THE SYNTHESIS OF TRANSISTOR AMPLIFIERS

THE SYNTHESIS OF TRANSISTOR AMPLIFIERS

Michael Kahn

New York City Community College

John M. Doyle

Electronic Aids, Inc.

HOLT, RINEHART AND WINSTON, INC.

New York · Chicago · San Francisco · Atlanta
Dallas · Montreal · Toronto · London · Sydney

PREFACE

This book provides one semester's work in transistor amplifiers at the junior college/technical institute level. It is not meant to be just another text in general electronics with the emphasis on transistors; rather, it is devoted entirely to the principles of amplification and the transistor is simply the active device through which these principles are illustrated. Thus, the chapters on oscillators, radio-frequency amplifiers, and so on, that one usually finds in a book of this type are conspicuous by their absence. We believe these things should be covered separately.

The book also differs from others in that no attempt is made to cover all of the so-called standard circuits. It is our belief that the continuing trend toward the use of integrated circuits will make this type of knowledge meaningless. Regardless of the form of the circuit, however, the concepts of bias, gain, distortion, and so forth, will remain unchanged, and these are the things on which we concentrate.

One final distinction is worthy of note. Most textbooks use the deductive approach when introducing new concepts. Equations are first treated in a literal manner and numerical data are not given until a later point. We feel that the inductive approach is best for a text at this level. Accordingly, when new concepts are introduced, typical numbers are used immediately in the analysis. General conclusions are then drawn from the numerical results. Finally, numerical examples are worked out to show how the values of components and device parameters affect a particular aspect of amplifier performance.

In dealing with the small signal equivalent circuit, we have taken the liberty of neglecting the feedback parameter h_{re}. This parameter has only a minimal effect on the value of amplifier gain in most instances, and thus its omission is not serious. However, the omission of h_{re} enables us to obtain a much clearer, simpler physical picture of the ac operation of the transistor, since the equivalent circuit can then be represented by two resistors and a controlled-current generator.

Due to its relatively limited application, the common-base amplifier is almost completely neglected. On the other hand, the emitter-follower circuit, one that is treated lightly in many texts, is discussed in considerable detail because of its vital importance as a buffer amplifier and impedance-matching device.

v

The assumed prerequisites are a working knowledge of ac and dc circuits, intermediate algebra, and trigonometry.

We have tried to minimize the complexity of algebraic derivations. When the derivations are extremely long they are omitted completely since only the final result is of interest to the reader.

The book begins with a review of volt–ampere characteristic curves, the load line, and transistor physics. An elementary transistor amplifier is introduced in chapter 4 together with a step-by-step development of the ac equivalent circuit. The reasons why this amplifier is not suited to practical applications are pointed out. Then, in chapters 5 through 9, attention is directed toward the modification of the basic amplifier—stabilization of the Q point, cascading stages for additional gain, gain-stabilizing techniques, and so on. The purpose of chapter 10, Amplifier Design, is twofold: First, it acquaints the reader with the procedures involved in the design of a practical amplifier to given specifications; second, it serves as a review of the basic concepts treated in the previous chapters.

Throughout the text, whether in the introduction or in the body of the chapter itself, we attempt to relate the material to the subject matter of both the preceding and succeeding chapters.

Brooklyn, New York   Michael Kahn
Baltimore, Maryland   John M. Doyle
January 1970

CONTENTS

THE SYNTHESIS OF TRANSISTOR AMPLIFIERS

1

VOLT–AMPERE
CHARACTERISTICS

In the study of passive electric circuits, you learned how to use a number of analytical techniques; e.g., Ohm's and Kirchhoff's laws, network theorems and simple vector algebra. Although you will continue to use these techniques in the study of active networks, i.e., networks containing amplifying devices, additional skills must be mastered. The development of these skills forms the substance of this chapter.

First, however, it is deemed essential to make a few brief general remarks about current.

1-1   CURRENT

Current is the movement of charge carriers. In the study of passive circuits it is convenient to use the concept of electron flow to indicate current direction. This results from the fact that, in the particular case of metallic conductors, the charge carriers are electrons. Using this concept, the direction of current is taken to be from the *most negative* point in a circuit *toward* the *most positive* point in the circuit.

In the study of semiconductor devices, two types of charge carriers, holes and electrons, must be considered. The hole is assumed to possess a *positive* charge while the electron is assumed to possess an equal *negative* charge. Thus, under the influence of an applied emf, the hole and electron currents move in opposite directions. To avoid the ambiguity which often results from this situation and, at the same time, to conform to the convention followed in the majority of serious textbooks, the concept of conventional current is used exclusively in this book. Thus, any reference to current always means the movement of charge carriers from the most positive point in a circuit to the most negative point.

1-2   THE VOLT–AMPERE CHARACTERISTIC OF A LINEAR DEVICE

Refer to Fig. 1.1. Except for the polarity of the source of emf, the two circuits, A and B, are assumed to be identical. Because the ganged rheostats, P_1 and P_2, control

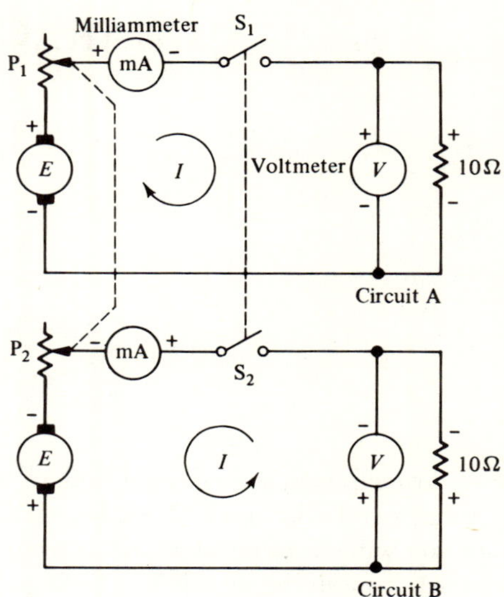

Figure 1.1

the total resistance of each circuit, any variation in the settings of the rheostats will change the magnitude of the current through, and voltage drop across, the 10 Ω load resistors. Notice that the polarity of E in circuit A is such as to send a current *down* through the 10 Ω resistor, while the opposite condition exists in circuit B; i.e., E is such as to send a current up through the 10 Ω resistor.

With the ganged switches S_1 and S_2 open, I and V are, of course, zero in both circuits. When the switches are closed, let us assume the rheostats are adjusted to

TABLE 1-1

Circuit A		Circuit B	
V	I	V	I
0	0	0	0
5	0.5	−5	0.5
10	1.0	−10	1.0
15	1.5	−15	1.5
20	2.0	−20	2.0
25	2.5	−25	2.5
30	3.0	−30	3.0

obtain 5-volt increments in the reading of the voltmeters, ranging from 0 to 30 V. The values of V and the corresponding values of I are shown in Table 1-1, and a plot of the tabular data is presented in Fig. 1.2. The curve of Fig. 1.2 is called a *characteristic*.

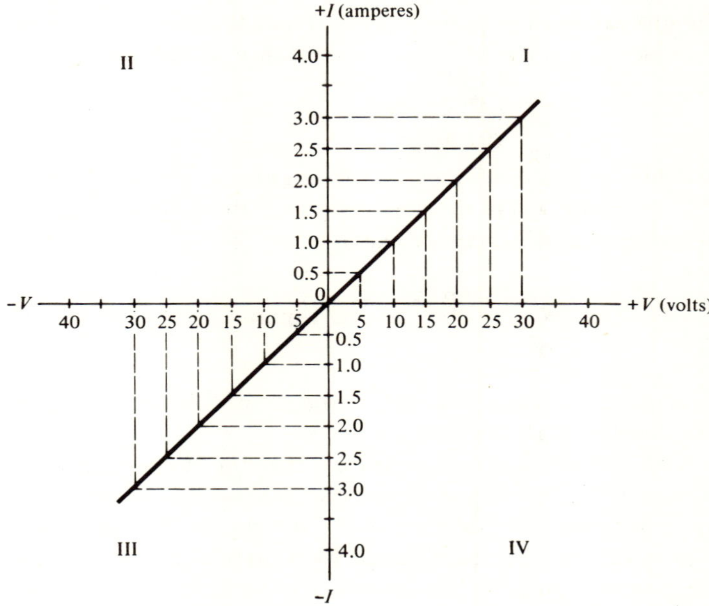

Figure 1.2

Ohm's law of constant proportionality tells us that the voltage drop across a resistor is directly proportional to the magnitude of current through the resistor. Thus, the volt–ampere characteristic of a resistor, as shown in Fig. 1.2, is a straight line and, accordingly, the device is said to be a *linear* device.

An important quantity which is defined for a curve, such as that shown in Fig. 1.2, is its *slope*. The slope gives a measure of the direction of a curve and is defined as the vertical distance from one point to another on the same curve, divided by the incremental horizontal distance from the first point to the second. Using the letter symbol m to represent slope, we have

$$m = \frac{y_2 - y_1}{x_2 - x_1} = \frac{\Delta y}{\Delta x} \tag{1-1}$$

where y_2 and y_1 are selected points on the y (vertical) axis
     x_2 and x_1 are selected points on the x (horizontal) axis
     Δ is the Greek letter delta, used to represent an increment

In Fig. 1.2 the y-scale represents amperes and the x-scale represents volts. Because the curve is a straight line, any two points can be selected. To illustrate the procedure, suppose we take the coordinates 10, 1 and 30, 3. Using these coordinates,

$$m = \frac{\Delta y}{\Delta x} = \frac{2 \text{ A}}{20 \text{ V}}$$

$$= 0.1 \text{ mho}$$

From the study of resistive networks, recall that the *mho* is the unit of *conductance* and that conductance is the reciprocal of resistance. Mathematically,

$$G = \frac{1}{R} \tag{1-2}$$

Thus, the slope of the volt–ampere characteristic represents the *conductance* of a device, while the reciprocal of the slope represents the *resistance* of the same device. In the case under consideration, for example,

$$R = \frac{1}{m} = \frac{1}{0.1 \text{ mho}}$$

$$= 10 \ \Omega$$

In constructing the curve of Fig. 1.2, V and I represent the voltage drop across, and current through, a 10 Ω resistor.

In view of the above, it is quite apparent that the steeper the slope of a volt–ampere characteristic, the larger the conductance of the device represented and the smaller its resistance. In the case of a short circuit, the volt–ampere characteristic of a device would be represented by a vertical line overlaying the *y*-axis since, for all practical purposes, the resistance of the device would be zero and the current would be infinite. At the other extreme, an open circuit would be represented by a straight horizontal line overlaying the *x*-axis—with the resistance of the device considered to be infinite and the current zero.

Two final points of interest can be drawn from an inspection of the curve of Fig. 1.2: (*1*) A volt–ampere characteristic is always constructed with the *x*-axis representing voltage and the *y*-axis representing current. (*2*) The magnitude of current through a resistor is the same for a given voltage drop across the resistor, without regard to the *polarity* of the voltage drop. For this reason, the resistor is a *bilateral* device.

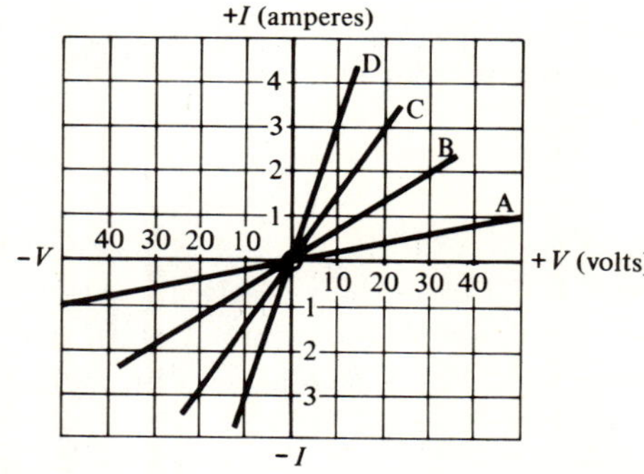

Figure 1.3

PRACTICE EXERCISES

Calculate the conductance and resistance of each of the four devices whose volt–ampere characteristics are shown in Fig. 1.3.

ANSWERS

DEVICE A: $G = 0.02$ mho, $R = 50$ Ω. DEVICE B: $G = 0.0667$ mho, $R = 15$ Ω. DEVICE C: $G = 0.15$ mho, $R = 6.67$ Ω. DEVICE D: $F = 0.3$ mho, $R = 3.33$ Ω.

1-3   THE VOLT–AMPERE CHARACTERISTIC OF A NONLINEAR DEVICE

Devices having a volt–ampere characteristic resembling that shown in Fig. 1.4 are encountered frequently in electronics. Such devices are called *diodes* because they

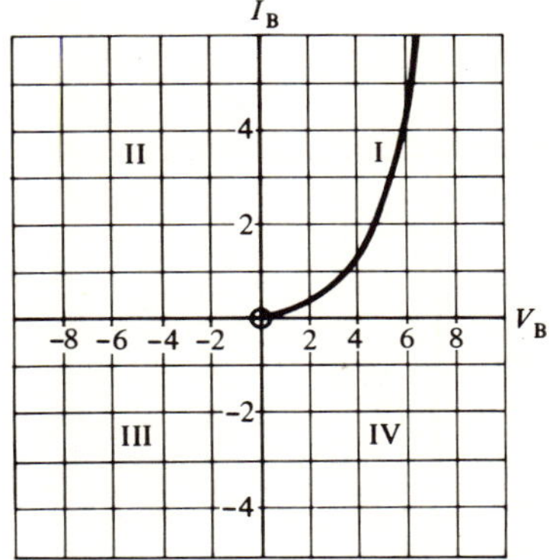

Figure 1.4

have two electrodes—an *anode* and a *cathode*. For the time being, our sole interest in the diode relates to its volt–ampere characteristic.

When the curve of Fig. 1.4 is compared with that of Fig. 1.2, several things become apparent immediately: (*1*)   The y (current) axis is labeled I_B to represent dc anode current and the x (voltage) axis is labeled V_B to represent dc anode voltage.   (*2*)   Unlike the resistor, which is a bilateral device, the diode is a *unilateral* device since, for all practical purposes, conduction occurs only when the anode is at a positive potential with respect to the cathode. When the anode is at a negative potential with respect to the cathode, the diode cannot conduct. (Actually, a very small current (measured in microamperes) exists, but for practical purposes, this current is considered to be negligible.)   (*3*)   The volt–ampere characteristic of the diode is *nonlinear* since the slope of the curve varies from point to point.   (*4*)   Because the slope of the curve is determined by the resistance of the device, the resistance of the diode also varies. As the anode becomes progressively more positive with respect to the cathode, the slope of the curve increases and the resistance, therefore, decreases. To limit I_B to some safe value, as specified by the manufacturer, a current-limiting resistor is placed in the external circuit.

Refer now to Fig. 1.5, A and B. The reader will recognize these circuits as being essentially the same as those presented in Fig. 1.1, with one exception; a diode is

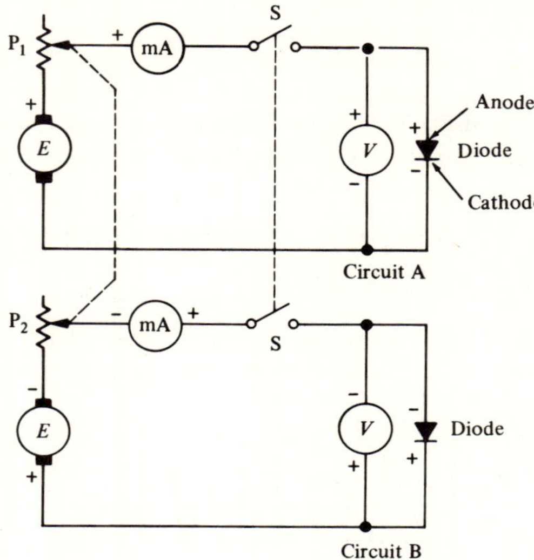

Figure 1.5

used in place of the 10 Ω resistor. Notice that the symbol for a diode combines the head of an arrow and a straight line. The arrow represents the anode, also called the *plate*, and the straight line represents the cathode.

In Circuit A the anode is positive with respect to the cathode. Under these conditions, the diode is said to be *forward biased*. As the forward bias on the diode is increased, by varying rheostat P_1, the diode *forward current* increases and its *forward resistance* decreases. The *forward characteristic* of the diode lies in quadrant I of Fig. 1.4.

In Circuit B the anode is negative with respect to the cathode. Under these conditions, the diode is said to be *reverse biased*. As the reverse bias on the diode is increased, by varying rheostat P_2, the diode *reverse current* is considered to be zero and its *reverse resistance* is considered to be infinite. In Fig. 1.4, the reverse characteristic of the diode lies along the negative portion of the *x*-axis. Bear in mind that the reverse-bias conditions noted above are *ideal* (a convenient fiction). Actually, a small reverse current does exist and the reverse resistance, although very high, is not infinite. Ideal conditions are assumed at this point to simplify the discussion.

1-4   DETERMINING THE STATIC AND DYNAMIC RESISTANCES OF A NONLINEAR DEVICE

As you will see later, both dc (biasing) and ac (signal) potentials are applied simultaneously to circuits which utilize nonlinear amplifying devices, such as the transistor.

In the absence of ac (signal) input, the voltage drop across, and current through, these devices remain constant. Because no change is taking place within the circuit, the dc operating conditions are said to be either *static* or *quiescent*. The *static resistance* of any nonlinear device used in this manner is, by Ohm's law, $R = V/I$.

When ac (signal) voltages are superimposed upon the dc (biasing) potentials, circuit conditions vary from instant to instant and operation is said to be *dynamic*. The *dynamic resistance* of the device is also determined by the application of Ohm's law, but the instantaneously varying voltages and currents are termed v and i, respectively.

Because both the static and dynamic resistances of a nonlinear device are important in the design of a practical amplifier circuit, let us see how they are calculated and in what ways they differ from one another.

Figure 1.6 shows the volt–ampere characteristic of a nonlinear device. To find the static resistance of the device, select a particular value of I_B and locate the corresponding point on the curve. In Fig. 1.6, the selected value of I_B is 1 mA.

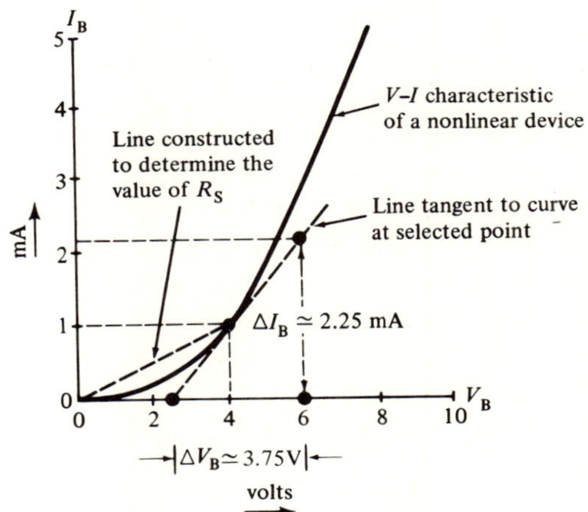

Figure 1.6

Next, draw a straight line from the origin of the graph to the selected point on the curve. The static resistance of the device, termed R_S, is equal to the reciprocal of the slope of this line. Using 1 mA as the selected value of I_B,

$$m = \frac{\Delta I_B}{\Delta V_B} = \frac{1 \times 10^{-3}\ \text{A}}{4\ \text{V}} = 2.5 \times 10^{-4}\ \text{mho}$$

and

$$R_S = \frac{1}{1 \times 10^{-4}\ \text{mho}} = 4{,}000\ \Omega$$

The calculations shown above can be avoided simply by recognizing that the static resistance of such a device is its resistance when passing a specified current. Then, it is only necessary to determine the corresponding value of voltage and to use the familiar expression $R = V_B/I_B$. In Fig. 1.6, for example, when $I_B = 1$ mA, $V_B = 4$ V and

$$R_S = \frac{4 \text{ V}}{1 \text{ mA}}$$

$$= 4{,}000 \ \Omega$$

To determine the dynamic resistance, R_D, of the same device at the same selected value of I_B, a line is drawn tangent to the curve at the selected value of I_B, as shown in Fig. 1.6. The incremental values of I_B and V_B are then determined by constructing a right triangle in which the line tangent to the selected point is the hypotenuse. Using this data,

$$m = \frac{\Delta I_B}{\Delta V_B} = \frac{2.25 \times 10^{-3} \text{ mA}}{3.75 \text{ V}} = 6 \times 10^{-4} \text{ mho}$$

and

$$R_D = \frac{1}{6 \times 10^{-4} \text{ mho}} = 1{,}667 \ \Omega$$

Notice that the dynamic resistance is considerably less than the static resistance for the same selected value of I_B. This is to be expected, since the slope of the line tangent to the curve at the selected point is steeper than the slope of the line drawn from the origin of the graph to the selected point. Of course, as the steepness of either line increases, the conductivity of the device increases, since the selected point is higher up on the curve, and the resistance under consideration decreases.

One final word about the static and dynamic resistances of a diode operating under conditions of reverse bias. Since I_B is assumed to be zero for all values of $-V_B$, both the static and dynamic resistances are considered to be infinite. In effect, we are considering the case of a pure resistor and this is the *only* case in which the dynamic and static resistances are identical.

1-5   THE LOAD LINE

In Fig. 1.7 a dc source of emf is used to forward bias a diode. If the volt–ampere characteristic of the diode is available, the determination of I_B (in this case, identical to I) is a simple matter since $V_B = E$.

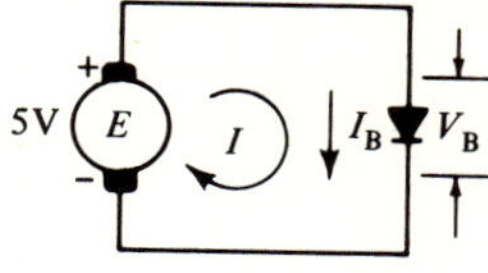

Figure 1.7

Suppose, however, a load resistor, R_L, is added to the circuit, as shown in Fig. 1.8. Now, $E \neq V_B$ ($\neq$ means "is not equal to"). Instead, $V_B = E - V_{RL}$. To determine the value of V_{RL}, however, I must be known. Since $I = E/R_T$, where R_T

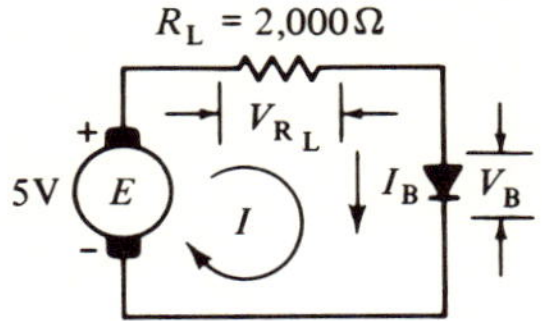

Figure 1.8

is the total resistance of the circuit, and we cannot determine R_T without knowing the static resistance of the diode ($R_S = V_B/I_B$), we are right back where we started from. Similar attempts to solve for I will also prove futile, but we can extend the application of an old friend, Kirchhoff's voltage law, to resolve the difficulty.

Kirchhoff's voltage law states that the sum of the voltage rises and voltage drops around a series circuit is equal to zero. Applying this law to the circuit of Fig. 1.8, we obtain:

$$-E + V_{\mathrm{RL}} + V_{\mathrm{B}} = 0 \qquad (1\text{-}3)$$

or

$$-E + IR_{\mathrm{L}} + V_{\mathrm{B}} = 0 \qquad (1\text{-}4)$$

where E = applied emf
$V_{\mathrm{RL}} = IR_{\mathrm{L}}$ = voltage drop across resistor R_{L}
V_{B} = voltage drop across the diode

Solving Eq. 1-4 for I we obtain

$$I = \frac{E}{R_{\mathrm{L}}} - \frac{V_{\mathrm{B}}}{R_{\mathrm{L}}} \qquad (1\text{-}5)$$

which may be rewritten as

$$I = \left(-\frac{1}{R_{\mathrm{L}}}\right) V_{\mathrm{B}} + \frac{E}{R_{\mathrm{L}}} \qquad (1\text{-}6)$$

Now, the general equation for a straight line is

$$y = mx + b \qquad (1\text{-}7)$$

Notice that Eqs. 1-6 and 1-7 have the same form. Equation 1-6 must, therefore, define a straight line. The slope of the line, m, is $-1/R_{\mathrm{L}}$, the x-axis of the plot is V_{B} and the y-axis is I_{B}. ($I_{\mathrm{RL}} = I_{\mathrm{B}}$ since this is a series circuit.)

To plot a straight line we need two points. We can obtain one point on the y-axis by setting V_{B} to zero. Under these conditions, $I_{\mathrm{B}} = E/R_{\mathrm{L}}$. We can obtain a second point on the x-axis by setting I_{B} to zero. Under these conditions, $V_{\mathrm{B}} = E$. In Fig. 1-8, $E = 5$ V and $R_{\mathrm{L}} = 2$ kΩ. Using these values, we can construct a straight line that intersects the volt–ampere characteristic of a diode, as shown in Fig. 1.9. A line constructed in this manner is called a *load line*. The point at which the load line intersects the volt–ampere characteristic of the diode is called the *quiescent point, operating point*, or *Q point*, and *specifies the dc operating conditions;* i.e., conditions with dc biasing voltages *only* applied.

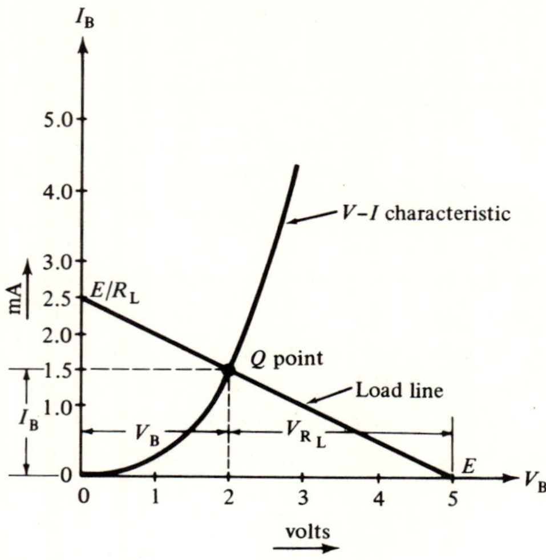

Figure 1.9

Notice that, when a vertical line is dropped from the Q point to the horizontal axis, *the voltage drop across the load resistor appears to the right of this line, while the voltage drop across the diode appears to the left of the line.* In this particular case, the voltage drop across the load resistor is 3 V and the voltage drop across the diode is 2 V. The dc diode current (in this case identical to the circuit current I) is 1.5 mA.

For the Q point indicated, the static resistance of the diode must be

$$R_S = \frac{V_B}{I_B} = 1{,}333 \ \Omega$$

Since the total circuit resistance is equal to the sum of the individual resistances,

$$R_T = R_L + R_S = 2{,}000 + 1{,}333 = 3{,}333 \ \Omega$$

The above calculation is easy to verify. At the Q point the circuit current is 1.5 mA. The applied voltage, E, is 5 V and,

$$R_T = \frac{E}{I_B} = \frac{5}{1.5 \times 10^{-3}} = 3{,}333 \ \Omega$$

Thus, for the Q point indicated in Fig. 1.9, the diode behaves as though it were a 1,333 Ω resistor. If the value of E or R_L is changed, the Q point is shifted, and the static resistance of the diode is changed.

We may think of the characteristic curve and the load line as depicting two relationships between V_B and I_B. The volt–ampere characteristic is determined by the diode alone while the load line is entirely dependent upon the external circuit (E and R_L). The one and only point at which both relationships are satisfied is at the intersection of the two curves. Essentially, we are solving two simultaneous equations by graphic methods, a technique learned in the study of intermediate algebra.

If, in the circuit of Fig. 1.8, the polarity of the source of emf is reversed, the diode becomes reverse biased and will not conduct. As shown in Fig. 1.10, the Q

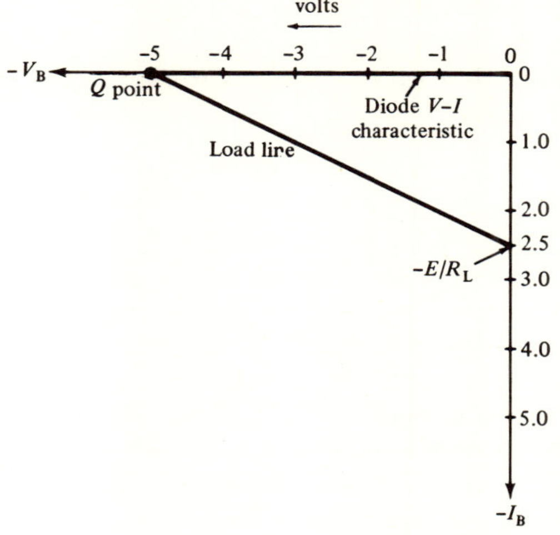

Figure 1.10

point then lies at the point of intersection between the load line and the V_B axis, because $I_B = 0$, $V_{RL} = 0$, and $V_B = E$.

PRACTICE EXERCISES

A series circuit containing a source of emf, a load resistor, and a diode is shown in Fig. 1.11(a). Using the V-I characteristic of the diode, Fig. 1.11(b), determine the Q point when

   (a) $R_L = 1$ kΩ and $E = 10$ V,
   (b) $R_L = 1$ kΩ and $E = 5$ V,
   (c) $R_L = 2$ kΩ and $E = 10$ V,

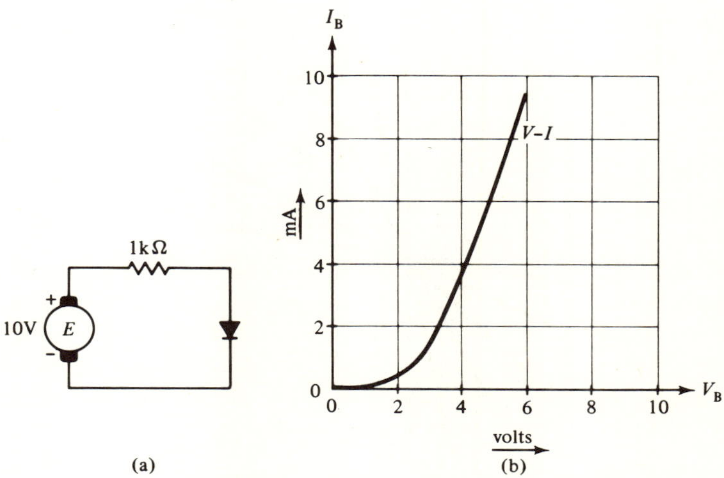

Figure 1.11

(d) R_L is infinite and $E = 10$ V, and

(e) $R_L = 0$ and $E = 10$ V.

Trace the *V-I* characteristic on to a sheet of work paper before starting to work on these problems.

ANSWERS

(a) $V_B = 4.5$ V, $I_B = 5.5$ mA;   (b) $V_B = 3.2$ V, $I_B = 1.8$ mA;   (c) $V_B = 3.8$ V, $I_B = 3.1$ mA;   (d) $V_B = 0$, $I_B = 0$;   (e) $V_B = 10$ V, $I_B \simeq$ infinite.

In the foregoing practice exercises, (a) and (b), R_L is the same, 1 kΩ, but the value of E is changed from 10 V to 5 V. Notice that, under these conditions, the load line moves *parallel* to itself. This is to be expected since the slope of the line is equal to the reciprocal of the resistance, which remains unchanged.

In (c), (d), and (e) above, the value of E is always the same, 10 V, but R changes from 2 kΩ to infinity, and then to zero. When R_L is infinite, the slope of the load line is zero and the Q point is located at the origin of the graph when I_B and V_B are both zero. When R_L is zero the slope of the load line is infinite (vertical) and the Q point would occur at some very high value of I_B (not shown on the graph).

The case in which R_L is infinite corresponds, of course, to an open circuit.

The case in which R_L is zero serves to indicate that the diode must always be operated with a current-limiting resistor. Otherwise, it will be destroyed by a current far in excess of its rating.

PROBLEMS

1.   The volt–ampere characteristics of five different devices (A through E) are drawn in Fig. 1.12.

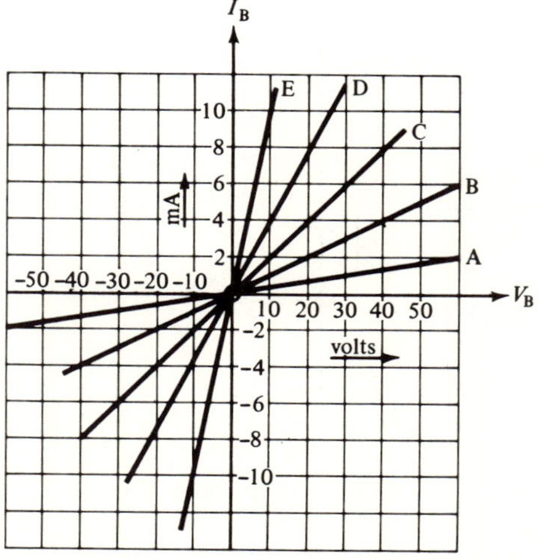

Figure 1.12

(a)   Compute the conductance and the resistance of each device.
(b)   Which curve has the largest slope?
(c)   Which device has the largest resistance?
(d)   What is the relationship between the slope of a characteristic curve and the resistance of the device?
(e)   What type of devices do these characteristics represent? Explain.

2.   Figure 1.13 shows the volt–ampere characteristic of a nonlinear device.
   (a)   Compute the values of both the static and dynamic resistance at point A.
   (b)   Repeat for point B.
   (c)   Repeat for point C.

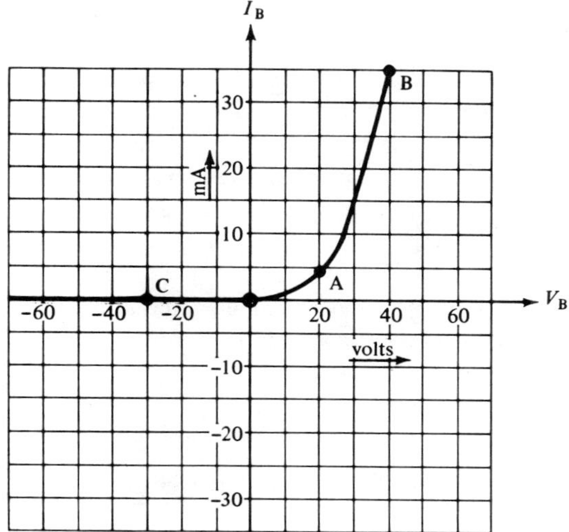

Figure 1.13

3.   The diode of Fig. 1.13 is used in the circuit of Fig. 1.14.
   (a)   Find the value of I_B if $E = 20$ V.
   (b)   Find the value of I_B if $E = 40$ V.
   (c)   In going from part (a) to part (b) the value of E doubled. Did I_B double? Explain.
   (d)   What value of E is required to make $I_B = 15$ mA?
   (e)   If the terminals of the diode in Fig. 1.14 are reversed and $E = 5$ V, what is the value of I_B?

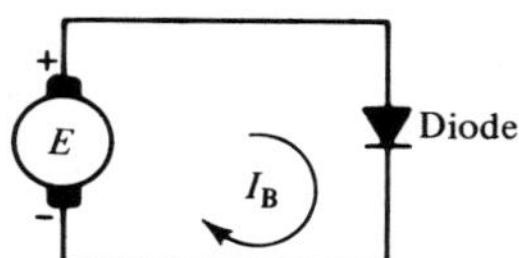

Figure 1.14

4.   The diode of Fig. 1.13 is used in the circuit shown in Fig. 1.15.
   (a)   Draw the load line and locate the Q point.

(b)   What is the value of I_B?
(c)   What is the value of V_B?
(d)   What is the value of V_{R_L}?
(e)   What is the value of the static resistance of the diode at the Q point?

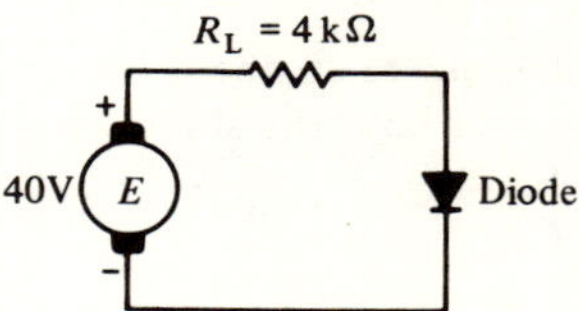

Figure 1.15

5.   (a)   Repeat problem 4 if $E = 60$ V.
    (b)   What is the relationship between the load lines of problems 4 and 5? Explain.

6.   Repeat problem 4 if the terminals of the diode in Fig. 1.14 are reversed.

7.   The diode of Fig. 1.13 is used in the circuit of Fig. 1.16. Assume that the meters are ideal (the resistance of the milliammeter is zero and that of the voltmeter is infinite).
    (a)   What are the readings of the milliammeter and the voltmeter?
    (b)   Repeat part (a) if the diode terminals are reversed.

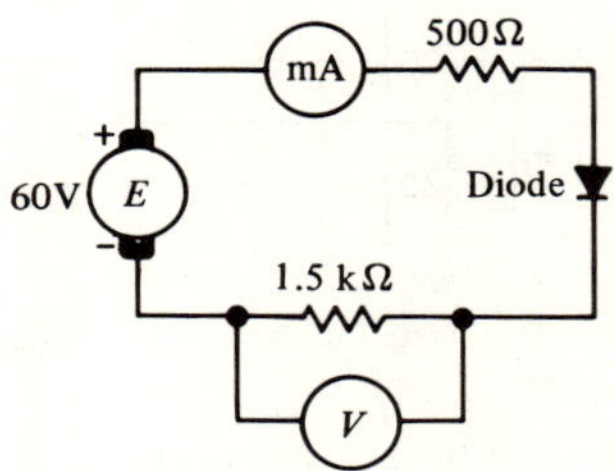

Figure 1.16

8.   The diode of Fig. 1.13 is used in the circuit of Fig. 1.17.
    (a)   What are the readings of the milliammeter and the voltmeter?
    (b)   Repeat part (a) if the diode terminals are reversed.

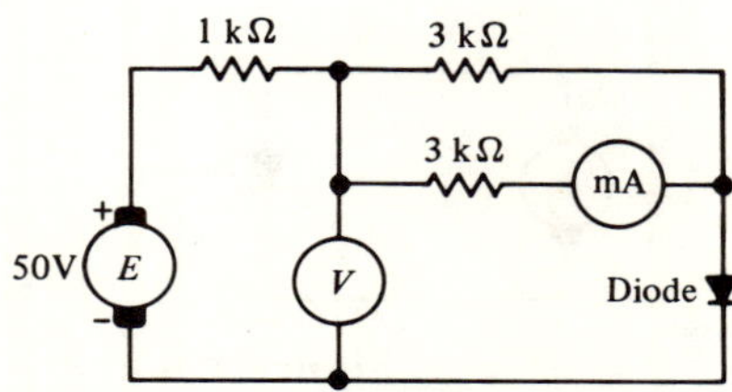

Figure 1.17

9.   The diode of Fig. 1.13 is used in the circuit of Fig. 1.18. The milliammeter reads 3.33 mA. What is the value of E?

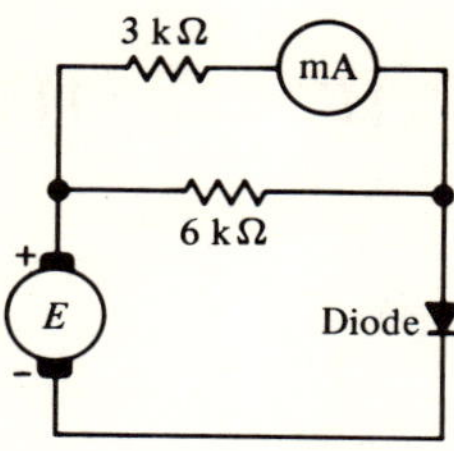

Figure 1.18

2

THE p–n JUNCTION DIODE

In chapter 1 we demonstrated methods used to analyze electronic circuits containing diodes. In this chapter, we examine the underlying physical theory of these and other solid-state devices, such as the transistor, as an understanding of this theory is essential before one can acquire a knowledge of how transistors operate. To avoid unnecessary difficulties, our discussion is essentially qualitative.

2-1   THE ATOM AND FREE ELECTRONS

From your study of basic dc theory you will recall that the atom is composed of a relatively massive positively charged nucleus, and negatively charged electrons, which revolve about the nucleus, as shown in Fig. 2.1. When the normal number of electrons are associated with an atom, the cumulative negative charge of the electrons is precisely equal to the positive charge of the nucleus. Hence, the net charge of a normal atom is zero.

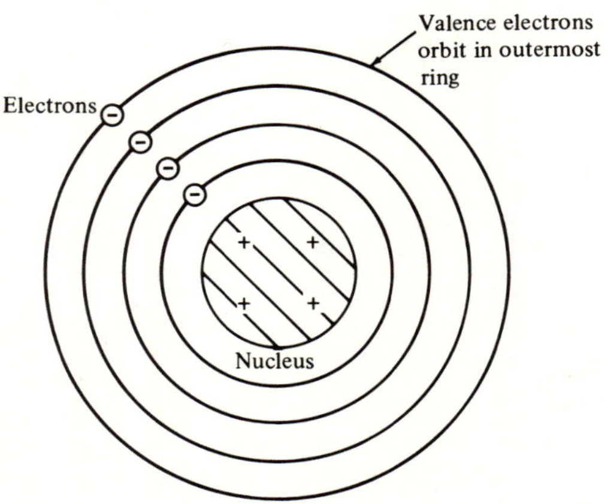

Figure 2.1

Electrons in the outer ring are known as *valence* electrons and are not bound as tightly to the nucleus as electrons in the other rings. For this reason, valence electrons frequently leave the parent atom and move through the material to another atom. When an atom acquires either a surplus or a deficiency of electrons it is said to be *ionized*. If there is a deficiency of electrons, the atom has a net positive charge and is called a *positive ion*. If, on the other hand, there is a surplus of electrons, the atom attains a net negative charge and becomes a *negative ion*.

Electrons which leave an atom are called *free* electrons and, in metal, it is the directed movement of these free electrons that constitutes current. Electrons become free by obtaining enough energy from an external source to allow them to break their bonds with an atom. This external energy may be derived from a source of EMF (electric energy), an increase in the temperature of the material (heat energy) or, with certain substances, by shining a light (radiant energy) on the material.

2-2 CRYSTALLINE STRUCTURES AND COVALENT BONDS

Germanium and silicon are two elements which have a crystalline structure, such as that shown in Fig. 2.2. Such elements are said to be *tetravalent* because they contain four valence electrons. In Fig. 2.2 each atom is neutral. The four black

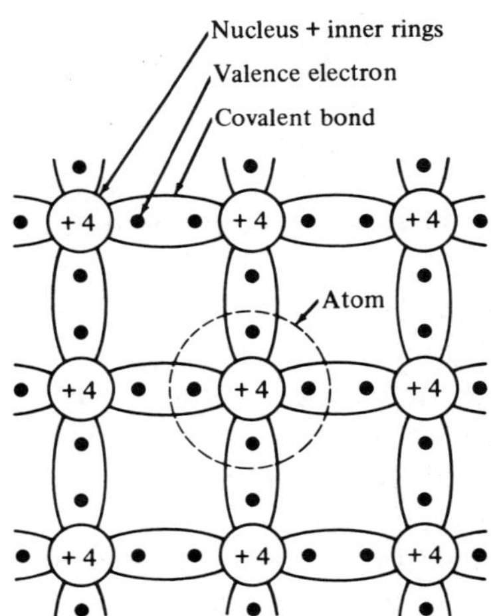

Figure 2.2

dots associated with each atom represent valence electrons. The circles marked $+4$ represent the remaining constituents of the atom; i.e., the nucleus plus all electrons not in the valence band. The charge of this portion of the atom must be $+4$ units to balance the -4 units negative charge of the valence electrons. Each valence electron shares a *covalent bond* with a valence electron from another atom and the entire structure is called a *crystal lattice*. The atoms are bound very tightly within this lattice structure.

2-3 ELECTRON-HOLE PAIRS

At temperatures close to absolute zero $(-273°C)$ valence electrons are held tightly in their covalent bonds and very large amounts of external energy are needed to free them from these bonds; consequently, there are few free electrons at these extremely low temperatures. Hence, the resistivity of the crystal is very high and, accordingly, its conductivity is very low.

As the temperature is increased, valence electrons gradually acquire additional energy. If the heating process is carried far enough, some of the valence electrons gain sufficient energy to become free electrons. Now, when a valence electron is free, it leaves behind a space in the lattice structure. We call this empty space a *hole*. See Fig. 2.3. The free electron and hole constitute a *electron-hole* pair. In an

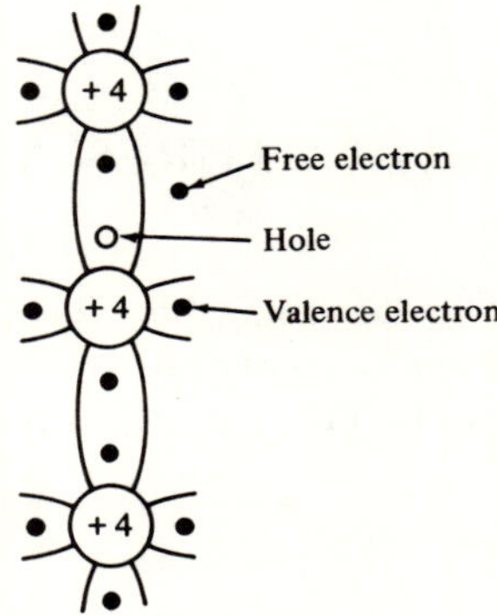

Figure 2.3

intrinsic (pure) germanium or silicon crystal, the number of free electrons is equal to the number of holes.

As the temperature is increased still further, more valence electrons acquire sufficient energy to break their bonds and more electron-hole pairs are formed. Once an electron is freed, it wanders through the crystal for some time, but eventually, drops into a *vacant* hole. The time interval during which an electron is free is called its *lifetime* and the process of filling a hole is called *recombination*. Once recombination occurs, neither the electron nor the hole are available as charge carriers. Typical lifetimes are in the order of microseconds.

Since current is defined as the movement of charge carriers, it is evident that the conductivity of silicon and germanium increases with increasing temperature. Although the conductivity of such materials at normal ambient temperatures is greater than that of an insulator, it is considerably less than that of a conductor, such as metal. For this reason, such substances are known as *semiconductors*.

Semiconductors exhibit a resistivity-temperature relationship opposite to that of metals. The resistivity of metals increases with increasing temperature. This increase is relatively gradual, however, compared to the decrease in resistivity of semiconductors with an equal increase in temperature. This phenomenon makes semiconductors very useful in temperature-control systems where a change in temperature initiates the operation of some circuit.

Refer now to Fig. 2.4. The six heavy dots in (a) represent valence electrons. The open circle in position 6 represents a hole. The valence electron has left this

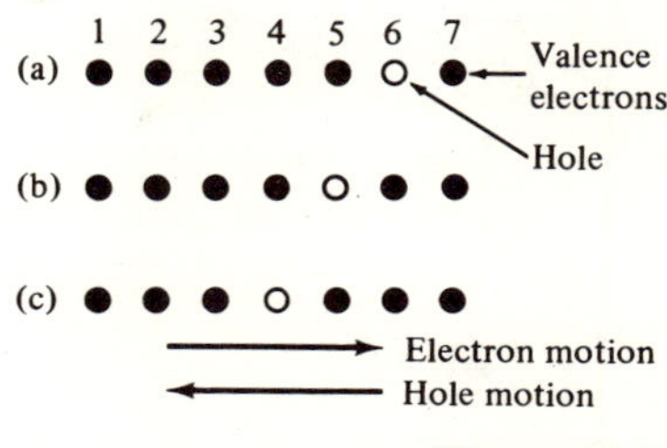

Figure 2.4

space and moved elsewhere in the crystal. Now let us assume that the electron in space 5 of (b) becomes a free electron and moves through the crystal until it fills the hole in space 6 of (a). As shown in (b), space 5 is now a hole. Figure 2.4 (c) shows what happens when the valence electron in space 4 moves to fill the hole in space 5 in (b). A study of the three parts of Fig. 2.4 reveals an important point. Although the electrons are moving from left to right, the holes *appear* to be moving from right to left. A hole does not actually move; it is simply a space fixed in the lattice, but it gives the appearance of moving opposite to the direction of an electron. Since electrons (negative charge) move to the right, we can consider the apparent motion of the holes to the left as a positive charge current. In other words, the holes act as carriers of electricity in a manner similar to electrons; therefore, the holes in the lattice *aid in the conduction of electricity* by providing paths through which the free electrons can travel. It is important to remember, however, that the crystal is neutral and its net charge is zero.

2-4   IMPURITY DOPING

Arsenic is a so-called *pentavalent* element because it has five valence electrons in its outer ring. If a very small amount of arsenic impurity is added chemically to intrinsic (pure) germanium or silicon, the germanium or silicon is said to be *doped*. During this chemical process, the impurity atoms become imbedded in the crystal lattice as shown in Fig. 2.5. The arsenic atom is shown inside the dashed circle. Four of the five valence electrons form covalent bonds with neighboring atoms of silicon or germanium, but the fifth electron is free. This additional free electron increases the conductivity of the material. If two arsenic atoms are added, there will be two extra free electrons in the crystal. In other words, for each pentavalent impurity atom added to the intrinsic crystal, there will be one extra free electron. It is important to remember that, in addition to the free electrons due to doping, there are also thermally generated electron-hole pairs present in the crystal.

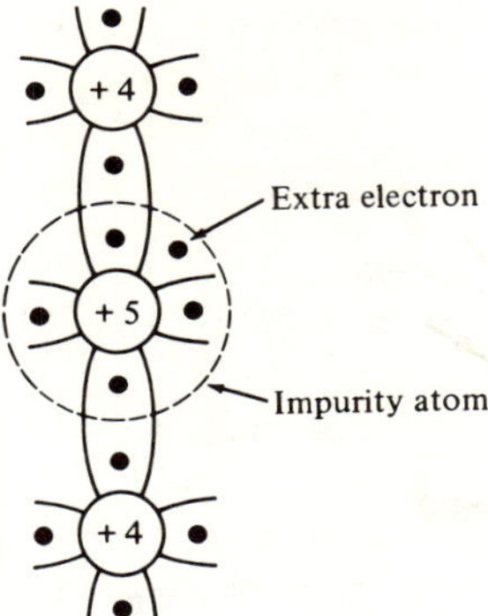

Figure 2.5

To illustrate the improvement in conductivity obtained by doping, let us assume that a small intrinsic crystal of germanium contains 100 million atoms and that, at room temperature (25°C), there is one electron-hole pair in the crystal. Now, suppose we add one arsenic atom for each one million germanium atoms. The one hundred arsenic atoms needed for this degree of doping would introduce 100 additional free electrons into the crystal. Since there was one free electron in the intrinsic crystal prior to doping, there are now 101 free electrons. In other words, the conductivity of the crystal has increased 101 times. It can be seen from the above discussion that, with only a very small amount of impurity added (one part in one million), the conductivity of the crystal is increased considerably. Of course, at higher temperatures, there will be more thermally generated electron-hole pairs present in the original crystal, so the percentage increase in conductivity will not be as great.

The crystal doped with pentavalent atoms will contain more free electrons than holes. Actually, the doped crystal will contain fewer holes than the intrinsic crystal, because recombination is statistically more probable.

As mentioned earlier, both free electrons and holes can be considered carriers of electricity. Because there are many more free electrons than holes when germanium or silicon is doped with a pentavalent atom, we call the free electrons *majority carriers* and the holes *minority carriers*. The pentavalent arsenic atom is called a *donor*, because it donates one free electron to the crystal. The doped crystal is called *n*-type germanium or silicon because the majority carriers are electrons (*n* for negative).

Indium is a so-called *trivalent* element because it has three valence electrons in its outer ring. Figure 2.6 shows what happens when germanium or silicon is doped with indium atoms. The indium (impurity) atom is shown within the dashed circle. The three valence electrons of indium form covalent bonds with neighboring germanium or silicon atoms. This leaves one hole in the lattice structure. Thus, for each trivalent impurity added to the intrinsic crystal, there will be one extra hole and this hole will act to increase the conductivity of the crystal in much the same manner as pentavalent doping.

Since there are many more holes than free electrons in a trivalent doped crystal,

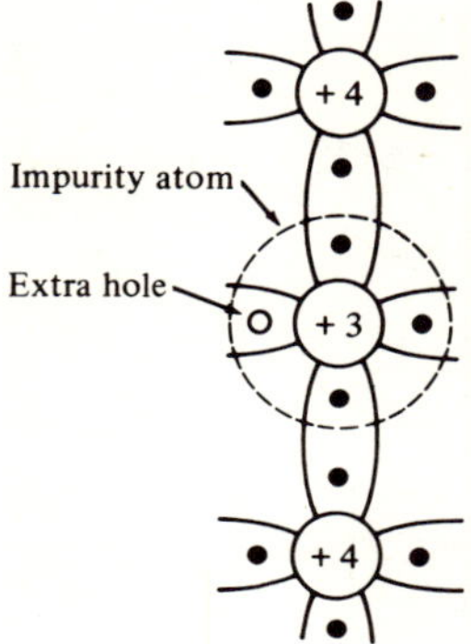

Figure 2.6

the holes are the majority carriers, and the free electrons are the minority carriers. The trivalent indium atom is called an *acceptor*, because the extra hole can accept an electron. This type of crystal is called *p*-type germanium or silicon because the majority carriers are holes (*p* for positive).

It is important to realize that the net charge of the doped crystal is still zero since the original intrinsic crystal is neutral and so also is the added impurity atom. Remember when a crystal is doped, its conductivity increases because there are more carriers present, but the crystal remains electrically neutral.

2-5   THE *p–n* JUNCTION

The union of *n*- and *p*-type materials, as shown in Fig. 2.7, produces a so-called *junction* diode. The junction is, of course, the point of union. Initially, let us assume that all of the carriers in the crystal are due to doping; i.e., there are no thermally generated electron-hole pairs. This is a fairly reasonable approximation in view of the discussion in the previous section. In the *n*-side of the crystal, each impurity atom is represented by a heavy dot and circle with a plus sign in its center. The dot represents the extra free electron contributed by the impurity atom. Similarly, in the *p*-side, each large circle with a minus sign in its center and the associated small circle represents one impurity atom. The small circle represents the extra hole. We know, of course, that there are many more germanium atoms in the crystal than impurity atoms, but since, in this case, they are assumed to contribute no charge carriers, they are not shown in Fig. 2.7 in order to avoid possible confusion.

At the instant the junction is formed, a redistribution of carriers in the vicinity of the junction takes place. Free electrons in the *n*-type germanium move across the junction and recombine with holes in the *p*-type material. Notice in Fig. 2.7 that recombination has taken place in the two columns of atoms nearest the junction. When this recombination occurs, the charge carriers, both holes and electrons, are eliminated. The region in which the recombination takes place is, therefore, referred to as the *depletion* region, i.e., a region depleted of charge carriers.

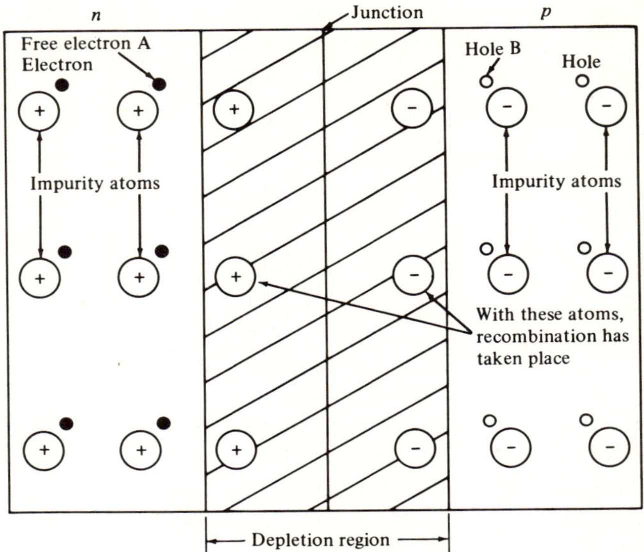

Figure 2.7

The loss of a free electron from each of the impurity atoms in the depletion region to the left of the junction makes these atoms *positively charged ions*. Similarly, the impurity atoms in the depletion region to the right of the junction have had their extra holes filled by free electrons from the *n*-type material and they are now *negatively charged ions*.

If the process of recombination described above were to continue, eventually there would be no free charge carriers and the entire substance would become an insulator. Fortunately, the process of recombination is limited. To understand why, in Fig. 2.7, notice the free electron on the *n*-side marked A, and the hole on the *p*-side marked B. As electron A approaches the junction in its attempt to recombine with hole B, two forces act on the electron. The positively charged ion in the depletion region of the *n*-type material exerts an *attractive force* that tends to pull electron A away from the junction. At the same time, the negatively charged ion in the depletion region of the *p*-type material exerts a force of *repulsion* that tends to push the electron away from the junction. Under these circumstances it becomes increasingly difficult for free electrons that are relatively distant from the junction to recombine with the holes on the opposite side. Thus recombination affects only those atoms in the immediate vicinity of the junction.

The depletion region is also known as the *space charge, transition,* or *barrier region.* The width of the depletion region is greatly exaggerated in Fig. 2.7 in relation to the size of the crystal, and the actual number of impurity atoms whose carriers undergo recombination across the junction is very, very small in comparison to the total amount of impurity atoms in the crystal. The positively and negatively charged ions which are formed as a result of recombination are located in the depletion region only. In the absence of thermally generated electron-hole pairs, there would be no carriers in the depletion region and the resistivity of the region would be infinite. Actually, however, some electron-hole pairs are always generated due to the ambient temperature and the role of these thermally generated charge carriers will be explained shortly.

2-6 DIODE ACTION

It was pointed out in the previous topic that the depletion region is limited to the immediate vicinity of the junction. The positive and negative ions in the depletion region form a barrier that acts to prevent recombination between the free electrons and holes which are farther removed from the junction.

In Fig. 2.8(a), a battery is connected across the junction diode in such a manner

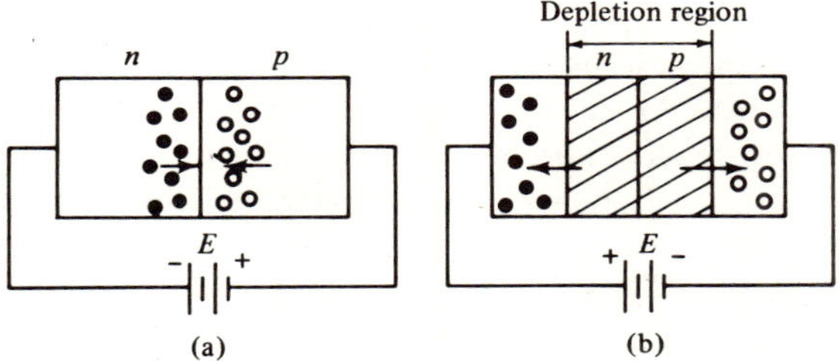

Figure 2.8

as to *forward* bias the diode; i.e., the negative terminal of the battery is connected to the *n*-section of the diode and the positive terminal of the battery is connected to the *p*-section. For simplicity, all ions and atoms are omitted in this illustration and only the charge carriers, free electrons and holes, are shown. Now, the potential at the negative terminal of the battery repels the free electrons in the *n*-section forcing them to *move toward the junction*. The potential at the positive terminal of the battery attracts thermally generated free electrons (minority carriers) in the *p*-type material and, as a result, holes in the *p*-type material *appear to move toward the junction*. The difference of potential that exists across the junction under these conditions forms a *junction barrier*. If the battery emf is about a tenth of a volt, very few of the free electrons from the *n*-side acquire sufficient energy to overcome the junction barrier and make their way through the *p*-type material to the positive terminal of the battery, so the resulting current is very small. As the battery emf is increased, however, the free electrons acquire a greater amount of energy and move toward the junction with greater force, thus overcoming the barrier caused by the ions in the depletion region. When the battery voltage exceeds a certain value (0.3 V for germanium and 0.6 V for silicon), the depletion region practically disappears and the current increases markedly.

In Fig. 2.8(b), the battery connection is reversed and the diode is *reverse* biased. Under these conditions, the positive terminal of the battery attracts the free electrons *away from the junction*. The negative terminal repels any minority free electrons in the *p*-material toward the junction so that the holes in the *p*-side appear to move *away from the junction*. This combined action produces a widening of the depletion region. With the assumption that no charge carriers exist in the depletion region under reverse-bias conditions, the resistivity of the region would be infinite, and there would be zero current through the entire circuit.

From the preceding discussion it should be apparent that the width of the depletion region is voltage sensitive. When the diode is forward biased, the depletion region is narrowed and, when the diode is reversed biased, the depletion region is widened. In effect, we have a variable resistance and a means of controlling the magnitude of current through the external circuit. The capability of the junction diode to control current in the external circuit makes the device extremely useful in a myriad of practical applications.

2-7   REVERSE CURRENT

Up to this point in our discussion, we have purposely neglected the presence of the thermally generated electron-hole pairs for the sake of simplicity and have assumed that all charge carriers present in the crystal were due to doping. Actually, at any given temperature, some electron-hole pairs always exist in the crystal. Consequently, there are always some *holes* in the *n*-type material and some *free electrons* in the *p*-type material. The number of these *minority carriers* is very small compared to the number of majority carriers, as shown in Fig. 2.9.

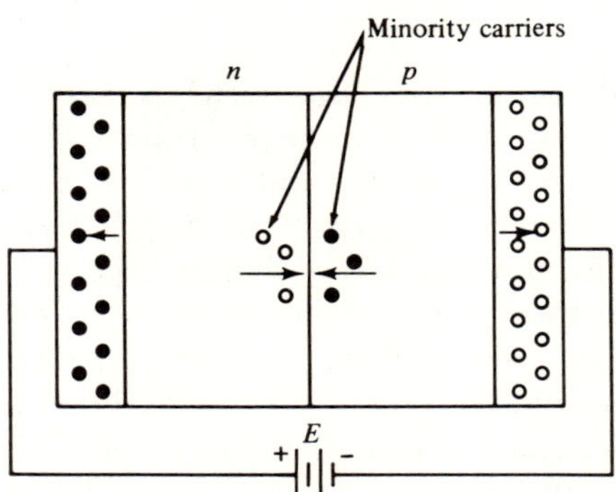

Figure 2.9

When reverse bias is applied to the diode, majority carriers move away from the junction, but *minority carriers move toward the junction*. In other words, as far as the *minority* carriers are concerned, the diode is forward biased. These minority carriers will cross the junction and produce a current through the external circuit. Thus, the junction diode is not an ideal diode as defined in chapter 1, and the reverse resistance of the diode is less than infinite. Because the reverse current is very small, however, usually a few microamperes, the reverse resistance is very high under *normal temperature conditions.* Pay particular attention to the last part of the preceding statement—under normal temperature conditions. If the temperature is increased, the number of thermally generated electron-hole pairs and, therefore, the number of minority carriers increases. With an increase in the number of minority carriers, the reverse current increases. In fact, for each 10°C increase in temperature, the reverse current in a germanium device doubles. For a silicon device the reverse current doubles for each 6°C increase in temperature.

If the temperature is permitted to exceed a value specified by the manufacturer, the reverse current may increase by many orders of magnitude, and this increase may destroy the device. This temperature sensitivity is a real problem in all semiconductor devices and a great deal more attention is given to the methods of controlling reverse current in succeeding chapters.

Because the reverse current is undesirable, it is often referred to as *leakage current*. In addition it is sometimes referred to as the *reverse saturation current* or, simply, saturation current. The letter symbol for reverse (saturation) current is I_{CO}.

TABLE 2-1

Temp. (°C)	$I_{CO}(\mu A)$
25	1
35	2
45	4
55	8
65	16
75	32
85	64
95	128

Table 2-1 indicates typical values of I_{CO} for a germanium device at various temperature increments, beginning at room temperature, 25°C. For the sake of simplicity, the figures shown in Table 2-1 are used throughout the remainder of this book. Keep in mind that these figures must be modified, as noted previously, for a silicon semiconductor device.

2-8   THE *p–n* DIODE CHARACTERISTIC

Figure 2.10 shows a typical characteristic curve for a real-life *p–n* junction diode. The characteristic in the forward direction shows the current increasing markedly after the voltage exceeds a few tenths of a volt. Notice that the slope of the curve in this region is rather steep, indicating that the resistance (both static and dynamic) of the diode in the forward direction is small. In the reverse direction,

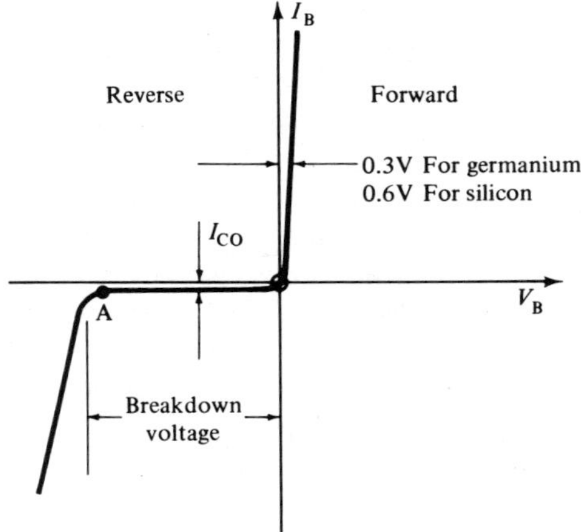

Figure 2.10

the current I_{CO} is small and practically independent of voltage up to point A. The slope of the I_{CO} curve is small indicating that both the static and dynamic resistances of the reverse biased diode are large, but not infinite.

Note that when the reverse voltage across the diode exceeds a particular value, point A in Fig. 2.10, the reverse current increases to very large values. The voltage at which this current begins to increase markedly is called the *breakdown voltage*. Beyond this voltage, the *p–n* junction no longer acts like a diode because, as the current increases, the resistance of the diode begins to decrease.

Actually there are two different mechanisms of breakdown. Refer to Fig. 2.11.

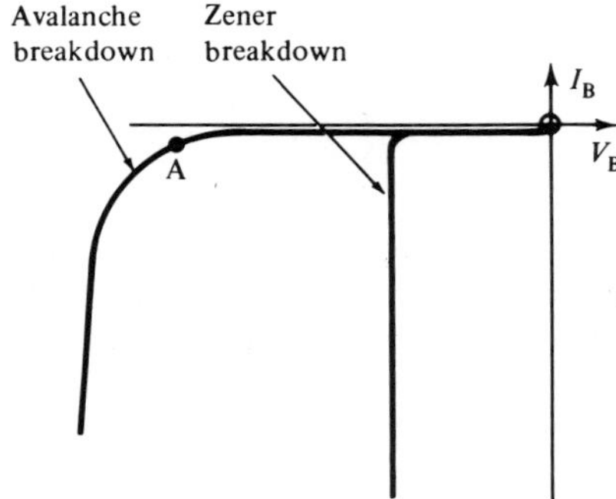

Figure 2.11

As the reverse voltage across the diode is increased, a minority carrier may acquire enough energy to knock an electron out of one of the covalent bonds. This creates two additional minority carriers—one free electron and one hole—and the reverse current increases slightly due to the presence of these additional carriers. If the reverse voltage is increased still further, more electrons are knocked out of their bonds, some by original thermally generated carriers, and some by newly freed electrons. Thus, as the reverse voltage is increased, the reverse current increases in *avalanche* fashion, each collision creating more carriers. *Avalanche breakdown* usually occurs at relatively high voltages. Generally, avalanche breakdown is an undesirable feature of diodes, as will become clear in succeeding chapters.

In some *p–n* diodes, the breakdown mechanism is entirely different. When the reverse voltage reaches some particular value, the force exerted by the voltage is enough to pull electrons from their covalent bonds in large numbers. There is an abrupt increase in current because of the presence of this large number of newly created carriers (holes and free electrons). This phenomenon is known as *Zener breakdown*, and usually occurs at lower voltages than avalanche breakdown.

In comparing the two types of breakdown, we see that avalanche breakdown is a gradual affair, caused by successive collisions, while Zener breakdown is abrupt with the current increasing sharply once the breakdown voltage is reached. Because the avalanche process is gradual, it is difficult to specify any one value as the breakdown voltage. We call the avalanche breakdown voltage that voltage at which the reverse current begins to increase markedly beyond the leakage value, point A in Fig. 2.11. Since the voltage at which Zener breakdown occurs is clearly defined, this value is easily determined by reference to the manufacturer's specification for a given diode.

The percentage of impurities added to the intrinsic crystal has an important effect on the type of breakdown that will occur in a diode. *Unless the dissipation capabilities of a diode are exceeded, reverse-current breakdown will not destroy the device.*

Both germanium and silicon diodes have the same general characteristic shown in Fig. 2.10. The forward portions of these characteristics are drawn on an expanded scale in Fig. 2.12. There are two major differences between germanium

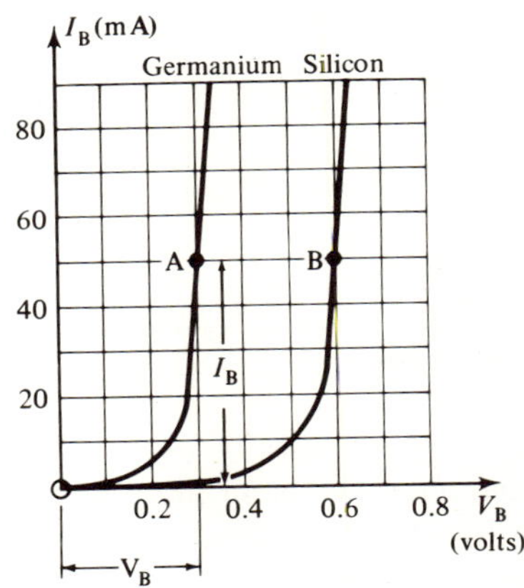

Figure 2.12

and silicon diodes. The first is the forward voltage needed to initiate heavy conduction. For germanium, this voltage is approximately *0.3 V* and, for silicon, approximately *0.6 V*.

Using Fig. 2.12, let us calculate the static resistance of the germanium diode at point A, and the silicon diode at point B. Since,

$$R_{\mathrm{S}} = \frac{V_{\mathrm{B}}}{I_{\mathrm{B}}}$$

we obtain

At point A:
$$R_{\mathrm{S}} = \frac{V_{\mathrm{B}}}{I_{\mathrm{B}}} = \frac{0.3 \text{ V}}{50 \text{ mA}} = 6 \text{ }\Omega$$

At point B:
$$R_{\mathrm{S}} = \frac{V_{\mathrm{B}}}{I_{\mathrm{B}}} = \frac{0.6 \text{ V}}{50 \text{ mA}} = 12 \text{ }\Omega$$

Thus, the forward resistance of the diode is quite small. This example also shows that the forward resistance of the silicon diode is larger than that of the germanium diode at the same current (50 mA). This is because the *silicon diode requires more voltage than the germanium diode to overcome the junction barrier and permit heavy conduction.*

The second major difference between germanium and silicon diodes concerns the leakage (minority carrier) current. The values of leakage current for a germanium diode are from 10 to 1000 times larger than those found in silicon diodes. Since leakage current increases drastically with an increase in temperature, silicon diodes and transistors are used exclusively in high-temperature applications. Germanium devices are practically useless at temperatures above 55°C because of their high leakage currents.

Figure 2.13 shows an expanded portion of the reverse characteristic of a diode

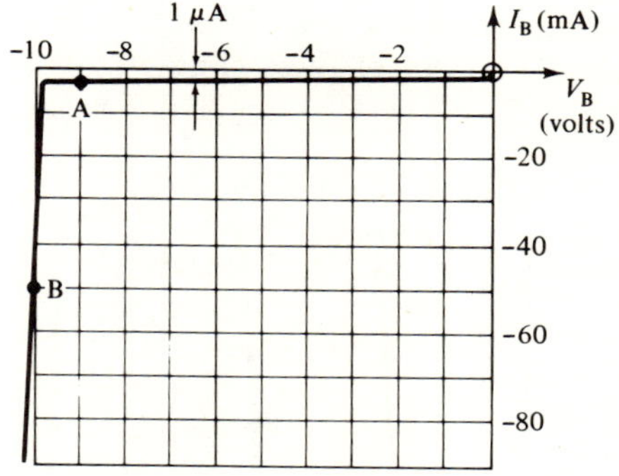

Figure 2.13

having a Zener breakdown voltage of 10 V and a leakage current of 1 µA. The leakage portion of the characteristic is, of course, greatly exaggerated. Let us determine the *static* resistance of the diode at point A in the leakage region and at point B in the breakdown region.

At point A:
$$R_{\mathrm{S}} = \frac{V_{\mathrm{B}}}{I_{\mathrm{B}}} = \frac{-9 \text{ V}}{-1 \text{ µA}} = 9\text{M}\Omega$$

At point B:
$$R_{\mathrm{S}} = \frac{V_{\mathrm{B}}}{I_{\mathrm{B}}} = \frac{-10 \text{ V}}{-50 \text{ mA}} = 200 \text{ }\Omega$$

Notice that the static resistance in the leakage region is quite large because the current is small. In the breakdown region, however, the static resistance of the diode is considerably smaller due to the increased current. The *dynamic* resistance in the breakdown region is very small, because the curve is steep.

Figure 2.14(a) shows the complete characteristic of a diode. Suppose this device

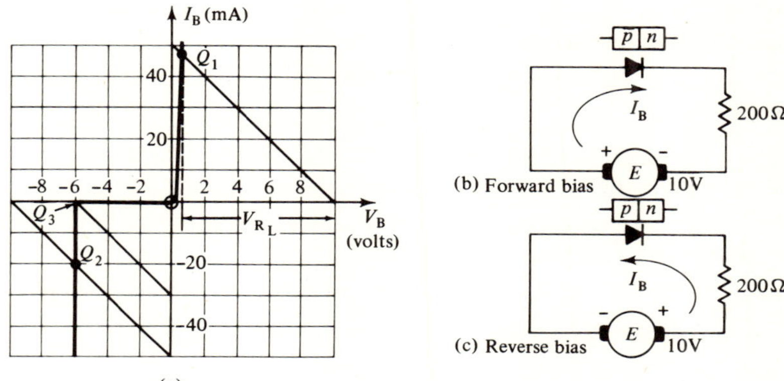

Figure 2.14

is connected as shown in (b). To determine the Q point, labeled Q_1, we must draw a load line on the forward characteristic. Reference to the load line indicates that

$$V_B = 0.5\,\text{V} \qquad \text{and} \qquad I_B = 47.5\,\text{mA}$$

Now suppose the diode is reverse biased as shown in Fig. 2.14(c). When a load line drawn on the reverse characteristic, the Q point is labeled Q_2,

$$V_B = -6\,\text{V} \qquad \text{and} \qquad I_B = -20\,\text{mA}$$

Notice that, when the diode is forward-biased, *most of the voltage is developed across the load resistor.* This is to be expected because the diode acts like a *very small resistance;* therefore, a smaller percentage of the supply voltage is dropped across the diode. When the diode is reverse biased, as shown in Fig. 2.14(c), it acts like a large resistance and we would expect a small reverse current. In this particular case, however, the supply voltage, E, is greater than the breakdown voltage, -6 V, and there is an appreciable reverse current, as shown at operating point Q_2. If the supply voltage is reduced from -10 to -6 V, the Q point becomes Q_3, as shown in Fig. 2.14(a). The current in the reverse direction is now the leakage current and is too small to estimate by reference to the characteristic shown.

Semiconductor diodes vary in physical size from fractions of an inch in length (switching diodes) to those which are more than one inch in diameter (high-power rectifiers).

QUESTIONS

1. What is the difference between a valence electron and a free electron?

2. What is meant by an intrinsic crystal?

3. What is the relationship between the number of free electrons and the number of holes in an intrinsic crystal?

4. Explain why the conductivity of an intrinsic crystal increases with increasing temperature.

5. Explain why the doping of a crystal increases its conductivity.

6. What is meant by the depletion region?

7. Why must a semiconductor diode be forward-biased by a few tenths of a volt before the diode current increases markedly?

8. Explain why the depletion region widens as the reverse voltage increases.

9. Explain why the leakage current increases with increasing temperature.

10. Distinguish between avalanche and Zener breakdown.

PROBLEMS

1. A germanium crystal containing 10^9 atoms has 5 electron-hole pairs at room temperature. The crystal is doped with one part trivalent impurity for each 10^6 germanium atoms.
    (a)   Are the majority carriers holes or free electrons? Explain.
    (b)   By what factor has the conductivity increased?

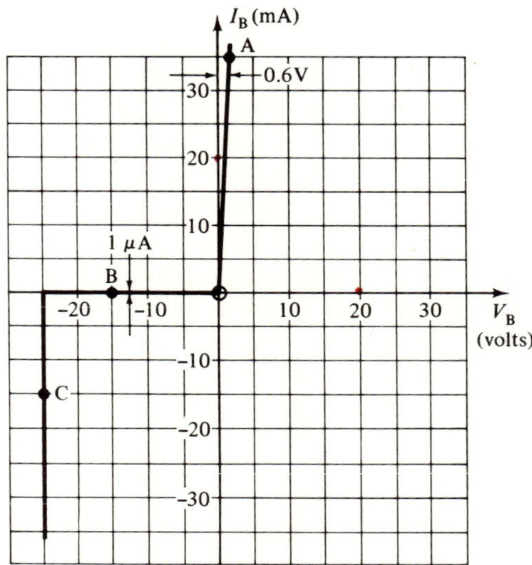

Figure 2.15

2.   Figure 2.15 shows the characteristic curve of a semiconductor diode.
    (a)   Compute the values of both the static and dynamic resistance at point A.
    (b)   Repeat for point B.
    (c)   Repeat for point C.
    (d)   Does this diode exhibit avalanche or Zener breakdown?

3.   The leakage current of a particular germanium diode is 0.2 μA at 25°C. What is its value at 65°C?

4.   The diode of Fig. 2.15 is used in the circuit of Fig. 2.16. The meters are ideal.
    (a)   Is the diode forward or reverse-biased?
    (b)   What is the reading of the milliammeter?
    (c)   What is the reading of the voltmeter?

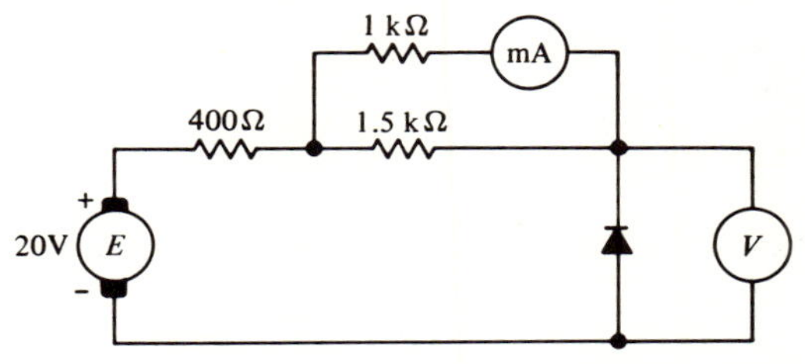

Figure 2.16

5.   Repeat problem 4 if the diode terminals are reversed.

6.   Repeat problem 4 if $E = 30$ V.

3

THE TRANSISTOR

In chapter 2 we studied the structure and dc behavior of a single-junction device, the diode. In this chapter, similar treatment is given to a two-junction device, the *transistor*.

3-1   RELATIVE DOPING AND THE DEPLETION REGION

We must return briefly to the p–n junction diode before proceeding with the study of the transistor. Refer to Fig. 3.1. Part (a) of this figure is a pictorial repre-

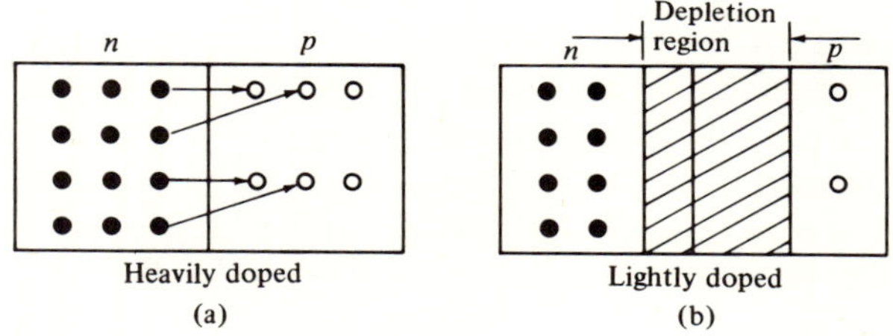

Figure 3.1

sentation of a junction diode at the instant of formation, i.e., before the depletion region is formed. Notice that, in this particular case, the n-type material is *more heavily* doped than the p-type material; specifically, there are twice as many donor impurities in the n-side than acceptor impurities in the p-side. Control over the relative amounts of doping is achieved readily during the manufacturing process.

Now, we know from the theory developed in chapter 2 that, initially, carriers in the vicinity of the junction recombine and a depletion region is formed. Because, in the case at hand, there are more free electrons in the n-side than there are holes in the p-side, the penetration of electrons into the p-side is deeper than the penetration of holes into the n-side, as shown in Fig. 3.1(b). As a result, the depletion region on the p-side is wider than on the n-side. This phenomenon always occurs when one side of the crystal is doped more heavily than the other. By controlling the relative amounts of impurity doping, manufacturers can make the depletion region extend to any desired depth into the relatively lightly doped material. Just remember that, in the event of unequal doping, the depletion region is widest on the side where the doping is lightest. Also bear in mind that the relative dimensions shown in Fig. 3.1(b) are considerably exaggerated. The actual width of the depletion region is much, much smaller than the overall dimensions of the crystal.

3-2   THE JUNCTION TRANSISTOR

Refer to Fig. 3.2(a). The crystal shown has its left and right ends doped with donor (n) impurities and its center doped with acceptor (p) impurites. A crystal

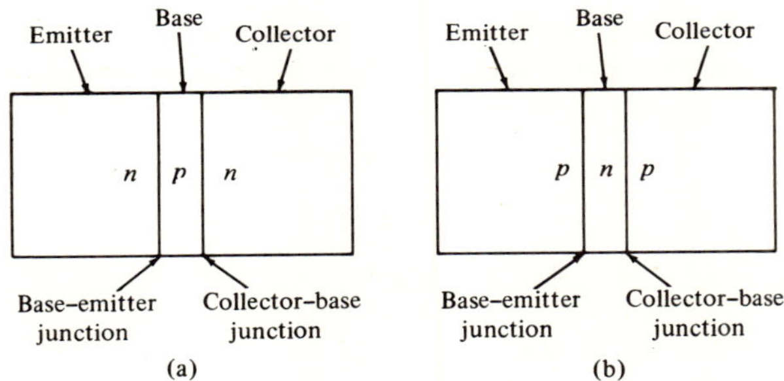

Figure 3.2

doped in this fashion is called an *npn* transistor. A transistor may also be of the *pnp* variety, with the ends doped with acceptor (p) impurities and the center doped with donor (n) impurities. A *pnp* transistor is shown in Fig. 3.2(b). Notice that, in either type transistor, the *emitter* and *collector* are doped with the same type impurity while the *base* is doped with the opposite-type impurity.

In Fig. 3.2, the junction between the emitter and base is called the *base-emitter* junction. In like manner, the junction between the collector and base is called the *collector-base* junction.

The schematic symbols for *npn*- and *pnp*-type transistors are shown in Figs. 3.3(a) and (b), respectively.

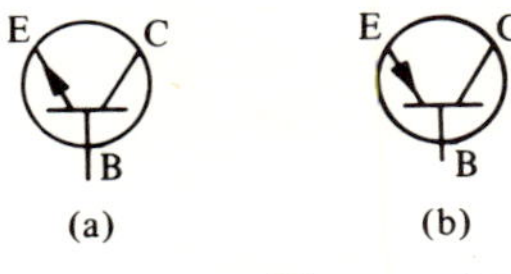

Figure 3.3

To permit connection into an external circuit, wire leads are attached to the emitter, base, and collector electrodes. To identify these leads, some transistors have a small tab formed on the lip of the protective casing, as shown in Fig. 3.4(a).

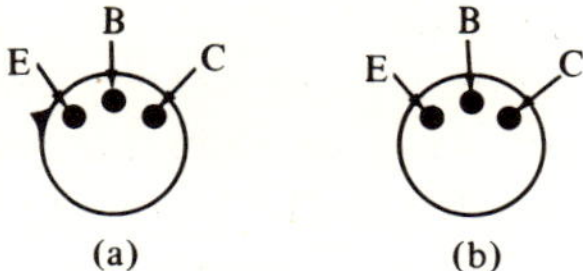

Figure 3.4

Reading clockwise from the tab, the leads are emitter, base, and collector, respectively. If the transistor does not have a tab, the emitter is held in the position shown in Fig. 3.4(b) and, again, the leads are read in a clockwise direction as emitter, base, and collector, respectively.

Physically, the base is much thinner than either the emitter or collector. In addition, the emitter and collector are doped very heavily in comparison to the base. The combined result of these two factors, i.e., relative widths and doping, is that the emitter and collector contain many more majority carriers than the base.

For an *npn* transistor, dc operating potentials are applied as shown in Fig. 3.5.

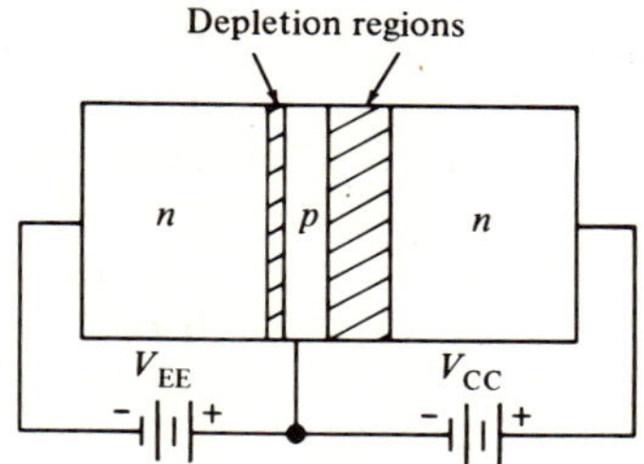

Figure 3.5

The dc source marked V_{EE} is connected between the emitter and base in such a manner as to *forward* bias the base-emitter junction. Hence, the width of the base-emitter depletion region is narrowed. A second dc source, marked V_{CC}, is connected between the collector and base in such a manner as to *reverse* bias the collector-base junction; consequently, the width of the collector-base junction is extended. *For normal operation*, a transistor is *always* biased in this manner; i.e., with the base-emitter junction forward biased and the collector-base junction reverse biased. To achieve these conditions in a *pnp* transistor, it is only necessary to reverse the polarities of the sources of emf. It should also be noted that, although batteries are shown in the illustration, the usual source of emf, except in portable applications, is an electronic power supply.

3-3　TRANSISTOR CURRENTS

A properly biased *npn*-type transistor is shown in Fig. 3.6. The emitter (*electron*) current, i_E, is made up of electrons crossing from the emitter region into the base and holes from the base crossing into the emitter. Some electrons from the emitter

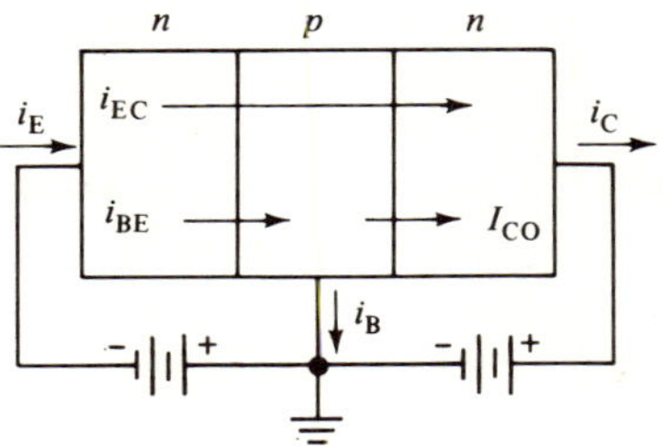

Figure 3.6

recombine with holes in the base region. Since the base is very narrow and lightly doped, however, the amount of recombination that takes place is small and the vast majority of electrons get to the collector junction. At the collector junction these electrons come under the influence of supply voltage V_{CC}, and are swept across the junction into the collector region.

Because the crystal must remain electrically neutral, when an electron is captured by a hole in the base region, an electron must flow *out* of the base lead. By the same reasoning, every hole from the base region that crosses into the emitter leaves an excess negative charge in the base region and this, too, forces an electron *down* the base lead. Thus the total base current, labeled i_B in Fig. 3.6, is made up of the current resulting from hole-electron recombination in the base and the movement of holes from the base into the emitter. It should be noted, however, that the portion of i_B due to recombination is quite small. The total electron and hole current across the base-emitter junction which contributes to the base current is termed i_{BE}. Now, if we label that part of the emitter current which reaches the collector i_{EC}, we can say that

$$i_E = i_{BE} + i_{EC} \qquad (3\text{-}1)$$

Let us now turn our attention to the collector-base section of the transistor, which is sometimes referred to as the *collector diode*. The leakage current of this effective diode is termed I_{CO} and is defined as the reverse bias collector current with the emitter open. Sometimes, I_{CO} is written I_{CBO}, but the meaning is identical. Current I_{CO} consists, of course, of thermally generated electrons flowing from the base to the collector and holes flowing from the collector to the base. Again, because the crystal must remain electrically neutral at all times, each thermally generated electron that passes from the base to the collector must be replaced by an electron flowing *up* the base lead into the transistor. This component of base current opposes i_B. In like manner, every thermally generated hole, in crossing from the collector to the base, places an extra positive charge in the base. To neutralize this positive charge, an electron flows *up* the base lead and this, too, is opposite to the direction of i_B. In Fig. 3.6 both of these components of base current are represented by I_{CO}.

Normally, a resistance ranging from about 10 to 200 Ω exists in the base lead and I_{CO} has no effect on i_{EC}. If the value of this resistance is much greater than about 200 Ω, however, the number of electrons moving *up* the base lead may become inadequate to neutralize the thermally generated holes moving into the base from the collector. Under these conditions, some of the holes may cross into the emitter region and recombine with electrons, thereby reducing i_{EC}.

For a *pnp*-type transistor, the analysis would be similar to that presented for the *npn* variety except that the majority and minority carriers as well as the polarities of the external sources of emf would be reversed.

Regardless of the type of transistor used, we can sum up all the transistor current relationships as follows:

$$i_E = i_{BE} + i_{EC}$$

$$i_C = i_{EC} + I_{CO} \tag{3-2}$$

$$i_B = i_{BE} - I_{CO} \tag{3-3}$$

and
$$i_E = i_B + i_C \tag{3-4}$$

It has been noted that most of the charge carriers injected into the base from the emitter reach the collector. The precise value of this quantity is given by the relationship

$$\alpha = \frac{i_{EC}}{i_E} \tag{3-5}$$

where α = the Greek letter alpha = *forward current transfer ratio*. Typical values of α range from about 0.9 to 0.99.

By transposition, Eq. 3-5 may be rewritten as

$$i_{EC} = \alpha i_E \tag{3-6}$$

and Eqs. 3-1 and 3-2 may be rewritten as

$$i_E = \alpha i_E + i_{BE} \tag{3-7}$$

and
$$i_C = \alpha i_E + I_{CO} \tag{3-8}$$

At this point, an example will serve to indicate typical values for the currents one encounters in dealing with transistor circuits.

EXAMPLE 1

A transistor has an alpha of 0.98 and, at room temperature, $I_{CO} = 10\ \mu\text{A}$. The emitter current is 10 mA. Calculate the collector and base currents.

SOLUTION

$$
\begin{aligned}
i_C &= \alpha i_E + I_{CO} \\
&= 0.98\,(10\ \text{mA}) + 10\ \mu\text{A} \\
&= 9.8\ \text{mA} + 0.01\ \text{mA} \\
&= 9.81\ \text{mA}
\end{aligned}
$$

and
$$i_B = i_E - i_C$$
$$= 10 \text{ mA} - 9.81 \text{ mA}$$
$$= 0.19 \text{ mA}$$
$$= 190 \ \mu\text{A}$$

From the example just given it should be noted that the collector current is just slightly smaller than the emitter current and that, at room temperature, leakage current is negligible.

The transistor configuration discussed thus far is known as the common-base (CB) configuration because the base terminal is common to both the emitter and collector circuits. Alternatively, this configuration is called the grounded-base configuration because the base is usually connected to ground.

Although the CB configuration is useful from the standpoint of illustrating basic transistor action, the practical application of this configuration is quite limited. The most common transistor configuration by far is the so-called common-emitter (CE) configuration.

3-4   THE RELATIONSHIP BETWEEN α AND β

Suppose, in the example just given, i_E is increased from 10 to 11 mA by increasing the value of V_{EE}. Since the value of I_{CO} does not change without a change in temperature, it remains the same, that is, 10 μA. This assumes, of course, that the value of V_{CC} is less than the breakdown voltage. Then, by Eq. 3-8,

$$i_C = \alpha i_E + I_{CO}$$
$$= (0.98)(11) + 0.01$$
$$= 10.78 + 0.01$$
$$= 10.79 \text{ mA}$$

From Eq. 3-4

$$i_B = i_E - i_C$$
$$= 11 - 10.79$$
$$= 0.21 \text{ mA}$$
$$= 210 \ \mu\text{A}$$

Notice that an increase in i_E causes a corresponding increase in both i_B and i_C. Refer to Table 3-1. The column labeled OLD lists the currents computed for the

TABLE 3-1

i (mA)	Old	New	Change
i_E	10	11	1.0
i_C	9.81	10.79	0.98
i_B	0.19	0.21	0.02

conditions of example 1. The NEW column contains the values found in the problem just solved, and the last column lists the CHANGE in the values of these currents.

Several important conclusions can be drawn from Table 3-1. First, notice that the change in i_E is equal to the sum of the changes in i_C and i_B, that is,

$$\Delta i_E = \Delta i_C + \Delta i_B \tag{3-9}$$

Second, when a transistor is connected in the CB configuration, the ratio of the change in i_C to the change in i_E is of great interest. Notice that this ratio, as previously defined, is alpha. To show the incremental changes in i_C and i_E, however, the equation for alpha is written as

$$\alpha = \frac{\Delta i_C}{\Delta i_E} \tag{3-10}$$

Finally, from the data in Table 3-1, we can determine another very important parameter called beta. Beta, represented by the Greek letter symbol β, is defined as the ratio of a change in collector current to the corresponding change in base current, that is,

$$\beta = \frac{\Delta i_C}{\Delta i_B} \tag{3-11}$$

The usefulness of beta will become apparent when the so-called common-emitter (CE) configuration is discussed.

Since α and β both represent current ratios, the units of measurement cancel and both quantities are expressed as pure numbers. Using the information contained in Table 3-1, $\alpha = 0.98$ and $\beta = 49$.

Alpha and beta may also be expressed in terms of one another as follows:

$$\beta = \frac{\Delta i_C}{\Delta i_B}$$

But,
$$\Delta i_B = \Delta i_E - \Delta i_C$$

so
$$\beta = \frac{\Delta i_C}{\Delta i_E - \Delta i_C}$$

Now, dividing $\Delta i_C / (\Delta i_E - \Delta i_C)$ by Δi_E gives

$$\beta = \frac{\dfrac{\Delta i_C}{\Delta i_E}}{\dfrac{\Delta i_E}{\Delta i_E} - \dfrac{\Delta i_C}{\Delta i_E}}$$

Then, since $\alpha = \Delta i_C / \Delta i_E$,

$$\beta = \frac{\alpha}{1 - \alpha} \tag{3-12}$$

Multiplying both members of Eq. 3-12 by $1 - \alpha$,

$$\beta(1 - \alpha) = \alpha$$
$$\beta - \beta\alpha = \alpha$$
$$\beta = \alpha + \beta\alpha$$
$$\beta = \alpha(1 + \beta)$$
$$\alpha = \frac{\beta}{1 + \beta} \tag{3-13}$$

Both alpha and beta are quantities associated with a particular transistor. If one of these quantities is known, either from direct measurement or by reference to the manufacturer's specification sheet, the other can be found by using either Eq. 3-12 or Eq. 3-13.

EXAMPLE 2

Find the β of a transistor whose $\alpha = 0.98$.

SOLUTION

From Eq. 3-12

$$\beta = \frac{\alpha}{1 - \alpha}$$
$$= \frac{0.98}{1 - 0.98}$$
$$= \frac{0.98}{0.02}$$
$$= 49$$

EXAMPLE 3

A certain transistor has an alpha of 0.99. Find its β.

SOLUTION

From Eq. 3-12

$$\beta = \frac{\alpha}{1 - \alpha}$$
$$= \frac{0.99}{1 - 0.99}$$
$$= 99$$

EXAMPLE 4

A certain transistor has an alpha of 0.995. Find its β.

SOLUTION

Using Eq. 3-12 once again, we would find that

$$\beta = 199$$

These three examples show that β increases rapidly as α approaches unity.

EXAMPLE 5

A transistor has a β of 95. Find its α.

SOLUTION

From Eq. 3-13

$$\alpha = \frac{\beta}{1+\beta}$$

$$= \frac{95}{1+95}$$

$$= 0.989$$

Obviously, α can never exceed unity, regardless of the value of β.

3-5   THE COMMON-EMITTER CONFIGURATION

Figure 3.7(a) shows an *npn* transistor connected in the common-emitter configuration. The emitter terminal is now common to both base and collector circuits and,

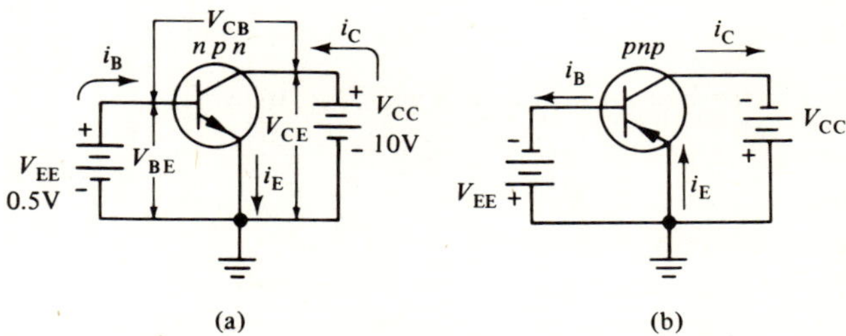

(a)   (b)

Figure 3.7

with the emitter grounded, all potentials are stated with respect to ground. Battery V_{EE} is connected directly across the base-emitter junction with the positive terminal tied to the base. Thus, the base-emitter junction is forward-biased and the base-to-emitter voltage is equal to V_{EE}. Battery V_{CC} is connected between the collector and emitter with the positive terminal tied to the collector. The collector-to-emitter voltage is, therefore, 10 V. The voltage across the collector-base junction, by Kirchhoff's voltage law, is equal to the difference between V_{CE} and

V_{BE}—about 9.5 V. The base is 0.5 V above the emitter (ground) and the collector is 10 V above the emitter; therefore, the collector is 9.5 V above the base. Because the n-type collector is more positive than the p-type base, the collector to base junction is reverse-biased. A voltmeter used to measure V_{CB} would, therefore, have to be connected with the positive lead going to the collector. Notice that, although the common-emitter configuration is wired differently than the common-base configuration, the base-emitter junction is still forward-biased and the collector-base junction is still reverse-biased. These are the conditions stipulated previously for normal operation of a transistor.

Figure 3.7(b) shows a *pnp* transistor connected in the common-emitter configuration. To maintain the same conditions existing in Fig. 3.7(a), the polarities of both batteries are reversed. Of course, if it is necessary to measure V_{CB}, the negative lead of the voltmeter would go to the collector.

Notice the direction of each current in Figs. 3.7(a) and (b). Because the base-emitter junction is forward-biased and the collector-base junction is reverse-biased, the (*conventional*) currents have the same direction as in the CB configuration and all of the current relationships developed for the CB circuit are applicable, including those for alpha and beta. Notice that the collector current is supplied by battery V_{CC} and the base current by battery V_{EE}. These two currents combine inside the transistor and emerge from the emitter as i_E. When i_E arrives at the ground terminal, it splits; i_B returns to V_{EE} and i_C to V_{CC}.

The collector-base junction is reverse-biased as long as V_{CC} is greater than V_{EE}. This poses no problem, since V_{EE} must be kept small to limit the amount of current forced into the base. If V_{EE} is excessive, i_B will be so great as to damage the transistor. In fact, as you will see shortly, a current-limiting resistor is always included in series with the base and supply voltage V_{EE}.

It is extremely important to know for certain the type of transistor to be used in a given circuit. If a transistor of unknown type is inserted into a circuit it may be damaged. Suppose, for example, a *pnp*-type transistor is inserted into a circuit that is biased for an *npn* unit. Under these conditions, the base-emitter junction would be reverse-biased and the collector-base junction would be forward biased—the reverse of normal operating conditions. The high forward bias applied to the collector-base junction would probably ruin the transistor, regardless of the circuit arrangement.

An ohmmeter containing a single 1.5 V battery can be used to determine the transistor type. Connect the ohmmeter across the collector-base junction with the positive lead of the meter touching the collector lead. If the ohmmeter reads a few thousand ohms or less, the collector-base junction is forward-biased and the transistor is of the *pnp* variety. If, on the other hand, the ohmmeter gives a reading of several hundred kilohms to a few megohms, the collector-base junction is reverse-biased and the transistor is an *npn* type.

In making the above noted test, do not use the low-resistance scale of the ohmmeter because excessive current might be drawn if the junction is forward-biased. To be completely safe, insert a 1,000-ohm resistor in series with the positive lead of the ohmmeter.

The ohmmeter test may be used to advantage in checking practical circuits where the condition of one or more transistors is suspect. Remove each transistor from the circuit and check each of its two junctions in both the forward and reverse directions. If the transistor is not damaged you will obtain a high resistance in one direction and a low resistance in the other direction for each junction.

3-6 THE COMMON-EMITTER OUTPUT CHARACTERISTICS

In chapter 1 it was shown that a characteristic curve is the plot of current vs. voltage for a particular device, and several characteristics were drawn for two-terminal devices. Because the transistor is a three-terminal device, we may choose which two electrodes to use in constructing the characteristic curve. For reasons which will be made clear shortly, the collector and emitter are selected; however, the effect of the base must be considered.

To obtain the characteristic curve of a transistor, use is made of the circuit shown in Fig. 3.8. The base-emitter junction is forward biased by supply V_{EE}.

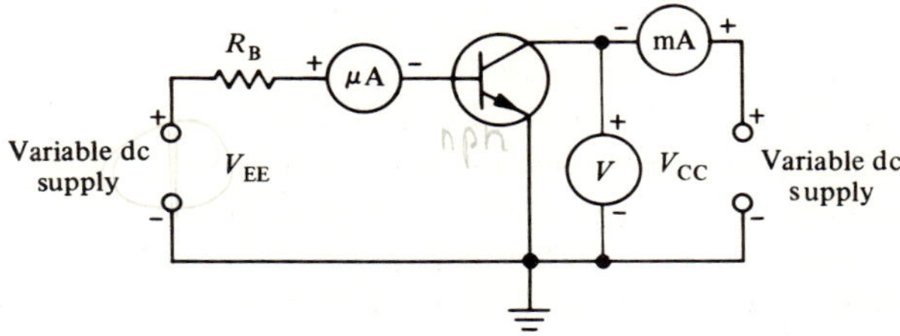

Figure 3.8

A resistor, labeled R_B, is inserted in the base lead to limit base current and a dc microammeter is used to measure this current. V_{EE} is made variable to supply a means of controlling the base current; a change in V_{EE} causes a change in i_B. The collector dc supply, V_{CC}, is also variable. The collector-to-emitter voltage is, of course, equal to V_{CC} since there is no resistor in the collector circuit. The magnitude of this voltage is indicated by a dc voltmeter. A dc milliammeter indicates the value of collector current.

To obtain information for plotting a family of characteristic curves, V_{EE} is set to some value and the resulting base current is recorded. For the sake of illustration, let $i_B = 0.05$ mA (50 μA). Supply V_{CC} is then varied in equal increments from zero volts to some larger value that is less than the breakdown voltage of the collector-base diode. Again, for the purpose of illustration, suppose V_{CC} is varied from 0 to 50 V in 5-volt increments. At each increment of V_{CC} the resulting i_C is recorded. During this procedure it may be necessary to readjust V_{EE} to keep i_B constant at 50 μA. The accumulated data is set down in tabular form. Then, V_{EE} is increased to provide increments in i_B of, perhaps, 50 μA, and, at each increment, the above described procedure is repeated.

The accumulated data are then plotted on a graph similar to that shown in

Fig. 3.9. Notice that i_C is plotted vs. v_{CE} and that a different characteristic curve results for each value of i_B. These curves are referred to collectively as a family of common-emitter output characteristics since they relate to the output circuit formed by supply V_{CC}, the collector and the emitter. A study of the curves shown

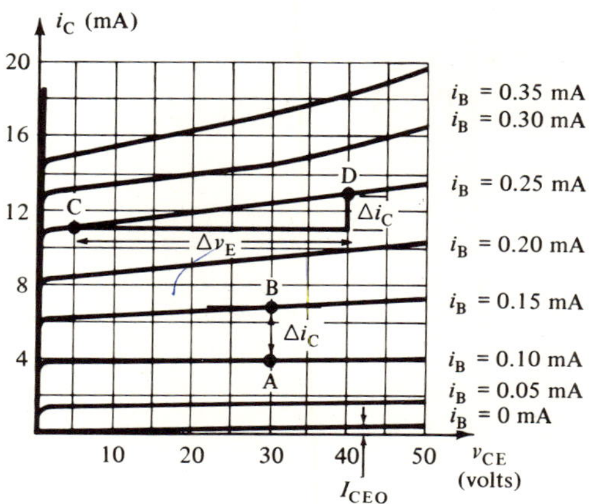

Figure 3.9

in Fig. 3.9 reveals that, as i_B increases, i_C also increases for a given value of V_{CE}. The collector current drawn when i_B is 250 μA and V_{CE} is 35 V, for example, is much larger than that which occurs when i_B is 50 μA and V_{CE} is 35 V. Another point to note is that, at the low values of i_B, i_C is practically independent of V_{CE}; e.g., when i_B is 50 μA, increasing V_{CE} from 5 to 50 V has almost no effect on i_C which remains essentially constant at 1.8 mA. At higher values of i_B, however, an increase in V_{CE} causes a noticeable increase in i_C. As a result, the slope of the curves at the higher values of i_B is much steeper than at the lower values of i_B.

It is important to note that the graph is a plot of i_C vs. v_{CE} and that only these two values can be read directly from the graph. The indicated values of i_B are constants that serve only to identify a given curve.

Each characteristic curve in Fig. 3.9 is similar to those studied in chapter 1, since the collector and emitter by themselves may be considered to be a diode having a nonlinear characteristic. The essential difference between the curves of Fig. 3.9 and those discussed earlier is that the base current determines which characteristic represents the combined behavior of the collector and emitter. An infinite number of characteristic curves could be drawn, of course, for a given transistor. For practical purposes, however, about seven or eight curves usually suffice.

Close observation of Fig. 3.9 reveals that there is also a curve drawn for the condition $i_B = 0$. To obtain this condition, V_{EE} is set to zero volts and this makes i_B zero. Alternatively, the base lead may be open circuited. Although the collector current under these conditions is very small, it is not zero; rather, a leakage current termed I_{CEO} (collector-to-emitter current with the base open) exists.

Since, under these conditions, i_B is zero, i_C must be equal to i_E; i.e., the current into the collector is the same as that coming out of the emitter. The magnitude of this current is very small, but, as we shall see later, its value is an important piece of information for the circuit designer to know.

Recalling the current relationships developed in preceding topics, we can perform a few simple algebraic manipulations to rearrange the terms in a more meaningful form as follows:

$$i_C = \alpha i_E + I_{CO}$$

but

$$i_E = i_C + i_B$$

Then,

$$i_C = \alpha(i_C + i_B) + I_{CO}$$
$$i_C = \alpha i_C + \alpha i_B + I_{CO}$$
$$i_C - \alpha i_C = \alpha i_B + I_{CO}$$
$$i_C (1 - \alpha) = \alpha i_B + I_{CO}$$

$$i_C = \left(\frac{\alpha}{1 - \alpha}\right) i_B + \left(\frac{1}{1 - \alpha}\right) I_{CO}$$

However,

$$\frac{\alpha}{1 - \alpha} = \beta$$

and

$$\frac{1}{1 - \alpha} = \frac{\beta}{\alpha} = \frac{\beta}{\dfrac{\beta}{\beta + 1}} = \beta + 1$$

Therefore,

$$i_C = \beta i_B + (\beta + 1)I_{CO}$$

and

$$i_C = \beta i_B + I_{CEO} \tag{3-14}$$

where

$$I_{CEO} = (\beta + 1)I_{CO}$$

If i_B is set to zero in Eq. 3-14, the equation becomes

$$i_C = I_{CEO} = i_E$$

which shows that, when $i_B = 0$, i_C and i_E result entirely from I_{CEO}. The physical reasons why I_{CEO} is much larger than I_{CO} are rather involved and, because they do not contribute to the practical understanding of transistor operation, no attempt is made here to explain these reasons. It is entirely sufficient to recognize that I_{CEO} is, indeed, larger than I_{CO}.

EXAMPLE 6

Let the leakage current I_{CO} for a given transistor be 1 μA at room temperature. If $\beta = 60$, find I_{CEO}.

SOLUTION

Using the equation just derived,

$$I_{CEO} = (\beta + 1)I_{CO} = 61 \times 1 = 61 \ \mu A$$

3-7 TRANSISTOR PARAMETERS

Parameters are quantities used to describe the behavior of various devices. We have already discussed in detail two major parameters of transistors; viz., alpha and beta. We defined β as the ratio of a change in i_C to a change in i_B. We now put one further restriction on this definition. Beta is fully defined as the ratio of a change in i_C to a change in i_B with the collector-to-emitter voltage, V_{CE}, held constant. As an equation,

$$\beta = \left(\frac{\Delta i_C}{\Delta i_B}\right)_{V_{CE}=K} \tag{3-15}$$

Notice that β, as defined above, is an ac quantity. This quantity is often referred to in the literature as h_{fe}. There is also a dc beta associated with a transistor and this is referred to as h_{FE}. More is said about the dc beta at a later point in the discussion. For the moment, let us concern ourselves with the ac beta, β, i.e., h_{fe} only.

Let us calculate β for the characteristic of Fig. 3.9 by assuming a change in i_B from 100 to 150 μA while V_{CE} is held constant at 30 V. Note that, for a constant value of V_{CE}, it is necessary to move along a vertical line on the graph, in this case, from point A, 100 μA, to point B, 150 μA. The incremental change in i_C between these two points is $7 - 4 = 3$ mA. Then,

$$\beta = \left(\frac{\Delta i_C}{\Delta i_B}\right)_{V_{CE}=30 \text{ V}} = \frac{7 - 4}{0.15 - 0.10} = \frac{3 \text{ mA}}{0.05 \text{ mA}} = 60$$

Now, let us compute the value of β when i_B is changed from 0 to 0.05 mA and V_{CE} is held constant at 30 V (assuming $I_{CEO} \simeq 0$).

$$\beta = \left(\frac{\Delta i_C}{\Delta i_B}\right)_{V_{CE}=30 \text{ V}} = \frac{1.8 - 0 \text{ mA}}{0.05 - 0 \text{ mA}} = 36$$

From these two examples it should be quite obvious that the β of a transistor is not a constant. In both examples the incremental change in base current is 0.05

mA; however, the incremental change in collector current is greater in the first example than in the second. The reason for this is that, up to a certain point, the curves are spaced farther apart as i_C increases. This is always true when the curves are drawn for equal increments of base current. For the transistor having the characteristic curves depicted in Fig. 3.9, β is maximum when i_C is about 5 mA. At both higher and lower values of i_C, the curves are closer to one another and the value of β decreases. Manufacturers generally specify the maximum β point.

As noted previously, a transistor also has a dc beta. By definition,

$$h_{\text{FE}} = \beta_{\text{DC}} = \left(\frac{I_C}{I_B}\right) \tag{3-16}$$

To calculate the dc beta we take the ratio of the *total* collector current at a given point on a characteristic curve to the *total* base current at the same point.

Using the characteristic curves of Fig. 3.9, the dc beta at point B is

$$h_{\text{FE}} = \beta_{\text{DC}} = \left(\frac{I_C}{I_B}\right) = \frac{7 \text{ mA}}{0.15 \text{ mA}} = 46.6$$

As will be seen later, both the ac and dc betas are important parameters in the design of an amplifier.

At this point it should prove worthwhile to develop other dc relationships for the CE amplifier. The ac case has already been considered.

In the CE amplifier, if the emitter is dc open-circuited, and a voltage is applied to the collector that reverse biases the collector-to-base junction, a small collector current exists. This current is called the reverse (leakage) current of the collector-to-base junction, and is denoted either as I_{CO} or I_{CBO}. This leakage current exhibits wide variance in transistors of similar types, and is extremely sensitive to temperature. At a specified temperature, however, I_{CO} is considered a constant for a particular transistor. For the time being we simply note that variations in I_{CO} have a very marked effect upon the operating point. More is said about this matter later.

From the previous discussion, it follows that

$$I_C = \alpha I_E + I_{CO} \tag{3-17}$$

at any particular collector potential.

In a CE amplifier, if the collector is dc open-circuited and a voltage is applied from emitter to base that reverse biases the base-emitter junction, a leakage current, designated either as I_{EO} or as I_{EBO}, occurs. This leakage current is called the reverse leakage current of the base-to-emitter junction, and it is also sensitive to temperature.

Again, in a CE amplifier, if the base is dc open circuited, and a voltage is applied from collector to emitter that reverse biases the collector-to-base junction, a collector current exists. This current is termed I_{CEO}. The relative magnitude of I_{CEO} is indicated by noting that $I_E = I_C + I_B$, and substituting for I_E in Eq. 3-17. Then,

$$I_C = \alpha(I_C + I_B) + I_{CO} \tag{3-18}$$

Rearranging

$$I_C = \frac{\alpha I_B}{1 - \alpha} + \frac{I_{CO}}{1 - \alpha} \tag{3-19}$$

If $\beta = \alpha/1 - \alpha$ is substituted into Eq. 3-19,

$$I_C = \beta I_B + (\beta + 1)I_{CO} \tag{3-20}$$

As noted above, when the base is dc open circuited, $I_C = I_{CEO}$. For any particular collector potential, therefore,

$$I_{CEO} = (\beta + 1)I_{CO} \tag{3-21}$$

Then Eq. 3-20 becomes

$$I_C = \beta I_B + I_{CEO} \tag{3-22}$$

In summary, I_{CO} is the collector-to-base leakage current when a normal collector potential is applied and the input circuit (emitter) is open for the CB configuration. For the CE configuration, collector current is I_{CEO} when the input circuit (base) is open. I_{CEO} is many times greater than I_{CO}.

Another parameter that is useful in the design of an amplifier is the *dynamic-output resistance* termed r_O. The so-called *h*-parameter that corresponds to r_O is $1/h_{oe}$. Just remember, r_O and $1/h_{oe}$ refer to the exact same thing—the dynamic output resistance of the transistor. By definition,

$$r_O = \frac{\Delta v_{CE}}{\Delta i_C} \tag{3-23}$$

From chapter 1 you will recall that the dynamic resistance of a device is equal to the reciprocal of the slope of a line drawn tangent to its characteristic curve at some point. In Fig. 3.9, a line tangent to any point on the flat portion of the curve marked $i_B = 0.25$ mA is tangent to every point along that portion of the curve. To find r_O for this curve, take the reciprocal of the slope of the line C–D as follows:

$$r_O = \frac{\Delta v_{CE}}{\Delta i_C} = \frac{40\text{ V} - 5\text{ V}}{13\text{ mA} - 11\text{ mA}} = \frac{35\text{ V}}{2\text{ mA}} = 17.5\text{ k}\Omega$$

Compare this with the value of r_O for the curve labeled $i_B = 0.05$ mA. Since the latter curve is almost horizontal, it will be found that r_O approaches infinity. Thus, the output resistance tends to decrease as i_B increases. The value of r_O for most transistors ranges from about 10 kΩ to 50 kΩ. As you will see, r_O is very important in practical amplifier circuits.

3-8 THE COMMON-EMITTER INPUT CHARACTERISTICS

The circuit used to obtain the data needed to plot the input characteristics of a transistor is shown in Fig. 3.10(a). Recall that the output characteristics were plots of i_C vs. v_{CE} with i_B held constant. The input characteristics are plots of i_B

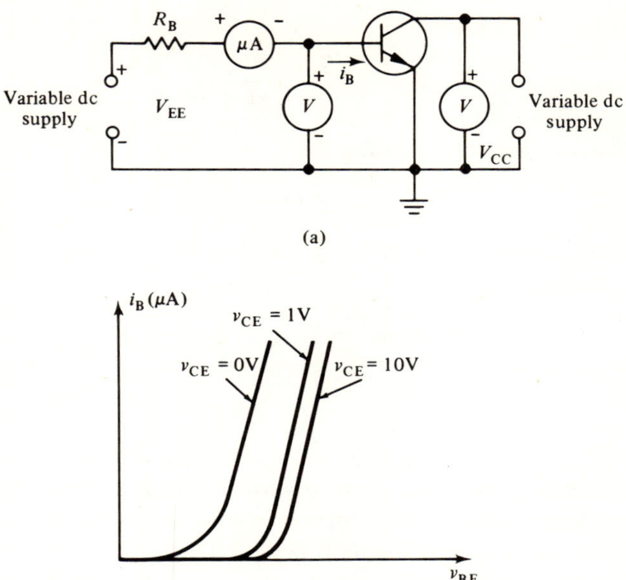

Figure 3.10

vs. v_{BE} with v_{CE} held constant. Initially, V_{CC} is set to some desired value of v_{CE}, say, 1.0 V. Then V_{EE} is varied in incremental steps over a given range. For each new value of V_{EE} the base current and base-to-emitter voltage are recorded. At each new value of V_{EE} it may be necessary to readjust V_{CC} to maintain a constant collector-to-emitter voltage. In the next run, V_{CC} is set to a new value, e.g., 2 V, and the above procedure is repeated. This is done for a number of different values of V_{CC} and the resulting data is tabulated.

Figure 3.10(b) shows the general slope of the characteristics for three values of v_{CE}; viz., 0, 1, and 10 V. Notice that these curves have the same general shape as those of the *pn* diodes studied previously. This is not surprising. Since the base-to-emitter junction is forward biased by V_{EE}, the base-emitter junction is, in effect, a forward-biased diode.

It can be seen from Fig. 3.10(b) that the behavior of the emitter-base diode depends upon the value of the collector-to-emitter voltage v_{CE}. For each value of v_{CE} there is a different characteristic curve. Between the curves for $v_{CE} = 0$ V and $v_{CE} = 1$ V, there is a relatively large space. Between $v_{CE} = 1$ V and $v_{CE} = 10$ V, however, the spacing is quite small. If curves were drawn for each one-volt increment between 1 and 10 V, the curves would be crowded very close to one another and all of the curves would have approximately the same slope. Since, as you will see, the transistor amplifier is usually operated at collector-to-emitter voltages greater than 1 V, the assumption may be made that the input characteristic is relatively *independent* of v_{CE}. In view of the fact that the characteristics of several transistors of the same type are never identical, this approximation is well justified. In addition, the use of this approximation acts to simplify to a considerable extent the methods of practical circuit analysis.

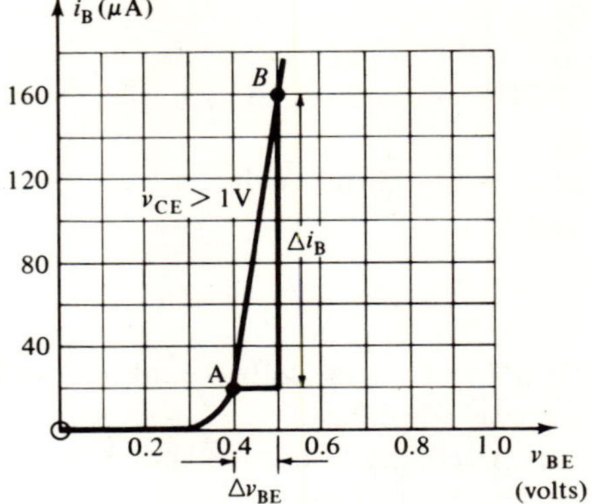

Figure 3.11

Figure 3.11 shows a single input characteristic that approximates all input characteristics when v_{CE} is greater than one volt. Even if v_{CE} is less than one volt, the approximation is quite good for most applications. Using this characteristic, we can derive the *dynamic-input resistance, r_I,* of a transistor. Manufacturers usually refer to r_I as h_{ie}. By definition,

$$r_I = \frac{\Delta v_{BE}}{\Delta i_B} \tag{3-24}$$

EXAMPLE 7

Using the input characteristic of Fig. 3.11, find the value of r_I between points A and B.

SOLUTION

Applying the definition of dynamic resistance, the reciprocal of the slope of the line A–B is found as follows:

$$r_I = \frac{\Delta v_{BE}}{\Delta i_B} = \frac{0.1 \text{ V}}{140 \ \mu\text{A}} = 710 \ \Omega$$

Notice that the input resistance of a transistor is quite small compared to its output resistance. Typical values of r_I range from about 500 Ω to 2 kΩ.

3-9   TEMPERATURE EFFECTS

A major problem in transistor amplifier design is the variation in device parameters with changes in temperature. Strictly speaking, all of the transistor parameters are temperature sensitive. From the practical point of view, however, only β and I_{CEO} are important.

Since the common-emitter amplifier is, by far, the most common transistor amplifier, β is the single most important transistor parameter. In a typical transistor, β varies over a two-to-one range in the temperature range from -55 to $+95°C$. The β of a type 2N738 transistor, for example, changes from 25 at $-55°C$ to about 60 at $95°C$.

The current from collector to emitter with the base open, I_{CEO}, is also temperature sensitive. You recall that

$$I_{CEO} = (\beta + 1)I_{CO}$$

This means that I_{CEO} is a function of both β and I_{CO} and must be taken into consideration when designing amplifiers.

Now, we know from the preceding discussion that β increases with temperature. We also know that I_{CO} doubles for about every $10°C$ increase in temperature. Thus, I_{CEO} will also increase with an increase in temperature. Of course, I_{CO} has the greater effect on I_{CEO} since it increases at a much more rapid rate than β.

There is another quantity that varies with temperature that has not yet been mentioned. In discussing the characteristics of a forward-biased diode and the input characteristics of a transistor, it was noted that, when the voltage across the forward-biased junction exceeds a few tenths of a volt, a large forward current commences. The voltage at which this current begins is about 0.3 V for germanium and about 0.6 V for silicon, at normal room temperatures. At higher temperatures, however, this voltage *decreases* at the rate of 2.5 mV per $°C$. Fortunately, by using suitable biasing methods this variation can be made negligible for most designs.

EXAMPLE 8

In a given transistor operating at a normal room temperature of $25°C$, a heavy forward current occurs when v_{BE} exceeds 0.4 V. Determine the value of v_{BE} at which the heavy forward current commences when the ambient temperature is $95°C$.

SOLUTION

$\Delta T = 95°C - 25°C = 70°C.$

The change in v_{BE} is $\left(\dfrac{2.5 \text{ mV}}{°C}\right)(70°C) = 175 \text{ mV} = 0.175 \text{ V}.$

Then, heavy forward current begins when

$$v_{BE} = 0.4 - 0.175 = 0.225 \text{ V}$$

3-10   BETA SPREAD

It has been shown that β varies as a result of changes in collector current and ambient temperature. There is also a third and even more crucial factor that tends to vary the beta of a transistor—manufacturing tolerances in transistor reproducibility. As a result of these tolerances, the β of a given type of transistor, operating at a specified collector current and temperature, will generally vary from one unit to another. This variation is referred to as the *beta spread* of a transistor.

Manufacturers usually give the minimum guaranteed β at a specified value of collector current and temperature. Sometimes, however, they give a typical value for a large number of samples. Consider the type 2N738 transistor, for example. The β of this unit varies from a minimum of 20 to a maximum of 50 when the collector current is 5 mA, the collector to emitter voltage is 5 V, and the unit is operating at room temperature. For this particular unit, a minimum β of 12 is guaranteed when the temperature is $-55°C$.

It is sometimes necessary to use a group of transistors of a given type having a smaller than normal β spread. Transistors of this type are said to be tightly specified. Since the manufacturer must select these units by actual measurement, a considerable cost factor is involved.

3-11   dc ANALYSIS OF THE COMMON-EMITTER CONFIGURATION

Refer to Fig. 3.12. The only difference between this circuit and those discussed earlier is the inclusion of a load resistor, R_L, in series with the collector. Now, V_{CE}

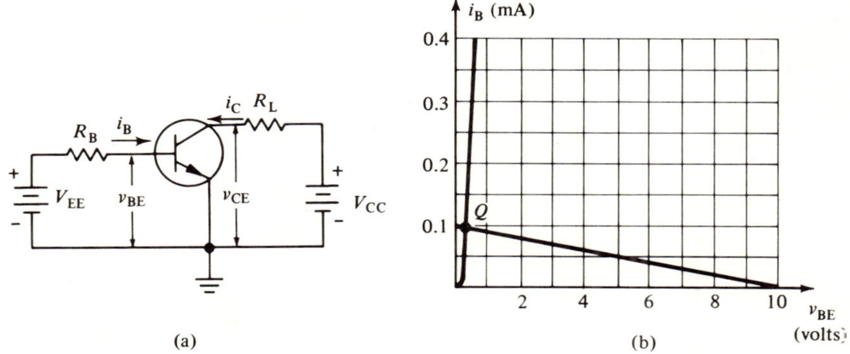

Figure 3.12

is no longer equal to V_{CC}; rather, $V_{CE} = V_{CC} - V_{RL}$. To analyze such a circuit, it is necessary to calculate the values of I_B, I_C, and V_{CE}.

There are a number of methods by which the circuit of Fig. 3.12(a) can be analyzed; the method selected usually depends upon the information available, the accuracy required, and the amount of time one has to spend on the calculations. A few of these methods are illustrated in the following example.

EXAMPLE 9

In the circuit of Fig. 3.12(a), let $V_{EE} = 10$ V, $R_B = 100$ kΩ, $V_{CC} = 50$ V, and $R_L = 5$ kΩ. Determine I_B, I_C, and V_{CE}.

SOLUTION

The input and output characteristics of the transistor used in this example are shown in Figs. 3.12(b) and 3.13, respectively. Let us first calculate I_B. We know

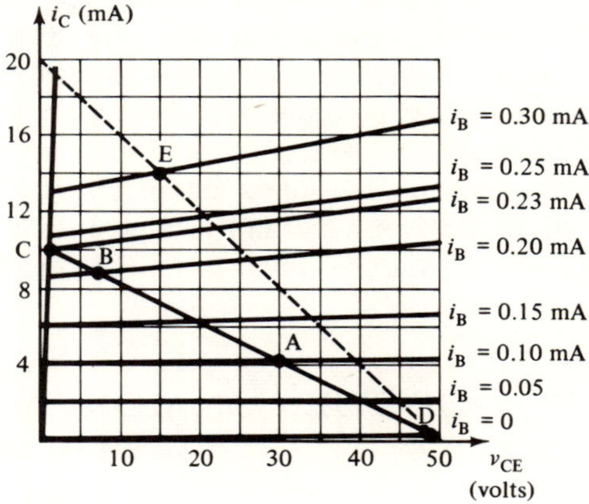

Figure 3.13

that the base-to-emitter junction of the transistor is forward-biased and acts like a forward-biased *pn* diode. The base and emitter portion of the transistor, therefore, can be depicted as a simple diode. See Fig. 3.14(a). In Fig. 3.14(a) the dashed line represents the transistor as seen from the base, *B*, and emitter, *E*, terminals. The base current passes through the series circuit formed by V_{EE}, R_B, and the diode.

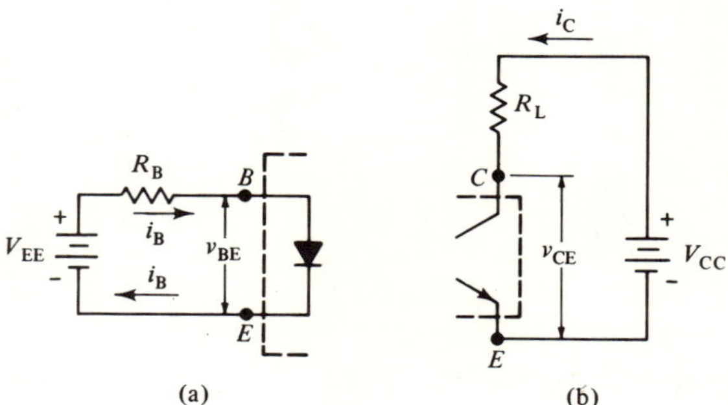

Figure 3.14

Thus, for the sake of analysis, Fig. 3.14(a) may be considered as a simple series circuit containing a battery, a resistor, and a nonlinear device. Recall from the studies in chapter 1 that such a circuit can be analyzed by the load-line technique, provided the characteristic of the nonlinear device is known. In this case, the characteristic is the input characteristic of the transistor, shown in Fig. 3.12(b). The load line is, therefore, constructed on the characteristic of Fig. 3.12(b) using

10,0 as one point (V_{EE} = 10 V), and 0,0.1 as the other point (10 V/100 kΩ = 0.1 mA).

The Q point, as indicated in the figure, is set at I_B = 0.096 mA (96 μA) and V_{BE} = 0.4 V. Most of the supply potential is dropped across R_B, 9.6 V, because the resistance of the base-emitter diode is relatively small.

The method just demonstrated for determining I_B is quite accurate; however, the input characteristic may not be available.

Here is a faster method for determining the value of I_B. We know that the base-to-emitter voltage rarely exceeds 0.6 or 0.7 V. We can, therefore, assume that V_{BE} remains relatively constant. Although V_{BE} actually increases slightly as I_B increases, this is a good assumption. If it is known that the forward voltage for a given transistor is, say, 0.4 V, the voltage drop across R_B is equal to V_{EE} − 0.4 V. Once the voltage drop across R_B is known, I_B can be found by the application of Ohm's law.

$$V_{R_B} = V_{EE} - V_{BE} = 10 - 0.4 = 9.6 \text{ V}$$

and
$$I_B = \frac{V_{R_B}}{R_B} = \frac{9.6 \text{ V}}{100 \text{ k}\Omega} = 0.096 \text{ mA} = 96 \ \mu\text{A}$$

This method is much faster than the first and requires no graphical construction. An even faster method is to neglect V_{BE} completely, and assume that the total supply voltage V_{EE} is dropped across R_B. With this assumption, R_B works out to be 100 μA. This result compares favorably with that obtained by using the more accurate methods. Circuit designers make frequent use of this approximation.

With I_B determined, the next step is to find I_C and V_{CE}. Refer now to Fig. 3.14(b). Note that V_{CC} is in series with R_L and the collector and emitter terminals of the transistor. Now, strictly speaking, this is not a series circuit because the current out of the emitter terminal is not equal to the current into the collector terminal. Since I_E is only slightly larger than I_C, however, we will assume that $I_C = I_E$.

As noted previously, the collector and emitter terminals of a transistor may be considered to be those of a nonlinear device whose characteristic curve is controlled by the value of base current. Whenever a dc source is in series with a resistor and a nonlinear device, and the characteristics of the device are known, the load line is a suitable analytical technique. The characteristics describing the behavior of the collector and emitter terminals are curves of I_C vs. V_{CE}, that is, the output characteristics of the transistor.

See Fig. 3.13. The solid load line is drawn on the collector characteristics with its extremities at V_{CE} = V_{CC} = 50 V and i_C = 10 mA. This load line intersects many characteristic curves; however, we know that the base current is approximately 0.1 mA, so the Q point is the intersection of the load line with the curve I_B = 0.1 mA (point A). The quiescent collector current and collector-to-emitter voltage are:

$$I_C = 4.2 \text{ mA} \quad \text{and} \quad V_{CE} = 29 \text{ V}$$

Our problem is now completely solved. We have found I_B, I_C, and V_{CE}. Whenever the values of I_C and V_{CE} are to be determined, the Q point is always located at the intersection of the load line with the appropriate base current characteristic.

EXAMPLE 10

Repeat the example just given, but make $V_{EE} = 20$ V.

SOLUTION

$$I_B = \frac{V_{EE}}{R_B} = \frac{20 \text{ V}}{100 \text{ k}\Omega} = 0.2 \text{ mA}$$

Since V_{CC} and R_L are unchanged, the load line is the same as that used previously. The Q point is the intersection of the load line with the 0.2 mA characteristic curve, that is, point B in Fig. 3.13, where $I_C = 8.6$ mA and $V_{CE} = 7$ V.

Notice that, although the base currents are much smaller than the collector currents, a change of 0.1 mA in the base current produces a change of 4.4 mA in the collector current. In other words, even though the base current is quite small, it exerts a considerable amount of control over the collector current. This is the basic principle of amplification and is elaborated upon in the next chapter. Notice in the last two examples that an increase in base current causes an increase in collector current and a decrease in the collector-to-emitter voltage.

EXAMPLE 11

In the circuit of Fig. 3.12(a), let $V_{EE} = 23$ V, $R_B = 100$ kΩ, $V_{CC} = 50$ V, and $R_L = 5$ kΩ. Determine I_B, I_C, and V_{CE}.

SOLUTION

$$I_B = \frac{V_{EE}}{R_B} = \frac{23 \text{ V}}{100 \text{ k}\Omega} = 0.23 \text{ mA}$$

The Q point is found at point C in Fig. 3.13. $I_C = 9.9$ mA and $V_{CE} = 0.5$ V.

EXAMPLE 12

Repeat the above problem, but make $V_{EE} = 30$ V.

SOLUTION

$$I_B = \frac{V_{EE}}{R_B} = \frac{30 \text{ V}}{100 \text{ k}\Omega} = 0.3 \text{ mA}$$

Notice in Fig. 3.13 that the load line does not intersect the 0.3 mA characteristic on its flat portion. At point C, the characteristics for base currents of 0.23, 0.25, 0.27, 0.3, mA, etc., are the same. With the load line drawn, base currents larger than 0.23 mA will not produce any further change in the Q point. Since, in this example, the base current is 0.3 mA, the Q point is the same as that found in the preceding example. For any base current greater than 0.23 mA the Q point will remain fixed at point C. When an increase in base current produces no further change in the Q point, the transistor is said to be *saturated*. The transistor can be brought out of saturation by either one of two methods. The first method is to decrease the base current to some value below 0.23 mA by either decreasing V_{EE} or increasing R_B. The second method involves decreasing R_L. If R_L is reduced to, say, 2.5 kΩ, the load line will shift to the position shown by the dashed curve in Fig. 3.13. Then, when $I_B = 0.3$ mA, the Q point is at point E and the transistor is no longer saturated.

EXAMPLE 13

Repeat the example given above, but remove V_{EE} from the base circuit and leave the base and emitter terminals open.

SOLUTION

The base current is zero. The original (solid) load line in Fig. 3.13 intersects the $I_B = 0$ curve at point D. Since the base is open, the current through the collector and emitter is I_{CEO}. At room temperature, this current is too small to be read directly from the curves. Thus, at room temperature we assume I_C is zero and $V_{CE} = V_{CC}$ (50 V, in this case). When I_B is zero, and the collector current is essentially zero, the transistor is said to be *cut-off*.

In Fig. 3.13 the portion of the characteristic below the $I_B = 0$ curve is called the cut-off region; that to the left of the vertical portion of the curves is the saturation region. The region between cut-off and saturation, where a change in I_B causes a change in I_C, is called the active region of the characteristic.

Problems similar to those presented in the above examples can also be solved without the use of characteristic curves, provided the approximate value of the dc beta of the transistor is known. Let us assume that the dc beta, that is h_{FE}, of the given transistor is 45. Since h_{FE} is the ratio I_C/I_B, once I_B is known, it is a simple matter to determine I_C.

EXAMPLE 14

In a CE amplifier circuit, let $V_{EE} = 10$ V, $V_{CC} = 50$ V, $R_B = 100$ kΩ, and $R_L = 5$ kΩ. Find I_B, I_C, and V_{CE}.

SOLUTION

$$I_B = V_{EE}/R_B = 10/10^5 = 0.1 \text{ mA}$$

From Eq. 3-22, $I_C = \beta I_B + I_{CEO}$. Neglecting I_{CEO},

$$I_C = h_{FE}I_B = 45 \times 0.1 \text{ mA} = 4.5 \text{ mA}$$
$$V_{CE} = V_{CC} - V_{RL} = 50 - 22.5 = 27.5 \text{ V}$$

The values for I_C and V_{CE} compare favorably with the values found in example 9 where identical values of V_{EE}, V_{CC}, R_B, and R_L were used. In that example, I_C was found to be 4.2 mA and V_{CE} was found to be 29 V.

EXAMPLE 15

Solve the problem of example 14 if $V_{EE} = 20$ V.

SOLUTION

$$I_B = V_{EE}/R_B = 20/10^5 = 0.2 \text{ mA}$$

$$I_C = h_{FE}I_B = 45 \times 0.2 \text{ mA} = 9.0 \text{ mA}$$

$$V_{CE} = V_{CC} - V_{RL} = 4.5 \text{ V}$$

EXAMPLE 16

Solve the problem of example 14 if $V_{EE} = 30$ V.

$$I_B = V_{EE}/R_B = 30/10^5 = 0.3 \text{ mA}$$

$$I_C = h_{FE}I_B = 45 \times 0.3 \text{ mA} = 13.5 \text{ mA}$$

$$V_{CE} = V_{CC} - V_{RL} = 50 - 67.5 = -17.5 \text{ V}$$

This example is identical to example 11 but uses a different technique of solution. In example 11, $I_C = 9.9$ mA and $V_{CE} = 0.5$ V.

When V_{CE} is computed to be negative, as in example 16, the transistor is saturated. The largest possible collector current is approximately 10 mA ($V_{CC}/R_L = 50/5{,}000 = 10$ mA) because V_{CE} cannot be smaller than 0 V. When I_C is computed to be greater than this value, the transistor has to be saturated and, in the usual case, V_{CE} is about 0 V.

This last example shows clearly that, when we use quick approximations to analyze a circuit, we must be prepared to evaluate the results realistically.

Here is a quick method of determining whether or not the transistor is operating in the saturation region. First, calculate the saturated collector current as follows:

$$I_{C(\text{sat})} = \frac{V_{CC}}{R_L}$$

Then, calculate I_C by using the relationship

$$I_C = h_{FE}I_B$$

If the collector current computed from the second expression is greater than that computed from the first, the transistor is operating in the saturation region. If, on the other hand, the first computed value is greater than the second, the transistor is operating in the active region as long as I_B is not zero.

It was noted earlier that h_{FE} varies with collector current, temperature, and beta spread. Thus, the solution for I_C using the relationship $I_C = h_{FE}I_B$ is, at best, a very rough approximation. In fact, it is only useful when the transistor is a tightly specified unit.

3-12   TRANSISTOR RATINGS

Now we come to the problem of the maximum ratings of a transistor. The quantities involved, i.e., current, power, voltage, and temperature, are values which should not be exceeded when using a particular transistor. If any of the maximum ratings are exceeded, the transistor will very likely be destroyed very quickly.

Many, but not all, manufacturers specify a maximum collector current. This information, if given, will appear along with the other ratings under a heading that reads something like this: Absolute Maximum Ratings at 25°C Free Air Temperature. Under no conditions should these ratings be exceeded if trouble-free operation is to be expected.

The next rating is that of collector power dissipation, termed P_C. The collector circuit is treated like a resistor and its power dissipation is equal to the product of V_{CE} and I_C. There is, of course, some power dissipated in the base circuit; however, since I_B and V_{BE} are relatively small compared to the corresponding quantities in the collector circuit, base-circuit dissipation is usually neglected. Just as with any resistor, supplying too much power to the transistor will raise the temperature of the device to the point where it will suffer physical damage due to overheating.

There is also a maximum allowable collector dissipation, termed $P_{C(max)}$. This rating is usually specified at an ambient temperature of 25°C. Along with $P_{C(max)}$ there is also a derating factor, expressed in mW/°C. The derating factor indicates how much the maximum allowable collector dissipation decreases for each °C rise in temperature above 25°C.

EXAMPLE 17

For a 2N738 transistor, $P_{C(max)}$ at 25°C ambient temperature is 0.5 W. The derating factor is 3.33 mW/°C. Find the maximum allowable dissipation of this unit at an operating temperature of 55°C.

SOLUTION

Since the temperature increases 30° above the reference temperature, the incremental change in P_C is

$$\Delta P_C = 30 \times 3.33 \text{ mW} = 100 \text{ mW}$$

Then,
$$P_{C(max)} = 0.5 - 0.1 = 0.4 \text{ W}$$

Some manufacturers also give $P_{C(max)}$ at the case temperature. Since the case is usually somewhat warmer than the ambient temperature, this rating is more conservative. The derating factor is also different for the case-temperature rating; e.g., in the case of the 2N738, the derating factor becomes 6.66 mW/°C. In this text the ambient rating is used. Further consideration is given to this problem in chapter 9.

Since $P_C = V_{CE}I_C$, the current can be made very large only if the voltage is kept very small, and vice versa. As long as the product of V_{CE} and I_C does not exceed $P_{C(max)}$, the transistor is operating within its maximum power rating.

Now, the equation of the product of two variables equal to a constant is graphed as an equilateral hyperbola.

$$(xy) = C$$

$$(V_{CE}I_C) = P_{C(max)} \tag{3-25}$$

For the 2N738, $P_{C(max)}$ is 0.5 W at 25°C. To obtain data to plot the curve, one variable is solved for in terms of the other. We may, for example, solve for I_C in terms of V_{CE} as shown below.

$$I_C = \frac{P_{C(max)}}{V_{CE}} \tag{3-26}$$

A chart, similar to that shown in Fig. 3.15(b) is then made. Values of V_{CE} are substituted into Eq. 3-26 and computed values are listed alongside them in the chart. Notice from the chart that, as V_{CE} increases from 20 to 50 V, I_C decreases from 25 to 10 mA. These results are then plotted and the curve appears as shown in Fig. 3.15(a), where the solid curve is for $P_{C(max)} = 0.5$ W. This curve is known as the

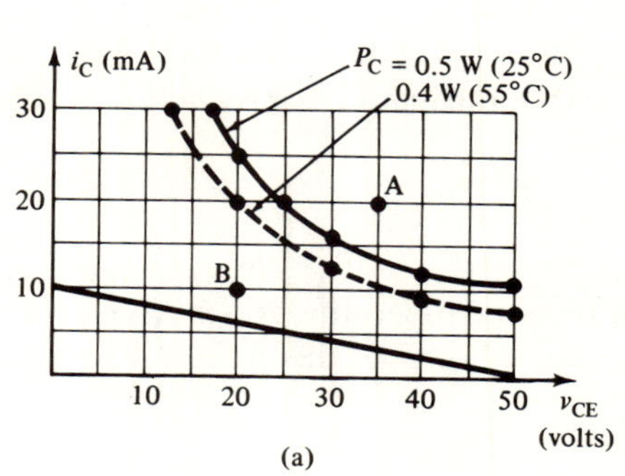

v_{CE} (V)	i_C (mA)
20	25
25	20
30	16.7
40	12.5
50	10

Figure 3.15

maximum dissipation hyperbola. The product of I_C and V_{CE} at any point along this curve is 0.5 W. The product of I_C and V_{CE} at any point below this curve is less than 0.5 W, while above the curve the product is greater than 0.5 W. Thus, once this curve is drawn, we can be sure that the transistor is being operated safely if the Q point is below the maximum dissipation hyperbola.

The load line for the problems worked out previously is also shown in Fig. 3.15(a). It should be noted that any Q point along this load line will be below the hyperbola.

The dashed curve shown in Fig. 3.15(a) is the maximum dissipation hyperbola for the same transistor at an operating temperature of 55°C where $P_{C(max)}$ is 0.4 W. Notice that this curve falls below the one for room temperature indicating that the range of operation on the characteristics is now limited at higher temperatures.

EXAMPLE 18

Find P_C at points A and B in Fig. 3.15(a).

SOLUTION

At point A, $P_C = V_{CE}I_C = 35$ V $\times$ 20 mA $= 0.7$ W.
At point B, $P_C = V_{CE}I_C = 20$ V $\times$ 10 mA $= 0.2$ W.
At point A, P_C is greater than 0.5 W. At point B, P_C is less than 0.4 W.

PROBLEMS

1. A transistor has an alpha of 0.97 and $I_{CO} = 5$ μA. The emitter current is 5 mA. Compute the values of the collector and base currents.

2. Repeat problem 1, except make the emitter current equal to 3 mA.

3. Find the β of a transistor whose $\alpha = 0.97$.

4. Find the β of a transistor whose $\alpha = 0.99$.

5. Find the α of a transistor whose $\beta = 50$.

6. Find the α of a transistor whose $\beta = 100$.

7. The value of I_{CO} for a particular transistor is 10 μA. If the $\beta = 50$, compute the value of I_{CEO}.

8. Using the curves of Fig. 3.9, compute the β of the transistor between the curves labeled $i_B = 0.15$ mA and $i_B = 0.20$ mA at $V_{CE} = 30$ V.

9. Repeat problem 8, using the curves labeled $i_B = 0.05$ mA and $i_B = 0.10$ mA.

10. Using the curves of Fig. 3.9, compute the dc β (h_{FE}) at $i_B = 0.1$ mA and $V_{CE} = 30$ V.

11. Repeat problem 10 at $i_B = 0.2$ mA.

12. Using the curves of Fig. 3.9, compute the value of r_O for the $i_B = 0.3$ mA curve.

13. Repeat problem 12 for the $i_B = 0.15$ mA curve.

14. The circuit of Fig. 3.12(a) has the following values:

$$V_{EE} = 10 \text{ V} \qquad V_{CC} = 40 \text{ V}$$
$$R_B = 200 \text{ k}\Omega \qquad R_L = 5 \text{ k}\Omega$$

   Using the curves of Fig. 3.13 and neglecting V_{BE}, find the values of I_B, I_C, and V_{CE}.

15. Repeat problem 14, except make $R_B = 100$ kΩ.

16. Repeat problem 14, except make $R_B = 50$ kΩ.

17. Repeat problem 14, except make $R_B = 40$ kΩ.

18. The circuit of Fig. 3.12(a) has the following values:

$$V_{EE} = 20 \text{ V} \qquad V_{CC} = 30 \text{ V} \qquad h_{FE} = \beta_{DC} = 50$$
$$R_B = 1 \text{ M}\Omega \qquad R_L = 5 \text{ k}\Omega$$

Compute the values of I_B, I_C, and V_{CE}.

19. Repeat problem 18, except make $R_B = 500 \text{ k}\Omega$.

20. Repeat problem 18, except make $R_B = 200 \text{ k}\Omega$.

21. Repeat problem 18, except make $R_B = 100 \text{ k}\Omega$.

22. $P_{C(max)} = 1 \text{ W}$ at 25°C for a particular transistor. The derating factor is 5 mW/°C. What is the value of $P_{C(max)}$ at 95°C?

4

THE ELEMENTARY
TRANSISTOR AMPLIFIER

The development of the transistor in the late 1940's opened up an entirely new area of electronics. Many uses were found for this new device, one of which was in an amplifier circuit. In this chapter we apply the analysis techniques developed in chapter 1, the general semiconductor device theory developed in chapters 2 and 3, and basic dc and ac circuits theory to the explanation of the operation of a simple one-stage transistor amplifier. We shall see that the transistor, when used in the proper circuit, has the ability to amplify, or magnify an ac voltage and current. The development of the amplifier made possible many modern electronic systems, such as the radio, audio system, television, radar, digital computer, etc. The emergence of the transistor amplifier made it possible to build these electronic systems in localities where space is at a premium, such as in airplanes, missiles, space capsules, etc., because of the inherently small size of the transistor.

In order to provide an insight into the actual workings of the circuit, we begin this chapter with a step-by-step explanation of the operation of the amplifier. Then follows a description of a method for analyzing the circuit. The method is useful for quick computations and approximates quite closely the actual situation in the circuit.

4-1 ANALYSIS OF A SINGLE-STAGE AMPLIFIER

Figure 4.1 shows the circuit of an elementary transistor amplifier. Notice that this circuit is similar in many respects to those studied in the previous chapter. The transistor is connected in the common-emitter configuration with the base-emitter junction forward biased and the collector-base junction reverse biased. The resistor, R_B, has been removed from the circuit and V_{EE} has been reduced to 0.6 V to limit the current to the base. The removal of the resistor and the reduction of the value of V_{EE} have been done to facilitate the explanation of the amplifier circuit. It is shown in the next chapter that the circuit of Fig. 4.1 is not a practical circuit for a number of reasons; the remaining chapters of this text are devoted to transforming the circuit of Fig. 4.1 into a practical, workable amplifier.

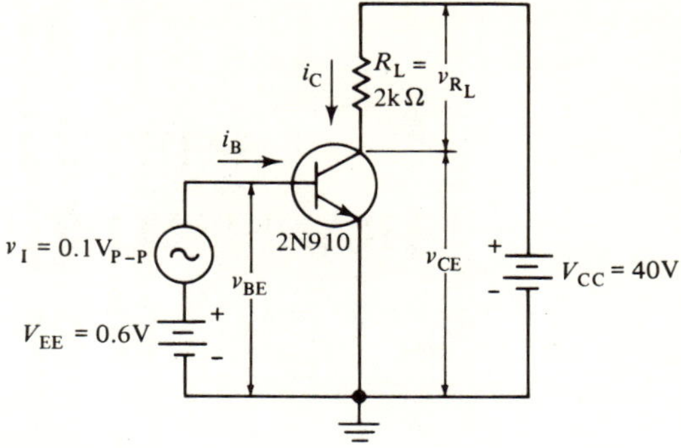

Figure 4.1

The major difference between the circuit of Fig. 4.1 and those of the last chapter is the inclusion of an ac voltage, v_I, in *series* with V_{EE} and the base lead. It is assumed that the internal resistances of both v_I and V_{EE} are zero. The ac *input* voltage, v_I, is an ordinary sine wave whose *peak-to-peak* value is 0.1 V. Figure 4.2(a)

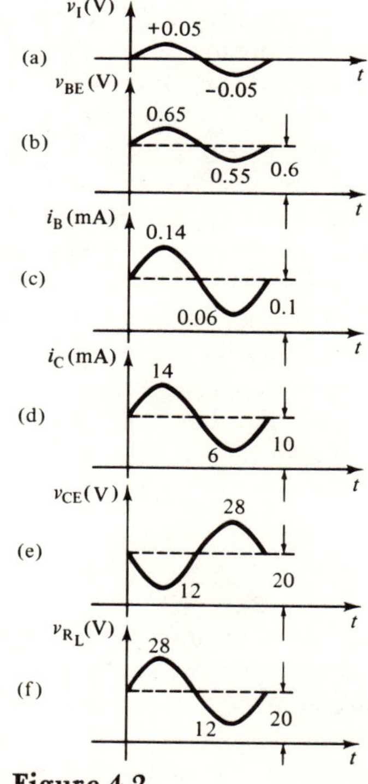

Figure 4.2

shows the waveform of v_I; notice that it reaches a maximum of *plus* 0.05 V and a minimum of *minus* 0.05 V. According to Kirchhoff's law, the base-to-emitter voltage, v_{BE}, is equal to the *sum of v_I and V_{EE}*. When a dc and an ac voltage are added, the result is the *complex wave* shown in Fig. 4.2(b). The 0.6 V battery, V_{EE}, has the effect of *shifting up the ac voltage;* we say that the ac voltage now *rides* on the dc level. The dc level is 0.6 V; the maximum value of v_{BE} is 0.65 V (0.05 V above the dc level), and the minimum value is 0.55 V (0.05 V below the dc level), as shown in Fig. 4.2(b). In other words, the insertion of the ac voltage into the base circuit has caused the base-to-emitter voltage to vary sinusoidally with time. We now see that this varying base-to-emitter voltage causes the other circuit quantities (base current, collector current, collector-to-emitter voltage) to vary with time as well. We shall examine the operation of the circuit at *five* instants of time: the instants at which the sine wave is at 0, 90, 180, 270, and 360 degrees.

Use will be made now of both Figs. 4.2 and 4.3. Figure 4.3 shows an approximate

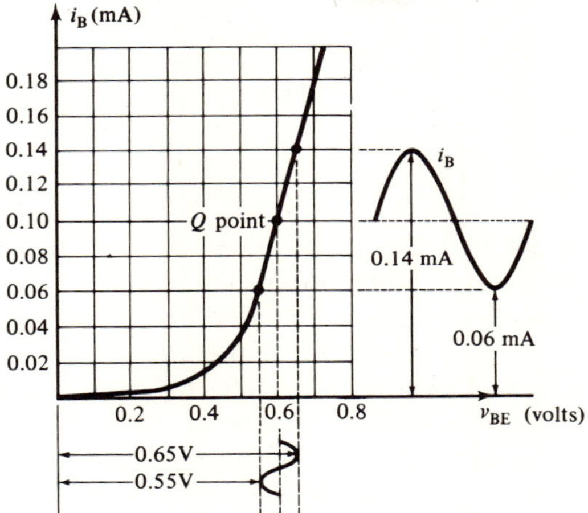

Figure 4.3

input characteristic for the 2N910 transistor (recall from chapter 3 that this is a plot of base current vs. base-to-emitter voltage). Observation of Fig. 4.2(b) reveals that at the zero degree instant of time, the value of v_{BE} is 0.6 V (the value of a sine wave at zero degrees is zero and thus v_I contributes no effect to the base-to-emitter voltage at this instant of time). Figure 4.3 indicates that the base current is equal to 0.1 mA (100 μA) when v_{BE} is 0.6 V. Thus i_B is 0.1 mA at the zero degree instant as shown in Fig. 4.2(c). Figure 4.4 shows the approximate collector characteristics for the 2N910 transistor with the load line drawn ($V_{CC} = 40$ V and $R_L = 2$ kΩ). With the base current equal to 0.1 mA, the Q point is exactly at the center of the load line; thus at the zero degree instant $i_C = 10$ mA and $v_{CE} = 20$ V, as shown in Figs. 4.2(d) and 4.2(e), respectively. Figure 4.2(f) shows the waveform of v_{R_L}; since the voltage across the collector-to-emitter terminals is 20 V at zero degrees, the voltage across the load resistor is also 20 V (the sum of these two voltages must add up to V_{CC}, or 40 V).

At instants of time between zero and 90 degrees, v_I and v_{BE} increase in value. At the 90 degree instant, v_{BE} has *increased* to 0.65 V. Figure 4.3 reveals that at this new value of v_{BE} the base current has *increased* to 0.14 mA; this is shown also

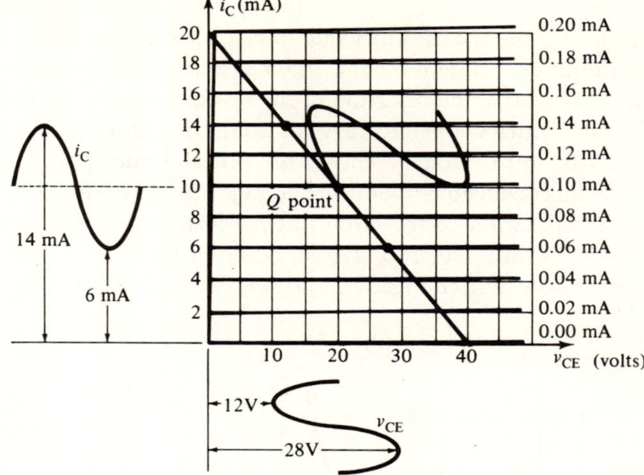

Figure 4.4

in Fig. 4.2(c). Figure 4.4 shows that an increase in base current from 0.1 mA to 0.14 mA results in an *increase* in collector current (from 10 mA to 14 mA) and a *decrease* in collector-to-emitter voltage (from 20 V to 12 V). If v_{CE} decreases to 12 V, v_{RL} must *increase* to 28 V in order that the sum of the two voltages remains equal to 40 V. This checks, since with the collector current equal to 14 mA and R_L equal to 2 kΩ, the voltage across the resistor is (14 mA)(2 kΩ), or 28 V. The values of all of these quantities are shown at the 90 degree instant in Fig. 4.2.

At instants of time between 90 and 180 degrees, v_{BE} decreases toward 0.6 V; thus i_B decreases toward 0.1 mA, i_C decreases toward 10 mA, v_{CE} increases toward 20 V, and v_{RL} decreases toward 20 V. At the 180 degree instant, the circuit conditions are the same as they were at the zero degree instant. At instants between 180 and 270 degrees, v_{BE} *decreases* from 0.6 V to 0.55 V (reaching 0.55 V at 270 degrees). Reference to Figs. 4.3 and 4.4 reveals that i_B *decreases* from 0.1 mA to 0.06 mA, i_C *decreases* from 10 mA to 6 mA, v_{CE} *increases* from 20 V to 28 V, and v_{RL} *decreases* from 20 V to 12 V. Finally at instants of time between 270 degrees and 360 degrees, the process again reverses and the circuit returns again to the conditions present at zero degrees. This process is repeated many times each second, depending, of course, on the frequency of v_I.

4-2   WAVEFORMS

Figures 4.2(a) through (f) show the waveforms of all of the currents and voltages developed in the amplifier of Fig. 4.1. Notice first that, with the exception of v_I, all of the waveforms are complex; they all consist of an ac voltage riding on a dc level. The second point of importance is that all of the currents and voltages are in phase with each other, *except* v_{CE}; the waveform of v_{CE} is *180* degrees out of phase with all of the other waveforms. This comes about because an *increase* in collector current causes a *decrease* in collector-to-emitter voltage; similarly a decrease in i_C causes an increase in v_{CE}. If the collector current increases, the voltage drop across R_L increases, and the collector-to-emitter voltage must therefore decrease. A third point in connection with the waveforms is that both the dc and ac portions of the waveforms can be treated *separately* when dealing with amplifier circuits. A circuit theorem known as the *Superposition Theorem* states that when dc and ac are applied to a circuit simultaneously (in this case the dc is supplied from V_{CC}, and the ac from v_I), the circuit can be analyzed as if first the dc were applied, and then the ac; the final results can be obtained by combining the ac portion with the dc. We have more to say about this concept later on in the chapter.

A fourth, and final point concerns the relationship between the waveforms of v_{CE} and v_{RL}, shown in Figs. 4.2(e) and 4.2(f), respectively. The two waveforms are out of phase by 180 degrees because when one voltage increases, the other must decrease so that their sum remains equal to V_{CC}. If indeed the sum of the two voltages must remain equal to V_{CC}, the *change* in v_{CE} must equal the *change* in v_{RL}. For example, consider the 90 degree instant in Figs. 4.2(e) and (f). V_{CE} has decreased from its dc level of 20 V to 12 V (for a net decrease of 8 V), while v_{RL} has increased from its 20 V dc level to 28 V (for a net increase of 8 V). In other words, both voltages have *changed by the same amount*, although in *opposite directions*. The *ac portion* of the v_{CE} waveform is 16 V_{P-P} (28 minus 12), and the ac portion of the v_{RL} waveform is also 16 V_{P-P}. Thus, we can say that *the ac portion* of the waveform of v_{CE} is equal to *the ac portion* of the waveform of v_{RL}. This particular point is very important as is shown later on. It should be noted that, although the ac portions of the two waveforms are equal, *the dc portions need not be*. They happen to be equal in the case of the amplifier of Fig. 4.1 (20 V) because the Q point was in the *center* of the load line; if this were *not* the case the dc levels of the two voltages would *not* be equal. However, the statement made above regarding the ac portions of the waveforms holds whether or not the dc levels are equal.

4-3   GAIN

We turn now to a discussion of the most important concept to be considered in this text: gain. Refer again to the waveforms of Fig. 4.2. We concern ourselves now with only *the ac portions* of the waveforms. From chapter 3 you may recall that the base circuit was considered to be the input, and the collector circuit, the output. Thus the ac portions of the voltages and currents developed in the base circuit are *inputs*, while those developed in the collector circuit are *outputs;* these inputs and outputs are given as follows:

Input voltage:	ac portion of v_{BE}	0.1 V_{P-P}	v_I
Input current:	ac portion of i_B	0.08 mA_{P-P}	i_I
Output voltage:	ac portion of v_{CE} (or v_{RL})	16 V_{P-P}	v_O
Output current:	ac portion of i_C	8 mA_{P-P}	i_O

The gain is defined as the ratio of the ac portion of the output to the ac portion of the input. There are three gains of importance: *voltage* gain, *current* gain, and *power* gain.

The voltage gain is defined by the following formula:

$$A_V = \text{Voltage Gain} = \frac{v_O(\text{ac})}{v_I(\text{ac})}$$

Computing the voltage gain for the circuit of Fig. 4.1, we get:

$$A_V = \frac{v_O(\text{ac})}{v_I(\text{ac})} = \frac{16\ \text{V}_{P-P}}{0.1\ \text{V}_{P-P}} = 160$$

In this case the input voltage, v_I, is equal to the ac portion of v_{BE}, or 0.1 V_{P-P}. But this is merely the value of the ac voltage inserted in series with V_{EE} and the base, v_I. Thus, the circuit of Fig. 4.1 has magnified, or amplified the ac input voltage 160 times, from 0.1 V_{P-P} (in the base circuit) to 16 V_{P-P} (in the collector circuit).

The current gain is given as follows:

$$A_I = \text{Current Gain} = \frac{i_O(\text{ac})}{i_I(\text{ac})}$$

Computing the current gain we get:

$$A_I = \frac{i_O(\text{ac})}{i_I(\text{ac})} = \frac{8\ \text{mA}_{P-P}}{0.08\ \text{mA}_{P-P}} = 100$$

Thus, not only the voltage, but the current also is increased; the ac portion of the collector current is 100 times as large as the ac portion of the base current.

The power gain is computed by taking the ratio of the output power to the input power. In order to compute the power, we must first obtain the rms, or effective values of all of the ac portions of the currents and voltages. It is known that the maximum value of a sine wave is equal one-half the peak-to-peak value, and the rms value is equal to 0.707 V_M, where V_M is the maximum value of the voltage (or current). The following algebraic manipulations put the expression for power computation in a more convenient form:

$$P = v_{\text{rms}}\, i_{\text{rms}} = \frac{v_m}{\sqrt{2}}\, \frac{i_m}{\sqrt{2}} = \frac{\dfrac{v_{P-P}}{2}\, \dfrac{i_{P-P}}{2}}{2} = \frac{(v_{P-P})(i_{P-P})}{8}$$

$$P = \frac{(v_{P-P})(i_{P-P})}{8}$$

This expression holds *only* for a *sine* wave; if some other type of waveform is used, a new expression will have to be derived. Computing the ac input and output powers, we get:

$$P_I = \frac{(v_{IP-P})(i_{IP-P})}{8} = \frac{(0.1\ \text{V})(0.08\ \text{mA})}{8} = 0.001\ \text{mW}$$

$$P_O = \frac{(v_{OP-P})(i_{OP-P})}{8} = \frac{(16\ \text{V})(8\ \text{mA})}{8} = 16\ \text{mW}$$

The power gain is defined as follows:

$$A_P = \text{Power Gain} = \frac{P_O(\text{ac})}{P_I(\text{ac})}$$

The power gain for the circuit of Fig. 4.1 is:

$$A_P = \frac{P_O(\text{ac})}{P_I(\text{ac})} = \frac{16 \text{ mW}}{0.001 \text{ mW}} = 16{,}000$$

The power gain can be computed from a knowledge of the voltage and current gains as follows:

$$A_V A_I = \frac{v_O(\text{ac})}{v_I(\text{ac})} \times \frac{i_O(\text{ac})}{i_I(\text{ac})} = \frac{v_O i_O}{v_I i_I} = \frac{P_O}{P_I} = A_P$$

$$A_P = A_V A_I$$

$$A_P = A_V A_I = (160)(100) = 16{,}000$$

Thus the power gain can be computed by multiplying the voltage gain by the current gain.

Now let us examine the concept of gain more closely. It is assumed from this point on that when the terms input and output are used, we are referring to the ac portions of the various quantities. Notice that all three output quantities (voltage, current, and power) are *larger* than the input quantities. The circuit of Fig. 4.1 is an *electronic amplifier;* it magnifies ac voltage, current, and power. The various gains are indications of the amount of amplification of the input quantities; the voltage has been amplified 160 times, the current, 100 times, and the power, 16,000 times. The circuit of Fig. 4.1 is called a *single-stage* amplifier because only *one* transistor is used; we shall see in chapter 6 that an amplifier may be composed of many stages (many transistors).

Referring again to Fig. 4.3, it can be seen that the small ac input voltage, v_I, when applied in series with the base and V_{EE}, causes a variation in i_B. Figure 4.4 shows that this variation in i_B causes a variation in i_C and v_{CE}. In other words, an amplifier converts a *small change* in input voltage or current to a *large change* in output voltage or current. The ac variation above and below the dc level *is* the change. For example, the 0.1 V_{P-P} variation about the 0.6 V dc level in Fig. 4.2(a) is the change in the input, and the 16 V_{P-P} variation is the much larger change in output about the 20 V dc level in Fig. 4.2(e). This property of an amplifier has made possible the development of many electronic systems, as mentioned previously. Almost every type of electronic system requires an amplifier of some sort to make the system function properly; very often a small voltage or current must be amplified.

The quantities A_V, A_I, and A_P are all ratios of similar quantities; e.g., voltage gain is the ratio of output voltage to input voltage. Thus, the gains are all dimensionless quantities; they have no units. Since gain is the ratio of output to input, the output can be computed if the input and the gain are known. Thus:

$$v_O = A_V v_I$$

We see in later sections that the gain of an amplifier is dependent upon the transistor parameters, so the gain can be computed if these parameters are known. Once the gain is known, the output can be computed by using the above expression.

It might be assumed by the reader that since there is a power gain (output power greater than input power), the law of conservation of energy is contradicted; this law states that energy cannot be created or destroyed, only changed. The analysis of the amplifier seems to indicate that power is somehow created since the output power is greater than the input power. This is *not* the case as we now illustrate. Figure 4.5(a) shows the schematic diagram of a step-down transformer. Assume

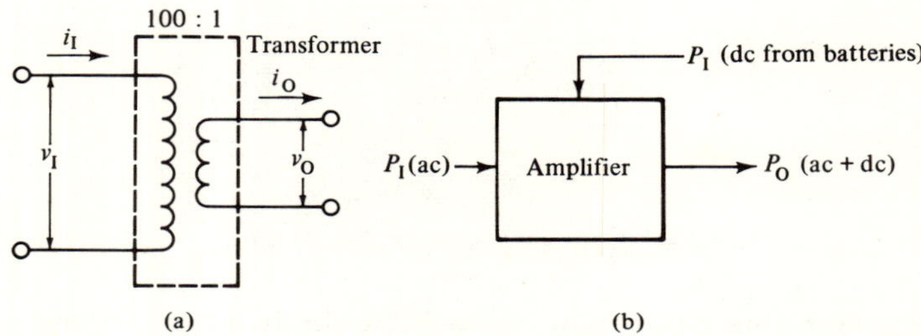

Figure 4.5

that the input voltage, v_I, is 0.1 V_{P-P} and the input current, i_I, is 0.08 mA_{P-P}. The input power to the transformer is thus:

$$P_I = \frac{(v_{I_{P-P}})(i_{I_{P-P}})}{8} = \frac{(0.1 \text{ V})(0.08 \text{ mA})}{8} = 0.001 \text{ mW}$$

If the transformer has a 100:1 step-down turns ratio, the voltage at the secondary (v_O) will be one one-hundredth of the input voltage (voltage is stepped down) and the secondary current, i_O, will be 100 times as large as the input current (current is stepped up); that is:

$$v_I = 0.1 \text{ V}_{P-P} \qquad v_O = 0.001 \text{ V}_{P-P}$$
$$i_I = 0.08 \text{ mA}_{P-P} \qquad i_O = 8 \text{ mA}_{P-P}$$
$$P_I = 0.001 \text{ mW} \quad \rightarrow P_O = 0.001 \text{ mW}$$

$$P_I = P_O \text{ (Assuming No Losses)}$$

In other words, the transformer steps the voltage down and the current up, or vice versa. The voltage is stepped down in the *same* ratio that the current is stepped up. This must be true in order that the output power be equal to the input power, as shown above. The output power from a transformer will always be equal to, or slightly less (due to losses) than the input power. Since no additional power is supplied to the transformer from some other source, the primary power must equal the secondary power. Thus the transformer can only step up the voltage *or* current, not both.

Figure 4.5(b) shows a schematic block diagram of an amplifier. In the previous discussions, we said that *the ac output power* of an amplifier was greater than *the ac input power;* nothing was said about dc power. In Fig. 4.5(b), it is shown that dc power is supplied to the amplifier from the V_{CC} battery; this must be considered input power since it must be *supplied to* the amplifier (power is also supplied from the V_{EE} battery). The power supplied by a battery is equal to the product of the battery voltage and the dc current flowing through the battery; thus:

$$P_{CC} = V_{CC}I_C = (40 \text{ V})(10 \text{ mA}) = 400 \text{ mW} \quad (V_{CC} \text{ Battery})$$

$$P_{EE} \doteq V_{EE}I_B = (0.6 \text{ V})(0.1 \text{ mA}) = 0.06 \text{ mW} \quad (V_{EE} \text{ Battery})$$

The total input power to the amplifier is composed of the power supplied by the two batteries and the ac input power; the ac output power is 16 mW.

$$P_I = P_I(\text{ac}) + P_{EE} + P_{CC} = 0.001 \text{ mW} + 0.06 \text{ mW} + 400 \text{ mW}$$

$$P_I = 400.061 \text{ mW} \simeq 400 \text{ mW}$$

$$P_O(\text{ac}) = 16 \text{ mW}; \textit{Less} \text{ than } 400 \text{ mW}$$

Thus it can be seen that the total input power to the amplifier is *greater* than the ac output power, and thus the law of conservation of energy is not contradicted; the remainder of the input power goes to perform other functions within the amplifier which will be discussed in more detail in chapter 9. Thus the amplifier can provide both current *and* voltage gain because power is supplied from an outside source (the batteries), whereas the transformer can provide either voltage *or* current gain in order that the law of conservation of energy be satisfied. The ability to provide both voltage and current gain is a distinct advantage as we will see later.

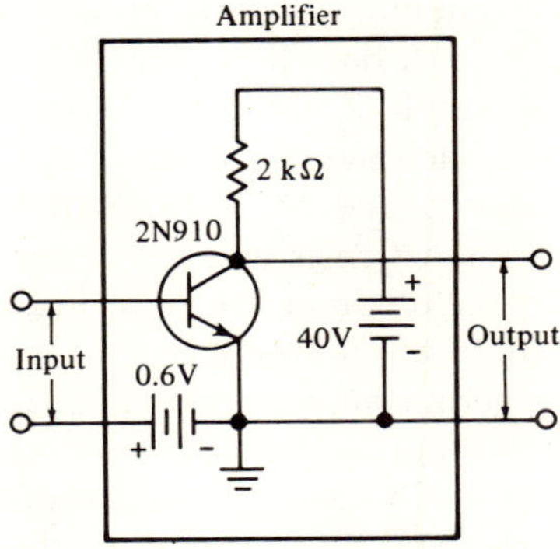

Figure 4.6

The amplifier of Fig. 4.1 is repeated in Fig. 4.6. Notice that the amplifier itself (within the box) is composed of the transistor, load resistor, and the two batteries; all of these components are necessary as we see later.

4-4   THE FUNCTION OF THE BATTERIES

It was pointed out in the last section that the batteries were necessary in the amplifier in order that the circuit provide current, voltage, and power gain. The importance of the presence of the batteries can be shown in another way. The waveform of the collector current in Fig. 4.2(d) is repeated in Fig. 4.7 for con-

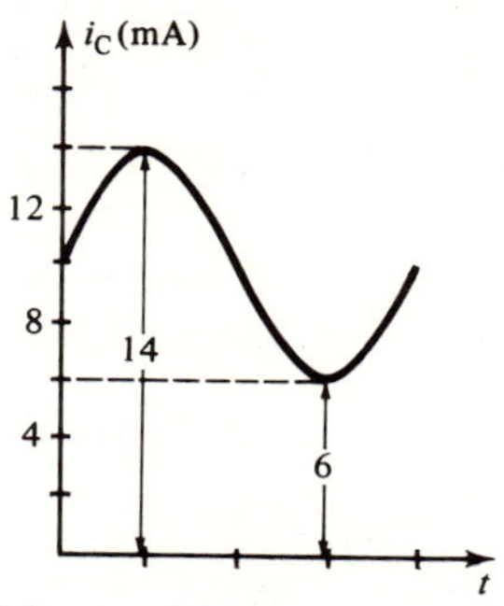

Figure 4.7

venience. Notice that because the waveform is complex, the values of current are always *above* the *t*-axis; i.e., the collector current is always *positive*. Figure 4.7 shows that the maximum value of current (at 90 degrees) is 14 mA, and the minimum value (at 270 degrees) is 6 mA. This means that the *collector current flows in the same direction for the full 360 degrees of the cycle;* we say that the collector current is *unidirectional*. Since the 2N910 is an *npn* transistor, the collector current flows from the positive terminal of V_{CC} through the collector to the emitter; if the transistor were a *pnp*, the current direction would be reversed. Observation of the waveforms of Fig. 4.2 reveals that all of the other voltages and currents are also uni-directional (except v_I). The collector current of Fig. 4.7 *does* vary with time (from a maximum of 14 mA to a minimum of 6 mA), but always flows in the same direction. It can be seen from Fig. 4.7 that the result of adding the V_{CC} battery is the establishment of the 10 mA dc level about which the ac collector current varies.

To illustrate the importance of the 0.6 V V_{EE} battery, let us remove it from the circuit. The base-emitter junction is now no longer forward-biased and the Q point is approximately at the origin as shown in Fig. 4.8(a). If a *p–n* junction is not forward biased, no current will flow through the junction. Because of the absence of V_{EE}, v_{BE} is no longer a complex wave (there is no dc level about which the ac can vary). The input voltage, v_I, is applied directly to the base and emitter terminals, and is, therefore, equal to v_{BE}; in other words, v_{BE} is an ordinary sine wave with no dc level. Figure 4.8(a) shows that when the input voltage makes the base positive with respect to the emitter (as at point A in the figure), a very small amount of base current flows; when the base becomes negative with respect to the emitter (second half of cycle), the junction is reverse biased and no current flows. The waveform of the resulting base current is shown at the right in Fig. 4.8(a). Notice first that even when current is flowing (during that portion of the cycle when the junction is forward-biased), the current is *very small*. Second, because the

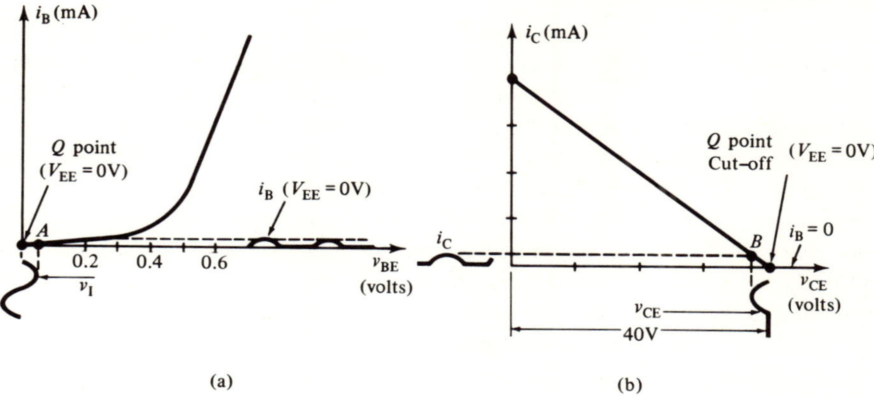

Figure 4.8

junction is actually reverse biased for one-half of the cycle, *no base current flows* for a full one-half cycle. The reason for this phenomenon is that with V_{EE} equal to zero, we are forced to operate in the *flat* portion of the input characteristic as shown in Fig. 4.8(a). Compare this situation with that shown in Fig. 4.3; notice in Fig. 4.3 that the Q point is on the *steep* portion of the characteristic because of the presence of the 0.6 V battery, and the same 0.1 V_{P-P} input voltage produces a *full* 0.08 mA$_{P-P}$ base current (current flow for the full 360 degrees of the cycle). Thus the presence of the small 0.6 V battery has a very considerable effect on the waveform of the base current as can be seen by a comparison of Figs. 4.3 and 4.8(a).

Let us now see what effect the waveform of the base current of Fig. 4.8(a) has on the waveforms of the collector current and collector-to-emitter voltage. With the V_{EE} battery missing, the quiescent base current is zero as explained with the aid of Fig. 4.8(a). This means that the Q point on the collector characteristics [Fig. 4.8(b)] is at the intersection of the load line with the characteristic for which $i_B = 0$; recall from chapter 3 that the current that flows through the collector (and emitter) when the base current is zero is called I_{CEO}. We assume, for the purpose of this explanation, that I_{CEO} for the 2N910 transistor is approximately equal to zero; thus the Q point is at *cut-off* (extreme lower right-hand portion of the load line) as shown in Fig. 4.8(b). During the half-cycle when the base current is flowing, the increase in base current will cause a corresponding increase in collector current and a decrease in collector-to-emitter voltage, shown at point B in Fig. 4.8(b). Since the base current remains at zero for the second half-cycle [Fig. 4.8(a)], the collector current remains at its quiescent value of zero mA, and the collector-to-emitter voltage remains at its quiescent value of V_{CC} (40 V in this case), during the second half-cycle.

Examination of the waveforms of i_C and v_{CE} in Fig. 4.8(b) reveals that, like the waveform of i_B, they too show variations for only one-half of a cycle; in addition, the ac portions of both waves are quite small. The reason for the existence of these waveforms is that the base current was very small and showed variations for only one-half of a cycle; this again stems from the fact that the 0.6 V battery was removed from the circuit. Thus, it can be seen that the V_{EE} battery has an important effect on not only the base current waveform, but the waveforms of i_C and v_{CE} as well. The effect of the absence of V_{EE} can be illustrated by comparing the waveforms of i_C and v_{CE} as shown in Fig. 4.8(b) (with V_{EE} removed) with those waveforms of i_C and v_{CE} shown in Fig. 4.4 (with V_{EE} present); notice that those waveforms shown in Fig. 4.4 have much larger ac portions than those of Fig. 4.8(b), and the variations occur over the full 360 degrees of the cycle.

The waveforms of i_C and v_{CE} shown in Fig. 4.8(b) are undesirable in an amplifier for *two* reasons. The first is explained as follows: The object of an amplifier is to magnify the ac voltage and current, *maintaining the same waveshape*. Notice in Fig. 4.8 that, while the input voltage, v_I, is a pure 0.1 V_{P-P} sine wave, the ac portions of the collector current and voltage are *not* sine waves; this effect is known as *distortion* of the waveform and is discussed in more detail later on in the text. The effect of a condition such as shown in Fig. 4.8 is to produce a garbled sound in an audio system and a distorted picture in a video system. The second undesirable feature concerns the gain. From the previous section you will recall that the voltage gain equals the ratio of the ac portion of the output voltage to the ac portion of the input voltage. Notice in Fig. 4.8(b) that the output voltage, v_{CE}, is quite small even though the input voltage is 0.1 V_{P-P}; thus the voltage gain is much *smaller* when the V_{EE} battery is not in the circuit. The waveforms of Fig. 4.8 and the discussion of this section indicate the importance of the presence in the amplifier circuit of both V_{EE} and V_{CC}.

In the discussion of this section a few approximations and assumptions have been made. The input characteristic shown in Fig. 4.8 is an approximation to the actual case; in reality the input characteristic is located partially *below* the v_{BE}-axis as shown in Fig. 4.9. A careful analysis using this characteristic curve will show that the waveforms produce variations for *less* than one-half of a cycle.

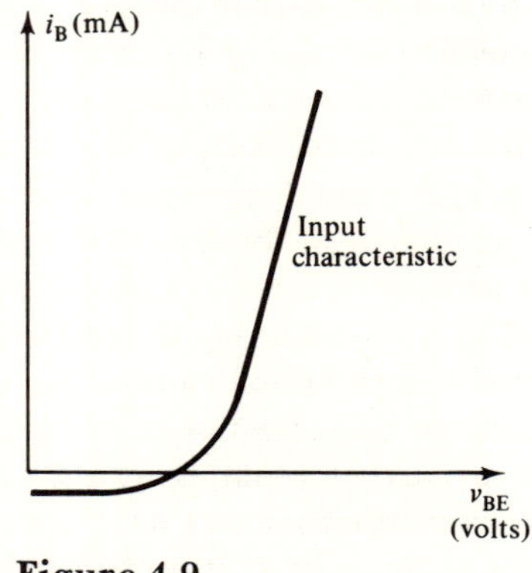

Figure 4.9

The purpose of the previous discussion has been to show the necessity for the *presence* of the V_{EE} battery, in order that a relatively large sinusoidal collector current and collector-to-emitter voltage be obtained from the amplifier. We now examine the effect of the *size* of the battery on the output waveforms. Let us assume that the V_{EE} battery voltage has been *increased* from 0.6 V to 0.75 V. The input characteristic, repeated in Fig. 4.10(a), shows that the Q point has shifted farther

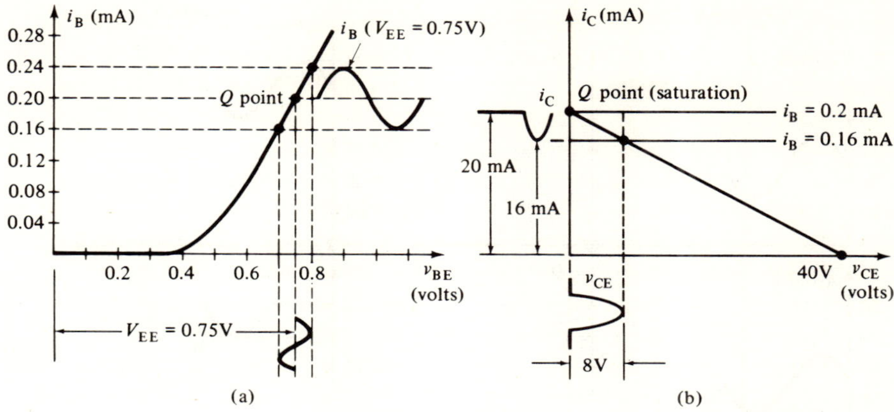

Figure 4.10

up the characteristic so that the new quiescent base current is 0.2 mA. If the same 0.1 V_{P-P} input voltage is applied, the base-to-emitter voltage will again be a complex wave, varying about the 0.75 V dc level. Figure 4.10(a) shows that the base current is also a complex wave, having a dc level of 0.2 mA and a peak-to-peak ac value of 0.08 mA (from 0.24 mA to 0.16 mA). Thus far the only difference between this case and the original one ($V_{EE} = 0.6$ V) is that the Q point has shifted on the input characteristic.

Figure 4.10(b) shows the collector characteristics. Notice that, with the quiescent base current equal to 0.2 mA, the Q point on the collector characteristics is *just at the saturation point;* v_{CE} at this point is approximately equal to zero volts and i_C is approximately equal to the collector saturation current of 20 mA. Figure 4.10(a) shows that, as the base-to-emitter voltage increases from 0.75 V to 0.8 V, the base current increases from 0.2 mA to 0.24 mA. Figure 4.10(b) shows that, although the base current increases to 0.24 mA, the collector current cannot increase beyond its saturation value of 20 mA, and the collector-to-emitter voltage cannot decrease below zero volts. Thus, during the half-cycle when v_I is going positive, both i_C and v_{CE} must remain at their quiescent values (20 mA and 0 V, respectively). During the half-cycle when v_I is going negative the base current decreases from 0.2 mA to 0.16 mA [Fig. 4.10(a)]. Figure 4.10(b) shows that, as the base current decreases below 0.2 mA, the transistor comes out of saturation; the collector current decreases to 16 mA and the collector-to-emitter voltage increases to 8 V.

Examination of the waveforms of i_C and v_{CE} in Fig. 4.10(b) reveals that there is a sinusoidal variation for only one-half of a cycle. The problem here is that the transistor is biased at saturation, and thus the current and voltage can vary in only *one* direction. In this case, the value of V_{EE} was too large because it provided a value of quiescent base current (0.2 mA) which biased the transistor at saturation. Thus it is important to select the value of V_{EE} so that the quiescent base current is neither too large nor too small.

Figure 4.11 summarizes the waveforms of i_B, i_C, and v_{CE} for three different conditions of bias: bias at the center of the load line, bias at cut-off, and bias at

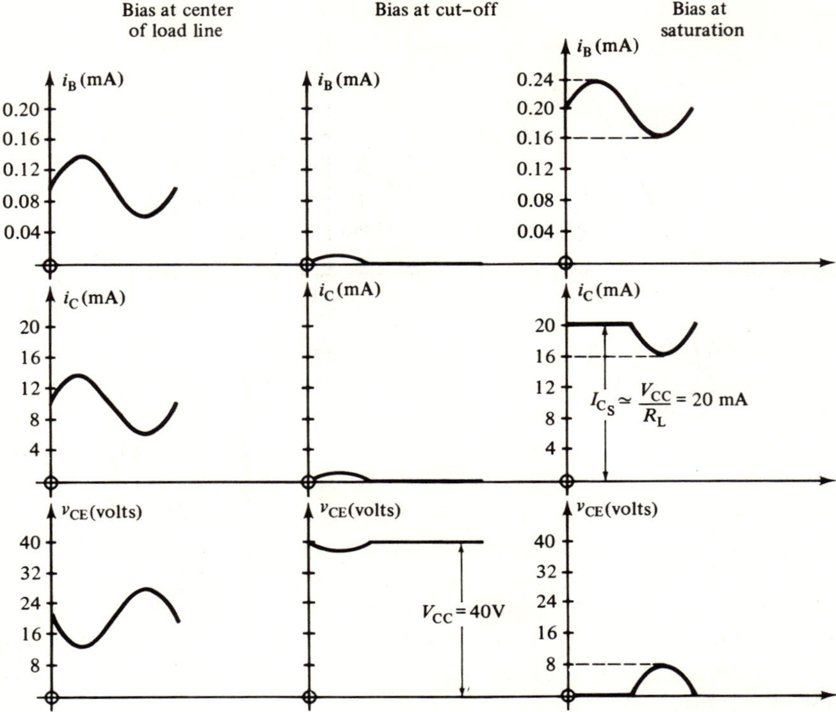

Figure 4.11

saturation. Notice that all three waveforms produce sinusoidal variations for the full 360 degrees of the cycle when the Q point is selected at the *center of the load line;* the quiescent collector-to-emitter voltage, V_{CE}, is selected to be one-half of V_{CC} (in this case, 20 V), and the quiescent collector current, I_C, is selected to be one-half of the collector saturation current (in this case, 10 mA). This provides the maximum amount of room on the characteristics for the ac portions of the currents and voltages to swing in either direction.

The problem involved with selecting the bias at cut-off is that the collector current can only become larger, not smaller, since it cannot go below zero (negative). Similarly, in a circuit of the type being studied, the collector-to-emitter voltage cannot increase above V_{CC} (this is not true in other types of amplifier circuits as will be shown later on in the text). Thus, the purpose of providing dc is to establish a dc level (quiescent collector current), so that the collector current and collector-to-emitter voltage can vary sinusoidally *above and below* this level, as shown in Fig. 4.11 when the Q point is selected at the center of the load line. If *too much* dc base current is supplied to the transistor, however, the transistor may be biased at saturation; here the collector current cannot increase above the saturation value with the result that the entire sine wave will not be obtained.

Thus, we have seen that both batteries are essential for the proper operation of the amplifier. The Q point is generally selected in the center of the load line (halfway between cut-off and saturation). The V_{EE} battery is selected to have a value which forward biases the base-emitter junction; the value of base current must be such as to set the Q point in the center of the load line. In the next chapter we see that we can eliminate the V_{EE} battery in favor of a more practical scheme. We can see now why the word *bias* has been used. The word generally implies a *one-way* attitude, or viewpoint. The base-emitter junction must be biased in the forward direction (one-way) in order that the Q point on the input characteristic be on the steep portion of the characteristic [see Fig. 4.10(a)].

4-5 THE CURRENT SOURCE

Before proceeding with the development of the transistor amplifier, it is instructive to consider the concept of the current source. The reader will recall from basic dc circuits theory that any voltage source of power (E), whether battery, power supply, or ac generator, can be represented schematically as a voltage generator, in series with an internal resistance, R_{IN}, as shown in Fig. 4.12(a); the complete voltage source is included within the dotted box in the figure. When a load, R_{L}, is connected across the output terminals, current will flow through the circuit; a voltage drop will exist across R_{IN} and therefore the load voltage, v_{L}, will be *less* than E. The only situation where $v_{\text{L}} = E$ is when the output terminals are open-circuited; if no resistor is connected across the terminals ($R_{\text{L}} = $ infinity), no current will flow, there will be no voltage drop across R_{IN}, and thus the full voltage, E, will appear across the output terminals (open circuit load voltage $= E$). If E, R_{IN}, and R_{L} are known, the voltage across the load terminals can be computed by application of the basic laws of circuit analysis. A simple example will review the basic techniques.

EXAMPLE 1

In the circuit of Fig. 4.12, $E = 10\ \text{V}_{\text{P--P}}$, $R_{\text{IN}} = 1.0\ \Omega$, and $R_{\text{L}} = 9\ \Omega$. Compute the load voltage, v_{L}, and the load current, i_{L}.

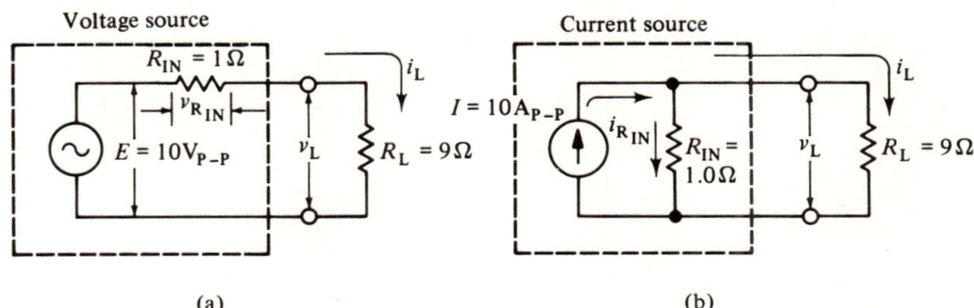

Figure 4.12

SOLUTION

This problem can be solved in either one of two ways: We can first solve for the current and then find the voltage by means of Ohm's law; or we can find the voltage directly by means of the voltage divider expression, and then find the current by Ohm's law. We illustrate both methods

(a)

$$i_{\mathrm{L}} = \frac{E}{R_{\mathrm{IN}} + R_{\mathrm{L}}} = \frac{10\ \mathrm{V}}{1\ \Omega + 9\ \Omega} = 1.0\ \mathrm{A}$$

$$v_{\mathrm{L}} = R_{\mathrm{L}}i_{\mathrm{L}} = (9\ \Omega)(1.0\ \mathrm{A}) = 9.0\ \mathrm{V}$$

(b)

$$v_{\mathrm{L}} = \frac{ER_{\mathrm{L}}}{R_{\mathrm{IN}} + R_{\mathrm{L}}} = \frac{10(9)}{1 + 9} = 9.0\ \mathrm{V}$$

$$i_{\mathrm{L}} = \frac{v_{\mathrm{L}}}{R_{\mathrm{L}}} = \frac{9\ \mathrm{V}}{9\ \Omega} = 1.0\ \mathrm{A}$$

It can be shown mathematically that a *voltage* source (voltage generator in *series* with an internal resistance) can be converted to a *current* source; the current source consists of a current generator in *parallel* with an internal resistance, as shown in Fig. 4.12(b). The current source is located within the dotted box in the figure. To effect the conversion *two* things must be done: First the value of the current, I, for the current source must be found by dividing the voltage, E, by the internal resistance, R_{IN}.

$$I = \frac{E}{R_{\mathrm{IN}}} = \frac{10\ \mathrm{V_{P-P}}}{1\ \Omega} = 10\ \mathrm{A_{P-P}}$$

The second operation is to place the internal resistance (which was in series with the voltage, E) in parallel with the current generator, I. The theory maintains that, as far as the load current and load voltage are concerned, the circuits of Figs. 4.12(a) and (b) are identical.

It was mentioned above that when a load is placed across the output terminals in Fig. 4.12(a), the load voltage will always be less than E by an amount equal to the drop across the internal resistance; e.g., in example 1, the load voltage was 9 V and the drop across the internal resistance was 1 V (1 A $\times$ 1 Ω), and the sum of these two voltages added up to E (10 V). The load voltage was equal to E only on open circuit. Similarly, an examination of Fig. 4.12(b) reveals that *only a portion* of I flows through the load; the remainder of this current will re-circulate through the source via the internal resistance which is now in parallel with the load. The source current, I, splits into two paths, part of it going to the load and the rest of it remaining in the source (flowing through the internal resistance). The percentage of I which flows through the load depends upon the relative values of R_{IN} and R_{L}; the *larger* the value of R_{L}, the *smaller* will be the percentage of I which flows through the load, with *more* current re-circulating through the source, and vice versa. Thus, just as v_{L} is less than E by an amount equal to the drop across R_{IN} in Fig. 4.12(a), i_{L} is less than I by an amount equal to the current flowing through R_{IN} in Fig. 4.12(b). Just as $v_{\mathrm{L}} = E$ only on *open* circuit in Fig. 4.12(a), $i_{\mathrm{L}} = I$ only on *short* circuit ($R_{\mathrm{L}} = 0$) in Fig. 4.12(b).

Mathematically speaking, the two circuits pictured in Fig. 4.12 are equivalent *at the load terminals;* either one may be used in a particular problem. To illustrate the equivalence of the two sources, let us compute the load voltage and current in Fig. 4.12(b) and compare them with the results of example 1.

EXAMPLE 2

Compute the load voltage and load current for the circuit of Fig. 4.12(b) and compare these values with the results of example 1.

SOLUTION

This problem can be attacked by one of two methods: The first is to combine R_{IN} and R_L in parallel, find the voltage across the parallel combination, R_P (see Fig. 4.13), and then find the load current by Ohm's law. The second method is to

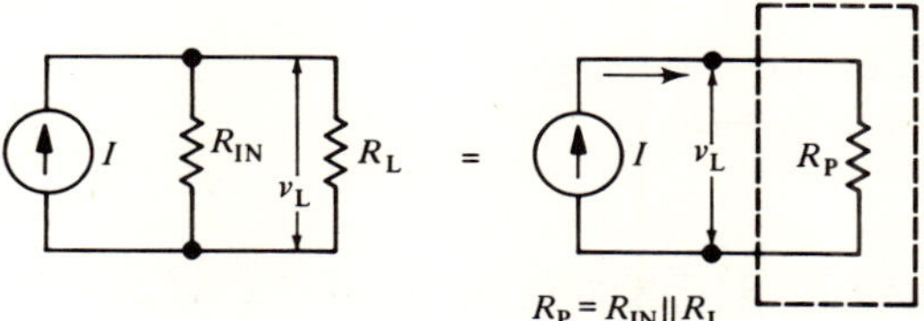

$$R_P = R_{IN} \| R_L$$

Figure 4.13

find the load current by means of the current-divider expression, and then find the load voltage by Ohm's law. Both methods will be illustrated.

(a)

$$R_P = \frac{R_{IN}R_L}{R_{IN} + R_L} = \frac{(1\ \Omega)(9\ \Omega)}{1\ \Omega + 9\ \Omega} = 0.9\ \Omega \ \text{(see Fig. 4.13)}$$

$$v_L = IR_P = (10\ \text{A})(0.9\ \Omega) = 9\ \text{V}$$

$$i_L = \frac{v_L}{R_L} = \frac{9\ \text{V}}{9\ \Omega} = 1.0\ \text{A}$$

(b)

$$i_L = \frac{IR_{IN}}{R_{IN} + R_L} = \frac{(10\ \text{A})(1\ \Omega)}{1\ \Omega + 9\ \Omega} = 1.0\ \text{A} \ \text{(Current Divider)}$$

$$v_L = i_L R_L = (1.0\ \text{A})(9\ \Omega) = 9\ \text{V}$$

Comparing the results of this example with those of example 1 it can be seen that both the load current (1 ampere) and the load voltage (9 V) are the *same for both circuits* [Figs. 4.12(a) and (b)]. This illustrates that the two sources (voltage and current) are *equivalent as far as conditions in the external circuit* are concerned; by external circuit is meant that portion of the circuit exclusive of the source (in the case of the circuits of Fig. 4.12, the external circuit is the load resistor, 9 Ω).

It should be pointed out that the source conversion can proceed in the opposite direction; a current source can be converted to a voltage source by multiplying I by R_{IN} to obtain E, and then placing R_{IN} in series with E. The reader should bear in mind when dealing with these sources that the source voltage, E, remains constant; only the load voltage changes depending upon the load resistance. Similarly the source current, I, remains constant; only the load current changes, depending upon the load resistance.

In analyzing a particular circuit, we are not interested in what is actually inside a particular source; we are concerned only with the equivalent circuit representation of the source. Either one can be used in a particular application with the same degree of legitimacy. The reason that we concern ourselves with current sources at this point is that transistor circuits lend themselves more readily to current source representations than to voltage source configurations; this will be illustrated in the following sections.

To summarize, the conversion of voltage sources to current sources proceeds as follows:

1. To compute I, divide E by R_{IN}.
2. Place R_{IN} in parallel with I.

4-6   THE SMALL SIGNAL ac EQUIVALENT CIRCUIT: APPROXIMATE MODEL

The analysis of the transistor amplifier of Fig. 4.1 as described in Secs. 4-1, 4-2, and 4-3 of this chapter was rather tedious and cumbersome. The procedure included the following steps:

1. Draw the load line on the collector characteristics and find the Q points on both input and collector characteristics.

2. Obtain the complete waveforms of v_{BE}, i_B, i_C, and v_{CE} from the characteristic curves.

3. Compute the various gains by taking the ratio of the ac portions of the output quantities to those of the input quantities.

Although the above procedure is correct, it is extremely time-consuming. The purpose of this section is to develop a simpler method of computing the various gains of the amplifier.

It was mentioned in Sec. 4-2 that it is desirable to treat the complex waveforms developed in the amplifier as if they were composed of *two* separate portions: an ac and a dc. For example, refer to the complex waveform of the collector current in Fig. 4.14(a); notice that it is composed of an 8 mA$_{P-P}$ sine wave riding on a 10 mA dc level. This waveform is equal to the *sum* of the two waveforms shown in Figs. 4.14(b) and (c); Fig. 4.14(b) shows just the 10 mA dc level, while Fig. 4.14(c) shows only the ac portion (an 8 mA$_{P-P}$ pure sine wave). For most of the remainder of this text, we treat the two portions of the complex waves separately; once the values of the two portions are ascertained, they can be combined to produce the entire complex wave.

Since the most important piece of information to know about an amplifier is its gain (how many times the voltage, current, or power is amplified), we are generally interested in knowing the magnitudes of the ac portions of the waveforms; it is by taking the ratios of these magnitudes that we compute the gain. Thus we should like to develop some type of system which enables us to compute the ac portion of

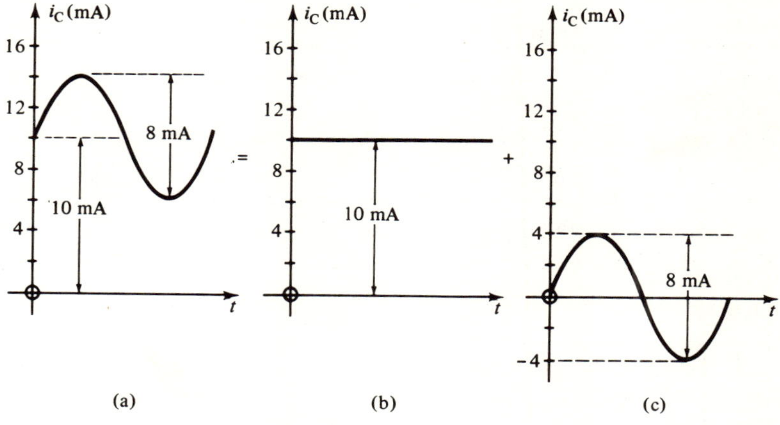

Figure 4.14

the waveform without the necessity of resorting to the laborious procedure described earlier. A circuit will now be developed which approximates *the ac operation of the amplifier*; it cannot be emphasized too strongly that *this circuit applies only to the ac portions of the waveforms, not the dc*.

From the beginning of this chapter you recall that a 0.1 V_{P-P} ac voltage was applied to the input of the amplifier; this produced a 0.08 mA_{P-P} base current. Figure 4.15 shows this voltage applied to the input terminals (base and emitter)

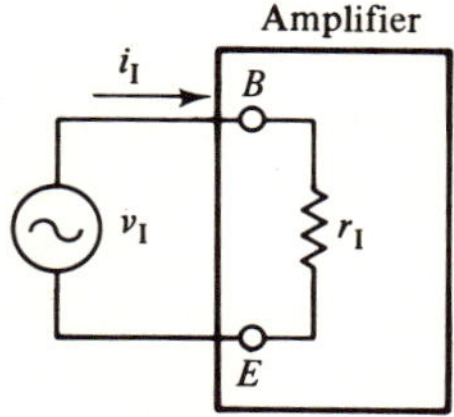

Figure 4.15

of the amplifier; notice that the V_{EE} battery is left out of the circuit since we are considering the effect of only the ac quantities. If a voltage is applied between two terminals, and current flows, we can say that there is a certain *resistance* (actually impedance) connected between the terminals. The value of resistance which allows 0.08 mA_{P-P} to flow through it when 0.1 V_{P-P} is across it can be found by dividing the voltage by the current. Thus, we can say that the input circuit of the transistor can be represented by an *input resistance*, r_I; for the amplifier of Fig. 4.1 the input resistance is:

$$\text{Input Resistance} = r_{\text{I}} = \frac{\text{ac Input Voltage}}{\text{ac Input Current}}$$

$$r_{\text{I}} = \frac{0.1 \text{ V}_{\text{P--P}}}{0.08 \text{ mA}_{\text{P--P}}} = 1.25 \text{ k}\Omega$$

In other words, as far as the 0.1 $V_{\text{P--P}}$ input source voltage is concerned, it is feeding a 1.25 kΩ resistor. In this case, we are not interested in whether or not a 1.25 kΩ resistor is actually inside the transistor; all we know or care about is that *the effect is the same as if a 1.25 kΩ resistor is connected between the base and emitter terminals inside the transistor.* As far as we are concerned, the transistor can contain one 1.25 kΩ resistor, or two 625 Ω resistors in series, or two 2.5 kΩ resistors in parallel, etc.

There is another way of computing the input resistance of the transistor. Figure 4.16(a) shows the input characteristic of the 2N910 transistor. Notice that, with

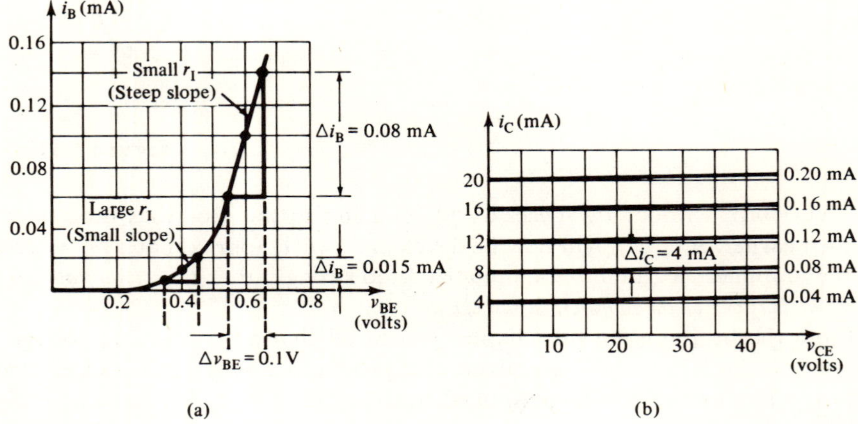

Figure 4.16

the bias set at 0.1 mA (base current), a 0.1 V change in v_{BE} produces a 0.08 mA change in i_{B}. The ratio of the change in voltage to the change in current is equal to the 1.25 kΩ input resistance, r_{I}. But this is nothing more than the dynamic resistance (reciprocal of the slope) on the straight line portion of the input characteristic (note the triangle in Fig. 4.16(a), used to compute the slope). In other words, we can say that *the input resistance of the amplifier is approximately equal to the reciprocal of the slope of the input characteristic;* in the next section it will be explained why the word *approximately* was used. It can be seen now why the transistor must be biased in the steep, linear portion of the input characteristic. It is in the steep portion (large slope) where the resistance is small; here, a given change in v_{BE} produces a relatively large change in i_{B} (a voltage applied across a small resistance produces a large current). In the flat portion of the curve (small slope), the resistance is large; the same change in v_{BE} produces a much smaller change in i_{B} (a voltage applied across a large resistance produces a small current).

You again recall from the analysis of the first three sections of this chapter that a 0.08 mA$_{\text{P--P}}$ base current produced an 8 mA$_{\text{P--P}}$ collector current; i.e., the ac portion of the base current was multiplied by 100 in the collector (current gain equals 100). Now refer to the collector characteristics in Fig. 4.16(b); we know that the β of a transistor is equal to the ratio of the change in collector current to the change in base current, with the collector-to-emitter voltage held constant. It can be seen in Fig. 4.16(b) that a 0.04 mA change in base current (from 0.08 to 0.12 mA) produces a 4 mA change in collector current (from 8 to 12 mA); thus the β of the 2N910 transistor is:

$$\beta = \frac{\Delta i_C}{\Delta i_B} = \frac{4 \text{ mA}}{0.04 \text{ mA}} = 100$$

We know from chapter 3 that curves are not equally spaced and therefore β will vary from point to point; however, for the sake of simplicity, we have assumed that all of the curves are equally spaced.

Thus it is the β of the transistor which is responsible for the amplification of the current; the current in the collector circuit is equal to β multiplied by the current in the base circuit. *The collector-emitter circuit of the transistor can therefore be represented by a current generator* as shown in Fig. 4.17; the value of this current generator is equal to βi_B, where i_B is the ac portion of the base current and β is the β of the transistor. There is no separate representation for the emitter current since i_E is only very slightly larger than i_C; the equivalent circuit of Fig. 4.17 assumes that $i_C = i_E$.

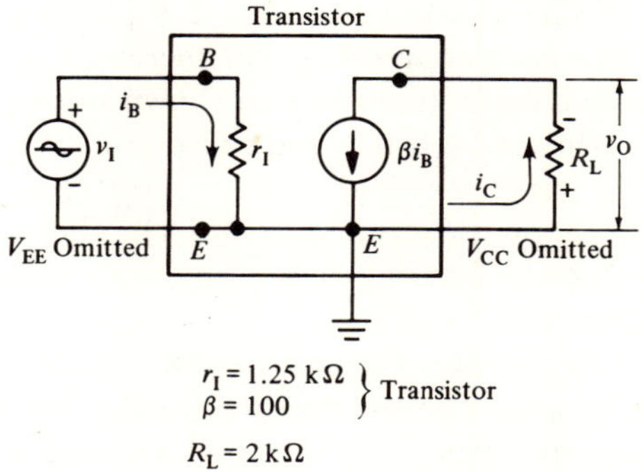

Figure 4.17

Figure 4.17 represents an approximate model for the small signal ac equivalent circuit of the basic amplifier of Fig. 4.1. The input voltage, v_I, is applied across the base and emitter (input) terminals of the transistor; between these two terminals the transistor acts like a resistor, r_1. When the voltage is applied across the terminals, a base current, i_B, flows into the transistor. This base current is multiplied by β in the collector circuit. The current generator, βi_B, in the collector circuit is called a *controlled generator;* i.e., the value of the current in this generator is controlled by (depends upon) the value of the base current. The collector of the transistor in Fig. 4.1 was connected directly to the load resistor, R_L, and R_L was connected to the positive terminal of V_{CC}; the negative terminal of V_{CC} was connected to the emitter (or ground). Since the circuit of Fig. 4.17 is an *ac equivalent circuit, all dc sources are omitted.* Thus V_{CC} is *not* shown in the circuit of Fig. 4.17; the load resistor is connected *directly* across the collector and emitter terminals, as far as the ac portions of the voltages and currents are concerned.

The transistor circuit itself is represented inside the box shown in Fig. 4.17; notice that it can be considered to be merely a resistor and a controlled current source. Both the resistance, r_{I}, and β are parameters which depend upon the particular transistor. It should be mentioned that this is just an approximate equivalent circuit; a more exact form is given in the next section. The approximate circuit was introduced at this point to give the reader an insight into the basic workings of a transistor amplifier. The effect of r_{I} and β on the gain of the amplifier is discussed later on in the chapter.

A few final points are now mentioned in connection with the circuit of Fig. 4.17. First, notice that the load resistor is connected directly across the collector and emitter terminals. This means that *the ac portion* of the collector-to-emitter voltage is equal to the ac portion of the voltage across the load resistor; this fact was mentioned earlier in Sec. 4-2, Waveforms. It is important to bear in mind that this holds true *only* for the ac portions, *not* the dc. The second point concerns the fact that the arrow for the current source in Fig. 4.17 is pointing *downward* in Fig. 4.17, with the result that the collector current flows *upward* through the load resistor as shown. The reason for this is explained as follows: When the input voltage is going positive (first half-cycle), the base current is increasing in the direction shown in the figure. We know from the analysis made earlier in the chapter that the collector-to-emitter voltage is 180 degrees out of phase with the input voltage. Since an increasing base current results in an increasing collector current, the collector current is increasing in the direction shown during the first half-cycle. Thus the voltage across the load resistor (and collector-emitter terminals) acquires the polarity shown; the upper terminal of R_{L} (collector) tends to become more *negative* with respect to the lower terminal (emitter). Therefore, when the input voltage is *increasing* during the first half-cycle, the output voltage, v_{O}, is *decreasing;* this explains the 180 degree phase shift.

The circuit of Fig. 4.17 is called the small signal circuit because, as is further explained in later chapters, the circuit ceases to be a good approximation for large inputs. It is called the ac equivalent circuit because it is valid for only the ac portion of the waveform. Finally, we shall see in the following section that the circuit of Fig. 4.17 is only an approximation.

4-7   THE SMALL SIGNAL ac EQUIVALENT CIRCUIT: EXACT MODEL

For the sake of completeness it is necessary at this point to introduce the exact model of the small signal ac equivalent circuit, shown in Fig. 4.18. The transistor itself is represented by the circuitry within the box; notice that this model is considerably more complicated than that shown in Fig. 4.17. This is known as the *hybrid equivalent circuit,* and h_{ie}, h_{re}, h_{fe}, and h_{oe} are called the h (or hybrid) *parameters.* The term "hybrid" is used because the parameters are measured in different units; some are measured in ohms, some in mhos, and some are dimensionless.

The input circuit of Fig. 4.18 shows that, between base and emitter, the transistor is represented by a resistance, h_{ie}, in series with a *controlled voltage generator*, $h_{re}v_O$. This small generator is dependent upon the output voltage, and tends to act in opposition to the input base current; in other words, we can say that the *input*

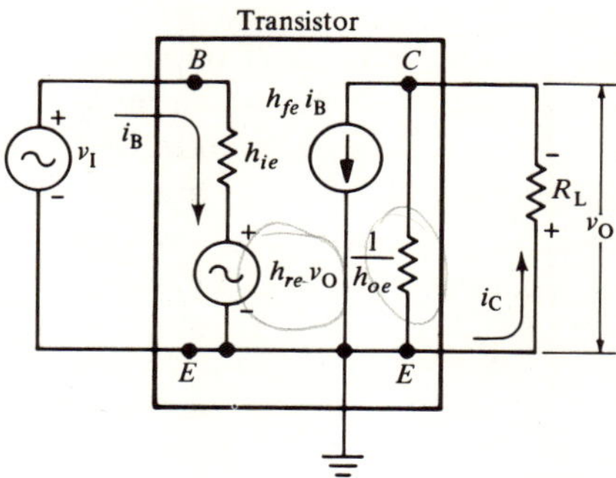

Figure 4.18

current depends in some measure upon the output voltage, just as the output voltage and current depend upon the input current. We shall soon see that the addition of this controlled voltage generator complicates considerably the analysis of the amplifier. The parameter, h_{ie}, is a *resistance*, measured in ohms; the subscript "*i*" means input, and "*e*" stands for common emitter configuration (there is a different set of parameters for the common base configuration, and one for the common collector). The parameter, h_{ie}, can be computed as the reciprocal of the slope of the input characteristic; the relationship between the *h*-parameter, h_{ie}, and the input resistance of the amplifier, r_I, is explained shortly. The controlled source, $h_{re}v_O$, is a voltage; since it is directly proportional to another voltage (v_O), h_{re} is a *dimensionless quantity* (no units). The "*r*" stands for *reverse*, since the input depends upon the output (in addition to the output's depending upon the input).

The collector circuit of Fig. 4.18 contains a *controlled current generator*, $h_{fe}i_B$; this current generator is dependent upon another current (i_B), and therefore the parameter h_{fe} is *dimensionless*. This parameter *is the same as the β of the transistor as* described in chapter 3 and earlier in this chapter; it is computed in the same manner as shown in chapter 3. The term "β" is used instead of h_{fe} because it is simpler to work with in equations; the reader should bear in mind, however, that the two terms are identical. The "*f*" in h_{fe} stands for *forward*, since it is through this particular parameter that the output depends upon the input.

There is a resistance, $1/h_{oe}$, in parallel with the current generator $h_{fe}i_B$ in Fig. 4.18. Since $1/h_{oe}$ is a resistance, h_{oe} is a *conductance*, measured in mhos; the "*o*" in h_{oe} stands for *output* because this parameter is directly across the output terminals. The parameter, $1/h_{oe}$, is computed as the reciprocal of the slope of the collector characteristics; the relationship between the parameter, $1/h_{oe}$, and the output resistance of the transistor, r_O, is explained shortly. Notice that the combination of $h_{fe}i_B$ and $1/h_{oe}$ in Fig. 4.18 constitute a current source *within the transistor;* therefore *not all of* $h_{fe}i_B$ will flow through the collector, because some portion re-circulates through the *internal resistance*, $1/h_{oe}$ (see Sec. 4-5). We have more to say about this phenomenon shortly.

The hybrid equivalent circuit is only one of a number of equivalent circuits developed to represent the ac operation of the transistor. This circuit was constructed by engineers, technicians, and circuit theorists after some years of studying the operation of the transistor. The results obtained from working with this circuit give a very exact indication of what is actually happening in the circuit; however, we shall find that it is more convenient to use an approximate representation of the circuit. The current and voltage polarities indicated in Fig. 4.18 are given for a particular instant of time; these will, of course, reverse during the next half-cycle since all of the quantities are considered to be pure ac sine waves (no dc).

Summarizing, the four *h*-parameters are as follows:

Parameter	Units	Measured from
h_{ie}	ohms	input characteristic
h_{re}	none	input characteristic
h_{fe}	none	collector characteristics
h_{oe}	mhos	collector characteristics

We now illustrate the application of the use of these parameters to a particular problem. It can be shown that the expressions for input resistance (r_I), output resistance (r_O), current gain (A_I), and voltage gain (A_V) can be stated as follows:

$$r_I = \frac{h_{ie} + R_L(h_{ie}h_{oe} - h_{re}h_{fe})}{1 + h_{oe}R_L}$$

$$r_O = \frac{1}{h_{oe} - \dfrac{h_{fe}h_{re}}{h_{ie} + R_g}}$$

$$A_I = \frac{h_{fe}}{1 + h_{oe}R_L}$$

$$A_V = \frac{-h_{fe}R_L}{h_{ie} + R_L(h_{ie}h_{oe} - h_{re}h_{fe})}$$

These equations are derived by applying Ohm's and Kirchhoff's laws to the circuit of Fig. 4.18, and then employing a considerable amount of algebraic manipulation. R_L is the collector load resistance (2 kΩ in the case of the circuit of Fig. 4.1); R_g is the internal resistance of the input generator, v_I, which was assumed to be equal to zero ohms. The minus sign in the expression for the voltage gain symbolizes the 180 degree phase shift. Let us compute the values of r_I, etc., for the amplifier of Fig. 4.1. The *h*-parameters for the 2N910 transistor are:

$$h_{ie} = 1.25 \text{ k}\Omega, \; h_{re} = 10^{-4}, \; h_{fe} = \beta = 100, \; h_{oe} = 20 \times 10^{-6} \text{ mhos (or 20 umhos)}$$

$$R_L = 2 \text{ k}\Omega, \; R_g = 0$$

It is convenient first to compute the value of the factor $h_{ie}h_{oe} - h_{re}h_{fe}$, since this appears a number of times in the expressions, thus:

$$h_{ie}h_{oe} - h_{re}h_{fe} = (1.25 \times 10^3)(20 \times 10^{-6}) - (10^{-4})(10^2) = 0.015$$

$$r_\mathrm{I} = \frac{h_{ie} + R_\mathrm{L}(h_{ie}h_{oe} - h_{re}h_{fe})}{1 + h_{oe}R_\mathrm{L}} = \frac{1.25\ \mathrm{k\Omega} + 2\ \mathrm{k\Omega}\ (0.015)}{1 + (20 \times 10^{-6})(2 \times 10^3)} = \underline{1.23\ \mathrm{k\Omega}}$$

$$r_\mathrm{O} = \frac{1}{h_{oe} - \dfrac{h_{fe}h_{re}}{h_{ie} + R_g}} = \frac{1}{20 \times 10^{-6} - \dfrac{(10^2)(10^{-4})}{1.25 \times 10^3 + 0}} = \underline{83.3\ \mathrm{k\Omega}}$$

$$A_\mathrm{I} = \frac{h_{fe}}{1 + h_{oe}R_\mathrm{L}} = \frac{100}{1 + (20 \times 10^{-6})(2 \times 10^3)} = \underline{96}$$

$$A_\mathrm{V} = \frac{-h_{fe}R_\mathrm{L}}{h_{ie} + R_\mathrm{L}(h_{ie}h_{oe} - h_{re}h_{fe})} = \frac{-100(2\ \mathrm{k\Omega})}{1.25\ \mathrm{k\Omega} + 2\ \mathrm{k\Omega}\ (0.015)} = \underline{-156}$$

Next let us compute the quantities assuming that h_{re} is approximately equal to zero. The equations become:

$$r_\mathrm{I} = h_{ie}$$

$$r_\mathrm{O} = \frac{1}{h_{oe}}$$

$$A_\mathrm{I} = \frac{h_{fe}}{1 + h_{oe}R_\mathrm{L}} \qquad \rightarrow h_{re} \simeq 0$$

$$A_\mathrm{V} = \frac{-h_{fe}R_\mathrm{L}}{h_{ie}(1 + h_{oe}R_\mathrm{L})}$$

Substituting into the equations, we obtain:

$$r_\mathrm{I} = h_{ie} = \underline{1.25\ \mathrm{k\Omega}}$$

$$r_\mathrm{O} = \frac{1}{h_{oe}} = \frac{1}{20 \times 10^{-6}} = \underline{50\ \mathrm{k\Omega}}$$

$$A_\mathrm{I} = \frac{h_{fe}}{1 + h_{oe}R_\mathrm{L}} = \frac{100}{1 + (20 \times 10^{-6})(2 \times 10^3)} = \underline{96}$$

$$A_\mathrm{V} = \frac{-h_{fe}R_\mathrm{L}}{h_{ie}(1 + h_{oe}R_\mathrm{L})} = \frac{-100(2\ \mathrm{k\Omega})}{(1.25 \times 10^3)(1.04)} = \underline{-154}$$

Notice that the exact value of the input resistance, r_I, was 1.23 kΩ, while the value computed assuming that $h_{re} = 0$ is 1.25 kΩ, or h_{ie}; the two values agree quite closely. We can now see the relation between h_{ie} and r_I. The parameter, h_{ie}, is actually the reciprocal of the slope of the input characteristic as explained in chapter 3; however, if we assume that h_{re} is approximately equal to zero, we can say that the *input resistance, r_I, is approximately equal to h_{ie}*. We make use of this approximation throughout the remainder of the text. Similarly, assuming $h_{re} = 0$, we computed the output resistance, r_O, to be equal to $1/h_{oe}$, or 50 kΩ; the exact value of r_O is 83.3 kΩ. Although it is true that 50 kΩ is not a good approximation to 83.3 kΩ, we can assume that the output resistance is approximately equal to $1/h_{oe}$ for the following reason: The ac equivalent circuit of the amplifier can now be represented as shown in

Fig. 4.19. Figure 4.19(a) shows the equivalent circuit when h_{re} is approximately equal to zero. Notice that the current from the $h_{fe}i_B$ generator splits up into two paths: one through the 2 kΩ load resistor, and the other through the resistor, $1/h_{oe}$, or 50 kΩ. Generally speaking, $1/h_{oe}$ is much greater than R_L (50 kΩ compared with

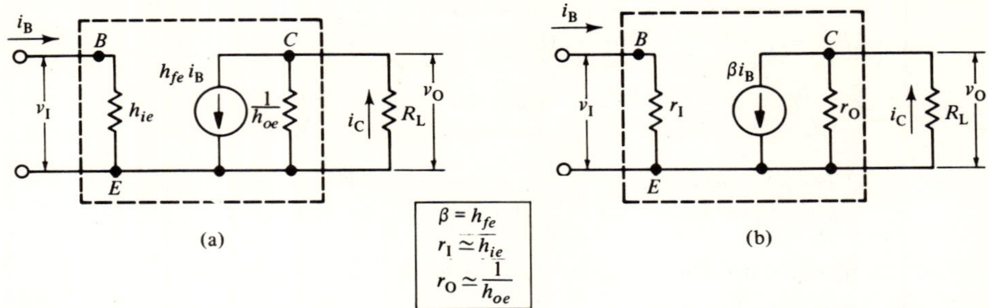

Figure 4.19

2 kΩ, in this case); therefore most of the source current will flow through the load resistor rather than recirculate through $1/h_{oe}$. Thus, if, as is usually the case, $1/h_{oe}$ is greater than 10 times the load resistor, we can neglect the effect of $1/h_{oe}$; i.e., we can assume that $1/h_{oe}$ is approximately infinite (or $h_{oe} = 0$). We therefore save the trouble of computing the output resistance by the formula; *we approximate the output resistance of the transistor, r_O, as $1/h_{oe}$*. The parameter $1/h_{oe}$ is actually the reciprocal of the slope of the collector characteristics as explained in chapter 3.

The circuits shown in Fig. 4.19 are used throughout the remainder of the text to represent the ac behavior of the transistor, subject to the approximations shown. Figure 4.19(b) is the same as Fig. 4.19(a), except that r_I is used in place of h_{ie}, r_O for $1/h_{oe}$, and β for h_{fe}.

Figure 4.20 is a chart showing the exact formulas, as well as certain approxi-

	Input resistance r_I	Output resistance r_O	Current gain A_I	Voltage gain A_V
Exact	$\dfrac{h_{ie} + R_L(h_{ie}h_{oe} - h_{re}h_{fe})}{1 + h_{oe}R_L}$	$\dfrac{1}{h_{oe} - \dfrac{h_{fe}h_{re}}{h_{ie} + R_g}}$	$\dfrac{h_{fe}}{1 + h_{oe}R_L}$	$\dfrac{-h_{fe}R_L}{h_{ie} + R_L(h_{ie}h_{oe} - h_{re}h_{fe})}$
$h_{re} \simeq 0$	h_{ie}	$\dfrac{1}{h_{oe}}$	$\dfrac{h_{fe}}{1 + h_{oe}R_L}$	$\dfrac{-h_{fe}R_L}{h_{ie}(1 + h_{oe}R_L)}$
$h_{re} \simeq 0$ $h_{oe} \simeq 0$	h_{ie}	∞	h_{fe}	$\dfrac{-h_{fe}R_L}{h_{ie}}$

Figure 4.20

mations. The approximations give results which are in good agreement with the exact expressions. For example, the voltage gain, A_V, computed using the exact expression was 156, and that computed from the expression where h_{re} was set equal to zero was 154; the voltage gain computed directly from the characteristics earlier in the chapter was 160. Thus the approximations are quite good. The equations and formulas given in the chart of Fig. 4.20 were stated only for the sake of completeness; we do not use these expressions in the text for a few important reasons. The first reason is that the formulas in the chart *hold only for a single-stage amplifier*, i.e., one transistor. We shall see later in the text that the addition of other stages to the circuit complicates the formulas considerably; thus it is more convenient to work with a simplified equivalent circuit like the one in Fig. 4.19(b). The designer who can adapt the simplified equivalent circuit to different situations, computing the various currents and voltages, is more versatile than the one who tries to solve circuit problems by memorizing long, unwieldy formulas; in addition, these simplified equivalent circuits yield more of an insight into the factors upon which the amplifier performance depends. The second reason for using the simpler circuits is that *all four of the h-parameters vary considerably with temperature, with Q point, and from device to device of the same type.* An indication of the extent of this variation is presented in chapter 10 when we consider the transistor specification sheet. Since the parameters vary so much, it is senseless to carry out long computations when we are not even sure to what extent the parameters are accurate.

When an engineer or technician designs a circuit, he generally does not work through long, involved formulas. He usually performs some rough, preliminary computations at his desk to get an idea of the approximate performance of the circuit; then he moves to the bench where he builds and tests the circuit to see if it operates as the computations predicted. If the circuit does not work exactly as anticipated (and the first time it usually does not), the designer makes the appropriate changes utilizing his knowledge of the theory. Since the parameters of the transistor are quite variable, and the resistors and capacitors have definite tolerances, and test equipment has its limitations, the design is usually finalized right at the bench with the actual circuit at the fingertips of the designer. We discuss the design process in more detail in chapter 10. For the foregoing reasons, we refrain from using long, involved formulas in this text and try instead to get a good approximation to the actual circuit performance, using the simplified equivalent circuit of Fig. 4.19(b) as the basic model.

One final point in connection with the equivalent circuit; this concerns $1/h_{oe}$. You will recall that $1/h_{oe}$ is the reciprocal of the slope of the collector characteristics. Figure 4.21(a) shows the load line drawn on the collector characteristics of a transistor; the base current, i_B (peak-to-peak) is varying between the two values, i_{B_1} and i_{B_2}. β is defined as the ratio of change in collector current to change in base current *when v_{CE} is held constant;* if v_{CE} were held constant the change in collector current would be from point 1 to point 2, and thus the peak-to-peak collector current would be equal to Z in the figure. The base current, however, must vary along the load line, and thus the collector current varies between points 1 and 3; the peak-to-peak collector current is *actually* equal to Y in the figure. Figure 4.21(c) shows that the *maximum amount of current generated is $\beta i_B = Z$*; however, only an amount equal to Y gets to the load while the remainder, X, re-circulates through $1/h_{oe}$.

Figure 4.21(b) shows the same set-up, except that the transistor has flatter-sloped

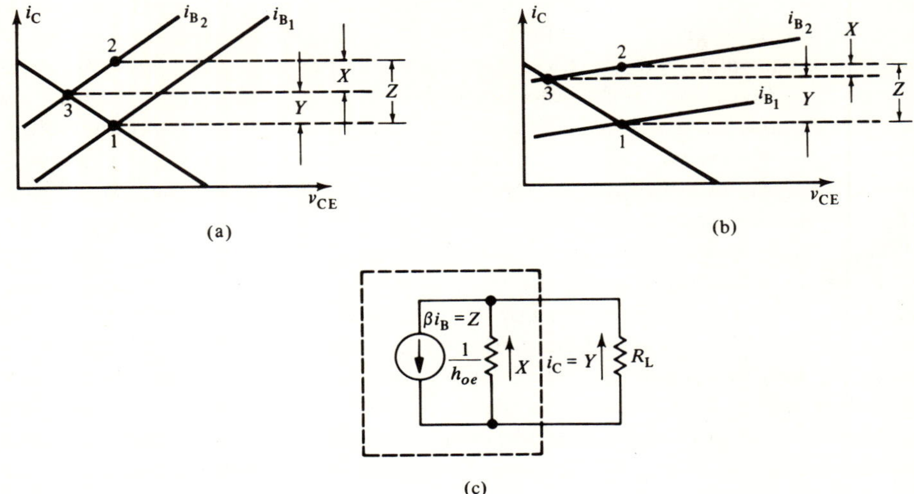

Figure 4.21

characteristics; notice that Y is a *greater* percentage of Z (X is much smaller). This means that *more* current gets to the load, and correspondingly *less* flows through $1/h_{oe}$. The reason for the action observed in Fig. 4.21(b) is that the slope of the characteristics is smaller (larger $1/h_{oe}$) than in Fig. 4.21(a); a larger resistor will always draw less current than a smaller one in parallel with it. It should be remembered when dealing with these equivalent circuits that $1/h_{oe}$ is not a physical resistor; it is just included in the circuit to account for the effect shown in Fig. 4.21.

4-8 APPLICATIONS OF THE EQUIVALENT CIRCUIT

In this section we illustrate the use of the equivalent circuit in solving circuit problems. Before continuing with examples of this procedure, it is necessary to define a few terms. We use lower-case letters to represent *the ac portions* of the voltages and currents in a transistor amplifier; e.g., v_I represents the ac input sine wave, i_B is the ac portion of the base current, etc. Since we are working with sine waves in this text, these lower-case letters will represent the *peak-to-peak* values of the quantities. Similarly, capital letters are used to represent *all dc* (quiescent) *currents* and voltages.

EXAMPLE 3

Using the equivalent circuit, compute the output quantities and gains for the amplifier of Fig. 4.1; $v_I = 0.1$ V.

SOLUTION

The equivalent circuit is re-drawn in Fig. 4.22. Since $r_O = 1/h_{oe} = 50$ kΩ, we neglect the effect of the output resistance since it is more than 10 times as large as the

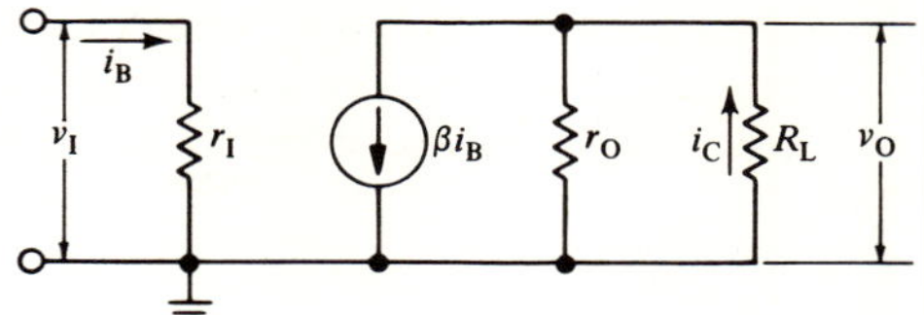

Figure 4.22

load resistance, $R_L = 2\ \text{k}\Omega$. Notice in Fig. 4.22 that v_I is applied directly across the input resistance, 1.25 kΩ. The first step is to compute the ac base current, i_B. By Ohm's law, we get:

$$i_B = \frac{v_I}{r_I} = \frac{0.1\ \text{V}}{1.25\ \text{k}\Omega} = 0.08\ \text{mA} = 80\ \mu\text{A}$$

We know that the current in the base is multiplied by β in the collector; thus:

$$\beta i_B = 100(0.08\ \text{mA}) = 8\ \text{mA}$$

Since we have neglected the effect of the output resistance, we can assume that the entire current from the controlled generator flows through the collector; thus:

$$r_0 \gg R_L$$
$$i_C = \beta i_B = 8\ \text{mA}$$

By Ohm's law, the output voltage, v_O, is:

$$v_O = i_C R_L = (8\ \text{mA})(2\ \text{k}\Omega) = 16\ \text{V}$$

The gains are computed as follows:

$$A_I = \frac{i_o}{i_I} = \frac{i_C}{i_B} = \frac{8\ \text{mA}}{0.08\ \text{mA}} = 100$$

$$A_V = \frac{v_O}{v_I} = \frac{16\ \text{V}}{0.1\ \text{V}} = 160$$

Note the agreement of these values with those computed earlier in the chapter by a more tedious procedure. Herein lies the beauty of the equivalent circuit: *it enables us to compute the ac quantities and gains quickly and easily.* Of course we neglected the effect of two of the h-parameters (h_{re} and h_{oe}); thus the answers are only approximately correct. However, the strength of this method becomes apparent when one is designing a circuit and wants to obtain an approximate idea of the circuit behavior without solving complex algebraic equations. It is important to note again that *this method tells us nothing about the dc quiescent values;* these must be computed by other methods. One approach is to use the load line and the characteristics as described in chapter 3 and the beginning of this chapter; other methods will be developed in chapter 5. In computing the ac quantities with this circuit, we have assumed that the internal resistance of the input voltage source was zero; this is never true. We see, in chapter 6, that the internal resistance of the input source has many important effects on the operation of the amplifier circuit.

EXAMPLE 4

Compute the output quantities and gains if the input voltage, v_I, is 0.001 V (1.0 mV).

SOLUTION

The input voltage is now *one one-hundredth* of the input voltage used in example 3. If this problem were attempted graphically, a difficulty would arise. In order to proceed graphically, the input voltage would have to be plotted on the input characteristic as in Figs. 4.3, 4.8(a), and 4.10(a); this enables us to plot the ac base current. However, the voltages and currents are *so small* that any readings taken from these curves would be highly inaccurate. Here is another reason for the popular use of the equivalent circuit; the *ac quantities and gains can be computed without the necessity of resorting to graphical procedures*. Since neither the batteries nor the transistor have been changed, the dc levels remain the same; the parameters of the transistor are still the same as in example 3. Proceeding as in example 3, we obtain:

$$i_\mathrm{B} = \frac{v_\mathrm{I}}{r_\mathrm{I}} = \frac{0.001 \text{ V}}{1.25 \text{ k}\Omega} = 0.8 \ \mu\text{A}$$

$$i_\mathrm{C} = \beta i_\mathrm{B} = 100(0.8 \ \mu\text{A}) = 80 \ \mu\text{A}$$

$$v_\mathrm{O} = i_c R_\mathrm{L} = (80 \ \mu\text{A})(2 \text{ k}\Omega) = 0.16 \text{ V}$$

$$A_\mathrm{I} = \frac{i_o}{i_\mathrm{I}} = \frac{i_\mathrm{C}}{i_\mathrm{B}} = \frac{80 \ \mu\text{A}}{0.8 \ \mu\text{A}} = 100$$

$$A_\mathrm{V} = \frac{v_\mathrm{O}}{v_\mathrm{I}} = \frac{0.16 \text{ V}}{0.001 \text{ V}} = 160$$

It can be seen from this example that all of the currents and voltages computed from example 4 are *one one-hundredth* of those of example 3 ($v_\mathrm{O} = 0.16$ V in example 4 compared with 16 V in example 3, etc.). The reason for this effect is that the input voltage used in example 3 was *100 times greater* than that in example 4. Note also, however, that the *gains* computed for both examples *are the same*. In other words, once the gain of an amplifier is known, the output voltage and current can be computed for any input voltage. All that need be done is to multiply the input quantity (voltage or current) by the gain to obtain the output quantity.

EXAMPLE 5

The voltage gain of a particular amplifier is 50. Compute the output voltage if the input voltage is (a) 0.01 V; (b) 0.1 V.

SOLUTION

Utilizing the general equation, we get:

(a)

$$A_V = \frac{v_O}{v_I}; \quad v_O = A_V v_I = 50(0.01 \text{ V}) = 0.5 \text{ V}$$

(b)

$$v_O = A_V v_I = 50(0.1 \text{ V}) = 5 \text{ V}$$

Thus, the gain of an amplifier is an important quantity.

EXAMPLE 6

Repeat example 5 if the voltage gain of the amplifier is 5.0.

SOLUTION

(a)

$$v_O = A_V v_I = 5(0.01 \text{ V}) = 0.05 \text{ V}$$

(b)

$$v_O = A_V v_I = 5(0.1 \text{ V}) = 0.5 \text{ V}$$

The results of example 5 indicate that, with the gain constant, a *greater input voltage will produce a greater output*. Example 6 shows that, for the same input, the *amplifier with the greater gain will produce the greater output;* e.g., in example 5(b) ($A_V = 50$) $v_O = 5$ V when $v_I = 0.1$ V, while in example 6(b) ($A_V = 5.0$) $v_O = 0.5$ V when $v_I = 0.1$ V. In other words, for the same input, an amplifier with 10 times as much gain will produce 10 times as much output.

There is, however, a limit to the amount of output which can be obtained from a particular amplifier. Figure 4.23 shows an *idealized* input characteristic for the 2N910 transistor; the term "idealized" is used because the input characteristic is *approximated* by the two straight line segments in Fig. 4.23(a) (the dotted curve represents the *actual* characteristic). The input characteristic is approximated as shown in Fig. 4.23(a) because it enables us to illustrate the important concepts which follow. We discuss the effect of the actual characteristic (dotted portion) on the waveforms in a later chapter. Since the sloping portion of the input characteristic is approximated as a straight line, we can assume that r_I is constant and equal to 1.25 kΩ (reciprocal of the slope) all along this portion of the characteristic.

Assume that $v_I = 0.25$ V (peak-to-peak); using either the graphical approach or the equivalent circuit we can compute the base current as 0.2 mA [see Fig. 4.23(a)]. Figure 4.23(b) shows the collector characteristics with the load line already drawn. With a peak-to-peak base current of 0.2 mA, the collector current

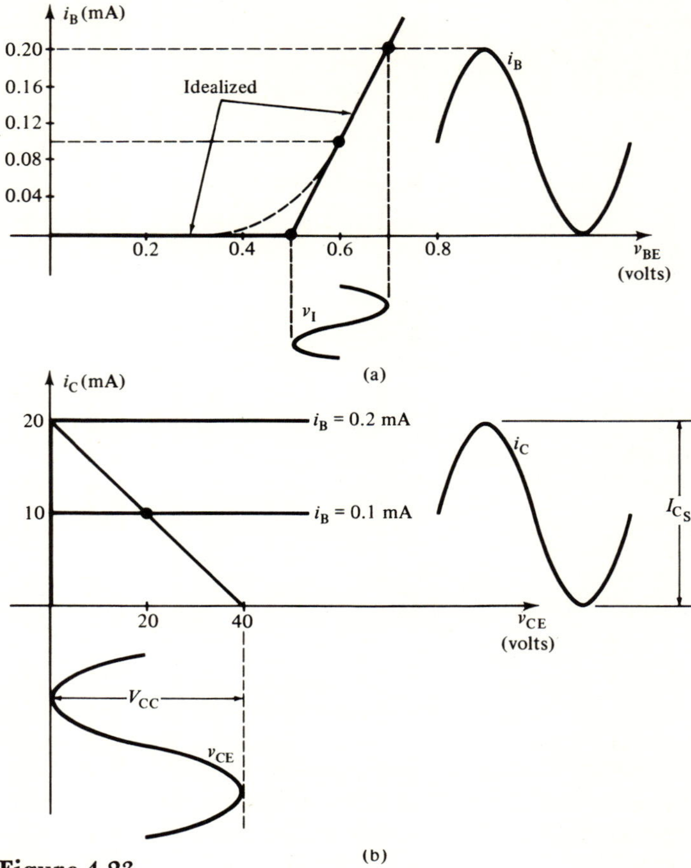

Figure 4.23

varies from a maximum of 20 mA (saturation) to a minimum of zero (cut-off), while the collector-to-emitter voltage varies from approximately zero (saturation) to 40 V (cut-off), as shown in Fig. 4.23(b). The waveforms shown in Fig. 4.23(b) (i_C and v_{CE}) represent the *maximum sinusoidal outputs* available from the amplifier of Fig. 4.1. This is so because the peak-to-peak collector-to-emitter voltage of an amplifier like the one in Fig. 4.1 cannot exceed the value of V_{CC} (in this case, 40 V), since v_{CE} cannot be greater than V_{CC} nor less than zero. Similarly, the peak-to-peak collector current cannot exceed the value of I_{Cs} (the saturation collector current, in this case equal to 20 mA), since i_C cannot be greater than I_{Cs} nor less than zero. The statements made above hold only for an amplifier with a resistive load; in chapter 9 we shall see that the situation is different with an inductive load.

Let us suppose now that the input voltage is increased above 0.25 V. The base current will increase (peak-to-peak) and therefore the ac portions of the collector current and the collector-to-emitter voltage will also *try* to increase. The word "try" is used because the ac portions of the waveforms cannot increase beyond the values shown in Fig. 4.23(b). Figure 4.24 shows the waveforms obtained when v_I is increased above 0.25 V. Notice that the waveforms are *not* sinusoidal; there is a definite *flattening* (distortion) at the 90 and 270 degree instants. When a voltage greater than 0.25 V is applied at the input of the amplifier of Fig. 4.1, the waveforms resembling those in Fig. 4.24 will appear because the ac portions of the

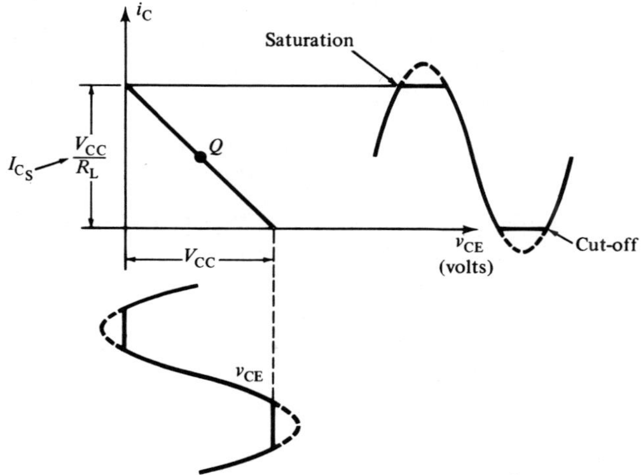

Figure 4.24

collector current and collector-to-emitter voltage will try to increase (dotted portions), but cannot because of saturation and cut-off. This phenomenon is known as *overdriving the amplifier;* i.e., the amplifier will *not* produce a sinusoidal output if more than a certain amount of input is applied. We say that the amplifier is *driven too hard.* In other words, the application of an ac sine wave to the input of an amplifier does *not* necessarily mean that the output will be sinusoidal. When using the equation,

$$v_O = A_V v_I$$

we cannot be sure that multiplying the input by the gain will give the output; we must first test to see if the amplifier can produce the predicted output.

EXAMPLE 7

The voltage gain of a particular amplifier is 50; V_{CC} is 20 V. Compute the output voltage if the input is (a) 0.1 V, and (b) 1.0 V.

SOLUTION

Applying the basic formula to part (a), we get:

$$v_O = A_V v_I = 50(0.1 \text{ V}) = 5 \text{ V}$$

Continuing with part (b) the formula *predicts* that the output will be:

$$v_O = A_V v_I = 50(1.0 \text{ V}) = 50 \text{ V}$$

However, since V_{CC} is 20 V, a sinusoidal output of 50 V is *impossible* with this amplifier; the output waveform will be severely distorted (non-sinusoidal). The maximum input which can be applied to this amplifier with the expectation that the output will be sinusoidal is found as follows: The maximum peak-to-peak output (provided the Q point is at the center of the load line) from the amplifier is 20 V; the gain is 50. Thus the maximum input which can be applied is:

$$v_{I(max)} = \frac{v_{O(max)}}{A_V} = \frac{20 \text{ V}}{50} = 0.4 \text{ V}$$

In other words, if the input voltage is so large that the amplifier is overdriven, the gain equations and formulas become meaningless. Therefore, the equations should be used with care. Of course, there are circuits where it is desirable that the output be non-sinusoidal (pulse circuits); however, in this text we are concerned primarily with sinusoidal waveforms, and therefore we want the output to be a pure undistorted sinusoidal waveform.

Up to this point the Q point has been located in the center of the load line. Let us see what the effect is on the output if the Q point is located at some point other than the center. Figure 4.25 shows the characteristics with the Q point located

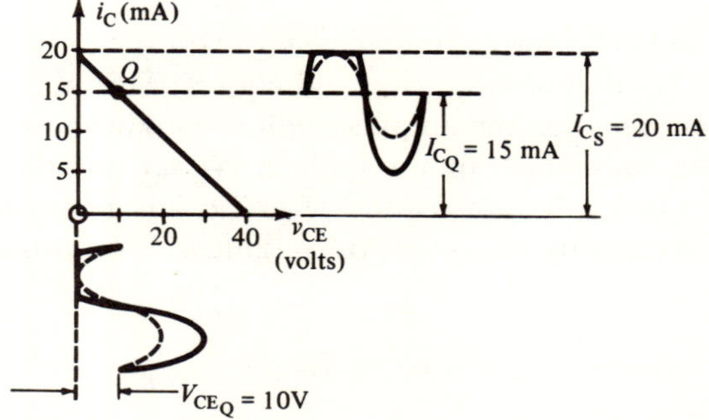

Figure 4.25

closer to the saturation point; the quiescent collector current is 15 mA (I_{C_Q}) and the quiescent collector-to-emitter voltage is 10 V (V_{CE_Q}). Notice that, because the Q point is located closer to *saturation*, the maximum undistorted peak-to-peak collector current obtainable from the amplifier is 10 mA (dotted curve); compare this with 20 mA in Fig. 4.23(b). Similarly, the maximum peak-to-peak collector-to-emitter voltage is 20 V (dotted curve); compare this with 40 V in Fig. 4.23(b). If the input is increased, the output will try to increase, but cannot do so because the transistor becomes saturated at 20 mA. Thus, the maximum peak-to-peak output of this amplifier is limited because the Q point is not located at the center of the load line.

Figure 4.26 shows what happens when the Q point is located at some point

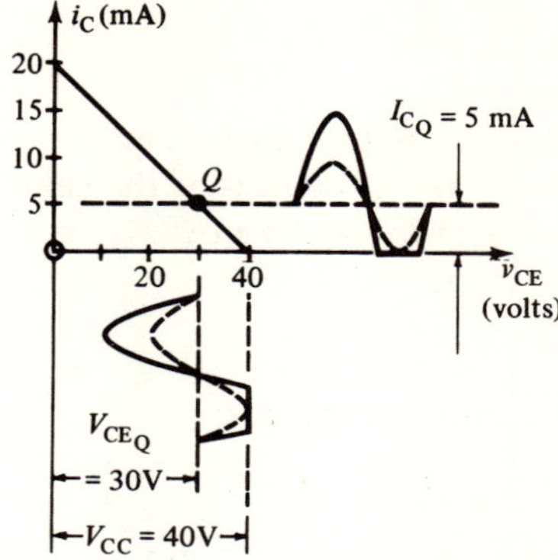

Figure 4.26

below the center of the load line. Here $I_{CQ} = 5$ mA and $V_{CEQ} = 30$ V. Now the maximum peak-to-peak output is limited by the *cut-off* point of the transistor. The dotted curves represent the maximum outputs obtainable; the solid lines show the distortion caused by overdriving the amplifier at cut-off.

The following chart summarizes the maximum undistorted outputs available from the amplifier:

Position of Q Point	Peak-to-Peak	
	$v_{O(\text{max})}$	$i_{O(\text{max})}$
Bias at Center	V_{CC}	$I_{Cs} = \dfrac{V_{CC}}{R_L}$
Bias above Center	$2V_{CEQ}$	$2(I_{Cs} - I_{CQ})$
Bias below Center	$2(V_{CC} - V_{CEQ})$	$2I_{CQ}$

Applying these to the amplifier of Fig. 4.1, we get:

BIAS AT CENTER:

$$v_{O(\text{max})} = V_{CC} = 40 \text{ V}$$
$$i_{O(\text{max})} = I_{Cs} = 20 \text{ mA}$$

BIAS AT 15 mA:

$$v_{O(\text{max})} = 2V_{CEQ} = 2(10 \text{ V}) = 20 \text{ V}$$
$$i_{O(\text{max})} = 2(I_{Cs} - I_{CQ}) = 2(20 - 15) = 10 \text{ mA}$$

BIAS AT 5 mA:

$$v_{O(\text{max})} = 2(V_{CC} - V_{CEQ}) = 2(40 - 30) = 20 \text{ V}$$
$$i_{O(\text{max})} = 2I_{CQ} = 2(5) = 10 \text{ mA}$$

These results show that in order to produce the maximum peak-to-peak output, the Q point should be in the center of the load line for a sine wave; any movement either way from the center limits the undistorted output. The farther the Q point moves from the center of the load line, the smaller is the maximum output

EXAMPLE 8

Find the maximum peak-to-peak output current and voltage of an amplifier whose $V_{CC} = 20$ V, $R_L = 5$ kΩ if (a) the Q point is at the center of the load line ($I_{CQ} = 2$ mA) and (b) the Q point is at $I_{CQ} = 3.5$ mA.

SOLUTION

(a) In the first case the maximum output is:

$$v_{O(max)} = V_{CC} = 20 \text{ V}$$

$$i_{O(max)} = I_{Cs} = \frac{V_{CC}}{R_L} = \frac{20 \text{ V}}{5 \text{ kΩ}} = 4 \text{ mA}$$

(b) Before proceeding, we must first compute the quiescent collector-to-emitter voltage, V_{CEQ}. From chapters 1 and 3 you recall that this can be done as follows:

$$V_{CEQ} = V_{CC} - I_{CQ}R_L = 20 \text{ V} - (3.5 \text{ mA})(5 \text{ kΩ}) = 20 \text{ V} - 17.5 \text{ V} = 2.5 \text{ V}$$

In this case the Q point is close to saturation; thus we must use the formulas which apply to this case. The maximum outputs are:

$$v_{O(max)} = 2(V_{CEQ}) = 2(2.5 \text{ V}) = 5 \text{ V}$$

$$i_{O(max)} = 2(I_{Cs} - I_{CQ}) = 2(4 \text{ mA} - 3.5 \text{ mA}) = 1 \text{ mA}$$

Thus, a comparison of parts (a) and (b) of example 8 shows that we can expect a greater maximum output when the Q point is in the center of the load line.

The amplifier circuit of Fig. 4.1 was discussed in this chapter because it is the easiest to explain. The purpose of this chapter has been to discuss the basic concepts involved in amplification; in order to avoid complicating the situation with other problems, the circuit of Fig. 4.1 was used since it is the simplest, most basic amplifier circuit. However, this particular circuit is rarely used because it is highly impractical for a number of very important reasons.

Chapter 5 illustrates the necessity of replacing the V_{EE} battery with a number of resistors connected in various configurations. This is done to stabilize the Q point because the circuit of Fig. 4.1 is very sensitive to changes in temperature, as well as replacement of transistors. We shall see in chapter 5 that stabilization of the Q point is one of the major problems in transistor design; the variation of the position of the Q point creates a number of serious difficulties.

In this chapter we just considered a single-stage amplifier, i.e., one transistor. In practice, amplifiers are usually connected together in various configurations to increase the gain. The problems involved in interconnecting these amplifier stages are considerable. Usually additional components have to be connected into the amplifier circuit to facilitate the interconnections of the amplifier stages; chapter 6 considers these problems in detail.

In addition to causing variations in the Q point, changing temperatures and transistor replacement contribute to changes in the gain of a particular amplifier. We see that the gain of the amplifier of Fig. 4.1 is highly dependent upon the transistor parameters; yet these parameters are temperature-sensitive. In chapter 7 methods are discussed which stabilize the amplifier gain; additional components have to be incorporated to effect this stabilization.

In the discussions of this chapter, nothing was mentioned about the frequency of the sine wave; chapter 8 shows that the gain of an amplifier depends upon the frequency of the input sine wave.

Chapter 9 discusses in more detail the problems of distortion which were only touched upon in this chapter. Additional changes must be made in the circuitry to reduce the effects of distortion.

In chapter 10, we consider a complete amplifier design utilizing the principles developed in chapters 1 through 9.

It can be seen from the above discussion that a considerable number of changes must be made in the circuit of Fig. 4.1 in order that the amplifier function in a practical application. The remainder of this text (chapters 5 through 10) is devoted to making the appropriate changes. Each of the remaining chapters deals with a particular facet of amplifier operation. By the time the reader reaches chapter 10, he will begin to appreciate the many problems involved in the analysis and design of transistor amplifiers.

PROBLEMS

1. Using the amplifier of Fig. 4.1 and the characteristics of Fig. 4.3 and Fig. 4.4, and setting $v_I = 0.05\ V_{(P–P)}$, sketch the waveforms of:

     (a) v_I (d) i_C
     (b) v_{BE} (e) v_{CE}
     (c) i_B (f) v_{R_L}

2. For the conditions of problem 1, compute:

     (a) A_V (b) A_I (c) A_P

3. A voltage source is composed of a 1 mV voltage generator in series with a 1 kΩ internal resistance. Convert the voltage source to a current source.

4. A current source is composed of a 0.1 mA current generator in parallel with a 10 kΩ internal resistance. Convert the current source to a voltage source.

5. In the approximate equivalent circuit of Fig. 4.17:

$$v_I = 1\ mV_{(P–P)} \qquad \beta = 100$$
$$r_I = 1\ k\Omega \qquad R_L = 10\ k\Omega$$

   Compute the peak-to-peak values of:

     (a) i_B (b) i_C (c) v_O (d) A_V (e) A_I

6. Repeat problem 5, except make $r_I = 500\ \Omega$. What effect does r_I have upon the gain?

7. Repeat problem 5, except make $\beta = 200$. What effect does β have upon the gain?

8. Repeat problem 5, except make $R_L = 20\ k\Omega$. What effect does R_L have upon the gain?

9. In the exact equivalent circuit of Fig. 4.18:

$$h_{ie} = 1 \text{ k}\Omega \qquad\qquad h_{fe} = 80$$
$$h_{re} = 2 \times 10^{-4} \qquad h_{oe} = 10 \times 10^{-6} \text{ mhos}$$

Compute:

(a) r_I (b) r_O (c) A_I (d) A_V

10. Repeat problem 9, except make $h_{re} = 0$. Comparing the results of problems 9 and 10, what effect does h_{re} have upon the various quantities?

11. The voltage gain of a particular amplifier is 100. Compute the output voltage if the input voltage is:

(a) 1 mV (b) 100 mV

12. The voltage gain of a particular amplifier is 100 and $V_{CC} = 40$ V. Assuming the Q point is in the center of the load line, compute the peak-to-peak output voltage if the input voltage is:

(a) 0.1 V (b) 1 V

13. What is the largest input voltage which can be applied to the amplifier of problem 12 without the output distorting?

14. Find the maximum peak-to-peak output current and voltage of an amplifier whose $V_{CC} = 30$ V and $R_L = 3$ kΩ if the Q point is in the center of the load line ($I_{CQ} = 5$ mA).

15. Repeat problem 14 if $I_{CQ} = 3$ mA.

16. Repeat problem 14 if $I_{CQ} = 7$ mA.

5

dc BIAS CIRCUITS AND STABILITY

It is the purpose of this chapter to study in detail the various methods used to bias the amplifier for dc operation. We will see that it is desirable to replace the V_{EE} battery by an alternate, more practical circuit. It is then shown that this alternate method (fixed bias) is also an undesirable means of biasing the amplifier under certain conditions. The reason is that the variation of transistor parameters with a change in the temperature causes a shift in the Q point of the circuit; this Q point shift is undesirable because it may lead to saturation (or cut-off) and distortion. Thus new bias circuits must be developed to replace the conventional methods described in chapters 3 and 4. In this chapter we are concerned exclusively with the problem of dc bias; the following chapters deal in more detail with additional aspects of the ac operation. We begin chapter 5 by illustrating the need for new bias circuits, utilizing some of the theory developed in chapters 3 and 4. Then we proceed to study these new circuits, applying the basic analytical techniques of chapters 1 and 3. Finally, we develop some simple methods for predicting the operation of the various circuits under specific conditions.

5-1 FIXED BIAS

As mentioned in chapters 3 and 4, the purpose of the V_{EE} battery was to forward bias the base-emitter junction of the transistor. This insured that a certain amount of dc current flowed to the base. A considerable amount of discussion was given in chapter 4 to the necessity for the presence of the battery in the amplifier circuit. Not only had the battery to be present, but its *value* was critical. As you will recall, the battery voltage had to be large enough so that the transistor was biased on the *steep, linear* portion of the input characteristic; this enabled the base current to vary both above *and* below the quiescent value of I_B (0.10 mA, in the amplifier of chapter 4). The proper quiescent value of I_B in turn set the Q point on the collector characteristics (in the center of the load line).

Batteries, however, are usually undesirable from the standpoint of practical circuit design. The problems of size (and space requirements), weight, cost, and the necessity of replacement or recharging encourage the designer to find an alternate means of supplying current. There is a fifth, more subtle, yet important reason for trying to replace the V_{EE} battery by a different type of circuit; this concerns the sensitivity of the Q point to the value of the battery voltage. Figure 5.1 shows the basic amplifier circuit of chapter 4, with the ac input voltage omitted;

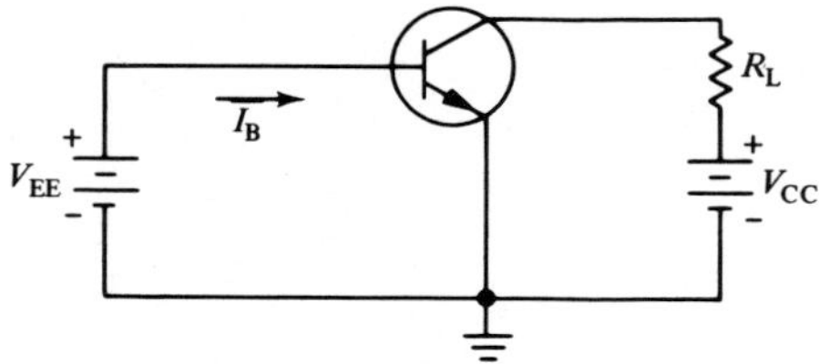

Figure 5.1

thus all of the voltages and currents are pure dc (no ac portions). Note that the V_{EE} battery forward biases the base-emitter junction. Figure 5.2(a) shows the input characteristic of the 2N910, and Fig. 5.2(b) shows the collector characteristics. If

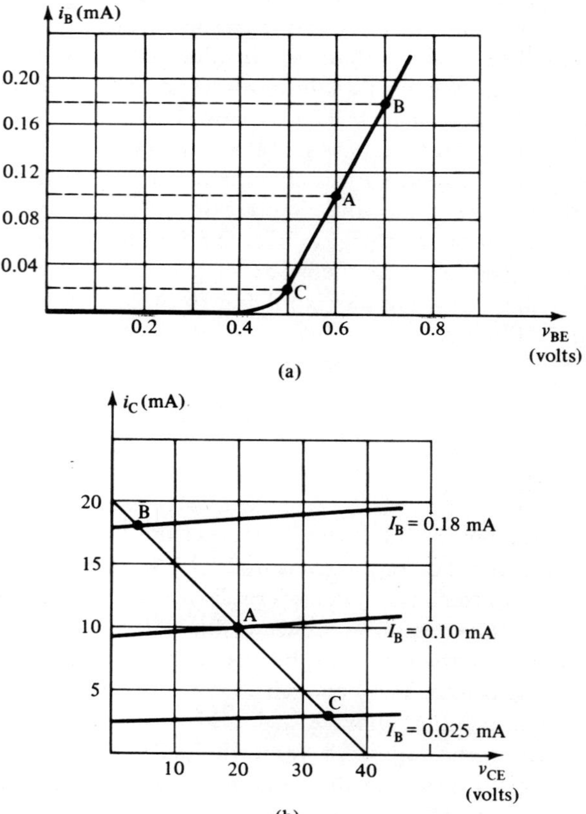

Figure 5.2

the battery voltage is exactly 0.6 V (point A in Fig. 5.2(a)), the quiescent base current is 0.1 mA, and the Q point on the collector characteristics is in the *center* of the load line (point A in Fig. 5.2(b)). Now suppose this 0.6 V battery is replaced by one having a voltage of 0.7 V (only 0.1 V larger). The quiescent base current is now 0.18 mA (point B) and the Q point on the collector characteristics has shifted to the *upper left-hand* portion of the load line (point B in Fig. 5.2(b)). Point B is very close to saturation; thus there is a possibility of distortion of the output waveforms. Suppose now that the voltage of the original 0.6 V battery has dropped 0.1 V to 0.5 V, due to aging of the battery. Point C in Fig. 5.2(a) shows that the quiescent base current is now only 0.025 mA, with the result that the Q point on the collector characteristics has shifted to the *lower right-hand* portion of the load line [point C in Fig. 5.2(b)]; point C is close to cut-off. The above results illustrate that a change of only a *few tenths of a volt* in V_{EE} causes a rather *large* change in the Q point; we say that the Q point is very *sensitive* to changes in the battery voltage. This sensitivity of the Q point to changes in the value of V_{EE} is undesirable, and is perhaps the primary reason why we should like to find an alternate biasing scheme. Thus, the following are the reasons why the use of the V_{EE} battery is undesirable:

1. Physical size may be prohibitive
2. Weight may be prohibitive
3. Cost may be excessive
4. Replacement or re-charging may be time-consuming and expensive
5. The Q point is highly sensitive to changes in the V_{EE} battery voltage

Figure 5.3(a) shows the same amplifier circuit, except that the V_{EE} battery has been increased to 40 V; of course a large resistor (400 kΩ) has been placed in series with the battery and the base lead to limit the size of the base current. From chapter 3, you may recall that in a circuit like the one in Fig. 5.3(a), the relatively small forward base-to-emitter voltage drop can be neglected in comparison with the 40 V battery voltage. Thus, the quiescent base current can be approximated as:

$$I_B \simeq \frac{V_{CC}}{R_B} = \frac{40 \text{ V}}{400 \text{ k}\Omega} = 0.10 \text{ mA}$$

In order to suppy dc current to the base of the transistor, it is necessary that terminal "X" in Fig. 5.3(a) be kept at a *more positive potential* than the emitter (more positive than ground, in this case). It makes no difference, however, where this positive potential comes from. Therefore, terminal "X" can be lifted from the positive terminal of the V_{EE} battery, as in Fig. 5.3(b), and connected to the positive terminal of the V_{CC} battery, as shown in Fig. 5.3(c). The positive terminal of the 40 V V_{CC} battery is 40 V more positive than the emitter, just as the positive terminal of the 40 V V_{EE} battery. As far as the 400 kΩ resistor is concerned, there is still 40 V across it, and therefore the dc base current is again approximately equal to 0.1 mA. Since the V_{EE} battery is no longer connected to any part of the circuit (Fig. 5.3(c)), it can be removed. Thus *we have replaced the V_{EE} battery by a resistor,* $R_B = 400$ kΩ.

Notice that the V_{CC} battery is still present; we have not been able to eliminate

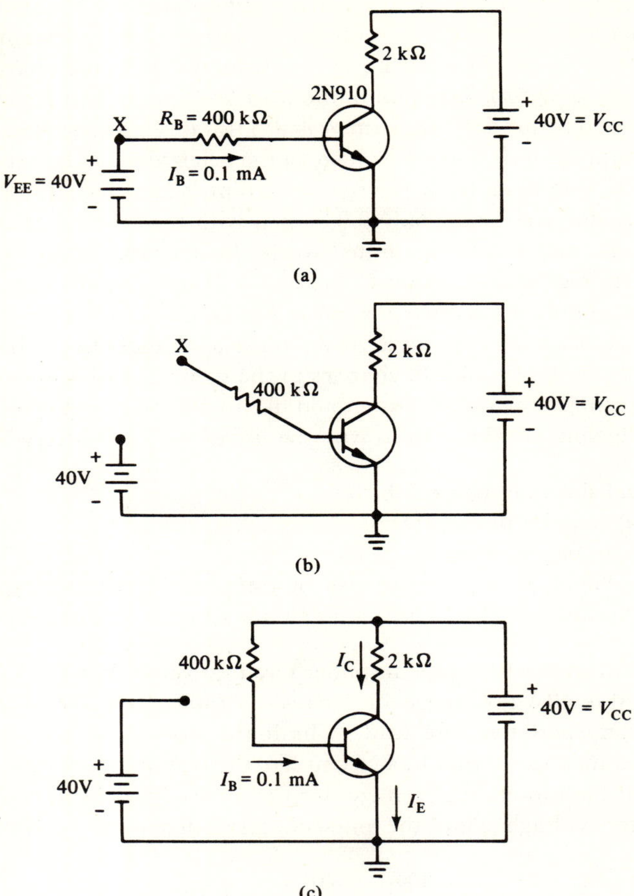

Figure 5.3

that one. Notice in Fig. 5.3(c) that the V_{CC} battery now must supply not only the collector current, but the base current as well; the currents are shown in the figure. The method of supplying dc current to the base pictured in Fig. 5.3(c) is known as the *fixed bias method*, because as we shall soon see, the base current is fixed by the value of the bias resistor, R_B. Figure 5.3(c) shows that the 40 V battery is connected between terminal "X" and ground; however, since the base-to-emitter drop is only a few tenths of a volt, we can consider that the base is approximately at *ground* potential. Therefore, we can say that the voltage across R_B is approximately equal to V_{CC}, and thus the same method can be used for computing the base current as was shown in chapter 3; i.e.:

$$I_B = \frac{V_{CC} - V_{BE}}{R_B}$$

For $V_{CC} \gg V_{BE}$: $\qquad I_B \simeq \dfrac{V_{CC}}{R_B}$ Fixed Bias

EXAMPLE 1

Compute the dc base current in the circuit of Fig. 5.3(c) if (a) $R_B = 400$ kΩ, (b) $R_B = 200$ kΩ, (c) $R_B = 800$ kΩ, (d) $R_B = 0$ Ω, and (e) $R_B = $ infinity.

SOLUTION

We neglect the small forward diode drop, V_{BE}, in the solution to example 1.
  (a) Utilizing the basic equation:

$$I_B = \frac{V_{CC}}{R_B} = \frac{40 \text{ V}}{400 \text{ k}\Omega} = 0.10 \text{ mA}$$

Thus the Q point is exactly in the center of the load line (point A in Fig. 5.2(b)).

(b) $\qquad\qquad I_B = \dfrac{V_{CC}}{R_B} = \dfrac{40 \text{ V}}{200 \text{ k}\Omega} = 0.20 \text{ mA}$

Notice that, since the value of the bias resistor was *halved*, the base current *doubled*. For the same circuit (2N910 transistor, $V_{CC} = 40$ V, $R_L = 2$ kΩ), a base current of 0.20 mA brings the Q point to saturation.

(c) $\qquad\qquad I_B = \dfrac{V_{CC}}{R_B} = \dfrac{40 \text{ V}}{800 \text{ k}\Omega} = 0.05 \text{ mA}$

Now that the bias resistor is doubled, the base current is halved. The Q point is now closer to cut-off.

(d) $\qquad\qquad I_B = \dfrac{V_{CC}}{R_B} = \dfrac{40 \text{ V}}{0 \text{ }\Omega} = \infty$

The base current, of course, *will not* become infinite; it *will* become so large that the transistor will fail. This example points up the necessity of always including some resistance in series with the base lead to protect the transistor.

(e) $\qquad\qquad I_B = \dfrac{V_{CC}}{R_B} = \dfrac{40 \text{ V}}{\infty} = 0 \text{ mA}$

If the base lead is left open ($R_B = $ infinity), the base current is, of course, equal to zero; in this case the collector current is equal to the relatively small I_{CEO}; i.e., the transistor is biased at cut-off.
  The above example illustrates the important effect of the value of R_B on both the base current and the Q point. Once the value of V_{CC} is selected, the value of R_B fixes the base current, and consequently, the Q point. A decrease in R_B increases the base current and moves the Q point toward saturation, while an

increase in the value of the resistor decreases the current and moves the Q point toward cut-off. If R_B is made too small the transistor will saturate; if it is made too large, it will be cut-off.

Comparing the circuit of Fig. 5.3(c) with that of Fig. 5.1, we can see the advantages of the fixed bias circuit. The bias resistor, R_B, which replaces the V_{EE} battery is smaller than the battery, it is lighter in weight, it is cheaper, and it does not have to be replaced or recharged; if the proper power rating is used, there is no reason why the resistor should ever have to be replaced. Most importantly, the Q point is not as sensitive to changes in the value of R_B as it was to changes in the value of V_{EE}. Figure 5.4(a) shows the effect on the value of the base current of a

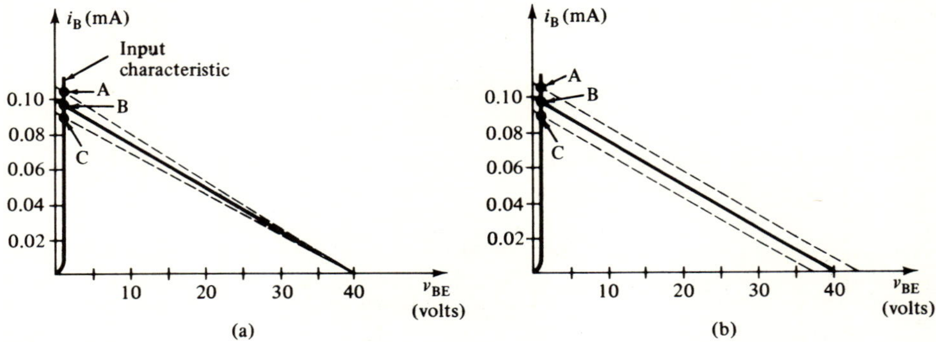

Figure 5.4

variation in the value of R_B. The solid load line is drawn on the input characteristics for $R_B = 400$ kΩ; the base current is found at the intersection of the load line and the input characteristic (point B). If R_B is slightly larger than 400 kΩ, the load line will have a smaller slope (lower dotted line) and intersect at point C. The upper dotted line (intersection with the characteristic at point A) is drawn for R_B slightly smaller than 400 kΩ. Notice that all three intersection points (A, B, and C) are located quite close to each other, indicating that the base current (and consequently, the Q point on the collector characteristics) is not very sensitive to changes in the value of R_B.

EXAMPLE 2

Assume that the tolerance of the 400 kΩ resistor is 5%. Compute the dc base current if (a) R_B is at its minimum possible value and (b) R_B is at its maximum possible value.

SOLUTION

The minimum and maximum possible values of R_B must first be ascertained.

(a)

$$0.05(400 \text{ k}\Omega) = 20 \text{ k}\Omega \qquad R_{B\text{min}} = 400 \text{ k}\Omega - 20 \text{ k}\Omega = 380 \text{ k}\Omega$$

$$I_{B\text{max}} = \frac{V_{CC}}{R_{B\text{min}}} = \frac{40 \text{ V}}{380 \text{ k}\Omega} = 0.105 \text{ mA}$$

(b)

$$R_{B_{max}} = 400 \text{ k}\Omega + 20 \text{ k}\Omega = 420 \text{ k}\Omega$$

$$I_{B_{min}} = \frac{V_{CC}}{R_{B_{max}}} = \frac{40 \text{ V}}{420 \text{ k}\Omega} = 0.095 \text{ mA}$$

The results of this example show that the variation in the value of R_B due to manufacturer's tolerance has very little effect on the Q point. Figure 5.4(b) shows the effect on the Q point of a variation in the value of V_{CC}; the three load lines are drawn for three different values of V_{CC}. The results of example 2 and Fig. 5.4 indicate that the fixed bias circuit is relatively insensitive to variations in V_{CC} and R_B; compare Fig. 5.4 with Fig. 5.2(a).

Summarizing, we can say that the fixed bias method is superior to the V_{EE} method for reasons of size, weight, cost, replacement, and sensitivity. By selecting the value of R_B, we can fix the base current (bias) at any particular value we choose.

5-2 INSTABILITY OF THE Q POINT

Although the fixed bias method is an improvement over the V_{EE} method of biasing, there are definite limitations to its use in a practical circuit; the study of these limitations forms the substance of this section. From chapter 3 you will recall that all of the transistor parameters and quantities vary somewhat with temperature, as well as with the particular unit being used. *Three* quantities in particular have a considerable effect on the Q point as we shall now see; these are:

1. I_{CO} (I_{CBO}), the reverse leakage current from collector to base with the emitter open-circuited.
2. β of the transistor.
3. V_{BE}, the forward voltage drop across the base-to-emitter junction.

As was explained in chapter 3, all three of these quantities vary with temperature in some particular fashion. It is the purpose of this section to illustrate the effect of the variation of these quantities with temperature, both collectively and separately, on the Q point. In order to simplify the presentation of these concepts we study, first, the effect of each of the quantities on the Q point, *separately;* i.e., we study the effect on the Q point of each of these quantities *assuming that the other two remain constant* with change in temperature. After studying the effect of each of these quantities separately, we combine these effects to see the *total resultant change* in Q point due to all of the factors operating simultaneously.

Figure 5.5 shows a 2N736 transistor using fixed bias. Let us first analyze the dc operation of this circuit at room temperature. Neglecting the forward drop across the base-emitter junction, we find the base current to be:

$$I_B = \frac{V_{CC}}{R_B} = \frac{20 \text{ V}}{500 \text{ k}\Omega} = 40 \text{ } \mu\text{A}$$

The β of the 2N736 varies over a wide range with temperature and component

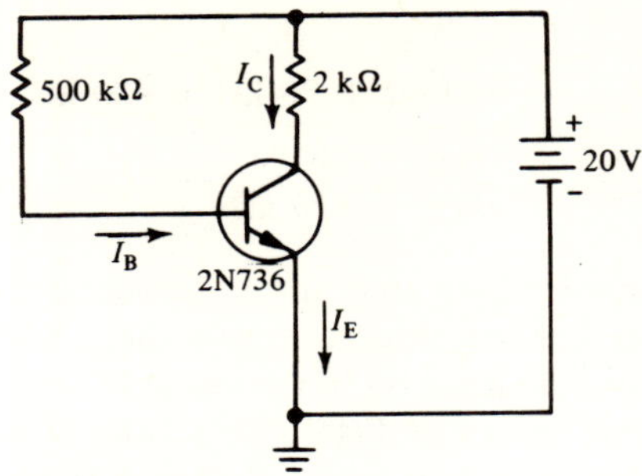

Figure 5.5

replacement. The typical value for this type transistor (at room temperature and $I_C = 5$ mA) is 125; let us assume that the transistor shown in Fig. 5.5 has a β of 125 at room temperature. Figure 5.6(a) shows the characteristics of this transistor at room temperature (25°C) with the load line drawn (from Fig. 5.5). With the dc base current fixed at 40 μA by the 500 kΩ bias resistor, the Q point is exactly in the center of the load line, as shown in Fig. 5.6(a); the quiescent collector current, I_{C_Q}, is 5 mA, and the quiescent collector-to-emitter voltage, V_{CE_Q}, is 10 V. Notice in Fig. 5.6(a) that the characteristic curves have been drawn with uniform spacing; i.e., the distance between the curve for $I_B = 0$ and $I_B = 40$ μA is the same as that between $I_B = 40$ μA and $I_B = 80$ μA. We know from the discussion in chapter 3 that the spacing between the curves is *not* uniform; however, we have drawn them as in Fig. 5.6(a) for simplicity. Thus the circuit of Fig. 5.5 will behave as shown in Fig. 5.6(a) at room temperature.

Let us now investigate the effect of the leakage current on the Q point. In doing this, we assume that all of the other quantities and parameters of the transistor remain constant with change in temperature. The manufacturer of the 2N736 specifies that the maximum value of I_{CO} at room temperature is 1.0 μA. From chapter 3 you will recall that the relationship between I_{CO} and I_{CEO} is the following:

$$I_{CEO} = (\beta + 1)I_{CO} \simeq \beta I_{CO} \qquad \text{since } \beta \gg 1$$

Therefore, I_{CEO} at room temperature for the transistor of Fig. 5.5 is:

$$I_{CEO} = \beta I_{CO} = 125(1 \ \mu A) = 125 \ \mu A = 0.125 \text{ mA}$$

Thus I_{CEO} for this transistor at room temperature is very small and can be considered equal to zero. This is shown in Fig. 5.6(a); the characteristic curve for $I_B = 0$ is almost collinear with the horizontal V_{CE}-axis.

The manufacturer also specifies that the maximum value of I_{CO} at *150°C* is *100 μA*. The term "maximum" is used here because some 2N736 units may have smaller values of I_{CO}; we are considering the maximum value because this presents the worst case, as we soon see. The fact that I_{CO} is much larger at 150° than at room temperature should not be surprising since we know from chapter 3 that I_{CO} increases markedly with an increase in temperature. It was also mentioned in chapter 3 that the leakage current approximately doubles for every 10°C increase

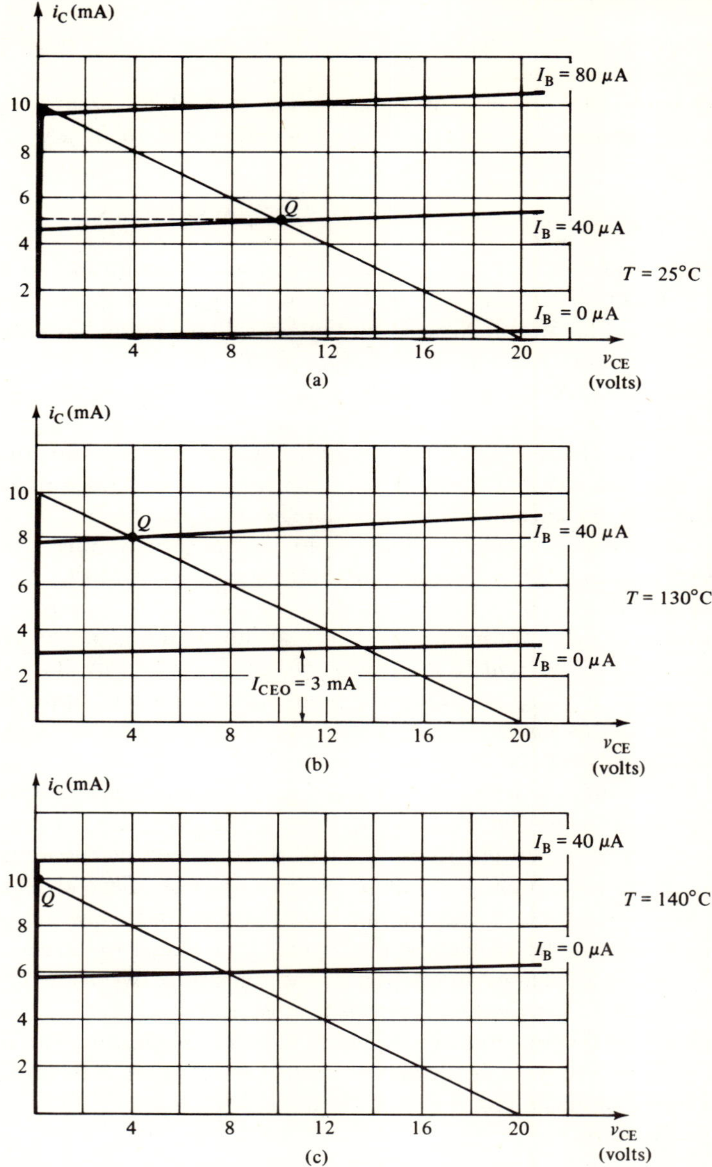

Figure 5.6

in temperature. Working backwards, then, as we *decrease* the temperature by 10°C, the leakage current should *halve*. Thus $I_{CO} = 50\ \mu A$ at 140°C and $I_{CO} = 25\ \mu A$ at 130°C. Before proceeding, it should be mentioned that these are only approximations. First, the rate of increase of I_{CO} with temperature (doubling for every 10°C increase) is *not the same at all temperatures;* the rate at room temperature is *different* from that at the high temperatures. Second, the rate of increase is different for germanium and silicon; thus the rate used above gives only a rough approximation to the actual situation.

Assuming, then, that we use the approximate rate of increase of I_{CO} with temperature, we can say that I_{CO} is approximately equal to 25 μA at 130°C. Therefore, I_{CEO} at 130° is:

$$I_{CEO} = \beta I_{CO} = 125(25\ \mu A) = 3125\ \mu A \simeq 3.0\ mA$$

Notice that I_{CEO} has changed from 0.125 mA at room temperature to approximately 3 mA at 130°. This increase has the effect of *shifting all of the characteristics upward*, as shown in Fig. 5.6(b). Notice that the $I_B = 40\ \mu$A curve has shifted from approximately 5 mA at room temperature to 8 mA at 130°C. Since the base current is fixed at 40 μA in Fig. 5.5, *the Q point has shifted* toward the upper left-hand portion of the load line; now (at 130°C) $I_{CQ} = 8$ mA and $V_{CEQ} = 4$ volts. Notice that all of the characteristic curves shift up by the same amount; the $I_B = 0$ curve is now approximately at 3 mA. The reason that this shift takes place is that the increasing leakage current causes an increase in I_{CEO} causing, in turn, an upward shift of the curves. Since I_B is fixed at 40 μA, the Q point must *shift upward along with the 40 μA curve.* If the temperature is increased now to 140°C, I_{CO} doubles to 50 μA and I_{CEO} doubles to 6 mA. Thus the curves now shift upward by 6 mA; the 40 μA curve is now at approximately 11 mA, as shown in Fig. 5.6(c). However, the saturation collector current of the circuit of Fig. 5.5 is 10 mA. Therefore the transistor *saturates* at 140 degrees, and the Q point is at the extreme upper left-hand end of the load line, as shown in Fig. 5.6(c). An increase in temperature above 140°C will obviously produce no further change in the Q point.

This shift in Q point due to change in temperature is an undesirable effect of the fixed bias circuit for a number of reasons. First, if the transistor saturates, as at 140°C, *only one-half* of the output sine wave will be produced by the amplifier, as explained in chapter 4; this will result in severe distortion. Second, even if the transistor does not saturate, the maximum sinusoidal output from the amplifier will be reduced (see chapter 4) because the Q point is no longer in the center of the load line. Finally, as explained in chapter 3, the spacing between the characteristic curves is not uniform; there is, for each transistor, a certain value of collector current at which the spacing, and consequently β, is at a maximum. The transistor is usually biased at this value of collector current. If the Q point shifts into a region of lower β, the gain of the amplifier will be reduced; this last point will be expanded upon in later chapters.

Since both I_{CO} and I_{CEO} are very small at room temperatures, the shift in the Q point in the vicinity of 25°C is minimal. Since the leakage current becomes even smaller at temperatures below 25°C, there is practically no shift in Q point *due to leakage current* at the lower temperatures (down to about -55°C). The words "due to leakage current" were emphasized because, as we shall soon see, the change in β at the lower temperatures *does* affect the Q point. Thus we conclude that the leakage current affects the Q point at only the higher temperatures.

The above illustrations indicate that the shift in Q point becomes more serious as the temperature increases. For example, if we keep taking one-half of I_{CO} as we drop in temperature by 10°C, I_{CO} at 100°C becomes approximately equal to 3 μA;

therefore $I_{\text{CEO}} = (125)(3\ \mu\text{A})$, or 0.375 mA. Thus if the temperature is increased from 25 to 100°C, the curves shift upward *by only 0.375 mA;* I_{C_Q} shifts from 5 mA (at 25°) to approximately 5.375 mA (at 100°), and V_{CE_Q} shifts from 10 V to 9.25 V. In other words, the 2N736 transistor, which is an *npn* double-diffused silicon mesa type, can be used with the fixed bias method up to a temperature of 100°C, with the *Q* point shifting only from 5 mA to 5.375 mA, due to change in leakage current. Of course, if the temperature is increased above 100°C, the shift in *Q* point becomes more noticeable, with the transistor saturating at some temperature between 130° and 140°. If a microammeter were placed in series with the base and a milliammeter with the collector, the microammeter reading would remain approximately constant at 40 μA, while the milliammeter reading would increase from 5 mA at 25°, to 10 mA at some temperature between 130° and 140°; the milliammeter reading would show no further increase as the temperature increased beyond 140°.

It was mentioned in chapters 2 and 3 that the leakage currents in germanium transistors are between 10 and 1000 times as large as those in silicon. Suppose the I_{CO} of a germanium transistor (whose β is 125) is 300 μA (0.3 mA) at 100°C. Therefore its I_{CEO} is (125)(0.3 mA), or 37.5 mA. This means that all of the curves will shift upward *by 37.5 mA* at 100 degrees; the collector current will try to shift from 5 mA to 42.5 mA. Of course it will shift only as far as 10 mA; therefore the germanium transistor will saturate at 100° whereas the silicon will *not*. For this reason, the useful temperature range over which a germanium transistor can operate is much more limited than for a silicon transistor. In practice, germanium transistors are rarely used at temperatures above *55°C;* they are used only in applications where the temperature does not vary over a wide range, such as in household equipment or on shipboard. The temperatures encountered in these particular environments are not particularly high when it is realized that 126°F is only 38°C. Silicon transistors, on the other hand, are suitable for high-temperature applications, such as in airborne equipment. Most electronic equipment designed for the military in recent years has been for use in airplanes, missiles, and orbiting satellites. Military specifications require that electronic equipment designed for these applications be able to function over a temperature range from -55°C to $+95$°C; the upper limit of 95° has recently been increased due to the high temperatures encountered when satellites re-enter the earth's atmosphere. It is plain from the above discussion that germanium transistors are wholly unacceptable for use in high temperature environments. We conclude that for high-temperature environments, we should like the leakage current of the transistor to be as small as possible, so that the *Q* point shift is minimal; for this reason, we use silicon transistors in these applications. Remember again that the *Q* point shifts discussed thus far are due only to a change in leakage current.

Let us now examine the effect on the *Q* point of the transistor β. The β of a transistor varies with collector current, temperature, and also with the particular device being used, as explained in chapter 3. We assume for the purposes of this discussion that the β remains constant with collector current, as shown by the uniform spacing of the curves in Figs. 5.6 and 5.7. The transistor specification sheet, to be studied in detail in chapter 10, lists curves showing the expected variation of

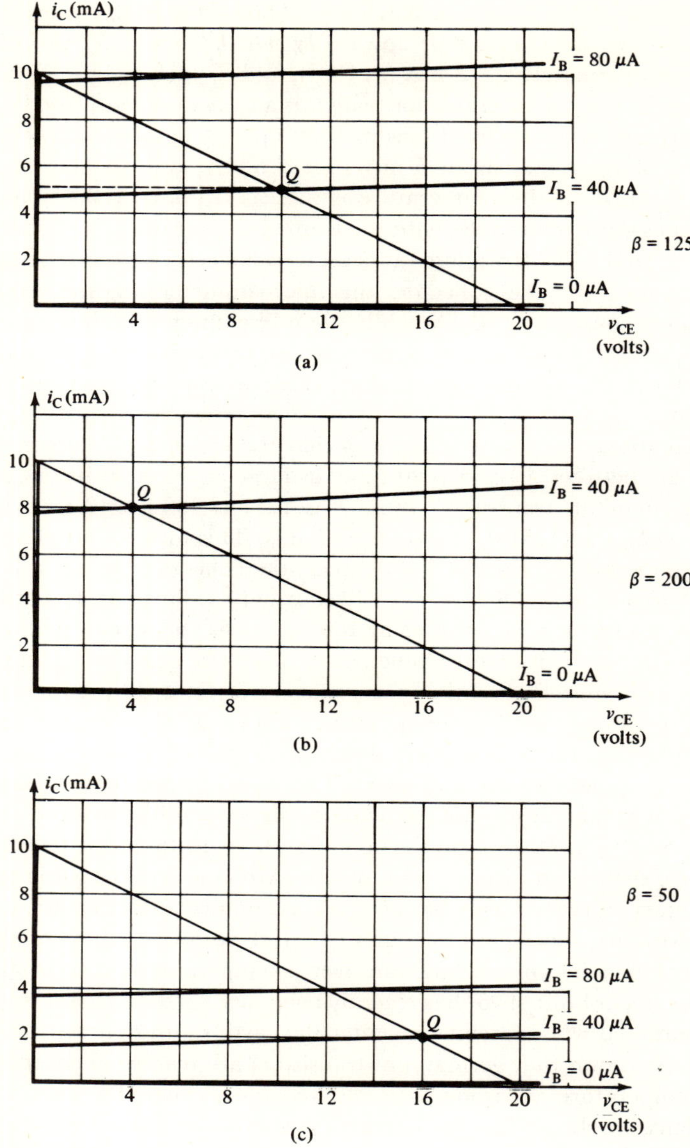

Figure 5.7

β with collector current and temperature; in addition, an indication of the β spread is given. The typical β of the 2N736 is 125 at room temperature and $I_C = 5$ mA. The β of the typical 2N736 is 70 at $-55°$, and is 260 at $+150°$. The β spread of the transistor at room temperature is from 80 to 200. Thus, considering the effects of both varying temperature and production variation, the curves indicate that the β of a particular 2N736 at any temperature can be as small as 40 (at $-55°C$) and can exceed 300 (at $+150°C$). This large variation in β can cause undesirable effects as we will now see.

We assume in the following analysis that the leakage current causes no shift in Q point. We will examine the effects of varying β on the Q point; in this analysis we are not concerned as to whether the change in β is caused by temperature or by replacement of the transistor with one having a different β. Figure 5.7(a) shows the characteristics of a 2N736 having a β of 125. Notice, that with the base current set at 40 μA, the Q point is in the center of the load line. Figure 5.7(b) shows the characteristics of a unit with a β of 200; note that the curves are spaced more widely apart because of the larger β. Since the base current is still equal to 40 μA, the Q point has shifted to the upper left; I_{CQ} is now equal to 8 mA and V_{CEQ} is 4 V. Figure 5.7(c) shows the curves for a transistor whose β is 50; note the closer spacing of the curves. The Q point has now shifted to the lower right-hand portion of the load line (closer to cut-off); $I_{CQ} = 2$ mA and $V_{CEQ} = 16$ V. *Thus the variation of β, whether due to changing temperature or transistor replacement, causes a shift in Q point just as does the changing leakage current.* Notice that the curves of Fig. 5.7(c) are drawn for a β of 50; since the minimum β of the 2N736 at room temperature is 70, these curves must be for a transistor at a temperature lower than room (lower than 25°C). Thus, unlike the change in Q point due to leakage current, the change in β due to temperatures below room can cause the Q point to shift closer to cut-off. Therefore, if the temperature is increased above, or decreased below room temperature, or if a transistor with a larger or smaller β than the typical value is inserted in the fixed bias circuit, the Q point can shift either toward saturation or toward cut-off, depending on whether β increases or decreases. This again is an undesirable situation for the reasons mentioned previously.

We come finally to the effect of V_{BE} on the Q point. We know that the voltage across the base-emitter terminals remains approximately constant at a few tenths of a volt if a large resistor is placed in series with the base lead, as in the case of fixed bias. This forward voltage is approximately 0.6 V for a silicon transistor and 0.3 V for germanium. It was shown in chapter 3 that this voltage is temperature sensitive, decreasing 2.5 mV for each centigrade degree increase. Let us assume that this voltage is 0.6 V at room temperature for the 2N736. If the temperature increases to 125° (total increase of 100°), this voltage will drop by:

$$\Delta V_{BE} = -2.5 \text{ mV}(°C) = -2.5 \text{ mV}(100°C) = -250 \text{ mV} = -0.25 \text{ V}$$

Thus the forward drop across the base-emitter terminals at 125 degrees is:

$$V_{BET=125°C} = V_{BET=25°C} - \Delta V_{BE}$$

$$V_{BET=125°C} = 0.60 \text{ V} - 0.25 \text{ V} = 0.35 \text{ V}$$

Now let us compute the effect on the Q point of the change in V_{BE}. Taking into account the base-to-emitter voltage, the base current can be expressed as:

$$I_B = \frac{V_{CC} - V_{BE}}{R_B}$$

The exact value of the base current at 25°C ($V_{BE} = 0.6$ V) is:

$$I_B = \frac{V_{CC} - V_{BE}}{R_B} = \frac{20 \text{ V} - 0.6 \text{ V}}{500 \text{ k}\Omega} = 38.8 \text{ } \mu\text{A}$$

The value of I_B at 125°C ($V_{BE} = 0.35$ V) is:

$$I_B = \frac{V_{CC} - V_{BE}}{R_B} = \frac{20 \text{ V} - 0.35 \text{ V}}{500 \text{ k}\Omega} = 39.3 \text{ } \mu\text{A}$$

Since the base-to-emitter voltage affects only the base current, the above computations show that the base current remains relatively constant over the temperature range. Therefore, we can say that the change in V_{BE} due to varying temperature has only a minimal effect on the base current, and consequently, on the Q point. We, thus, neglect the effect of the variation of V_{BE} on the Q point in the fixed bias circuit. It should be mentioned that there are circuits where even a change in I_B of the amount found above is intolerable. In addition, there are other types of circuits in use where the change in V_{BE} causes a much larger shift in Q point; this point is illustrated later on. Of course, strictly speaking, the resistance of the bias resistor varies somewhat with temperature, but this, like V_{BE}, has a negligible effect on the Q point.

The shift in Q point in the fixed bias circuit is due primarily to two factors: I_{CO} and β. We can say that the total change in the Q point is equal to the sum of the changes caused by the leakage current and β. Stated algebraically:

$$\Delta I_{C_T} = \Delta I_{C_L} + \Delta I_{C_\beta}$$

In this equation ΔI_{C_T} is the *total change* in quiescent collector current, ΔI_{C_L} is the change in quiescent collector current due to *leakage*, and ΔI_{C_β} is the change due to β. Thus the quiescent collector current increases at higher temperatures due to the cumulative effects of two factors: The curves *shift upward* because of the increase in I_{CO} (and I_{CEO}), and they also *spread farther apart* because of the increase in β. We return to a more detailed discussion of these concepts shortly.

The graphical explanations of this section are meant to give the reader an insight into the actual mechanism responsible for the instability of the Q point; it is important to keep the picture of the characteristic curves in mind when considering the operation of an actual circuit under changing environmental conditions. However, graphical techniques are usually time-consuming and laborious, and are therefore undesirable when it is necessary to analyze a particular circuit in a minimum amount of time. The engineer or technician working at the bench is usually interested in obtaining an approximate idea of the circuit performance by methods which require only a few short computations on a pad of paper. We now consider an example of these methods.

From chapter 3 you will recall that the collector current can be expressed as follows:

$$I_C = \beta I_B + (\beta + 1)I_{CO}$$
$$I_C = \beta I_B + I_{CEO}$$

We know that β, in addition to varying with temperature and production unit, varies as well with the collector current, reaching a maximum at some particular value of I_C (see chapter 3). In using the above equations, we have to assume that β remains relatively *constant with collector current;* thus the results obtained using these equations will be only approximate, but will yield a reasonably good approximation to the actual situation. Therefore, for any particular temperature and production unit, we assume that the *curves are uniformly spaced.* Let us now apply the equations to some specific examples.

EXAMPLE 3

Compute the quiescent collector current at room temperature for the circuit of Fig. 5.5, if the β of the particular unit at room temperature is 125 and the maximum leakage current at room temperature is 1.0 μA.

SOLUTION

Since fixed bias is used, and $R_B = 500$ kΩ, I_B is still 40 μA. Substituting into the equation we get:

$$I_{C_Q} = \beta I_{B_Q} + I_{CEO} = 125(40 \ \mu A) + 125(1 \ \mu A)$$
$$I_{C_Q} = 5 \text{ mA} + 0.125 \text{ mA} \simeq 5 \text{ mA}$$

Thus the leakage current is small enough to be neglected at room temperature, since I_{CEO} is very small in comparison with βI_B.

EXAMPLE 4

Repeat example 3 at room temperature if β of the unit is 200.

SOLUTION

$$I_{C_Q} = \beta I_B + I_{CEO} = 200(40 \ \mu A) + 200(1 \ \mu A)$$
$$I_{C_Q} = 8 \text{ mA} + 0.2 \text{ mA} \simeq 8 \text{ mA}$$

Note that the quiescent collector current has increased because of the larger β; note also that I_{CEO} has also increased because it too depends upon β. However, the contribution of the term due to I_{CEO} to the total collector current is still small enough to be neglected at room temperature (0.2 mA is small in comparison with 8 mA).

EXAMPLE 5

Assume that the temperature of the environment in which the circuit of Fig. 5.5 is located increases to 150°C. Compute the quiescent collector current if the β increases to 250 and I_{CO} increases to 100 μA.

SOLUTION

The equation "predicts" that the quiescent collector current will be:

$$I_{C_Q} = \beta I_{B_Q} + I_{CEO} = 250(40\ \mu A) + 250(100\ \mu A)$$
$$I_{C_Q} = 10\ mA + 25\ mA = 35\ mA$$

The term "predicts" is used because, in the circuit of Fig. 5.5, $V_{CC} = 20$ V and $R_L = 2$ kΩ; thus the collector saturation current, I_{Cs}, is equal to:

$$I_{Cs} = \frac{V_{CC}}{R_L} = \frac{20\ V}{2\ k\Omega} = 10\ mA$$

Therefore, although the current will try to increase to 35 mA, it will be able to reach a value of only 10 mA, and the transistor will be saturated at 150°C.

The above examples illustrate a convenient method for computation of the dc operating conditions in a fixed bias circuit. All that need be known are the β and the I_{CO} of the transistor at the particular temperature of the environment. In a practical situation, this analytical technique is preferred to the graphical procedure because of the time-savings involved. It should be emphasized that this method of computation is applicable only to the fixed bias circuit; other methods will be developed for different circuits.

We now illustrate one final problem for use in comparison with other bias circuits to be studied later on in this chapter.

EXAMPLE 6

Compute the quiescent collector current in the circuit of Fig. 5.5, both analytically and graphically, at room temperature ($\beta = 125$ and $I_{CO} = 1.0\ \mu A$), and at 100°C ($\beta = 200$ and $I_{CO} = 3.0\ \mu A$).

SOLUTION

We first compute the values of the quiescent collector currents at the two temperatures analytically, as follows:

At 25°C:

$$I_{C_Q} = \beta I_{B_Q} + I_{CEO} = 125(40\ \mu A) + 125(1\ \mu A) = 5.125\ mA \simeq 5\ mA$$
$$V_{CE_Q} = V_{CC} - I_{C_Q}R_L = 20\ V - (5\ mA)(2\ k\Omega) = 10\ V$$

At 100°C:

$$I_{C_Q} = \beta I_{B_Q} + I_{CEO} = 200(40\ \mu A) + 200(3\ \mu A) = 8.6\ mA$$
$$V_{CE_Q} = V_{CC} - I_{C_Q}R_L = 20\ V - (8.6\ mA)(2\ k\Omega) = 2.8\ V$$

Note that the increase in I_{C_Q} at the higher temperature is accompanied by a decrease in V_{CE_Q}.

Turning now to the graphical solution, we see that the curves of the transistor at room temperature ($\beta = 125$) are drawn in Fig. 5.8(a). With $I_B = 40\ \mu A$, the

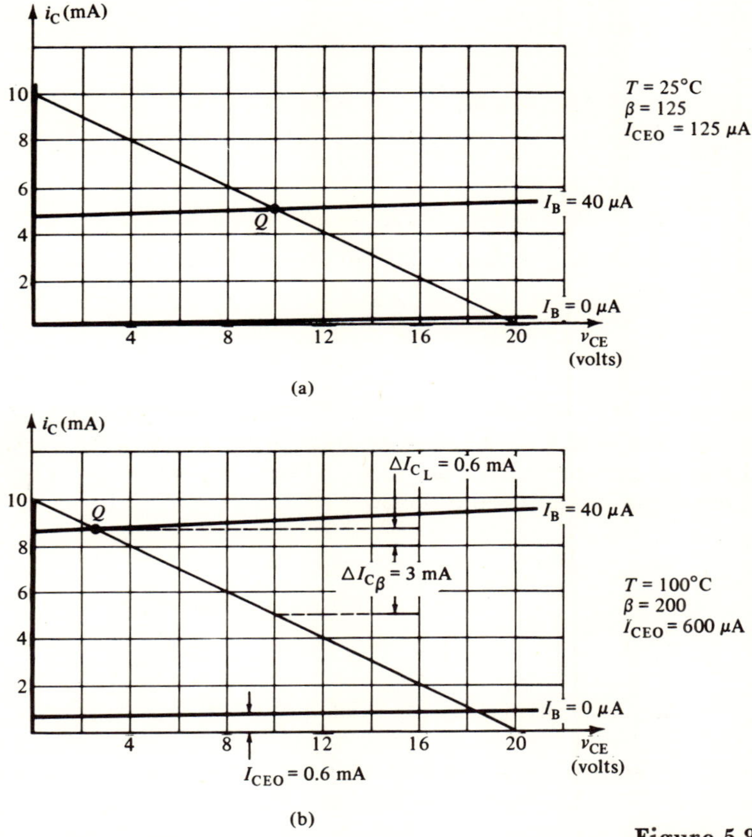

Figure 5.8

Q point is in the center of the load line as shown; thus the graphical results agree with those obtained analytically above since the Q point in Fig. 5.8(a) shows that $I_{C_Q} = 5$ mA and $V_{CE_Q} = 10$ V. Note that I_{CEO} is only 125 μA at room temperature and therefore the curve for $I_B = 0\ \mu A$ is almost collinear with the horizontal axis.

Figure 5.8(b) shows the characteristics of the same transistor at 100°C ($\beta = 200$ and $I_{CO} = 3\ \mu A$). The characteristic curves of Fig. 5.8(b) differ from those of Fig. 5.8(a) in *two* important respects: First, those in Fig. 5.8(b) *have shifted upward* due to the increase in I_{CO} (and I_{CEO}). The curve for $I_B = 0\ \mu A$ in Fig. 5.8(b) no longer lies along the horizontal axis as it does in Fig. 5.8(a); $I_{CEO} = 0.6$ mA in Fig. 5.8(b) (at 100°C). Second, the curves in Fig. 5.8(b) *are spaced more widely apart* than those in Fig. 5.8(a) due to the increase in β. Note that the spacing between the $I_B = 0\ \mu A$ and the $I_B = 40\ \mu A$ curves is much greater in Fig. 5.8(b) than in Fig. 5.8(a). Thus the new Q point (at 100°C) is located closer to saturation as shown in Fig. 5.8(b). The graphical results agree with those obtained analytically since the Q point in Fig. 5.8(b) is located at $I_{C_Q} = 8.6$ mA and $V_{CE_Q} = 2.8$ V.

The results of example 6 confirm what has been said before; namely, that the quiescent collector current increases at higher temperatures because the curves shift upward (due to increase in I_{CO}) and also because they spread farther apart (due to increase in β). The total *change* in quiescent collector current in example 6 was:

$$\Delta I_{C_T} = 8.6 \text{ mA} - 5 \text{ mA} = 3.6 \text{ mA}$$

Repeating the analytical results obtained above:

$$25°\text{C:} \qquad I_{C_Q} = \beta I_{B_Q} + I_{CEO} \simeq 125(40 \ \mu\text{A}) = 5 \text{ mA}$$

$$100°\text{C:} \qquad I_{C_Q} = \beta I_{B_Q} + I_{CEO} = 200(40 \ \mu\text{A}) + 200(3 \ \mu\text{A}) = 8.6 \text{ mA}$$

$$= \quad 8 \text{ mA} \quad + \quad 0.6 \text{ mA} \quad = 8.6 \text{ mA}$$

$$\beta I_{B_Q} \qquad\qquad I_{CEO}$$

The above results show that the quiescent collector current increased by 3 mA (from 5 mA to 8 mA) due to the increase in β (from 125 to 200), and by an additional 0.6 mA due to the increase in I_{CO}. Thus the *total change* in quiescent collector current can be expressed as follows:

$$\Delta I_{C_T} = \Delta I_{C\beta} + \Delta I_{C_L} \qquad \Delta I_{C\beta}: \text{ due to } \beta$$

$$\Delta I_{C_L}: \text{ due to leakage}$$

$$\Delta I_{C_T} = 3 \text{ mA} + 0.6 \text{ mA} = 3.6 \text{ mA}$$

We return to the results of example 6 later on in the chapter when we compare the fixed bias circuit with other methods of biasing.

We conclude from this section that the use of the fixed bias circuit is severely limited by the temperature range over which the circuit must operate, as well as by the β spread of the particular transistor type being used. A change in temperature and/or the replacement of the transistor by one having a different β will cause a shift in the Q point which can have undesirable effects on the operation of the circuit as an amplifier.

5-3   COLLECTOR-TO-BASE BIAS

It was shown in the last section that the Q point is very sensitive to changes in temperature and transistor replacement in the fixed bias circuit. We now consider an alternate method of biasing the transistor for dc operation. Recall that the purpose of bias is to supply a dc current to the base of the transistor; this means that the base-emitter junction must be forward biased. In order to forward bias this junction, the base must be maintained at a more *positive* potential than the emitter when an *npn* transistor is used (base more negative than emitter with *pnp*).

Remember also that the collector-base junction is reverse biased in an amplifier; with an *npn* transistor this means that the collector is at a more *positive* potential than the base (see chapter 3). Therefore, instead of connecting the base through a resistor to the *positive terminal of* V_{CC} as in the case of the fixed bias circuit, we can connect the lead to the *collector* of the transistor as shown in Fig. 5.9; this is called the *collector-to-base* method of biasing. Base current is now supplied to the transistor via the bias resistor, R_B, which is connected between the collector and base. Notice in Fig. 5.9 that the current which flows through the 2 kΩ load resistor splits

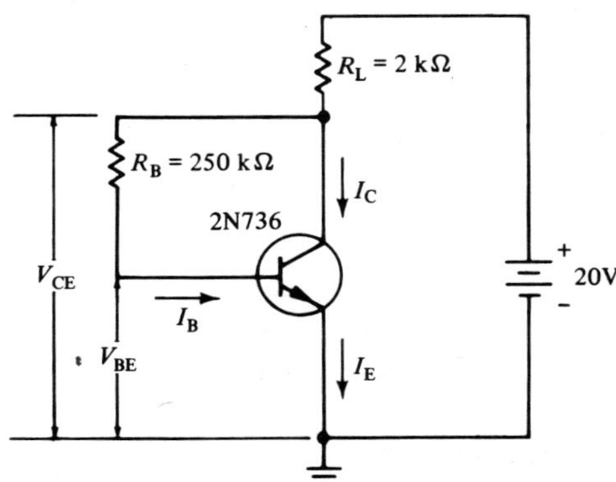

Figure 5.9

up at the collector; part of the current (I_C) flows to the collector and the remainder (I_B) flows through R_B to the base. These two currents recombine within the transistor to become emitter current, I_E. An examination of the circuit reveals that it is the emitter current, and not the collector current, which flows through the 2 kΩ load resistor. However, since both the collector and emitter currents are much larger than the base current, we can assume that I_E is approximately equal to I_C; thus we can say that the current in the 2 kΩ resistor is approximately equal to I_C.

Notice in Fig. 5.9 that the voltage drop across the bias resistor, V_{R_B}, is equal to the difference between V_{CE} and V_{BE}. Since V_{BE} is generally only a few tenths of a volt, we can neglect this voltage in comparison with V_{CE}; thus:

$$V_{R_B} = V_{CE} - V_{BE} \simeq V_{CE} \qquad V_{CE} \gg V_{BE}$$

The voltage V_{R_B} is directly across the bias resistor, R_B; therefore, if both this voltage and the value of R_B are known, the current through the resistor (same as the current through the base) can be computed by Ohm's law as follows:

$$I_B = \frac{V_{R_B}}{R_B} = \frac{V_{CE} - V_{BE}}{R_B} \simeq \frac{V_{CE}}{R_B}$$

$$I_B = \frac{V_{CE}}{R_B} \qquad \text{Collector-to-Base Bias}$$

Let us compute the value of R_B needed to bias the transistor of Fig. 5.9 at the same point as the transistor of Fig. 5.5. You will recall that at room temperature, the β of the 2N736 was 125, I_{Bq} was 40 μA, I_{Cq} was 5 mA, and V_{CEq} was 10 V; the Q point was in the center of the load line. Assuming that we are using the same 2N736 at room temperature in Fig. 5.9, we desire that the Q point again be in the center of the load line; this means that the base current must be equal to 40 μA. Since it is desired that the Q point be in the center, V_{CEq} must again equal 10 V (one-half of V_{CC}). Using the equation derived above for the collector-to-base bias circuit, we find that the value of R_B required is:

$$R_B = \frac{V_{CE}}{I_B} = \frac{10 \text{ V}}{40 \ \mu\text{A}} = 250 \text{ k}\Omega$$

In other words, the collector-to-base bias circuit of Fig. 5.9 with $R_B = 250$ kΩ will have the same Q point as the fixed bias circuit of Fig. 5.5 with $R_B = 500$ kΩ, assuming that the same 2N736 is used at room temperature in both circuits.

Now let us examine the behavior of the circuit of Fig. 5.9 under varying environmental conditions. In Fig. 5.10(a), the *lower* solid characteristic curve is for

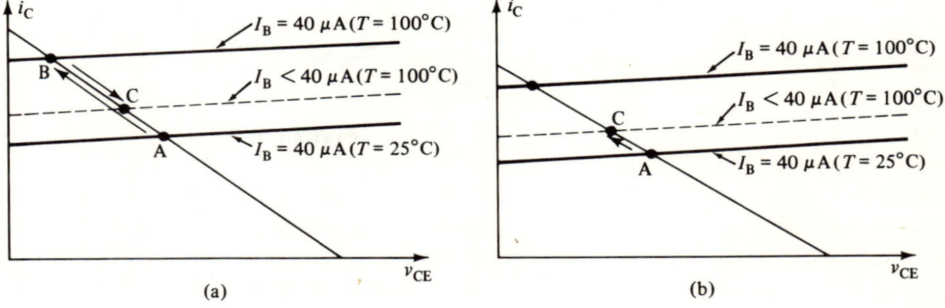

Figure 5.10

$I_B = 40$ μA at room temperature; at room temperature the Q point is at point A. Now let us say that the temperature increases to some value, say 100°C. We know from the discussion in the previous section that the curves will shift upward and spread apart; the characteristic for $I_B = 40$ μA at 100°C is the *upper* solid curve in Fig. 5.10(a). If the base current were *fixed* at 40 μA as in the *fixed* bias circuit, the Q point would shift upward along with the curves and the new Q point (at 100°C) would be at point B. This does *not* happen, however, in the collector-to-base bias circuit because an important mechanism prevails in this type of circuit. We know that the curves will shift upward at the higher temperatures and therefore the Q point will tend to shift from A to B (the use of the term "tend" will be explained shortly). If the Q point shifted from A to B, the collector current would *increase* and the collector-to-emitter voltage would *decrease*. However, we know that the base current is equal to the collector-to-emitter voltage divided by the bias resistor;

$$I_B = \frac{V_{CE}}{R_B}$$

therefore, since V_{CE} *decreases* as the Q point shifts from A toward B, the *base current must decrease as well*. If the base current decreases to a value *less than 40 μA*, the Q point cannot possibly be at point B since the Q point was defined as the intersection of the load line with the proper base characteristic curve; in other words, since the base current is now less than 40 μA, the Q point must be someplace between A and B in Fig. 5.10(a). We can say that, while the *upward shift of the curves tends to move the Q point from A to B, the decreasing base current which accompanies this upward shift tends to move the Q point from B back toward A*. Actually, the new Q point is at C, somewhere on the load line between A and B; notice that C is at the intersection of the load line with the dashed curve (I_B equals some value less than 40 μA at 100°C).

The reader may get the impression from Fig. 5.10(a) that the Q point shifts from A to B and then back to C; this actually does *not* happen. We say that the Q point tends to shift from A to B, but only moves as far as C. The reason for this action is that as soon as the temperature increases, I_C tends to increase and V_{CE} tends to decrease. But as soon as V_{CE} decreases, the base current decreases and the Q point moves back toward its original position (point A). Therefore the Q point never actually reaches point B. Figure 5.10(b) shows the actual situation; the Q point shifts from A ($I_B = 40$ μA) at room temperature to C (I_B less than 40 μA) at 100°C.

The above discussion reveals that for the same transistor and Q point, the collector-to-base bias circuit provides more stable operation under varying environmental conditions than does the fixed bias circuit. In other words, although the Q point *does shift* with the collector-to-base circuit, the *amount of shift* (or change in collector current) *is less* with this circuit than with the fixed bias circuit; we deal, shortly, with the exact amount of shift and the factors upon which this shift depends. Although the collector-to-base bias circuit provides a more stable Q point than does the fixed bias circuit, it *does not give perfect stabilization;* i.e., there will always be *some shift* in Q point with an increase in temperature. It is shown later that this circuit, in addition to providing stabilization under varying environmental conditions, also stabilizes the circuit against Q point shifts due to transistor replacement.

Figure 5.10(a) illustrates schematically the reason for the superiority of the collector-to-base bias circuit over the fixed bias circuit. In the fixed bias circuit, the base current (bias) is *fixed* by the values of V_{CC} and R_B; since neither of these values change noticeably with an increase in temperature, the *base current remains fixed at 40 μA as the temperature changes*, and the Q point shifts along with the 40 μA curve from A to B [see Fig. 5.10(a)]. However, in the collector-to-base bias circuit, the *tendency for the collector current to increase initiates a tendency for the base current to decrease to offset the increase in collector current*. In other words, the *base current does not remain fixed* in the collector-to-base bias circuit; *it readjusts itself so as to try to keep the collector current constant*. Therefore, the Q point shifts from A to C in Fig. 5.10(a) instead of from A to B.

This phenomenon (or mechanism) which enables the collector-to-base bias circuit to readjust itself under varying conditions is known as *negative feedback*. The concept of negative feedback is extremely important, not only in electronics, but in other fields as well. Basically, all systems employing negative feedback operate in the same manner. An external factor tends to produce an unwanted change in some quantity in the system. This change initiates a change in a *second* quantity in the system; this second quantity changes in such a manner as to offset the original unwanted change, and thus promotes stability. Figure 5.11 illustrates these con-

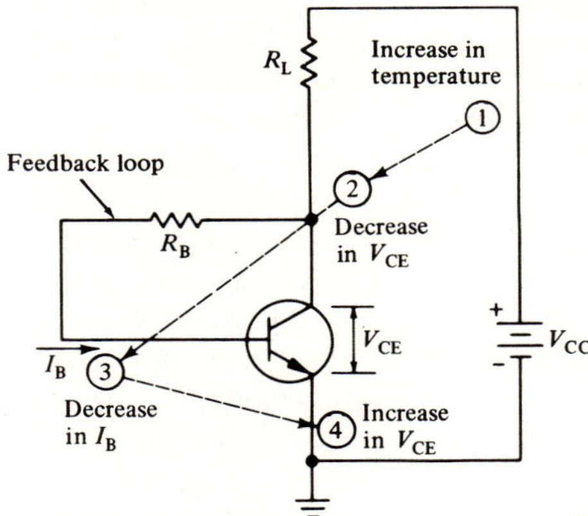

Figure 5.11

cepts using the collector-to-base bias circuit. The external factor tending to produce the original unwanted change in the system is the increase in temperature (point "1" in the figure), and the original unwanted change is the decrease in V_{CE} (point "2"). The decrease in V_{CE} tends to initiate a decrease in I_B, the second quantity in the system (point "3"). This change in I_B (decrease) causes an increase in V_{CE} back toward its original value (point "4"). The key to the operation of any feedback stabilization system lies in the mechanism whereby the change in the original quantity (decrease in V_{CE}) produces a change in a second quantity (decrease in I_B) such as to offset the original change. The circuit thus has a built-in intelligence; the change in the original quantity *feeds back* this information to the second quantity. The change in the conditions in the collector circuit is fed back to the base circuit by means of the *feedback loop*, consisting of the bias resistor, R_B. We study other types of negative feedback systems later on in the text; they all have some type of feedback loop. The above system is an example of *negative dc feedback* since this scheme is used to stabilize the dc operation of the circuit; we will see that this system is also used to stabilize the ac operation.

As mentioned above, feedback systems are employed in applications other than electronic ones. A common example is the thermostat. The thermostat is connected to the electrical heating system of a house; it provides for a comfortable room temperature, say 75°F. Now suppose the outside air temperature increases considerably during a warm afternoon (external factor); this tends to increase the air temperature *inside* the house (original unwanted change in the system). This increase in air temperature inside the house feeds back information to the thermostat to reduce the internal electrical heating (second quantity). The air temperature inside the house is thus reduced to a more comfortable level. The feedback loop in this case consists of the increased air temperature of the house acting on the thermostat, which in turn reduces the internal heating so as to keep the room temperature constant. It is important to emphasize that, although feedback systems provide stabilization, the *correction is never perfect;* there will always be some change in the system. The designer is concerned with the *amount of change,* as we see in the following sections of this chapter.

5-4 ANALYSIS OF THE COLLECTOR-TO-BASE BIAS CIRCUIT

Now that it has been shown that the collector-to-base bias circuit provides some measure of stability against temperature variation and transistor replacement, we turn our attention to the analysis of this circuit. The analysis is not quite as simple as that of the fixed bias circuit as we now see. Consider the circuit of Fig. 5.12 (the bias resistor is 500 kΩ). Let us now analyze this circuit; i.e., let us find the

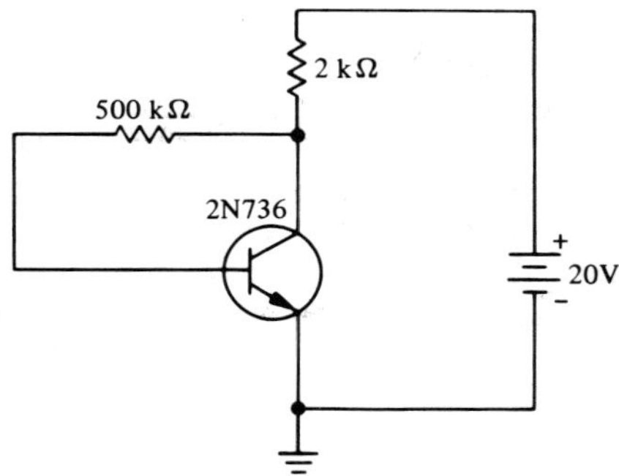

Figure 5.12

quiescent values of all of the currents and voltages. The first step, as usual, is to draw the load line on the characteristics; this load line depends upon V_{CC} (20 V) and R_L (2 kΩ). We know that the Q point is found at the intersection of the load line and the proper base current characteristic (this gives us the quiescent value of I_C (I_{C_Q}) and V_{CE} (V_{CE_Q})). Therefore, once the load line is drawn, we must find the quiescent value of the base current in order to determine which intersection of base current curve with the load line gives the Q point; i.e., is it the 40 μA curve? 60 μA? In the fixed bias circuit this was a simple matter since the quiescent base current was found as V_{CC}/R_B. However, we know from the theory developed in the preceding section that the quiescent base current in the collector-to-base bias circuit is equal to:

$$I_{B_Q} = \frac{V_{CE_Q}}{R_B}$$

We know the value of R_B but we cannot compute the quiescent value of I_B until we first find V_{CE_Q}. However, *it is precisely this quantity (V_{CE_Q}) which we are trying to find from the intersection of the load line and the base current characteristic.* In other words, we must know the value of I_{BQ} before we can find the Q point as the intersection of the load line and the characteristic curve; however, the value of I_{BQ} depends, in turn, upon the value of V_{CEQ}, which we cannot know until the Q point is found. Therefore, this circuit cannot be analyzed by straightforward methods. We could, of course, attempt a solution by trial-and-error; e.g., suppose we *assume* that V_{CEQ} is equal to 10 V (Q point in the center of the load line). Therefore, the quiescent base current must be:

$$I_{\mathrm{BQ}} = \frac{V_{\mathrm{CEQ}}}{R_{\mathrm{B}}} = \frac{10 \text{ V}}{500 \text{ k}\Omega} = 20 \ \mu\mathrm{A}$$

If the β of the unit is 125, the quiescent collector current must be:

$$I_{\mathrm{CQ}} \simeq \beta I_{\mathrm{BQ}} = 125(20 \ \mu\mathrm{A}) = 2.5 \text{ mA}$$

The quiescent collector-to-emitter voltage is therefore:

$$V_{\mathrm{CEQ}} = V_{\mathrm{CC}} - I_{\mathrm{CQ}}R_{\mathrm{L}} = 20 \text{ V} - (2.5 \text{ mA})(2 \text{ k}\Omega) = 15 \text{ V}$$

But this cannot be since we assumed that V_{CEQ} was equal to 10 V; i.e., our assumption was *incorrect.* We could continue this procedure until the correct Q point is found, but this is time-consuming and tedious. We now illustrate a graphical method which yields a rapid solution involving no guess-work.

We know that the base current can be found by the following expression:

$$I_{\mathrm{B}} = \frac{V_{\mathrm{CE}}}{R_{\mathrm{B}}}$$

$$V_{\mathrm{CE}} = I_{\mathrm{B}}R_{\mathrm{B}}$$

We construct a chart like the one shown in Fig. 5.13(a). In one column we list values of base current for which curves are drawn for the particular transistor (0 μA, 20 μA, 40 μA, etc.) ; in the second column we multiply each of these values of I_{B} by R_{B} to obtain corresponding values of V_{CE} (0 V, 10 V, 20 V, etc.). In other words, each value of V_{CE} in the second column is gotten by multiplying the corresponding value of I_{B} by R_{B} (in this case, 500 kΩ). Figure 5.14 shows the characteristics of the 2N736 ($\beta = 125$) at room temperature with the load line drawn. The values in the chart of Fig. 5.13(a) are now plotted on the characteristics of Fig. 5.14 in the following manner: The top row of Fig. 5.13(a) has $I_{\mathrm{B}} = 0 \ \mu$A and $V_{\mathrm{CE}} = 0$ V; a point is plotted on the characteristics of Fig. 5.14 on the $I_{\mathrm{B}} = 0$ curve at $V_{\mathrm{CE}} = 0$ V (at the origin). This is *point A* in Fig. 5.14. The second row in the chart in Fig 5.13(a) has $I_{\mathrm{B}} = 20 \ \mu$A and $V_{\mathrm{CE}} = 10$ V; therefore *point B* is plotted on the curves of Fig. 5.14 on the $I_{\mathrm{B}} = 20 \ \mu$A curve at $V_{\mathrm{CE}} = 10$ V. *Point C* is plotted in the same manner. The three points (A,B,C) are connected to form a curve; this is known as the *bias curve. The intersection of the bias curve with the load line gives the Q point.*

$$V_{CE} = I_B \ (500 \ k\Omega)$$

	$I_B \ (\mu A)$	$V_{CE} \ (V)$
A	0	0
B	20	10
C	40	20

(a)

$$V_{CE} = I_B \ (125 \ k\Omega)$$

	$I_B \ (\mu A)$	$V_{CE} \ (V)$
A	0	0
D	20	2.5
E	40	5.0
F	60	7.5
G	80	10.0

(b)

Figure 5.13

Notice in Fig. 5.14 that the Q point indicates that the quiescent collector-to-emitter voltage, V_{CEQ}, is approximately equal to 13.5 V. If this is the case, the quiescent base current must be:

$$I_{BQ} = \frac{V_{CEQ}}{R_B} = \frac{13.5 \ V}{500 \ k\Omega} = 27 \ \mu A$$

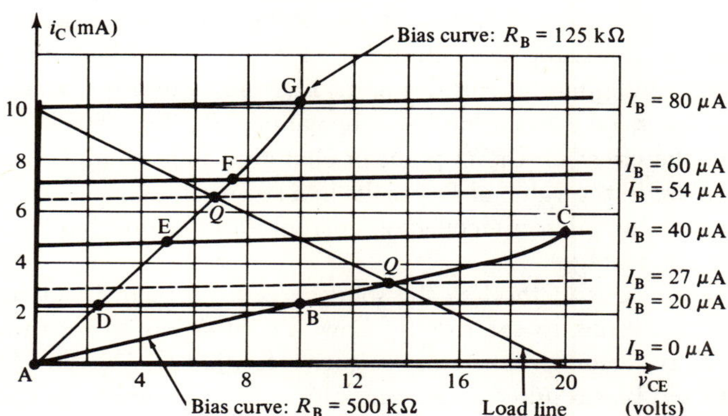

Figure 5.14

Note that the Q point (intersection of the load line with bias curve A–B–C) falls someplace between the $I_B = 20\ \mu A$ and the $I_B = 40\ \mu A$ curves, but closer to the $I_B = 20\ \mu A$ curve. This checks with the above computation which indicates that the quiescent base current is $27\ \mu A$ (between $20\ \mu A$ and $40\ \mu A$, but closer to $20\ \mu A$); the curve for $I_B = 27\ \mu A$ is shown dotted in Fig. 5.14. The above procedure can be used as a rough check on the accuracy of the construction. The bias curve method of analysis is a rapid, relatively simple method which can be used to determine the values of the quiescent currents and voltages in the collector-to-base bias circuit.

It is important to note that the bias curve is *not* a straight line in general; therefore, the designer should plot *at least three points* to obtain a reasonably accurate bias curve. Since we are interested in the intersection of the bias curve with the load line, *at least one point* should be plotted on either side of the load line (points B and C in Fig. 5.14).

Figure 5.13(b) shows the chart constructed if the bias resistor is changed from 500 kΩ to 125 kΩ. The bias curve for this new circuit ($R_B = 125$ kΩ) is drawn on the characteristics in Fig. 5.14; this curve is represented by the letters A–D–E–F–G. Since neither the transistor, nor the load resistor (R_L), nor the value of V_{CC} has been changed, the load line and the characteristics remain as before. However, since R_B *has* been changed, there is a new bias curve (A–D–E–F–G) and, consequently, a new Q point. To check this construction, we see that this new Q point indicates that $V_{CEQ} = 6.75$ V. Therefore,

$$I_{B_Q} = \frac{V_{CEQ}}{R_B} = \frac{6.75\ \text{V}}{125\ \text{k}\Omega} = 54\ \mu A$$

Figure 5.14 illustrates that this new Q point falls on a curve somewhere between $40\ \mu A$ and $60\ \mu A$, but closer to $60\ \mu A$, which checks with the above computation; the curve for $I_B = 54\ \mu A$ is shown dotted in Fig. 5.14.

Figure 5.14 illustrates that, while the load line depends upon V_{CC} and R_L, *the bias curve depends upon R_B and the characteristic curves.* Once V_{CC}, R_L, R_B, and the transistor characteristics are fixed, the Q point is also fixed. A change in either the value of R_B, or the transistor characteristics, or both, results in a change in the bias curve, and, consequently, the Q point. The only quantity which we have changed in the discussion thus far has been the value of R_B (from 500 kΩ to 125 kΩ); Fig. 5.14 shows the effect on the Q point of changing the value of R_B. It can be seen that a decrease in the value of R_B results in a steeper bias curve, which in turn causes an increase in quiescent collector current (and a decrease in quiescent collector-to-emitter voltage).

Let us now apply this method to the solution of a problem involving varying environmental conditions to see exactly how the situation is improved by the incorporation of collector-to-base bias (negative feedback). We saw in Sec. 5-2 that the use of fixed bias resulted in a large shift in Q point when the temperature of the surrounding environment was increased. Let us now study the behavior of the collector-to-base bias circuit under the same conditions to see the stabilizing effect of the feedback resistor, R_B. Recall from Sec. 5-3 that we found that a 250 kΩ

resistor placed between the collector and base terminals would bias the transistor at the center of the load line. Figure 5.15 shows the construction of a chart for

$$V_{\text{CE}} = I_{\text{B}}\, R_{\text{B}}$$
$$V_{\text{CE}} = I_{\text{B}}\, (250 \text{ k}\Omega)$$

I_{B} (μA)	V_{CE} (V)
0	0
20	5
40	10
60	15
80	20

Figure 5.15

plotting the bias curve with $R_{\text{B}} = 250$ kΩ. Figure 5.16(a) shows the characteristics of the 2N736 ($\beta = 125$) at room temperature with the load line drawn. Using the values appearing in the chart of Fig. 5.15, a bias curve ($R_{\text{B}} = 250$ kΩ) is plotted on the curves of Fig. 5.16(a); the Q point is at the intersection of the bias curve and the load line exactly in the center of the load line ($I_{\text{C}_{\text{Q}}} = 5$ mA and $V_{\text{CE}_{\text{Q}}} = 10$ V). This should not be surprising since the value of 250 kΩ was selected for R_{B} in Sec. 5-3 specifically for the purpose of biasing the transistor in the center of the load line. Figure 5.16(b) shows the characteristics of the same transistor at 100°C; notice that the curves have shifted upward (due to the increase in I_{CO}) and spread farther apart (due to the increase in β from 125 to 200). Since neither V_{CC} nor R_{L} has changed, the load line has remained unchanged; the chart in Fig. 5.15 also remains unchanged since R_{B} is still equal to 250 kΩ. However, since the curves have shifted, the bias curve plotted in Fig. 5.16(b) on the characteristics at 100°C is different from that plotted in Fig. 5.16(a) at room temperature. Therefore, the Q point has shifted, as can be seen from a comparison of Figs. 5.16(a) and (b). Figure 5.16(a) shows that $I_{\text{C}_{\text{Q}}} = 5$ mA at room temperature, while Fig. 5.16(b) shows that $I_{\text{C}_{\text{Q}}}$ is some value between 6 mA and 7 mA at 100°C. The results of Fig. 5.16 illustrate the major ideas concerning negative feedback which were brought out in the preceding section. Although the varying environmental conditions *did* produce a change in the system (a change in the Q point from 5 mA to 6.5 mA), *the change was not as much as when no feedback at all was used*; reference to Figs. 5.8(a) and (b) reveals that *when fixed bias* was used with the *same transistor* under the *same environmental conditions,* the Q point shifted from 5 mA to 8.6 mA. In other words, while *perfect correction is not obtained, the collector-to-base bias circuit provides a significant improvement over the fixed bias circuit.* Notice in Fig. 5.16(b) that the Q point is located approximately half-way between the 20 μA and 40 μA curves, while in Fig. 5.16(a) it is located at exactly 40 μA. This illustrates that the base current *decreases* at the higher temperatures (to about 30 μA, in this case)

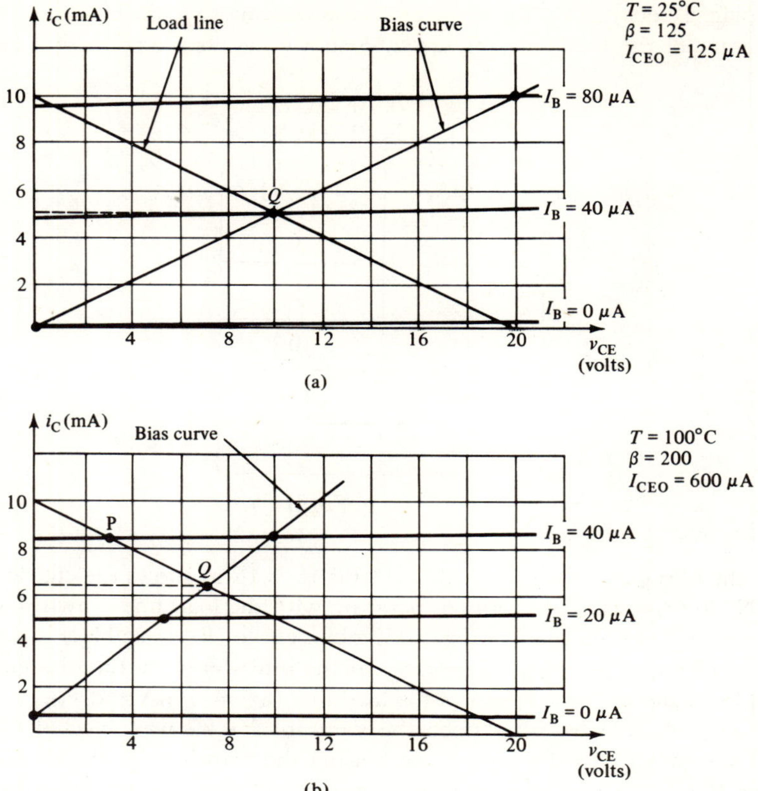

Figure 5.16

to partially offset the effect of the *increasing* collector current. If meters are placed in series with the base and collector leads, respectively, the reading of the base current meter will decrease (from 40 μA to about 30 μA) as the temperature increases from 25°C to 100°C, and the reading of the collector current meter will increase (from 5 mA to about 6.5 mA). If the base current remained fixed at 40 μA as the temperature increased, as in the case of fixed bias, the Q point would have shifted to point P (8.6 mA) in Fig. 5.16(b).

Just as with the case of the fixed bias circuit, we should like to develop algebraic expressions for use in analyzing the collector-to-base bias circuit, since graphical procedures are unwieldy. The graphical methods illustrated above can be, and are, used to analyze various circuits. However, the primary purpose for describing them in this text has been to give the reader an insight into the actual processes which occur within the transistor circuit. The derivation of the equations which describe the behavior of the collector-to-base bias circuit involves a considerable amount of algebraic manipulation, and is therefore not shown. If the basic concepts of Ohm's and Kirchhoff's laws are applied to the circuit and the equations are re-arranged algebraically, the quiescent collector current is given by the following equation:

Collector-to-Base Bias

$$I_{C_Q} = \frac{\beta(V_{CC} - V_{BE})}{R_B + R_L(\beta + 1)} + \frac{(\beta + 1)(R_L + R_B)(I_{CO})}{R_B + R_L(\beta + 1)}$$

If we assume that V_{CC} is much greater than V_{BE}, and also that β is much greater than 1.0, the equation can be approximated as follows:

$$I_{C_Q} = \frac{\beta V_{CC}}{R_B + \beta R_L} + \frac{\beta(R_L + R_B)I_{CO}}{R_B + R_L\beta}$$

Notice that this equation is somewhat similar to that used for the fixed bias circuit:

$$I_{C_Q} = \beta I_{B_Q} + (\beta + 1)I_{CO} \qquad \text{Fixed Bias}$$

Both equations are composed of two terms, the second term depending primarily upon the leakage current. We shall see that most bias circuits are associated with equations which have the form of those shown above.

Let us apply, now, the equation to the specific problem studied earlier in this section. We compute the quiescent collector current in the circuit of Fig. 5.9 at room temperature; $V_{CC} = 20$ V, $R_L = 2$ kΩ, $R_B = 250$ kΩ, the β of the transistor is 125, and $I_{CO} = 1.0$ μA. Substituting into the equation we get:

$$I_{C_Q} = \frac{\beta V_{CC}}{R_B + \beta R_L} + \frac{\beta(R_L + R_B)I_{CO}}{R_B + \beta R_L}$$

$$I_{C_Q} = \frac{125(20 \text{ V})}{250 \text{ k}\Omega + 125(2 \text{ k}\Omega)} + \frac{125(2 \text{ k}\Omega + 250 \text{ k}\Omega)(0.001 \text{ mA})}{250 \text{ k}\Omega + 125(2 \text{ k}\Omega)}$$

$$I_{C_Q} = 5 \text{ mA} + 0.0625 \text{ mA} \simeq 5 \text{ mA}$$

We can see that the term due to the leakage current (0.0625 mA) is negligible; the quiescent collector current is 5 mA, the same value obtained by the graphical procedure of Fig. 5.16(a).

Next we compute the quiescent collector current at 100°C; the leakage current is now 3.0 μA and β is 200. The quiescent collector current is now found as:

$$I_{C_Q} = \frac{200(20 \text{ V})}{250 \text{ k}\Omega + 200(2 \text{ k}\Omega)} + \frac{200(2 \text{ k}\Omega + 250 \text{ k}\Omega)(0.003 \text{ mA})}{250 \text{ k}\Omega + 200(2 \text{ k}\Omega)}$$

$$I_{C_Q} = 6.15 \text{ mA} + 0.231 \text{ mA}$$

$$I_{C_Q} = 6.38 \text{ mA}$$

Figure 5.16(b) showed that the quiescent current at 100°C was somewhere between 6 mA and 7 mA; the algebraic solution shows it to be 6.38 mA. Thus the algebraic method agrees quite well with the graphical procedure, and in addition provides a rapid, accurate means of computing the dc operating conditions in the collector-to-base bias circuit.

The total change in quiescent collector current, ΔI_{C_T}, for the collector-to-base bias circuit is:

$$\Delta I_{C_T} = 6.38 \text{ mA} - 5.0 \text{ mA} = 1.38 \text{ mA}$$

Compare this with 3.6 mA (8.6 mA − 5.0 mA) for the fixed bias circuit under the same conditions. Re-writing the expression for the quiescent collector current at 100°C:

$$I_{C_Q} = 6.15 \text{ mA} + 0.23 \text{ mA} = 6.38 \text{ mA}$$

We see that the current increased by 1.15 mA due to the change in β (from 125 to 200) and by approximately 0.23 mA due to the change in leakage current. Therefore the total change in collector current may be expressed as follows:

$$\Delta I_{C_T} = \Delta I_{C_\beta} + \Delta I_{C_L}$$

$$\Delta I_{C_T} = 1.15 \text{ mA} + 0.23 \text{ mA} = 1.38 \text{ mA}$$

Comparing this with the results of the fixed bias circuit, we can see again the superiority of the collector-to-base bias circuit as far as stability is concerned.

Using the general expression relating base and collector current:

$$I_{C_Q} = \beta I_{B_Q} + I_{CEO}$$

We can solve for I_{B_Q} as follows:

$$\beta I_{B_Q} = I_{C_Q} - I_{CEO}$$

$$I_{B_Q} = \frac{I_{C_Q} - I_{CEO}}{\beta}$$

We now compute the quiescent base current at 100°C; we know from the results obtained above that $I_{C_Q} = 6.38$ mA, $\beta = 200$, and $I_{CEO} = 0.6$ mA. Substituting into the equation, we get:

$$I_{B_Q} = \frac{I_{C_Q} - I_{CEO}}{\beta} = \frac{6.38 \text{ mA} - 0.6 \text{ mA}}{200} = 28.9 \ \mu\text{A}$$

Remember that Fig. 5.16(b) showed that the base current decreased at the higher temperature (100°C) to about 30 μA to partially offset the increase in collector current; the results of the above computation indicate excellent agreement with the graphical procedure.

EXAMPLE 7

Using the algebraic method, compute the quiescent collector current in the collector-to-base bias circuit; $V_{CC} = 20$ V, $R_L = 2$ kΩ, $R_B = 500$ kΩ, $\beta = 125$. The transistor is at room temperature.

SOLUTION

Since the transistor is at room temperature, we neglect the effect of the leakage current; therefore, we drop the second term in the expression for I_{C_Q}. Substituting into the equation, we get:

$$I_{C_Q} = \frac{\beta V_{CC}}{R_B + \beta R_L} = \frac{125(20 \text{ V})}{500 \text{ k}\Omega + 125(2 \text{ k}\Omega)} = 3.33 \text{ mA}$$

The quiescent base current is found as follows:

$$I_{B_Q} = \frac{I_{C_Q} - I_{CEO}}{\beta} \simeq \frac{I_{C_Q}}{\beta} = \frac{3.33 \text{ mA}}{125} = 26.6 \text{ } \mu\text{A}$$

Compare these results with those obtained from the intersection of the load line and the bias curve A–B–C in Fig. 5.14. Figure 5.14 shows that the quiescent base current is approximately equal to 27 μA, whereas the results of the above computations show it to be 26.6 μA.

EXAMPLE 8

Repeat example 7 if $R_B = 125 \text{ k}\Omega$.

SOLUTION

The quiescent collector current is:

$$I_{C_Q} = \frac{\beta V_{CC}}{R_B + \beta R_L} = \frac{125(20 \text{ V})}{125 \text{ k}\Omega + 125(2 \text{ k}\Omega)} = 6.67 \text{ mA}$$

The quiescent base current is:

$$I_{B_Q} = \frac{I_{C_Q}}{\beta} = \frac{6.67 \text{ mA}}{125} = 53.3 \text{ } \mu\text{A}$$

Compare these results with the intersection of the load line and bias curve A–D–E–F–G in Fig. 5.14; note the agreement. We use the algebraic method from this point on in the text to analyze the collector-to-base bias circuit.

The stabilizing action of this circuit can be shown in another way. Let us first repeat the expression for the quiescent collector current at room temperature.

$$I_{C_Q} = \frac{\beta V_{CC}}{R_B + \beta R_L} \qquad \text{Collector-to-Base Bias}$$

Notice that there is a term proportional to β in both the numerator (βV_{CC}) and the denominator (βR_L) of the fraction. Therefore, if β increases, both the numerator *and* the denominator will increase; this tends to keep the value of the fraction constant. Of course, the value of the fraction *will* change as β changes, but not as much as if no feedback were used. Compare this with the case of the fixed bias circuit; the equations are as follows:

$$I_{C_Q} = \beta I_{B_Q} + I_{CEO} \simeq \beta I_{B_Q}$$

$$I_{B_Q} = \frac{V_{CC}}{R_B}$$

$$I_{C_Q} = \beta \frac{V_{CC}}{R_B}$$

$$I_{C_Q} = \frac{\beta V_{CC}}{R_B} \qquad \text{Fixed Bias}$$

Notice that in this equation, the quiescent collector current is *directly proportional to* β; if β doubles, I_{C_Q} doubles, etc.; there is no stabilization.

5-5   THEVENIN'S THEOREM

Before proceeding with the development of another type of bias circuit, it is instructive to illustrate the use of a very important circuit theorem. This theorem states that any linear circuit can be represented by a *voltage* generator in *series* with an impedance, or a *current* generator in *parallel* with the impedance. Figure 5.17(a) shows a simple circuit composed of a 10 V battery and three 1 kΩ resistors

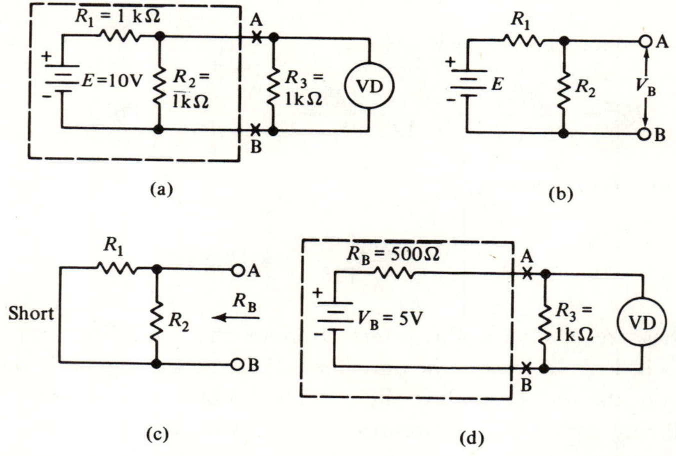

(a)   (b)

(c)   (d)

Figure 5.17

connected in a series-parallel arrangement; a dc voltmeter, VD, is connected across the parallel combination of R_2 and R_3. The computation of the voltage, V_{AB}, is a relatively simple matter which can proceed by any one of a number of methods. One method involves first combining the two resistors R_2 and R_3 in parallel to obtain the equivalent parallel combination:

$$\frac{R_2 R_3}{R_2 + R_3} = \frac{(1\ \text{k}\Omega)(1\ \text{k}\Omega)}{1\ \text{k}\Omega + 1\ \text{k}\Omega} = 0.5\ \text{k}\Omega = 500\ \Omega$$

The circuit is now composed of the same 10 V battery in series with R_1 (1 kΩ) and a 500 Ω resistor between terminals A and B (the parallel combination of R_2 and R_3). Since the circuit is now composed of a voltage in series with two resistors, the voltage divider can be applied to compute the voltage across the 500 Ω resistor (voltage across terminals A and B):

$$V_{AB} = \frac{(10 \text{ V})(0.5 \text{ k}\Omega)}{1 \text{ k}\Omega + 0.5 \text{ k}\Omega} = 3.33 \text{ V}$$

In other words, if this circuit were wired as shown in Fig. 5.17(a), the voltmeter would read 3.33 V (assuming we use a high sensitivity voltmeter). This is a relatively simple problem which the reader has no doubt come across in his study of basic circuits theory. The problem can be solved in another way using Thevenin's theorem.

Thevenin's theorem states that the portion of the circuit within the dotted box of Fig. 5.17(a) (composed of a battery and two resistors) *can be replaced by one battery in series with one resistor.* In other words, the portion of the circuit to the *left* of terminals A and B can be replaced by a simpler circuit such that the conditions (voltages and currents) to the *right* of the terminals remain unchanged. The procedure for converting the portion of the circuit of Fig. 5.17(a) inside the dotted box to a simpler form is composed of *two* steps:

First we must compute the *open circuit voltage,* V_B, across terminals A and B. To do this we must remove that portion of the circuit which we do *not* wish to replace; in this case it is the portion of the circuit to the right of terminals A and B, namely the resistor, R_3, and the voltmeter. Figure 5.17(b) shows the procedure for computing V_B; note that R_3 has been removed. V_B for this particular circuit can be computed by the voltage divider as follows:

$$V_B = \frac{10 \text{ V}(1 \text{ k}\Omega)}{1 \text{ k}\Omega + 1 \text{ k}\Omega} = 5 \text{ V}$$

The second step involves computation of the *Thevenin impedance* (in this case, resistance), R_B. This is the *resistance seen looking into terminals A and B with all voltage generators shorted and all current generators opened.* Figure 5.17(c) illustrates the procedure for computing the Thevenin impedance; note that the 10 V battery has been shorted. Figure 5.17(c) illustrates that the equivalent impedance to the left of terminals A and B is equal to the parallel combination of R_1 and R_2.

$$R_B = \frac{R_1 R_2}{R_1 + R_2} = \frac{(1 \text{ k}\Omega)(1 \text{ k}\Omega)}{1 \text{ k}\Omega + 1 \text{ k}\Omega} = 0.5 \text{ k}\Omega = 500 \text{ }\Omega$$

Thevenin's theorem now states that if V_B is connected in series with R_B, as shown in Fig. 5.17(d), the circuits of Figs. 5.17(a) and 5.17(d) are *identical* as far as that portion of the circuit to the *right* of terminals A and B is concerned. In other words, it makes no difference whether we connected the circuit within the dotted box of Fig. 5.17(a) to R_3, or the circuit within the dotted box of Fig. 5.17(d) (a 5 V battery in series with a 500 Ω resistor); the voltage measured (or computed) across

R_3 (terminals A and B) would be the *same* in both cases. To check, we merely apply the voltage divider to the circuit of Fig. 5.17(d) as follows:

$$V_{AB} = \frac{5\ \text{V}(1\ \text{k}\Omega)}{1\ \text{k}\Omega + 0.5\ \text{k}\Omega} = 3.33\ \text{V}$$

Thus, a voltmeter will read 3.33 V if it is connected across terminals A and B in Fig. 5.17(a), *or* if it is connected across the same terminals in Fig. 5.17(d). This indicates that the two circuits within the dotted boxes are equivalent as far as the resistor, R_3, is concerned.

Thevenin's theorem can also be applied to ac circuits. The problem illustrated above is relatively simple, and therefore Thevenin's theorem need not be applied in this case as an aid in simplifying the solution; this particular circuit was used to illustrate the basic concepts involved in the application of the theorem. This theorem becomes extremely useful when one is analyzing more complex circuits, such as the Wheatstone bridge. We have introduced the theorem at this point because we need it to aid in analyzing the bias circuit to be studied in the following section.

5-6   EMITTER BIAS

We now study another method of biasing the transistor amplifier, namely, the *emitter bias circuit* shown in Fig. 5.18. Although this circuit looks considerably more

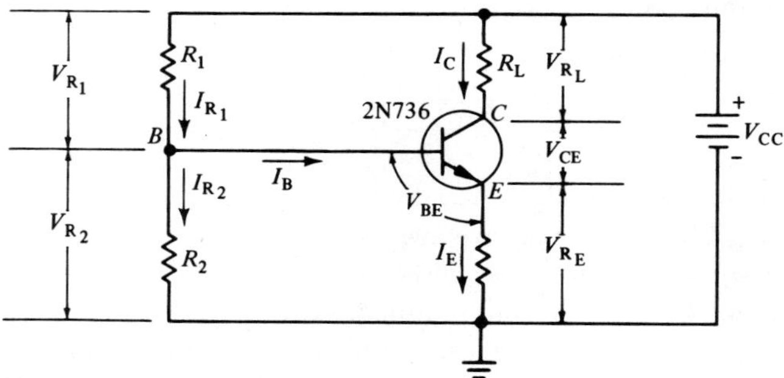

Figure 5.18

complicated than those discussed thus far, we will show that it is merely a variation of the other circuits (fixed bias and collector-to-base bias). Note first that an *emitter resistor*, R_E, has been inserted in series with the emitter lead between the emitter terminal and ground. This means that the *emitter terminal of the transistor is no longer at ground potential*. Emitter current, I_E, flows through the emitter resistor and thus there is a voltage drop, V_{RE}, across this resistor; therefore the emitter terminal is above ground by an amount equal to the voltage drop across the emitter resistor. The second point to note is that, if the base-emitter junction is forward biased (we will show that it is), the base-to-emitter voltage, V_{BE}, is approximately

0.6 V (with the base more positive than the emitter since the 2N736 is an *npn* transistor); thus the base-to-ground voltage (which is the same as the voltage drop across resistor R_2, V_{R_2}) is a few tenths of a volt *larger* than V_{RE}. In other words, the base is at a *higher* (more positive) potential than the emitter. Third, the base current is supplied through the voltage divider consisting of R_1 and R_2. It can be seen from Fig. 5.18 that the sum of the voltages across R_1 and R_2 is equal to V_{CC}. It is evident from the current directions in the figure that the current through R_1 (I_{R_1}) splits up into *two* paths at the base; part goes to the base (I_B) and the remainder goes to R_2 (I_{R_2}). Finally, an important point must be made about the collector circuit. Now that the emitter resistor has been included in the circuit, V_{CC} is equal to the sum of the *three* voltages shown in the collector circuit of Fig. 5.18, namely, V_{RL}, V_{CE}, and V_{RE}. Since I_B is generally much smaller than either I_C or I_E, as shown earlier in the text, we shall continue to assume that I_C *is approximately equal to* I_E. Therefore, for purposes of analysis, we can also assume that R_L and R_E are connected in *series* (although they actually are not) since they have *approximately the same current* flowing through them; thus when the load line is drawn, R_L and R_E are treated as a *single resistor* equal in value to $R_L + R_E$.

Let us see if this circuit provides stabilization. Assume that, due to increasing temperature, the curves shift upward and spread apart; the currents, I_E and I_C, will *tend* to *increase* and the Q point will tend to shift toward saturation, as explained earlier in the chapter. If the emitter current, I_E, increases, the voltage across the emitter resistor will also *increase*, since by Ohm's law, the following relationship is true:

$$V_{RE} = I_E R_E$$

Now it is known from Kirchhoff's voltage law that the voltage across R_2 is equal to the sum of the base-to-emitter drop (V_{BE}) and V_{RE}; stated algebraically:

$$V_{R_2} = V_{BE} + V_{RE}$$

Neglecting the effect of the 0.6 V drop from base to emitter, we can see from the above equation that, if V_{RE} increases as mentioned above, V_{R_2} must also *increase*. If V_{R_2} increases, then, according to Ohm's law I_{R_2} must *increase*.

$$I_{R_2} = \frac{V_{R_2}}{R_2}$$

It was also mentioned above that the sum of the voltages across R_1 and R_2 is equal to V_{CC}; this is a consequence of Kirchhoff's voltage law as can be seen from Fig. 5.18.

$$V_{R_1} + V_{R_2} = V_{CC}$$

Therefore, if V_{R_2} increases, V_{R_1} must *decrease*. But if V_{R_1} decreases, I_{R_1} must also *decrease*, since by Ohm's law, the following relationship is true:

$$I_{R_1} = \frac{V_{R_1}}{R_1}$$

Now by Kirchhoff's current law, the relationship between I_{R_1}, I_{R_2}, and I_B is given as follows:

$$I_{R_1} = I_B + I_{R_2}$$

$$I_B = I_{R_1} - I_{R_2}$$

Therefore, if, due to an increase in temperature, I_E and I_C tend to increase, it follows from the above reasoning that I_{R_2} *increases* and I_{R_1} *decreases;* thus the base current, I_B, must *decrease* to offset the *increase* in I_C and I_E. We can say that the emitter bias circuit provides a measure of temperature stabilization by means of negative feedback similar to the collector-to-base bias circuit. If meters were placed in series with the collector and base, respectively, that in series with the collector would show *some* increase in reading as the temperature is increased, while that in series with the base would show a decrease.

Figure 5.19(a) shows an emitter bias circuit with values included for the various

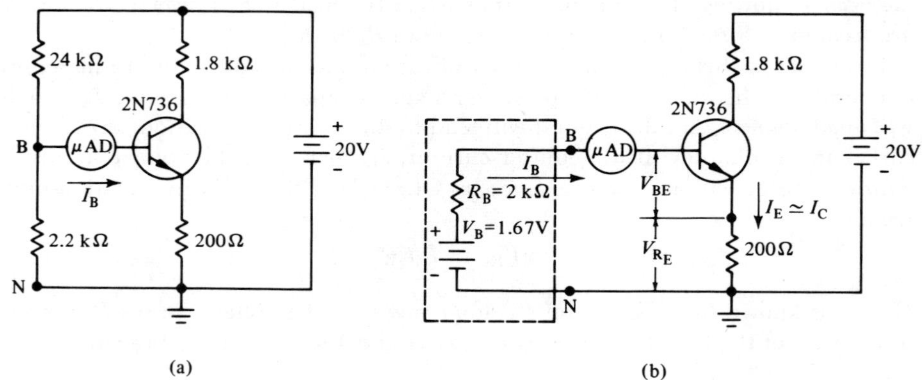

Figure 5.19

components. If the characteristics were at hand, the first step in analysis would be to draw the load line. Next, the quiescent value of base current would have to be ascertained in order to locate the Q point at the intersection of the load line and the proper curve. In order to find the value of I_{B_Q}, we must first know both I_{R_1} and I_{R_2}, since the base current is equal to the difference in these two values. In order to find I_{R_2}, we must know the value of V_{R_2}; but this in turn depends upon the value of V_{R_E}. Since it is desired that V_{R_E} be known, we must first know the value of I_{E_Q}, or I_{C_Q}. However, *it is precisely these values* (I_{E_Q} *and* I_{C_Q}) *which we are trying to ascertain by attempting to locate the Q point.* In other words, this situation is similar to that of the collector-to-base bias circuit in that a direct graphical solution is impossible.

The analysis of this circuit can proceed by one of two methods: a graphical solution utilizing a bias curve, or an algebraic method. Although the algebraic method is used in practice and is illustrated shortly, we discuss, first, the graphical solution because it gives a valuable insight into the operation of the circuit.

The first step in analyzing the circuit of Fig. 5.19(a) is to simplify the voltage divider portion of the circuit (consisting of the 2.2 kΩ and the 24 kΩ resistors) by means of Thevenin's theorem. We replace the voltage divider by its Thevenin equivalent. To find first the Thevenin open circuit voltage, V_B, we remove that portion of the circuit to the right of terminals B and N in Fig. 5.19(a), namely the transistor, R_E, and R_L; this is equivalent to saying that we *open the lead* in series with the base and the microammeter, μAD. The open circuit voltage, V_B, between B and N is merely the voltage across the 2.2 kΩ resistor with the base open. This can be found by the voltage divider expression as follows:

$$V_B = \frac{R_2 V_{CC}}{R_1 + R_2} = \frac{2.2\ \text{k}\Omega(20\ \text{V})}{2.2\ \text{k}\Omega + 24\ \text{k}\Omega} = 1.67\ \text{V}$$

The Thevenin impedance is that impedance seen looking back into terminals B and N with the 20 V battery shorted. This is the parallel combination of the two resistors, R_1 and R_2, or:

$$R_B = \frac{R_1 R_2}{R_1 + R_2} = \frac{(2.2\ \text{k}\Omega)(24\ \text{k}\Omega)}{2.2\ \text{k}\Omega + 24\ \text{k}\Omega} = 2\ \text{k}\Omega$$

Now if the open circuit voltage, V_B, is placed in series with the Thevenin impedance, R_B, the circuit of Fig. 5.19(b) results. The portion of the circuit within the dotted box of Fig. 5.19(b) replaces the voltage divider of Fig. 5.19(a). The portion of the circuit to the *right* of terminals B and N is equivalent in both diagrams; i.e., the μAD will read the *same* value of current in *both* circuits. In other words, as far as the transistor *collector* circuit is concerned, *it makes no difference whether the base circuit is wired as in Fig. 5.19(a) or (b)*. It should be mentioned at this point that the circuit of Fig. 5.19(a) is the one which is used in practice, since it requires only *one* battery; *the circuit of Fig. 5.19(b), while it will function properly, is discussed here only to facilitate our analysis.*

To obtain information to plot the bias curve, we must write the loop equations for the base and collector circuits of Fig. 5.19(b). Note that since V_B (1.67 V) is *greater* than V_{BE} (0.6 V), the base-emitter junction is forward biased. Utilizing Kirchhoff's voltage law, we write the loop equation for the base (and emitter) circuit of Fig. 5.19(b) as follows:

$$V_B = I_B R_B + V_{BE} + I_E R_E$$

$$1.67 = 2I_B + 0.6 + 0.2I_C \qquad \text{Assuming } I_E \simeq I_C$$

$$1.07 = 2I_B + 0.2I_C$$

In this equation, the currents are expressed in *milliamperes* and the resistances in *kilohms*. Applying Kirchhoff's law to the collector circuit, we obtain:

$$V_{CC} = V_{CE} + I_C R_L + I_E R_E = V_{CE} + I_C(R_L + R_E)$$

$$20 = V_{CE} + I_C(1.8\ \text{k}\Omega + 0.2\ \text{k}\Omega) = V_{CE} + 2I_C$$

We re-write the two equations as follows:

$$1.07 = 2I_B + 0.2I_C$$

$$20 = V_{CE} + 2I_C$$

Using the method of simultaneous equations studied in elementary algebra, we eliminate I_C from both equations (not shown here) and obtain the following relationship:

$$V_{CE} = 9.3 + 20I_B \qquad (I_B \text{ in milliamperes})$$

This gives the necessary information for plotting the bias curve. If I_B is to be expressed in *microamperes*, the equation must be changed as follows:

$$V_{CE} = 9.3 + 0.02I_B \qquad (I_B \text{ in microamperes})$$

Figure 5.20 shows the chart which is constructed for plotting the bias curve; for

$$V_{CE} = 0.02I_B + 9.2$$

I_B (μA)	V_{CE} (V)
0	9.2
20	9.6
40	10.0
60	10.4
80	10.8

Figure 5.20

each value of base current, a corresponding value of V_{CE} is computed from the equation. Figure 5.21(a) shows the characteristics of the 2N736 at room temperature ($\beta = 125$) with both the load line and bias curve drawn. Note that the load line is drawn as if the *total resistance in the collector circuit is 2 kΩ* (the 1.8 kΩ load resistor plus the 200 Ω emitter resistor); the end-point for the load line on the vertical i_C-axis is found as:

$$I_{C(V_{CE}=0)} = \frac{V_{CC}}{R_L + R_E} = \frac{20 \text{ V}}{1.8 \text{ k}\Omega + 0.2 \text{ k}\Omega} = 10 \text{ mA}$$

The intersection of the bias curve with the load line gives the Q point; the quiescent collector current, I_{CQ}, is 5 mA, as shown in Fig. 5.21(a). The values of R_E, R_L, R_1, and R_2 were deliberately chosen to set the Q point in the center of the load line. Figure 5.21(b) shows the characteristics of the same transistor at 100°C ($\beta = 200$), with the load line and bias curve drawn. Notice that *the Q point of this circuit at 100°C*

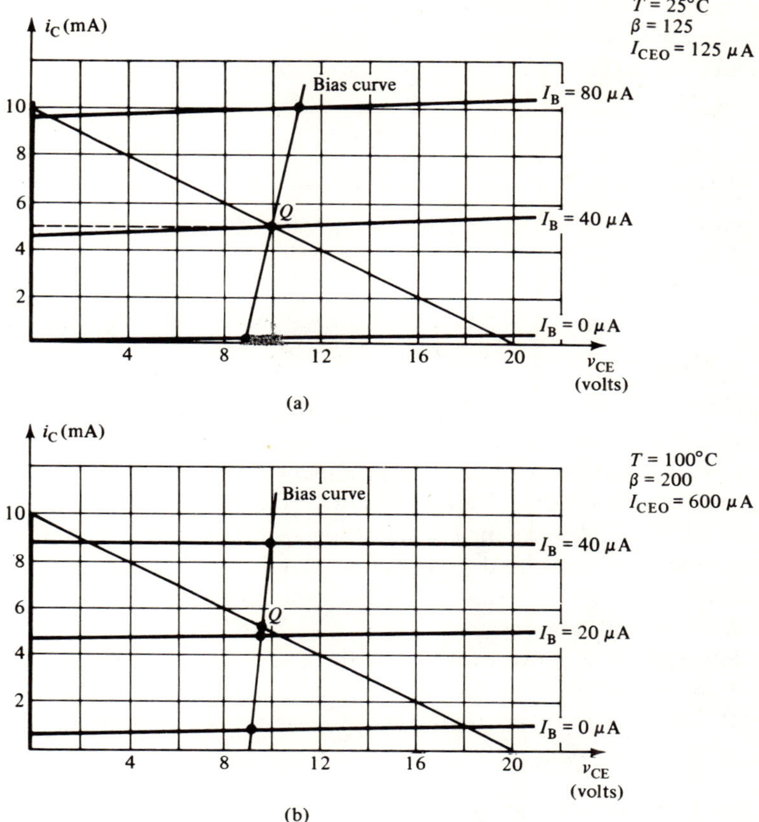

Figure 5.21

has moved only very slightly from its position at room temperature (Fig. 5.21(a)); the quiescent collector current at 100°C is only slightly greater than 5 mA. Therefore we can say that this circuit exhibits excellent temperature stability as compared with those circuits discussed earlier in this chapter. It should be emphasized that this does *not* mean that the emitter bias circuit generally yields better stability than the collector-to-base bias circuit (although they both *do* generally yield better stability than the fixed bias circuit); it merely means that *this particular* emitter bias circuit (that shown in Fig. 5.19(a)) yields more stable operation than the particular collector-to-base bias circuit shown in Fig. 5.9 ($R_B = 250$ kΩ). We discuss the factors upon which the *amount* of stability depends shortly.

Notice in Fig. 5.21(a) that the quiescent base current is 40 μA (at room temperature), while in Fig. 5.21(b) it is slightly larger than 20 μA (at 100°C). This shows again that the base current decreases at the higher temperatures, when negative feedback is used.

We turn now to the algebraic method of analysis for the emitter bias circuit. Just as with the case of the collector-to-base bias circuit, the algebraic manipulations necessary for the derivation of an expression for the quiescent collector current are not shown. If Kirchhoff's laws are applied to the circuit of Fig. 5.19(b), and the equations are properly manipulated, the following expression results:

$$\boxed{I_{C_Q} = \frac{\beta(V_B - V_{BE})}{R_B + \beta R_E} + \frac{(R_B + R_E)\beta I_{CO}}{R_B + \beta R_E}} \qquad \text{Emitter Bias}$$

In this equation it is assumed that β is much greater than 1.0, and the expressions for V_B and R_B are given as follows:

$$V_B = \frac{V_{CC}R_2}{R_1 + R_2} \qquad R_B = \frac{R_1 R_2}{R_1 + R_2}$$

Note again that the equation (similar to that for the other bias circuits) is composed of two terms, the second depending upon the leakage current. Note also that we have *not* neglected the base-to-emitter drop, V_{BE}, in the equation; the reason is that the 0.6 V base-to-emitter drop is *not* necessarily negligible in comparison with V_B (in this case, 1.67 V). We see in the following section that the variation in the value of V_{BE} with temperature does *not* have a negligible effect on the Q point in the emitter bias circuit; we assume for the moment, however, that V_{BE} remains constant with varying temperature.

Let us now apply the above equation to the analysis of the circuit of Fig. 5.19 at room temperature; $V_{CC} = 20$ V, $R_1 = 24$ kΩ, $R_2 = 2.2$ kΩ, $R_L = 1.8$ kΩ, $R_E = 200\ \Omega$, $\beta = 125$, $I_{CO} = 1.0\ \mu$A, and $V_{BE} = 0.6$ V. Substituting into the equation, we get:

$$I_{C_Q} = \frac{\beta(V_B - V_{BE})}{R_B + \beta R_E} + \frac{(R_B + R_E)\beta I_{CO}}{R_B + \beta R_E}$$

$$= \frac{125(1.67 - 0.6)}{2\ \text{k}\Omega + 125(0.2\ \text{k}\Omega)} + \frac{(2\ \text{k}\Omega + 0.2\ \text{k}\Omega)(125)(0.001)}{2\ \text{k}\Omega + 125(0.2\ \text{k}\Omega)} \qquad \begin{aligned} V_B &= 1.67\ \text{V} \\ R_B &= 2\ \text{k}\Omega \end{aligned}$$

$$= 5\ \text{mA} + 0.01\ \text{mA} \simeq 5.0\ \text{mA}$$

Notice that this result agrees quite well with the results of the graphical solution of Fig. 5.21(a); the term due to the leakage current (0.01 mA) is negligible. Repeating this operation to find the quiescent collector current at 100°C ($\beta = 200$, $I_{CO} = 3.0\ \mu$A), we get:

$$I_{C_Q} = \frac{200(1.07)}{2\ \text{k}\Omega + 200(0.2\ \text{k}\Omega)} + \frac{(2.2\ \text{k}\Omega)(200)(0.003)}{2\ \text{k}\Omega + 200(0.2\ \text{k}\Omega)}$$

$$= 5.09\ \text{mA} + 0.03\ \text{mA}$$

$$= 5.12\ \text{mA}$$

This indicates that the Q point shifts only very slightly as the temperature varies between room temperature and 100°C; the circuit of Fig. 5.19 thus exhibits excellent stability.

The total change in quiescent collector current, ΔI_{C_T}, for the emitter bias circuit is:

$$\Delta I_{C_T} = 5.12 \text{ mA} - 5.0 \text{ mA} = 0.12 \text{ mA}$$

Compare this with the 3.6 mA change (8.6 mA $-$ 5.0 mA) for the fixed bias circuit, and the 1.38 mA change (6.38 mA $-$ 5.0 mA) for the collector-to-base bias circuit. Re-writing the expression for the quiescent collector current at $100°C$:

$$I_{C_Q} = 5.09 \text{ mA} + 0.03 \text{ mA}$$

This shows that the quiescent collector current increased by 0.09 mA due to the increase in β, and by 0.03 mA due to the change in I_{CO}. Therefore, the total change in quiescent collector current may be expressed as follows:

$$\Delta I_{C_T} = \Delta I_{C_\beta} + \Delta I_{C_L}$$
$$\Delta I_{C_T} = 0.09 \text{ mA} + 0.03 \text{ mA} = 0.12 \text{ mA}$$

Compare this with the similar expressions for the fixed bias and collector-to-base bias circuits.

We know that the quiescent base current at room temperature is 40 μA; let us now find this value at $100°C$. At the higher temperature, $\beta = 200$, $I_{C_Q} = 5.12$ mA, and $I_{CEO} = 0.6$ mA. Substituting into the equation derived in the preceding section, we find the quiescent base current at $100°C$ to be:

$$I_{B_Q} = \frac{I_{C_Q} - I_{CEO}}{\beta} = \frac{5.12 \text{ mA} - 0.6 \text{ mA}}{200} = 0.0226 \text{ mA}$$
$$= 22.6 \text{ } \mu\text{A}$$

This indicates again that the base current decreases at the higher temperatures to offset the increase in collector current; recall that the construction of Fig. 5.21(b) reveals the base current at $100°C$ to be slightly greater than 20 μA.

We can now complete the analysis of the circuit of Fig. 5.19, since we know the quiescent collector current; i.e., we now have enough information to compute all of the voltages and currents in the emitter bias circuit. We analyze the circuit at room temperature. Since $I_{C_Q} = 5$ mA, and we have assumed that I_E is approximately equal to I_C, we say that $I_{E_Q} = 5$ mA. Therefore, by Ohm's law we can compute the voltage drop across the emitter resistor as:

$$V_{R_E} = I_E R_E = (5 \text{ mA})(0.2 \text{ k}\Omega) = 1 \text{ V}$$

Similarly, we can compute the voltage drop across the load resistor as:

$$V_{R_L} = I_C R_L = (5 \text{ mA})(1.8 \text{ k}\Omega) = 9 \text{ V}$$

Since the sum of V_{R_L}, V_{CE}, and V_{R_E} must add up to V_{CC} (20 V, in this case), V_{CE_Q} must be equal to:

$$V_{CE_Q} = V_{CC} - (V_{R_L} + V_{R_E}) = 20 \text{ V} - (9 \text{ V} + 1 \text{ V}) = 10 \text{ V}$$

These values are shown on the schematic diagram in Fig. 5.22(a). Notice that, since we assumed that R_L and R_E are in series, the *voltages across the two resistors are in the same ratio as the values of the resistors;* i.e., the ratio of V_{RL} to V_{RE} is 9 to 1 (9 V to 1 V) just as the ratio of R_L to R_E is 9 to 1 (1.8 kΩ to 200 Ω). Since the base-to-emitter drop is 0.6 V, the voltage drop across R_2 is 1.6 V, and that across R_1 is 18.4 V, shown as follows:

$$V_{R_2} = V_{BE} + V_{RE} = 0.6 \text{ V} + 1 \text{ V} = 1.6 \text{ V}$$

$$V_{R_1} = V_{CC} - V_{R_2} = 20 \text{ V} - 1.6 \text{ V} = 18.4 \text{ V}$$

The currents in these two resistors are found by Ohm's law as follows:

$$I_{R_2} = \frac{V_{R_2}}{R_2} = \frac{1.6 \text{ V}}{2.2 \text{ k}\Omega} = 0.727 \text{ mA} = 727 \ \mu\text{A}$$

$$I_{R_1} = \frac{V_{R_1}}{R_1} = \frac{18.4 \text{ V}}{24 \text{ k}\Omega} = 0.765 \text{ mA} = 765 \ \mu\text{A}$$

The quiescent base current is therefore equal to:

$$I_{BQ} = I_{R_1} - I_{R_2} = 765 \ \mu\text{A} - 727 \ \mu\text{A} = 38 \ \mu\text{A}$$

This checks approximately with the previous result, $I_{BQ} = 40 \ \mu\text{A}$; all of these values are shown in Fig. 5.22(a).

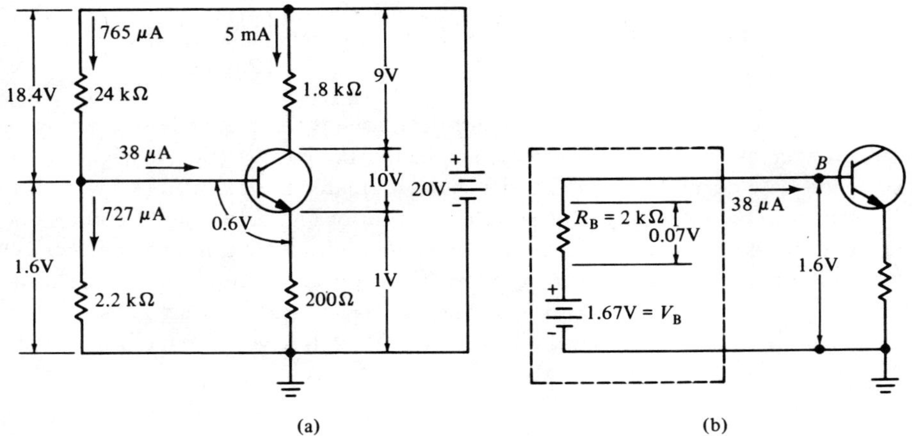

Figure 5.22

One important point should now be mentioned in connection with this circuit. Figure 5.22(b) shows the same circuit with the Thevenin equivalent circuit replacing the divider arrangement. Note that, while the *Thevenin voltage,* V_B, is equal to 1.67 V, the *actual* voltage from base to ground (voltage across R_2) is 1.6 V; in other words, *the voltage V_B is not equal to the voltage across R_2.* Figure 5.22(b) shows that the actual base terminal is *outside* the dotted box which contains the Thevenin equivalent circuit. Since I_B flows through the Thevenin impedance, R_B, there will be a voltage drop across this resistor equal to:

$$V_{R_B} = I_B R_B = (0.038 \text{ mA})(2 \text{ k}\Omega) = 0.076 \text{ V}$$

This 0.07 V drop accounts for the difference between V_B and V_{R_2}. The *actual voltage* across R_2 is 1.6 V, *not* 1.67 V; V_{R_2} can be found in either one of two ways:

1. $V_{R_2} = V_{BE} + V_{RE} = V_{BE} + I_E R_E$

2. $V_{R_2} = V_B - I_B R_B = V_B - R_B \left(\dfrac{I_C - I_{CEO}}{\beta} \right)$

These methods can be applied once the value of the quiescent collector current is ascertained.

Let us repeat the expression for the quiescent collector current at room temperature:

$$I_{C_Q} = \frac{\beta(V_B - V_{BE})}{R_B + \beta R_E}$$

Note again in this equation that there is a term proportional to β in both the numerator *and* the denominator; therefore, as β increases, both the numerator and the denominator increase and this tends to keep the collector current constant. Suppose for a moment that:

$$\beta R_E \gg R_B$$

If this is true, the expression can be approximated as:

$$I_{C_Q} \simeq \frac{\beta(V_B - V_{BE})}{\beta R_E} = \frac{V_B - V_{BE}}{R_E} \simeq I_{C_Q} \qquad \text{for } \beta R_E \gg R_B$$

This shows that *the quiescent collector current at room temperature is practically independent of β*; in other words, if the inequality stated above is true, the Q point will remain approximately constant even though units with differing β's are inserted in the circuit. This inequality requires that *both β and R_E be large, and R_B be small;* however, there are certain limitations on the values of these resistors and parameters, as we shall see.

The above approximation is also useful from the standpoint of a rapid analysis of the emitter bias circuit. The inequality mentioned above is valid in the case of the circuit of Fig. 5.19 since $R_B = 2 \text{ k}\Omega$ and βR_E is:

$$\beta R_E = 125(0.2 \text{ k}\Omega) = 25 \text{ k}\Omega$$

$$25 \text{ k}\Omega \gg 2 \text{ k}\Omega$$

$$\beta R_E \gg R_B$$

Substituting into the approximate form of the equation, we get:

$$I_{C_Q} = \frac{V_B - V_{BE}}{R_E} = \frac{1.67 \text{ V} - 0.6 \text{ V}}{0.2 \text{ k}\Omega} = 5.35 \text{ mA}$$

This agrees quite well with the value of 5 mA found using the exact expression. As a general rule of thumb, we use the *approximate* form of the equation if βR_E *is greater than ten times* R_B; i.e.,

$$\beta R_E > 10 R_B$$

This is generally the case since these are the conditions for good stability. If the inequality is not true, the more exact form of the equation must be used.

$$I_{C_Q} = \frac{\beta(V_B - V_{BE})}{R_B + \beta R_E}$$

It can be seen from the approximate form of the equation that V_B has an important effect on the Q point.

EXAMPLE 9

Compute the value of I_{C_Q} at room temperature if the values and parameters of the emitter bias circuit are: $R_1 = 12 \text{ k}\Omega$, $R_2 = 2.2 \text{ k}\Omega$, $R_L = 1.8 \text{ k}\Omega$, $R_E = 200 \text{ }\Omega$, $V_{CC} = 20 \text{ V}$, $\beta = 125$, and $V_{BE} = 0.6 \text{ V}$.

SOLUTION

The first step is to compute V_B and R_B.

$$V_B = \frac{R_2 V_{CC}}{R_1 + R_2} = \frac{(2.2)(20)}{12 + 2.2} = 3.1 \text{ V}$$

$$R_B = \frac{R_1 R_2}{R_1 + R_2} = \frac{12(2.2)}{12 + 2.2} = 1.86 \text{ k}\Omega$$

Next, we check to see if the inequality is valid, as follows:

$$\beta R_E = 125(.2 \text{ k}\Omega) = 25 \text{ k}\Omega$$

$$25 \text{ k}\Omega \gg 1.86 \text{ k}\Omega$$

Since βR_E is greater than ten times R_B, we can use the approximate formula to compute I_{C_Q}:

$$I_{C_Q} = \frac{V_B - V_{BE}}{R_E} = \frac{3.1 \text{ V} - 0.6 \text{ V}}{0.2 \text{ k}\Omega} = \frac{2.5 \text{ V}}{0.2 \text{ k}\Omega} = 12.5 \text{ mA}$$

This formula predicts that the quiescent collector current will be 12.5 mA; however, this is *greater* than the collector saturation current, which is:

$$I_{C_s} = \frac{V_{CC}}{R_L + R_E} = \frac{20 \text{ V}}{1.8 \text{ k}\Omega + 0.2 \text{ k}\Omega} = 10 \text{ mA}$$

Therefore the transistor is saturated, and the quiescent collector current is 10 mA. The reason for this increase in current (from 5 mA to 10 mA) is the *increase* in the value of V_B due to the change in the value of R_1 from 24 kΩ to 12 kΩ. In other words, the *halving* of the value of R_1 has changed the voltage divider ratio in such a manner as to *increase* V_B.

EXAMPLE 10

Repeat example 9 if R_2 is halved to 1.1 kΩ.

SOLUTION

Compute V_B and R_B.

$$V_B = \frac{R_2 V_{CC}}{R_1 + R_2} = \frac{1.1(20)}{12 + 1.1} = 1.68 \text{ V}$$

$$R_B = \frac{R_1 R_2}{R_1 + R_2} = \frac{12(1.1)}{12 + 1.1} = 1 \text{ k}\Omega$$

Since the inequality is again valid, we compute the value of I_{CQ} as:

$$I_{CQ} = \frac{V_B - V_{BE}}{R_E} = \frac{1.68 \text{ V} - 0.6 \text{ V}}{0.2 \text{ k}\Omega} = \frac{1.08 \text{ V}}{0.2 \text{ k}\Omega} = 5.4 \text{ mA}$$

Thus, halving the value of R_2 (from 2.2 kΩ to 1.1 kΩ) restores the value of I_{CQ} to its original value of approximately 5 mA. Examples 9 and 10 illustrate the importance of the value of V_B on the Q point; in particular, the examples illustrate that, since the base circuit is supplied from a resistive divider, it is the *ratio* of these resistors which determines the Q point. We discuss this concept further in the following section, and again in chapter 10.

5-7 STABILITY FACTOR

We know from the previous discussions in this chapter that the Q point will shift when the temperature of the environment is changed, or when a transistor with a different β is inserted in the circuit. We also saw that the incorporation of negative feedback (as embodied in the collector-to-base bias and emitter bias circuits) *reduces* the amount of Q point shift. We turn our attention now to the *amount* of Q point shift; i.e., we study the factors upon which the extent of the change in quiescent collector current depends. We saw that the Q point shifts because of the change in the values of the leakage current (I_{CO}) and β; strictly speaking, the change in the value of V_{BE} with changing temperature also has an effect on the Q point as we soon see.

We begin our discussion with the definition of a new quantity known as the *stability factor*, S, which is given as follows:

$$S = \frac{\Delta I_{CL}}{\Delta I_{CO}}$$

Note that it is the ratio of the change in collector current due to leakage current, divided by the change in leakage current. We soon see that S is a quantity which depends first upon the particular circuit configuration (fixed bias, collector-to-base bias, or emitter bias), and also on the values of the transistor parameters and the various resistors in the circuit. Re-arranging the basic equation, we get:

$$\Delta I_{CL} = S(\Delta I_{CO})$$

Now we know from previous discussions that we should like ΔI_{CL} to be *as small as possible* for good stability. This means that we should like both S and ΔI_{CO} also to be small numbers. The way to keep ΔI_{CO} small is to use silicon transistors, particularly if the circuit is to be used at very high temperatures, since silicon transistors have much smaller leakage currents than germanium. However, once the transistor is selected, and the temperature range over which it must operate is specified, the factor, ΔI_{CO}, is fixed (since ΔI_{CO} depends upon the particular transistor type, as well as the maximum temperature at which it will be used). Therefore, the only way of reducing ΔI_{CL} is to make the stability factor as *small* as possible.

The expression for the stability factor is different for each type of bias circuit; the derivation involves some algebraic manipulation and calculus and is not shown. These expressions are given as follows for the three types of bias circuits studied in this chapter:

$$(1) \qquad S = \beta \qquad\qquad \text{Fixed Bias}$$

$$(2) \qquad S = \frac{1 + \dfrac{R_L}{R_B}}{\dfrac{1}{\beta} + \dfrac{R_L}{R_B}} \qquad\qquad \text{Collector-to-Base Bias}$$

$$(3) \qquad S = \frac{1 + \dfrac{R_E}{R_B}}{\dfrac{1}{\beta} + \dfrac{R_E}{R_B}} \qquad\qquad \text{Emitter Bias}$$

In all of the above expressions it is assumed that β is much greater than 1.0, which is generally true. The β in the expressions is that value of β at the *higher* temperature.

We now compute the stability factor and the value of ΔI_{CL} for each of the three bias circuits studied in this chapter. We assume that I_{CO} is approximately equal to zero at room temperature, and that $I_{CO} = 3.0$ μA at $100°C$; therefore, $\Delta I_{CO} = 3.0$ μA for all three circuits. Taking the fixed bias circuit first, we know that β of the transistor was 200 at $100°C$; thus:

$$S = \beta = 200$$

$$\Delta I_{CL} = S(\Delta I_{CO}) = 200(3.0 \ \mu A) = 600 \ \mu A = 0.6 \text{ mA}$$

Compare this value of ΔI_{CL} with that found for the fixed bias circuit at the end of Sec. 5-2. It can be seen that the stability factor for the fixed bias circuit is independent of the resistors, and depends only upon the β of the transistor.

Next let us compute S and ΔI_{CL} for the collector-to-base bias circuit; remember that R_{B} for that particular circuit was 250 kΩ and R_{L} was 2 kΩ.

$$S = \frac{1 + \dfrac{R_{\mathrm{L}}}{R_{\mathrm{B}}}}{\dfrac{1}{\beta} + \dfrac{R_{\mathrm{L}}}{R_{\mathrm{B}}}} = \frac{1 + \dfrac{2}{250}}{\dfrac{1}{200} + \dfrac{2}{250}} = 76.8$$

$$\Delta I_{\mathrm{CL}} = S(\Delta I_{\mathrm{CO}}) = 76.8(3.0\ \mu\mathrm{A}) = 231\ \mu\mathrm{A} = 0.23\ \mathrm{mA}$$

Thus there is a *smaller* shift in Q point due to leakage current in the collector-to-base bias circuit than in the fixed bias circuit; this is because the stability factor is *smaller* for the collector-to-base bias circuit (76.8) than for the fixed bias circuit (200).

Finally, we compute the values of S and ΔI_{CL} for the emitter bias circuit; the values of the resistors in the particular circuit are: $R_1 = 24$ kΩ, $R_2 = 2.2$ kΩ, $R_{\mathrm{E}} = 200$ Ω. We already found that $V_{\mathrm{B}} = 1.67$ V (if $V_{\mathrm{CC}} = 20$ V) and $R_{\mathrm{B}} = 2$ kΩ. Proceeding we find that:

$$S = \frac{1 + \dfrac{R_{\mathrm{E}}}{R_{\mathrm{B}}}}{\dfrac{1}{\beta} + \dfrac{R_{\mathrm{E}}}{R_{\mathrm{B}}}} = \frac{1 + \dfrac{0.2}{2}}{\dfrac{1}{200} + \dfrac{0.2}{2}} = 10.5$$

$$\Delta I_{\mathrm{CL}} = S(\Delta I_{\mathrm{CO}}) = 10.5(3.0\ \mu\mathrm{A}) = 31.5\ \mu\mathrm{A} = 0.031\ \mathrm{mA}$$

Thus this circuit shows the smallest shift in Q point because it has the smallest value of S (10.5). Therefore, we conclude that, *for the same transistor and temperature range, the circuit with the smallest stability factor is the most stable circuit.*

The next step is to examine the factors upon which the stability factor depends, since we should like to make this quantity as small as possible. Since $S = \beta$ for the fixed bias circuit, the only way to reduce S is to reduce β; but we shall see in the next chapter that a reduction in β causes a reduction in the gain of the amplifier. Therefore, if good stability is desired, the fixed bias circuit is never used.

The stability factor for the collector-to-base bias circuit can be reduced by reducing the value of the bias resistor, R_{B}. Assume that $\beta = 200$, $R_{\mathrm{L}} = 2$ kΩ, and R_{B} is *reduced* from 250 kΩ to 125 kΩ; the stability factor now is:

$$S = \frac{1 + \dfrac{R_{\mathrm{L}}}{R_{\mathrm{B}}}}{\dfrac{1}{\beta} + \dfrac{R_{\mathrm{L}}}{R_{\mathrm{B}}}} = \frac{1 + \dfrac{2}{125}}{\dfrac{1}{200} + \dfrac{2}{125}} = 47.5$$

Notice that a reduction in R_{B} from 250 kΩ to 125 kΩ reduces the stability factor from 76.8 to 47.5. However, as was illustrated in Sec. 5-4, *a change in the value of R_{B} causes a change in the value of I_{CQ}; in particular, a reduction in R_{B} increases the value of I_{CQ}.* For example, computing the value of I_{CQ} at room temperature first with $R_{\mathrm{B}} = 250$ kΩ, we get:

$$I_{\mathrm{CQ}} = \frac{\beta V_{\mathrm{CC}}}{R_{\mathrm{B}} + \beta R_{\mathrm{L}}} = \frac{125(20)}{250 + 125(2)} = 5\ \mathrm{mA}$$

Computing next the value of I_{C_Q} with $R_B = 125$ kΩ, we get:

$$I_{C_Q} = \frac{\beta V_{CC}}{R_B + \beta R_L} = \frac{125(20)}{125 + 125(2)} = 6.67 \text{ mA}$$

This illustrates the important effect of the value of R_B on the Q point. Therefore, if we want to obtain smaller and smaller values of S, we have to decrease the value of R_B; but this causes an increase in I_{C_Q}, moving the Q point closer to saturation. In the limiting case, when $R_B = $ zero, the Q point is at saturation, since:

$$I_{C_Q} = \frac{\beta V_{CC}}{R_B + \beta R_L} = \frac{\beta V_{CC}}{0 + \beta R_L} = \frac{V_{CC}}{R_L}$$

$$I_{C_Q} = \frac{V_{CC}}{R_L} = \frac{20 \text{ V}}{2 \text{ k}\Omega} = 10 \text{ mA} = I_{Cs}$$

The stability factor with R_B set equal to zero Ω is:

$$S = \frac{1 + \dfrac{R_L}{R_B}}{\dfrac{1}{\beta} + \dfrac{R_L}{R_B}} = \frac{R_B + R_L}{\dfrac{R_B}{\beta} + R_L} = \frac{0 + R_L}{\dfrac{0}{\beta} + R_L} = 1.0$$

This is the *smallest* possible value that the stability factor can have. However, setting $R_B = $ zero Ω saturates the transistor; the transistor cannot be biased at saturation when it is to be used in an amplifier, as explained in chapter 4. Therefore, we cannot obtain very small values of S (by making R_B very small) and still maintain the Q point in the center of the load line, using the basic collector-to-base bias circuit of Fig. 5.9. A modification must be made in the basic circuit; this is shown later on in the chapter. Concluding, we say that a *reduction in the value of R_B decreases* (improves) *the stability factor.*

Turning finally to the emitter bias circuit, we now see that the values of R_E and R_B (R_1 in parallel with R_2) have an important effect on the stability factor. Let us compute S for the same emitter bias circuit ($V_{CC} = 20$ V, $R_1 = 24$ kΩ, $R_2 = 2.2$ kΩ, $\beta = 200$), but change R_E from 200 Ω to 100 Ω. The value of S is:

$$S = \frac{1 + \dfrac{R_E}{R_B}}{\dfrac{1}{\beta} + \dfrac{R_E}{R_B}} = \frac{1 + \dfrac{0.1}{2}}{\dfrac{1}{200} + \dfrac{0.1}{2}} = 19.1$$

Decreasing R_E from 200 to 100 Ω, has caused an increase in S from 10.5 to 19.1; therefore we should like R_E to be large for good stability. Let us compute S for the same circuit ($R_E = 100$ Ω), except halve the values of R_1 (from 24 kΩ to 12 kΩ) and R_2 (from 2.2 kΩ to 1.1 kΩ); the new value of R_B is:

$$R_B = \frac{R_1 R_2}{R_1 + R_2} = \frac{(12)(1.1)}{12 + 1.1} = 1 \text{ k}\Omega$$

Computing the new stability factor, we get:

$$S = \frac{1 + \dfrac{R_E}{R_B}}{\dfrac{1}{\beta} + \dfrac{R_E}{R_B}} = \frac{1 + \dfrac{0.1}{1}}{\dfrac{1}{200} + \dfrac{0.1}{1}} = 10.5$$

Thus, halving the value of R_B restores S to its previous value. It can be seen from the formula and from the results of the above computations that S depends upon the *ratio* of R_E to R_B; *we should like R_E to be large and R_B to be small for good stability.* In later chapters we see that there is a limit on how large we can make R_E and how small we can make R_B. Once R_E and R_B are chosen for the desired stability factor in a particular design, R_1 and R_2 are selected to give the desired Q point; this will be expanded upon in chapter 10.

As you may recall, there is also a shift in the Q point due to the change in β, $\Delta I_{C\beta}$. If the equations are manipulated algebraically, $\Delta I_{C\beta}$ can be found by use of the following relationship:

$$\Delta I_{C\beta} = I_{C_1} \frac{(\Delta\beta)S_2}{\beta_1\beta_2}$$

In this equation, $\Delta I_{C\beta}$ is the change in collector current due to a change in β, I_{C_1} is the original collector current, $\Delta\beta$ is the change in β, β_1 is the original β, β_2 is the final β, and S_2 is the stability factor with β set equal to β_2. Since we should like $\Delta I_{C\beta}$ to be as small as possible, we should like $\Delta\beta$ and S_2 to be small, and β_1 and β_2 to be large. However, once the transistor is selected, and the temperature range is specified, β_1, β_2, and thus $\Delta\beta$ are all fixed; therefore, we return again to the requirement that the stability factor be as small as possible.

Let us now compute $\Delta I_{C\beta}$ for all three bias circuits. In each case $\beta_1 = 125$, $\beta_2 = 200$, $\Delta\beta = 75$ $(200 - 125)$, $I_{C_1} = 5$ mA, and the stability factors for the three circuits are 200, 76.8, and 10.5, respectively. Solving first for the fixed bias circuit, we get:

$$\Delta I_{C\beta} = I_{C_1} \frac{(\Delta\beta)S_2}{\beta_1\beta_2} = (5) \frac{75(200)}{125(200)} = 3 \text{ mA}$$

Compare this with the value of $\Delta I_{C\beta}$ found for the fixed bias circuit in Sec. 5-2. Next, for the collector-to-base bias circuit, $\Delta I_{C\beta}$ is:

$$\Delta I_{C\beta} = I_{C_1} \frac{(\Delta\beta)S_2}{\beta_1\beta_2} = \frac{(5)(75)(76.8)}{125(200)} = 1.15 \text{ mA}$$

If the computations are worked carefully, it will be found that the value of I_{C_Q} at room temperature (I_{C_1}) for the emitter bias is actually less than 5 mA; it is equal to 4.95 mA. Thus we must use this value in the equation.

$$\Delta I_{C\beta} = I_{C_1} \frac{(\Delta\beta)S_2}{\beta_1\beta_2} = 4.95 \frac{75(10.5)}{125(200)} = 0.156 \text{ mA}$$

The results of these last three computations indicate that *the circuit with the smallest stability factor gives the smallest shift in Q point due to a change in β.*

Throughout this chapter we have neglected the effect of the variation of V_{BE} with temperature; since this quantity *does* vary as the temperature is changed, it has some effect on the Q point. We showed in Sec. 5-2 that this variation had a negligible effect on the Q point in the fixed bias circuit. However, this is not so for all bias circuits. Equations can be derived which enable us to compute the change in Q point due to a change in the value of V_{BE}; the equations themselves are shown as follows:

$$\text{(1)} \qquad \Delta I_{Cv} = -(\Delta V_{BE}) \frac{S_2}{R_B} \qquad\qquad \text{Fixed Bias}$$

$$\text{(2)} \qquad \Delta I_{Cv} = -(\Delta V_{BE}) \frac{S_2}{R_B + R_L} \qquad\qquad \text{Collector-to-Base Bias}$$

$$\text{(3)} \qquad \Delta I_{Cv} = -(\Delta V_{BE}) \frac{S_2}{R_B + R_E} \qquad\qquad \text{Emitter Bias}$$

In each of these equations, ΔI_{Cv} is the change in the quiescent collector current due to a change in V_{BE} expressed in milliamperes, ΔV_{BE} is the change in V_{BE} expressed in volts, S_2 is the stability factor at the higher temperature ($\beta = \beta_2$), and the resistors are expressed in kilohms. Let us now apply these equations. We know that V_{BE} decreases 2.5 mV for each centigrade degree increase in temperature. Since the total change in temperature is $75°$ (from 25 to 100), the change in V_{BE} over this temperature range is:

$$\Delta V_{BE} = (2.5 \text{ mV/C°})(\Delta T) = (2.5)(75) = 187.5 \text{ mV}$$
$$\Delta V_{BE} = -0.1875 \text{ V}$$

The minus sign is used to indicate that V_{BE} *decreases* as the temperature increases. The value of ΔV_{BE} (-0.1875 V) is the same for all three bias circuits, since the same transistor is used over the same temperature range. Let us first compute ΔI_{Cv} for the fixed bias circuit; $\beta_2 = S_2 = 200$ and $R_B = 500$ kΩ:

$$\Delta I_{Cv} = -(\Delta V_{BE}) \frac{S_2}{R_B} = -(-0.1875 \text{ V}) \frac{200}{500} = +0.075 \text{ mA}$$

This indicates that the variation in the value of V_{BE} with temperature has a negligible effect on the Q point in the fixed bias circuit. Notice that the two *negative* quantities, when multiplied together, yield a *positive* increment for ΔI_{Cv}. This serves to indicate that a *decrease* in the value of V_{BE} (with increasing temperature) causes an *increase* in the value of the quiescent collector current. For example, take the case of the fixed bias circuit; the expression for the quiescent collector current at room temperature is given by the following equation:

$$I_{CQ} = \frac{\beta(V_{CC} - V_{BE})}{R_B}$$

It can be seen by observation of this equation that a decrease in V_{BE} causes an increase in I_{CQ}; this can be shown to be true for the other two circuits as well. In other words, the change in V_{BE} has the same effect on the Q point as the change in β and the change in I_{CO}; as the temperature increases, the change in *all three* of these quantities tend to *increase* the quiescent collector current.

Next we compute the value of ΔI_{Cv} for the collector-to-base bias circuit; $S_2 = 76.8$, $R_B = 250$ kΩ, and $R_L = 2$ kΩ:

$$\Delta I_{Cv} = -(\Delta V_{BE}) \frac{S_2}{R_B + R_L} = -(-0.1875 \text{ V}) \frac{76.8}{252} = 0.057 \text{ mA}$$

The results illustrate that the change in V_{BE} also has a negligible effect on the Q point of this circuit. This is *not* generally true for the collector-to-base bias circuit, however; there are situations where the value of ΔI_{Cv} may be larger, and thus we do not make any generalizations.

Finally, computing ΔI_{Cv} for the emitter bias circuit ($S_2 = 10.5$, $R_1 = 24$ kΩ, $R_2 = 2.2$ kΩ, $R_E = 200$ Ω), we get:

$$\Delta I_{Cv} = -(\Delta V_{BE}) \frac{S_2}{R_B + R_E} = -(-0.1875 \text{ V}) \frac{10.5}{2.2} = 0.895 \text{ mA}$$

Thus the shift in Q point due to the variation in V_{BE} is *not* negligible in the emitter bias circuit. It can be shown mathematically that *decreasing* the value of R_B (to obtain a smaller stability factor) will *not* improve the situation (decrease ΔI_{Cv}) in the emitter bias circuit; only by *increasing* the value of R_E can ΔI_{Cv} be decreased. Thus, in general, we cannot neglect the effect of the variation in the value of V_{BE}.

The total shift in Q point, ΔI_{CT}, is composed of *three* parts:

$$\Delta I_{CT} = \Delta I_{CL} + \Delta I_{C\beta} + \Delta I_{Cv}$$

Each of these three individual shifts tends to increase I_{CQ} at the higher temperatures; all three taken together have a cumulative effect on the value of I_{CQ}. Let us now compute this total shift for the emitter bias circuit studied in this chapter. We found that:

$$\Delta I_{CL} = 0.031 \text{ mA}, \quad \Delta I_{C\beta} = 0.156 \text{ mA}, \quad \Delta I_{Cv} = 0.895 \text{ mA}$$

Therefore, for this particular circuit, the total shift is actually:

$$\Delta I_{CT} = \Delta I_{CL} + \Delta I_{C\beta} + \Delta I_{Cv} = 0.031 + 0.156 + 0.895 = 1.082 \text{ mA}$$
$$\Delta I_{CT} = 1.082 \text{ mA}$$

If I_{C_1} is the quiescent value at the lower temperature, and I_{C_2} is the value at the higher temperature, we can say that:

$$I_{C_2} = I_{C_1} + \Delta I_{CT} = 4.95 \text{ mA} + 1.082 \text{ mA} = 6.032 \text{ mA}$$
$$I_{C_2} = 6.032 \text{ mA}$$

Therefore, the actual value of I_{C_Q} at 100°C for the emitter bias circuit discussed in this chapter is 6.03 mA.

EXAMPLE 11

A particular emitter bias circuit has the following values and parameters at room temperature: $V_{CC} = 30$ V, $R_1 = 95$ kΩ, $R_2 = 5$ kΩ, $R_L = 5$ kΩ, $R_E = 250$ Ω, $\beta = 80$, $V_{BE} = 0.6$ V, I_{CO} is approximately equal to zero.

(a) Compute the quiescent collector current at room temperature.

SOLUTION

All that is needed is substitution into the basic equation. First we must find V_B and R_B:

$$V_B = \frac{R_2 V_{CC}}{R_1 + R_2} = \frac{5(30)}{95 + 5} = 1.5 \text{ V}$$

$$R_B = \frac{R_1 R_2}{R_1 + R_2} = \frac{(95)(5)}{95 + 5} = 4.75 \text{ k}\Omega$$

Since we are assuming that the leakage current at room temperature is zero, the expression for the quiescent collector current at room temperature is:

$$I_{C_Q} = \frac{\beta(V_B - V_{BE})}{R_B + \beta R_E}$$

Substituting into the equation, we get:

$$I_{C_Q} = \frac{80(1.5 - 0.6)}{4.75 + 80(0.25)} = 2.91 \text{ mA}$$

The collector saturation current for this circuit is:

$$I_{Cs} = \frac{V_{CC}}{R_L + R_E} = \frac{30 \text{ V}}{5 \text{ k}\Omega + 0.25 \text{ k}\Omega} = 5.71 \text{ mA}$$

Therefore the transistor is *not* saturated, and the Q point is approximately in the center of the load line, since 2.91 is approximately one-half of 5.71.

(b) Compute the quiescent collector current at 95°C if β at this temperature is 160 and $I_{CO} = 5$ μA.

SOLUTION

There are two ways to attack this problem: One is a direct substitution into the general equation; the second involves computing the individual shifts, and then adding them up to find the total shift, and finally, the new value of I_{C_Q}. In this particular problem, it is probably simpler to substitute directly into the equation since this is a straight analysis problem; we distinguish between *analysis* and *design* in chapter 10. The method of finding the individual shifts by means of the stability factor is more useful when we are actually engaged in designing a circuit. Both methods will be illustrated here for the sake of completeness. Since the total change in temperature is 70°C (from 25 to 95), the change in V_{BE} is:

$$\Delta V_{BE} = (2.5 \text{ mV/C}°)(70°\text{C}) = -0.175 \text{ V}$$

Therefore the value of V_{BE} at 95°C is:

$$V_{BE}(95°\text{C}) = 0.6 \text{ V} - 0.175 \text{ V} = 0.425 \text{ V}$$

Substituting into the general equation, we find the quiescent collector current in the circuit at 95°C to be:

$$I_{C_Q} = \frac{\beta(V_B - V_{BE})}{R_B + \beta R_E} + \frac{(R_E + R_B)\beta I_{CO}}{R_B + \beta R_E}$$

$$= \frac{160(1.5 - 0.425)}{4.75 + 160(0.25)} + \frac{(4.75 + 0.25)(160)(0.005)}{4.75 + 160(0.25)} = 3.93 \text{ mA}$$

Thus the total shift is:

$$\Delta I_{C_T} = I_{C_2} - I_{C_1} = 3.93 - 2.91 = 1.02 \text{ mA}$$

Now, proceeding with the alternate method of solution, we must first find the stability factor:

$$S = \frac{1 + \dfrac{R_E}{R_B}}{\dfrac{1}{\beta} + \dfrac{R_E}{R_B}} = \frac{1 + \dfrac{0.25}{4.75}}{\dfrac{1}{160} + \dfrac{0.25}{4.75}} = 17.9$$

We know that $\Delta I_{CO} = 5 \text{ }\mu\text{A}$, $\Delta\beta = 80$, $I_{C_1} = 2.91$ mA, $\beta_1 = 80$, $\beta_2 = 160$, and $\Delta V_{BE} = -0.175$ V. Computing the three increments, we get:

$$\Delta I_{C_L} = S(\Delta I_{CO}) = (17.9)(0.005 \text{ mA}) = 0.0895 \text{ mA}$$

$$\Delta I_{C_\beta} = I_{C_1} \frac{(\Delta\beta)S_2}{\beta_1\beta_2} = \frac{(2.91)(80)(17.9)}{80 \times 160} = 0.325 \text{ mA}$$

$$\Delta I_{C_V} = -(\Delta V_{BE}) \frac{S_2}{R_B + R_E} = -(-0.175)\frac{17.9}{4.75 + 0.25} = 0.626 \text{ mA}$$

The total change in quiescent collector current is:

$$\Delta I_{C_T} = \Delta I_{C_L} + \Delta I_{C_\beta} + \Delta I_{C_V} = 0.0895 + 0.325 + 0.626 = 1.04 \text{ mA}$$

This agrees quite well with the value of 1.02 mA found using the straight substitution method above. Thus both methods yield the same results. We shall return to further use of these equations in chapter 10.

One last problem is yet to be studied concerning the collector-to-base bias circuit. We saw earlier in this section that it is necessary to decrease the value of R_B in order to improve the stability factor. For example, using the same basic circuit with $\beta = 200$, let us find the value of R_B necessary to give a stability factor of 10.0. Substituting into the equation, and manipulating algebraically, we find:

$$S = \frac{1 + \dfrac{R_L}{R_B}}{\dfrac{1}{\beta} + \dfrac{R_L}{R_B}}; \quad 10 = \frac{1 + \dfrac{2}{R_B}}{\dfrac{1}{200} + \dfrac{2}{R_B}}; \quad R_B = 19 \text{ k}\Omega$$

It is left to the reader to work out the algebra. Note again that a reduction in the value of R_B (from 250 kΩ to 19 kΩ) is necessary in order to reduce the value of S (from 76.8 to 10.0). However, if the value of R_B is changed, the Q point is *also* changed. The *new* value of quiescent collector current at room temperature ($\beta = 125$, $R_L = 2$ kΩ, $R_B = 19$ kΩ, $V_{CC} = 20$ V), neglecting the 0.6 V V_{BE} drop, is:

$$I_{CQ} = \frac{\beta V_{CC}}{R_B + \beta R_L} = \frac{125(20)}{19 + 125(2)} = 9.3 \text{ mA}$$

The collector saturation current is:

$$I_{Cs} = \frac{V_{CC}}{R_L} = \frac{20 \text{ V}}{2 \text{ k}\Omega} = 10 \text{ mA}$$

Therefore, if a 19 kΩ resistor is inserted in the circuit for R_B, the new Q point is very close to saturation.

One method of resolving this difficulty is to insert a resistor, R_2, between base and emitter, as shown in Fig. 5.23(a). The base current (I_B) is now equal to the difference between the current in the 19 kΩ resistor (I_1) and the current in $R_2(I_2)$. *The resistor, R_2, diverts some current away from the base, thus decreasing I_B, and consequently*

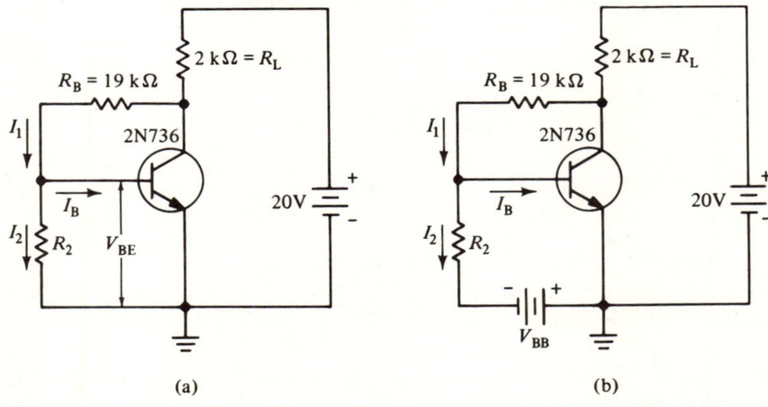

(a)   (b)

Figure 5.23

reducing the value of I_{CQ} so that the Q point is closer to the center of the load line. However, the use of the circuit of Fig. 5.23(a) creates *two new problems.*

The first problem involves the value of R_2 necessary to draw enough current away from the base to restore the Q point to the center of the load line. If the circuit equations are analyzed, it can be shown that the quiescent collector current at room temperature for the circuit of Fig. 5.23(a) is given by the following equation:

$$I_{CQ} = \frac{\beta V_{CC}}{R_B + \beta R_L} - \frac{\beta V_{BE}\left(1 + \frac{R_B}{R_2}\right)}{R_B + \beta R_L}$$

$$5\text{ mA} = \quad 9.3\text{ mA} \quad - \quad 4.3\text{ mA}$$

It can be seen above that the expression for I_{CQ} is composed of *two* terms; the value of I_{CQ} is equal to the *difference* in these two terms. We know that the first term is equal to 9.3 mA, and we desire that I_{CQ} be equal to 5 mA; therefore, the second term must be equal to 4.3 mA. It should be clear that this *second term is directly proportional to the current (I_2) diverted through R_2.* If R_2 is set equal to infinity (open circuit), the equation reduces to the original form, i.e., when there is no R_2 present in the circuit; thus:

$$\underline{R_2 = \infty}$$

$$I_{CQ} = \frac{\beta V_{CC}}{R_B + \beta R_L} - \frac{\beta V_{BE}\left(1 + \frac{R_B}{\infty}\right)}{R_B + \beta R_L}$$

$$= \frac{\beta V_{CC} - \beta V_{BE}}{R_B + \beta R_L} = \frac{\beta(V_{CC} - V_{BE})}{R_B + \beta R_L}$$

Now, to find the value of R_2 necessary to restore I_{CQ} to 5 mA, we must set the second term equal to 4.3 mA:

$$4.3\text{ mA} = \frac{\beta V_{BE}\left(1 + \frac{R_B}{R_2}\right)}{R_B + \beta R_L}$$

Substituing the proper values into the equation ($\beta = 125$, $V_{BE} = 0.6$ V, $R_B = 19$ kΩ, $R_L = 2$ kΩ) and solving algebraically for R_2, we get:

$$4.3 = \frac{125(0.6)\left(1 + \frac{19}{R_2}\right)}{19 + 125(2)}$$

$$1 + \frac{19}{R_2} = \frac{(4.3)(269)}{75}$$

$$R_2 = 1.32\text{ k}\Omega$$

We shall see in the next chapter that it is undesirable to place a resistor as small as

1.32 kΩ across the base and emitter terminals of an amplifier; it has the effect of *reducing the gain*.

The second problem concerns the stability of the circuit of Fig. 5.23(a). We know that the base current is given by:

$$I_\mathrm{B} = I_1 - I_2$$

We know also that I_2 is equal to:

$$I_2 = \frac{V_\mathrm{BE}}{R_2}$$

Now, if the temperature is increased, V_BE will decrease and therefore, I_2 will also decrease. If I_2 *decreases*, however, at the higher temperatures, this tends to cause the base current (I_B) to *increase*. But remember that, in order for the circuit to provide negative feedback, the base current must *decrease* at the higher temperatures in order to compensate partially for the increase in collector current. It is true that I_1 decreases at the higher temperatures because of the negative feedback path through the 19 kΩ bias resistor. But if both I_1 *and* I_2 decrease, I_B tends to remain constant; this is still unsatisfactory since I_B must decrease for proper stabilizing action. Therefore, the circuit of Fig. 5.23(a) is a source of instability due to the variation in V_BE.

Let us now examine quantitatively the stability of the circuit. Taking first the original circuit (without R_2); $S = 76.8$, $\Delta I_\mathrm{CO} = 3\ \mu\mathrm{A}$, $R_\mathrm{B} = 250$ kΩ, $R_\mathrm{L} = 2$ kΩ, $\beta_1 = 125$, $\beta_2 = 200$, $\Delta V_\mathrm{BE} = -0.1875$ V. We shall find the individual Q point shifts, and then add them together to find the total shift:

$$\Delta I_{\mathrm{C_L}} = S(\Delta I_\mathrm{CO}) = 76.8(0.003\ \mathrm{mA}) = 0.231\ \mathrm{mA}$$

$$\Delta I_{\mathrm{C}\beta} = I_{\mathrm{C_1}} \frac{(\Delta\beta)S_2}{\beta_1\beta_2} = 5\,\frac{(75)(76.8)}{(125)(200)} = 1.15\ \mathrm{mA}$$

$$\Delta I_{\mathrm{Cv}} = -(\Delta V_\mathrm{BE})\frac{S_2}{R_\mathrm{B} + R_\mathrm{L}} = -(-0.1875)\frac{76.8}{250 + 2} = 0.057\ \mathrm{mA}$$

$$\Delta I_{\mathrm{C_T}} = \Delta I_{\mathrm{C_L}} + \Delta I_{\mathrm{C}\beta} + \Delta I_{\mathrm{Cv}}$$

$$= 0.231 + 1.15 + 0.057 = 1.438\ \mathrm{mA}$$

Next let us compute the Q point shifts for the circuit of Fig. 5.23(a) with $R_2 = 1.32$ kΩ. It can be shown algebraically that, while the expressions for Q point shifts due to leakage and β remain the same for this circuit, the expression for the shift due to V_BE is modified as follows:

$$\Delta I_{\mathrm{Cv}} = -(\Delta V_\mathrm{BE})\frac{S_2}{R_\mathrm{B} + R_\mathrm{L}}\left(1 + \frac{R_\mathrm{B}}{R_2}\right)$$

Now repeating the operation for the circuit of Fig. 5.23(a) with $S = 10$, $\Delta I_\mathrm{CO} = 3\ \mu\mathrm{A}$, $R_\mathrm{B} = 19$ kΩ, $R_\mathrm{L} = 2$ kΩ, $R_2 = 1.32$ kΩ, $\beta_1 = 125$, $\beta_2 = 200$, and $\Delta V_\mathrm{BE} = -0.1875$ V, we get:

$$\Delta I_{C_L} = S(\Delta I_{CO}) = 10(0.003 \text{ mA}) = 0.03 \text{ mA}$$

$$\Delta I_{C_\beta} = I_{C_1} \frac{(\Delta\beta)S_2}{\beta_1\beta_2} = 5 \frac{(75)(10)}{(125)(200)} = 0.15 \text{ mA}$$

$$\Delta I_{C_V} = -(\Delta V_{BE}) \frac{S_2}{R_B + R_L} \left(1 + \frac{R_B}{R_2}\right) = -(-0.1875) \frac{10}{19 + 2} \left(1 + \frac{19}{1.32}\right)$$

$$\Delta I_{C_V} = 1.375 \text{ mA}$$

$$\Delta I_{C_T} = \Delta I_{C_L} + \Delta I_{C_\beta} + \Delta I_{C_V}$$

$$= 0.03 + 0.15 + 1.375 = 1.55 \text{ mA}$$

Thus the total shift for the circuit of Fig. 5.23(a) (1.55 mA) *is actually greater* than that for the original circuit (1.43 mA). Recall that the purpose of reducing R_B from 250 kΩ to 19 kΩ was to reduce the stability factor from 76.8 to 10; in other words, we wanted a more stable circuit. It can be seen that reducing S has given us *smaller* shifts due to *leakage* (0.03 mA compared with 0.231 mA) and β (0.15 mA compared with 1.15 mA); but it has caused a *much larger* shift due to V_{BE} (1.375 mA compared with 0.057 mA). In other words, instead of making the circuit *more* stable on the whole, we have actually made it *less* stable. The reason for this poor stability is the small value of R_2; it can be seen in the expression for ΔI_{C_V} that a *small* value of R_2 yields a *large* value for ΔI_{C_V}. Thus the circuit of Fig. 5.23(a) is unsatisfactory for two reasons: The small value of R_2 reduces the gain, and it also causes poorer stability.

The situation is improved enormously by the addition of a second battery, V_{BB}, in series with R_2 as shown in Fig. 5.23(b). Neglecting the V_{BE} drop (if V_{BB} is much larger than V_{BE}), the quiescent collector current at room temperature is given as follows:

$$I_{C_Q} = \frac{\beta V_{CC}}{R_B + \beta R_L} - \frac{\dfrac{\beta V_{BB}R_B}{R_2}}{R_B + \beta R_L}$$

$$5 \text{ mA} = 9.3 \text{ mA} - 4.3 \text{ mA}$$

If we make V_{BB} equal to 20 V, and we again set the second term equal to 4.3 mA, we find the necessary value of R_2 as follows:

$$4.3 = \frac{\dfrac{\beta V_{BB}R_B}{R_2}}{R_B + \beta R_L} = \frac{\dfrac{125(20)(19)}{R_2}}{19 + 125(2)}$$

$$R_2 = \frac{(2500)(19)}{(4.3)(269)} = 41 \text{ k}\Omega$$

Thus, the inclusion of a 20 V battery for V_{BB} in Fig. 5.23(b) enables us to make $R_2 = 41$ kΩ and *divert the same amount of current away from the base*, as in the case of the circuit of Fig. 5.23(a) when $R_2 = 1.32$ kΩ and *no* second battery is included. Now that the inclusion of a 20 V V_{BB} enables us to obtain $I_{C_Q} = 5$ mA with $R_2 = 41$ kΩ, we have a *much larger* resistor connected across the base and emitter terminals; this does *not* have as serious an effect on the gain as if a 1.32 kΩ resistor were connected (see chapter 6). In addition, we obtain an improvement in the stability of the circuit. The stability factor remains unchanged by the addition of the V_{BB} battery to the circuit; thus ΔI_{C_L} and ΔI_{C_β} are still equal to 0.03 mA and 0.15 mA, respectively. However, ΔI_{Cv} is reduced considerably, as follows:

$$\Delta I_{Cv} = -(\Delta V_{BE}) \frac{S_2}{R_B + R_L}\left(1 + \frac{R_B}{R_2}\right) = -(-0.1875)\frac{10}{19 + 2}\left(1 + \frac{19}{41}\right)$$

$$\Delta I_{Cv} = 0.13 \text{ mA}$$

This shows that the *larger* value of R_2 made possible by the addition of the V_{BB} battery yields a *smaller* value of ΔI_{Cv}. The total Q point shift is now equal to:

$$\Delta I_{C_T} = \Delta I_{C_L} + \Delta I_{C_\beta} + \Delta I_{Cv}$$

$$= 0.03 + 0.15 + 0.13 = 0.31 \text{ mA}$$

This is smaller than the shift obtained with the circuit of Fig. 5.23(a) (1.55 mA) or that obtained with the original collector-to-base bias circuit (1.43 mA). Thus the inclusion of the second battery and R_2 enables us to improve the stability while at the same time maintaining the Q point in the center of the load line.

The disadvantage of the circuit of Fig. 5.23(b) is the requirement of a second source of power; the circuit requires both a positive (V_{CC}) and a negative (V_{BB}) voltage with respect to ground (if a *pnp* transistor is used, the two supplies are reversed). The requirement of a second source necessitates the allocation of more space, as well as cost. However, many large electronic systems provide two sources of power specifically for circuits like the one in Fig. 5.23(b); there are other types of circuits which also require two power sources. If a small system is being designed, it is usually more economical to use the emitter bias circuit since this requires only one power source.

It should be mentioned that there are other methods of stabilizing the amplifier against changes in temperature; these circuits use various diodes and *thermistors* (resistors whose resistance changes with temperature). However, for most general purpose work involving mass production techniques, the circuits discussed in this chapter yield the most economical, trouble-free performance.

PROBLEMS

1.   A fixed bias circuit has $V_{CC} = 20$ V. Compute the value of the base current if:

    (a) $R_B = 100$ kΩ
    (b) $R_B = 200$ kΩ
    (c) $R_B = 500$ kΩ
    (d) $R_B = $ infinity

2.   A fixed bias circuit has the following values:

$$V_{CC} = 20 \text{ V} \qquad \beta = 100 \text{ and } I_{CO} = 1 \text{ } \mu\text{A at } 25°\text{C}$$
$$R_L = 5 \text{ k}\Omega \qquad \beta = 200 \text{ and } I_{CO} = 50 \text{ } \mu\text{A at } 95°\text{C}$$
$$R_B = 1 \text{ M}\Omega$$

    (a)   Compute the values of I_{CQ} and V_{CEQ} at 25°C.
    (b)   Compute the values of I_{CQ} and V_{CEQ} at 95°C.

3.   A fixed bias circuit has the following values:

$$V_{CC} = 40 \text{ V} \qquad \beta = 100 \text{ and } I_{CO} = 1 \text{ } \mu\text{A at } 25°\text{C}$$
$$R_L = 5 \text{ k}\Omega \qquad \beta = 150 \text{ and } I_{CO} = 10 \text{ } \mu\text{A at } 55°\text{C}$$
$$R_B = 1 \text{ M}\Omega$$

    (a)   Compute the values of I_{CQ} and V_{CEQ} at 25°C.
    (b)   Compute the values of I_{CQ} and V_{CEQ} at 55°C.
    (c)   Compute the values of ΔI_{CL}, $\Delta I_{C\beta}$, and ΔI_{CT}.

4.   A fixed bias circuit has $V_{CC} = 40$ V, $R_L = 5$ kΩ, $R_B = 1$ MΩ, $\beta = 100$, and $V_{BE} = 0.5$ V at 25°C. Assuming that both β and I_{CO} remain constant with temperature, compute the value of I_{CQ} at 25°C and 95°C. What effect does V_{BE} have upon the value of I_{CQ}?

5.   A fixed bias circuit has the following values:

$$V_{CC} = 30 \text{ V} \qquad R_B = 500 \text{ k}\Omega$$
$$R_L = 3 \text{ k}\Omega \qquad \beta = 50$$

    (a)   Compute the value of I_{CQ}.
    (b)   Compute the value of I_{BQ}.

6.   Repeat problem 5 if the transistor is replaced by one having a β of 100. What effect does the doubling of β have upon the value of I_{CQ}? on I_{BQ}?

7.   A collector-to-base bias circuit has the following values:

$$V_{CC} = 30 \text{ V} \qquad R_B = 333 \text{ k}\Omega$$
$$R_L = 3 \text{ k}\Omega \qquad \beta = 50$$

    (a)   Compute the value of I_{CQ}.
    (b)   Compute the value of I_{BQ}.

8.   Repeat problem 7 if the transistor is replaced by one having a β of 100. What effect does the doubling of β have upon the value of I_{CQ}? on I_{BQ}? Compare the results of problems 7 and 8 with those of problems 5 and 6.

9.   An emitter-bias circuit has the following values:

$$V_{CC} = 30 \text{ V} \qquad R_1 = 90 \text{ k}\Omega$$
$$R_L = 5 \text{ k}\Omega \qquad R_2 = 10 \text{ k}\Omega$$
$$R_E = 1 \text{ k}\Omega \qquad \text{Neglect } V_{BE}$$

Compute the value of I_{CQ} if $\beta = 100$.

10.   Repeat problem 9 if $\beta = 200$. Explain why I_{CQ} does not double when β doubles.

11.   Repeat problem 9 if $R_1 = 9$ kΩ and $R_2 = 1$ kΩ.

12.   Repeat problem 10 if $R_1 = 9$ kΩ and $R_2 = 1$ kΩ.

13. I_{CO} for the transistor in the circuit of problem 9 increases from 1 μA to 50 μA due to an increase in temperature.
    (a) Compute the stability factor; use $\beta = 200$.
    (b) Compute the value of ΔI_{C_L}.

14.   Repeat problem 13 for the circuit of problem 11.

15.   Compute the value of ΔI_{C_β} in problems 9 and 10.

16.   Compute the value of ΔI_{C_β} in problems 11 and 12.

6

AMPLIFIERS IN CASCADE

In the last chapter we saw that it was necessary to add a number of resistors to the basic amplifier circuit in order to maintain a stable Q point; at that point we were concerned only with the dc behavior of the amplifier. Now we deal again (as in chapter 4) with the ac operation of the amplifier circuit. In particular, we apply the basic techniques developed in chapter 4 (solution by means of the ac equivalent circuit) to the analysis of the more practical amplifier circuits discussed in chapter 5. Just as we were concerned *only with dc operation* in chapter 5, we deal *primarily with ac operation* in chapter 6.

We first study the methods of coupling (connecting) the ac source voltage (the one which is to be amplified, v_I) to the amplifier; this involves some important concepts. Another point to be studied is the effect of the various resistive components (R_L, R_E, R_1, R_2, R_B) and the transistor parameters (β, r_I, r_O) on the gain of the amplifier. We know that resistors such as R_1, R_2, R_E, and R_B had to be added to the circuit to stabilize the Q point; however, these resistors have very significant effects on the gain and therefore they must be selected with great care. We shall also see that the requirement that the resistors be in the circuit necessitates the addition of other components to the circuit, namely capacitors. Finally, we shall see that amplifiers must be connected in cascade (output of first stage to input of second) in order to increase the gain; this involves the consideration of some very important concepts.

6-1 CAPACITOR COUPLING

Let us begin with the study of the fixed bias circuit since it is the simplest. Figure 6.1(a) shows a fixed bias circuit using the 2N911, an *npn* silicon transistor. In this analysis we shall neglect temperature effect as well as those caused by transistor replacement, and assume that the β of the transistor remains constant at 50. Let us first analyze the circuit for dc operation: The quiescent base current is:

$$I_{B_Q} = \frac{V_{CC}}{R_B} = \frac{20 \text{ V}}{1 \text{ M}\Omega} = 20 \text{ }\mu\text{A}$$

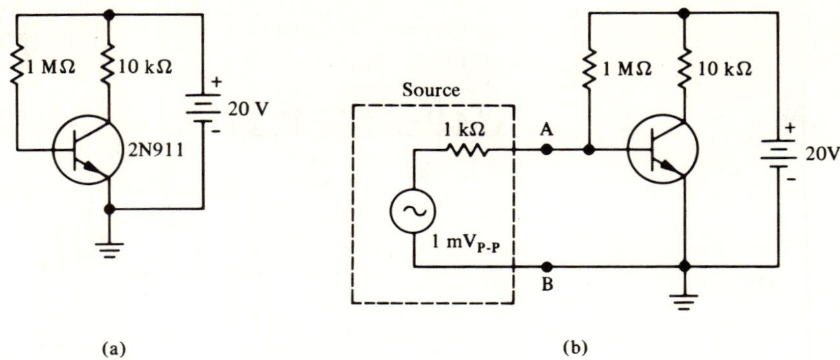

Figure 6.1

We know that the quiescent collector current is given by:

$$I_{Cq} = \beta I_{Bq} + I_{CEO}$$

Assuming that the value of I_{CEO} at room temperature is approximately equal to zero, the quiescent collector current is:

$$I_{Cq} \simeq \beta I_{Bq} = 50\,(20\ \mu A) = 1\ mA$$

The saturation collector current is:

$$I_{Cs} = \frac{V_{CC}}{R_L} = \frac{20\ V}{10\ k\Omega} = 2\ mA$$

Therefore the Q point is in the center of the load line.

Now suppose we should like to connect the ac source shown within the dotted box in Fig. 6.1(b) to the amplifier; i.e., we should like to amplify the small current and voltage from the source. One method is to connect the source *directly* to the input of the amplifier; in other words, the two terminals of the source (terminals A and B in Fig. 6.1(b)) are connected directly to the base and emitter terminals of the transistor, as shown in Fig. 6.1(b). This is known as *direct coupling, or dc coupling*, since the source terminals are connected directly to the amplifier input. The term "couple" means the same as connect. Note that the source is a 1.0 mV peak-to-peak sine wave voltage with an internal resistance of 1 kΩ. This can represent some type of microphone whose voltage is too small to be fed directly to a speaker to obtain a reasonable amount of sound; the voltage must first be amplified. At this point, we actually are not interested in the exact physical nature of the source; all that we are concerned with is that the source consists of a 1.0 mV voltage in series with a 1 kΩ resistance. From this point on, unless otherwise specified, all ac voltages and currents specified are *peak-to-peak values*.

Now let us proceed to the analysis of the circuit of Fig. 6.1(b). A portion of the circuit of Fig. 6.1(b) is re-drawn in Fig. 6.2(a). Note that the circuit is drawn with respect to the base and emitter terminals of the transistor; the collector circuit has been omitted from the diagram. Figure 6.2(a) shows that *both an* ac (1.0 mV) *and a* dc (20 V) are applied to the base and emitter terminals of the transistor. A little thought will reveal that the circuit of Fig. 6.2(a) can be re-drawn as shown in Fig. 6.2(b); the circuits of Figs. 6.2(a) and (b) are identical.

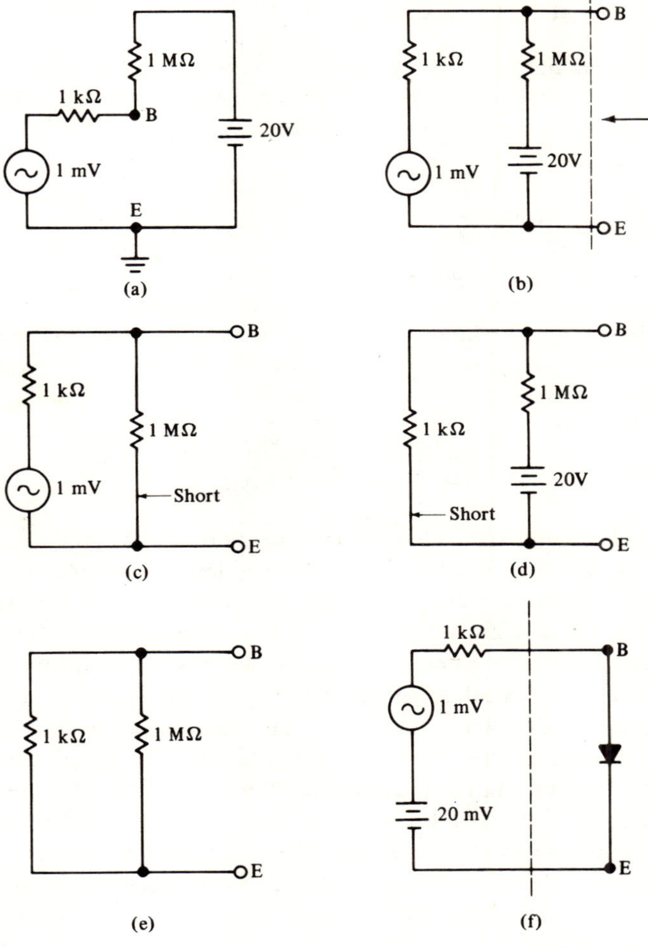

Figure 6.2

To analyze the circuit we must replace that portion of the circuit of Fig. 6.2(b) to the *left* of terminals B and E by its Thevenin equivalent. To do this we must first find the Thevenin open circuit voltage, V_{TH}. Since there are *two* sources (an ac and a dc) we can use the *superposition theorem* to compute V_{TH}. This states that we first *short* one of the sources, say the 20 V battery, and find the voltage across terminals B and E *due to the 1.0 mV ac voltage alone;* then we short the 1.0 mV voltage and find the voltage across terminals B and E *due to the 20 V battery alone.* The *sum* of these two voltages is the open circuit Thevenin voltage, V_{TH}.

Figure 6.2(c) shows the circuit with the 20 V battery shorted. The voltage across terminals B and E due to the 1.0 mV ac voltage (voltage across the 1 MΩ resistor) can be found by the voltage divider as follows:

$$V_{\mathrm{TH}} = \frac{(1 \text{ mV})(1 \text{ M}\Omega)}{1 \text{ M}\Omega + 1 \text{ k}\Omega} \simeq \frac{1 \text{ mV } (1 \text{ M}\Omega)}{1 \text{ M}\Omega} = 1.0 \text{ mV (ac)}$$

Figure 6.2(d) shows the circuit with the 1.0 mV ac voltage shorted. The voltage across terminals B and E due to the 20 V battery (voltage across the 1 kΩ resistor) can also be found by the voltage divider as follows:

$$V_{\mathrm{TH}} = \frac{20 \text{ V } (1 \text{ k}\Omega)}{1 \text{ M}\Omega + 1 \text{ k}\Omega} \simeq \frac{20 \text{ V } (1 \text{ k}\Omega)}{1 \text{ M}\Omega} = 0.02 \text{ V (dc)}$$

In other words, V_{TH} is a *complex waveform whose dc level* is 0.02 V (20 mV) and *whose ac portion* is a 1.0 mV sine wave. Figure 6.2(e) shows the circuit for computing the Thevenin impedance, R_{TH}; note that all sources are shorted. The Thevenin impedance, R_{TH}, is equal to the parallel combination of the 1 kΩ and the 1 MΩ resistances, or:

$$R_{\mathrm{TH}} = \frac{(1 \text{ k}\Omega)(1 \text{ M}\Omega)}{1 \text{ k}\Omega + 1 \text{ M}\Omega} \simeq \frac{(1 \text{ k}\Omega)(1 \text{ M}\Omega)}{1 \text{ M}\Omega} = 1 \text{ k}\Omega$$

Therefore, the Thevenin equivalent of that portion of the circuit to the left of terminals B and E in Fig. 6.2(b) is shown to the left of the same two terminals in Fig. 6.2(f). Note that the Thevenin equivalent consists of *three* units: a 20 mV *dc* voltage, a 1.0 mV *ac* voltage (the two combine to comprise a complex wave), and a 1 kΩ resistance. The transistor between the base and emitter terminals is represented by a diode; recall from chapter 3 that since this is an *npn* transistor, the diode is directed as shown in Fig. 6.2(f). The polarity of the 20 mV dc voltage indicates that the base-emitter diode is forward-biased. Thevenin's theorem states that, as far as the input terminals (base and emitter) of the transistor are concerned, it makes no difference whether we wire the circuit as in Fig. 6.2(a) or as in Fig. 6.2(f); both circuits are equivalent at the left of terminals B and E.

Figure 6.3(a) shows the input characteristic of the transistor (characteristic of the base-emitter diode) in the forward direction. Since the 2N911 is a silicon transistor, heavy conduction begins at approximately 0.6 V of forward bias, as shown. The shaded area of Fig. 6.3(a) is that portion of the graph where the input characteristic is *flat;* i.e., *below* 0.6 V, very little base current flows through the transistor. Figure 6.3(b) shows the shaded portion of Fig. 6.2(a) on a *vastly expanded* scale; note that only the flat portion (almost horizontal) and *not* the steep portion of the curve is shown in Fig. 6.3(b). Now let us neglect for a moment the 1 mV ac portion of the complex wave applied to the input terminals; thus the circuit of Fig. 6.2(f) consists of just a 20 mV dc voltage in series with a 1 kΩ resistor. A load line (A–B) is drawn on the input characteristic of Fig. 6.3(b), intersecting the v_{BE}-axis at 20 mV and the i_{B}-axis at 20 mV/1 kΩ, or 20 μA. The load line intersects the *flat* portion of the characteristic at the Q point, Q.

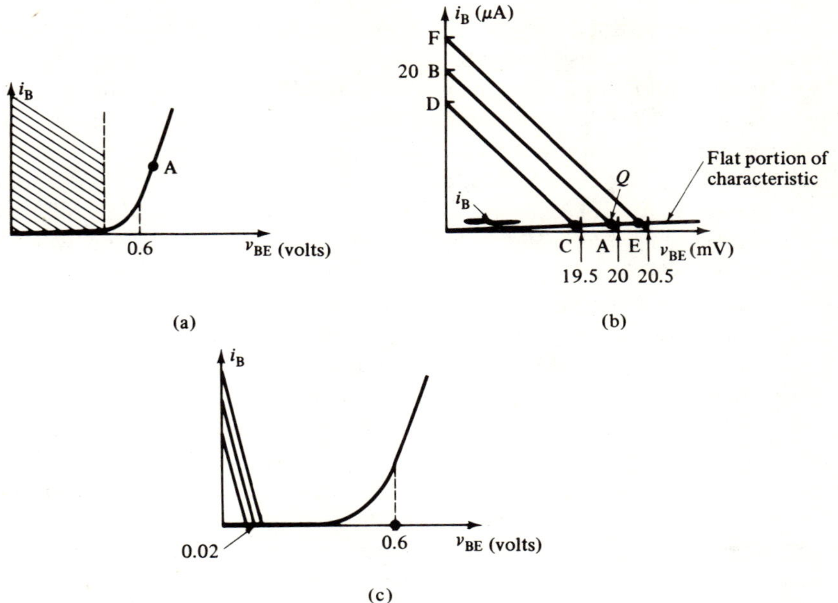

Figure 6.3

Now notice what this means. The transistor is biased in the *flat* portion of the input characteristic; recall from chapter 4 that it is important that the transistor be biased in the *steep* portion of the curve, such as at point A in Fig. 6.3(a). The value of quiescent base current computed for the transistor at the beginning of this section was 20 μA, *before the ac source was connected*. Now that the source *is* connected, the quiescent base current is *much less* than 20 μA, as can be seen from the location of the Q point in Fig. 6.3(b). If the base current is very small, the collector current is also small, and the collector-to-emitter voltage is large (close to the value of V_{CC}, 20 V); in other words, the direct connection of the ac source to the transistor input terminals biases the transistor close to *cut-off*. It was explained in Sec. 4-4 that biasing the transistor at cut-off can result in severe distortion of the waveform.

The reason for this phenomenon is that most of the dc current flowing through the 1 MΩ bias resistor in Fig. 6.1(b) flows, after splitting at the base, *through the source instead of into the base of the transistor*. The dc current finds a path through the relatively small 1 kΩ internal source resistance, and therefore much less current flows into the base with the result that the quiescent base current is very small, as shown in Fig. 6.3(b). It can be shown that the situation is improved markedly if the internal resistance of the source is much larger than 1 kΩ; in this case less current is diverted through the internal source resistance, and consequently, more current flows to the base. However, the value of the internal resistance of the source is usually beyond the control of the designer. If a large resistance is *intentionally* inserted in series with the source terminals to improve the situation pictured in Fig. 6.3(b), the amount of ac current fed to the base of the transistor will *decrease* considerably, with the result that there will be a large reduction in gain.

Returning now to a consideration of the 1 mV ac voltage in series with the 20 mV dc voltage and the 1 kΩ resistor of Fig. 6.2(f), we see that the ac and dc voltages combine to produce a complex wave. Since the ac voltage is 1 mV peak-to-peak, the complex voltage varies from 0.5 mV *above* the 20 mV dc (20.5 mV) to 0.5 mV *below* the 20 mV dc (19.5 mV). This has the effect of causing the *load line to move in synchronism with the ac voltage* as shown in Fig. 6.3(b). When the value of the complex wave is 19.5 mV, the load line is line C–D, and when the complex voltage is 20.5 mV, the load line moves to line E–F. From chapter 1 you may recall that if the voltage changes (from 19.5 mV to 20.5 mV) but the resistance remains the same (1 kΩ), the load line moves parallel to itself; note in Fig. 6.3(b) that the three load lines (A–B, C–D, and E–F) are all parallel to each other. The diagram in Fig. 6.3(b) is obviously *not* drawn to scale.

Now as the load line varies, the base current also varies. However, since we are operating on the flat portion of the input characteristic, the *total variation in the base current is very small*, as can be seen from Fig. 6.3(b). This is the same situation depicted in Sec. 4-4, Fig. 4.8, when the transistor is biased at cut-off. Figure 6.3(c) shows the three load lines of Fig. 6.3(b) drawn on the input characteristic to the same scale as Fig. 6.3(a); this figure illustrates again that the transistor is biased on the flat portion of the input characteristic.

We conclude, then, that we *cannot* couple the source directly to the transistor input terminals as shown in Fig. 6.1(b) for *two* reasons: The first is that the small internal resistance of the source causes a *change* in the dc conditions of the fixed bias circuit; when the source is connected, the Q point moves from the center of the load line (as it was in Fig. 6.1(a) *before* the source was connected) to cut-off. The second reason is that, since the transistor is biased in the flat portion of the curve, the input resistance is very *large*, and therefore practically *no* ac current is delivered to the base; the transistor amplifier provides little, if any, gain.

The next question to ask is: How *do* we couple the source to the amplifier? There are actually *two* methods which can be used. One is coupling by means of a capacitor and the other is transformer coupling; we shall study transformer coupling in chapter 7. The basic circuit using *capacitor coupling* is shown in Fig. 6.4. Notice that a 10 μF capacitor is connected between the source and the base. Let us now see how this circuit is an improvement over the direct coupling

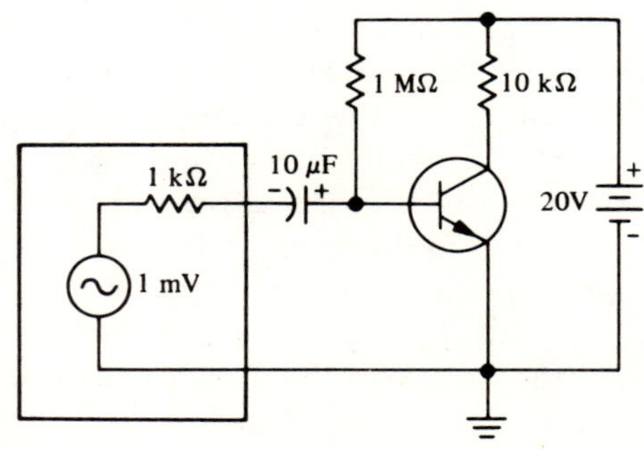

Figure 6.4

method. The first concept, which can be seen immediately, is that *all* of the dc current flowing through the 1 MΩ bias resistor *must* flow into the base of the transistor; *no portion* of this current can flow through the source since the capacitor acts like an *open circuit* to dc current (it has infinite capacitive reactance). Therefore, we say that the capacitor *isolates*, or separates, the source from the amplifier *as far as dc conditions are concerned.*

Now let us analyze the circuit of Fig. 6.4. Figure 6.5(a) shows the circuit of

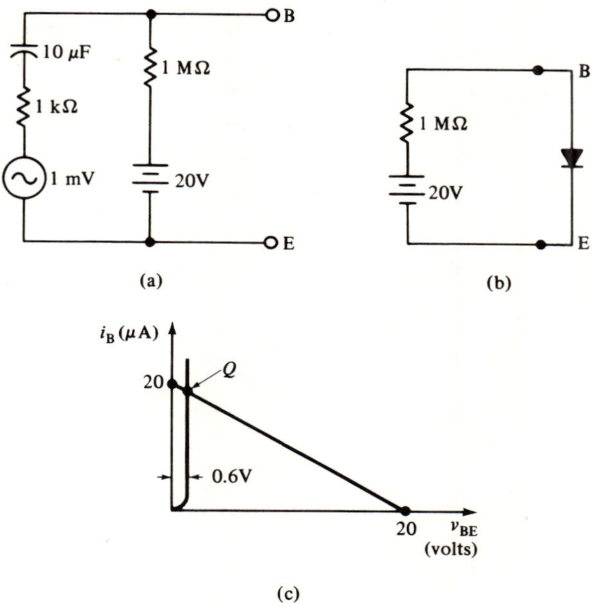

Figure 6.5

Fig. 6.4 re-drawn with respect to the base and emitter terminals, similar to the manner in which the circuit of Fig. 6.1(b) was redrawn in Fig. 6.2(b). Proceeding with the superposition theorem as before, we must first find the Thevenin open circuit voltage, V_{TH}, due to the 20 V battery. We know that the 10 μF capacitor will conduct no dc current; thus no dc current will flow through the *left*-hand branch in Fig. 6.5(a) and the entire branch can be *omitted* for this computation as shown in Fig. 6.5(b). All that remains now is the 20 V battery in series with the 1 MΩ resistor. The load line is drawn on the input characteristic in Fig. 6.5(c), and the Q point is indicated. Note in Fig. 6.5(c), that the Q point is in the *steep* portion of the input characteristic, and that the quiescent base current is approximately equal to 20 μA. In other words, the connection of the source to the amplifier through the capacitor has *no* effect on the dc conditions of the amplifier; contrast this with Fig. 6.3(b) when direct coupling was used.

The next step is to find the Thevenin open circuit voltage due to the 1.0 mV ac source voltage; to do this we must short the 20 V battery, as shown in Fig. 6.6(a). Now under ac conditions, the capacitive reactance is no longer infinite; i.e., it has some finite value. Let us say that the frequency of the 1.0 mV source voltage is 1.0 kHz. Computing the reactance of the 10 μF capacitor, we get:

$$X_C = \frac{1}{2\pi f C} = \frac{.159}{10^3 \times 10 \times 10^{-6}} = 15.9 \ \Omega$$

If the 1 kΩ internal source resistance is added vectorially to the 15.9 Ω capacitive reactance to find the impedance of the left-hand branch in Fig. 6.6(a), we find:

$$Z = \sqrt{(1000)^2 + (15.9)^2} = \sqrt{1,000,253} \simeq 1000 \ \Omega$$

In other words, since the resistance is so much larger than the reactance, we can neglect completely the effect of the capacitive reactance. Thus the circuit of

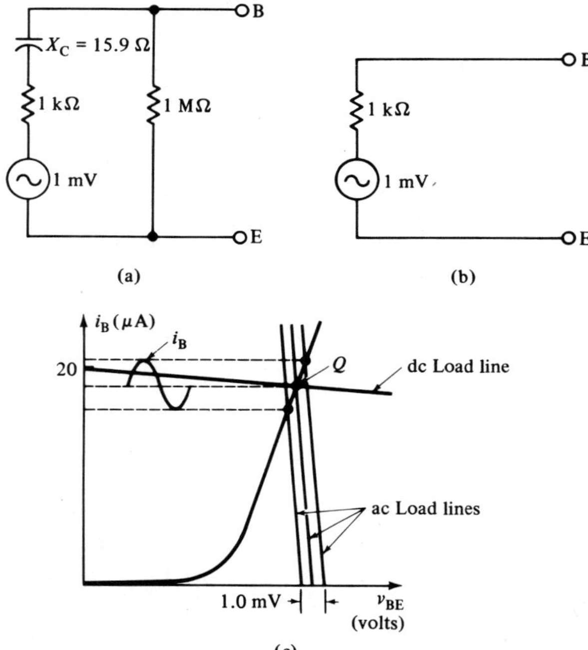

Figure 6.6

Fig. 6.6(a) reduces to that of Fig. 6.2(c). The Thevenin open circuit voltage due to the ac source is again, by the voltage divider:

$$\frac{(1 \text{ mV})(1 \text{ M}\Omega)}{1 \text{ M}\Omega + 1 \text{ k}\Omega} \simeq \frac{(1 \text{ mV}) \ 1 \text{ M}\Omega}{1 \text{ M}\Omega} = 1 \text{ mV}$$

The Thevenin impedance for ac conditions is:

$$R_{TH} = 1 \text{ k}\Omega \ || \ 1 \text{ M}\Omega \simeq 1 \text{ k}\Omega$$

Thus the Thevenin equivalent circuit for ac conditions consists of the 1.0 mV source voltage in series with a 1 kΩ resistance as shown in Fig. 6.6(b).

Figure 6.6(c) shows the input characteristic of Fig. 6.5(c) on an expanded scale. *The dc load line* is the load line drawn *for dc conditions* using the 20 V battery and the 1 MΩ resistor; this establishes the Q point and is the same load line shown in Fig. 6.5(c). Now we just saw that *under ac conditions,* we can consider the input terminals of the amplifier to be fed by a 1.0 mV ac voltage in series with a 1 kΩ resistor, as shown in Fig. 6.6(b). *Three ac load lines* are shown in Fig. 6.6(c). Note first that all three ac load lines have a steeper slope than the dc load line. From chapter 1 you may recall that the slope of the load line is equal to the reciprocal of the resistance:

$$m = -\frac{1}{R}$$

In other words, a *large* resistance would result in a load line with a *small* slope. Therefore, since the dc load line was drawn using a 1 MΩ resistance, it has a much smaller slope than the ac load lines which were drawn using a 1 kΩ resistance. All three ac load lines are parallel to each other since they were all drawn using a 1 kΩ resistance. The second point to note is that there are *three* ac load lines; these represent the maximum, zero, and minimum values of the 1.0 mV ac voltage (note in Fig. 6.6(c) that the horizontal distance between the two outer ac load lines is 1.0 mV). These ac load lines (there are actually an infinite number, one for each instant of time) can be viewed as a *single* ac load line moving in synchronism with the 1.0 mV ac voltage.

Thus there are *two* load lines; there is a dc load line drawn using the V_{CC} voltage and the R_B resistance (20 V and 1 MΩ, respectively, in this case) which establishes the Q point, and an ac load line which is drawn using *the ac Thevenin open circuit voltage and impedance* (1.0 mV and 1 kΩ, respectively, in this case). Note in Fig. 6.6(c) that the variation in the position of the ac load line produces a variation in the base current at the steep portion of the input characteristic. Since the input resistance is small in this portion of the curve, a reasonable amount of ac base current is produced, as illustrated by the ac base current waveform shown at the left of Fig. 6.6(c).

Figure 6.7(a) illustrates the effect of the location of the Q point on the base current waveform. Ac load lines A–B and C–D are located at the *flat* portion of the input characteristic (where r_I is large), and thus the base current waveform labeled "1" results. Ac load lines E–F and G–H (same separation and slope as A–B and C–D) are located at the *steep* portion of the curve (where r_I is small), and thus the base current waveform labeled "2" results. The fact that waveform "2" is significantly larger than waveform "1" indicates the importance of biasing at the steep portion of the curve. Ac load lines C–D and E–F produce waveform "1" in Fig. 6.7(b). Ac load lines A–B and G–H produce waveform "2." A–B and G–H have a *wider* separation than C–D and E–F, which means that there is a *larger* ac voltage associated with them. All other things being equal, a larger ac *voltage* will produce a larger ac *current,* and that explains why waveform 2 is larger than waveform 1 in Fig. 6.7(b).

The discussion of the preceding paragraphs has shown that capacitor coupling has eliminated the two major problems associated with direct coupling. First, the capacitor prevents any dc current from flowing through the source (we say that it "blocks" the dc), and therefore the Q point is determined by the transistor, the

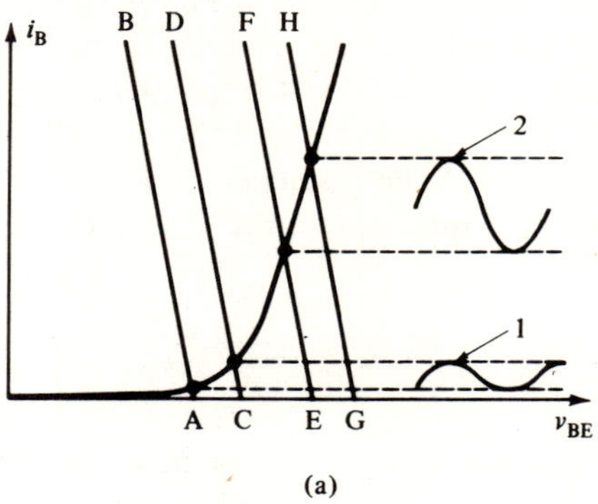

(a)

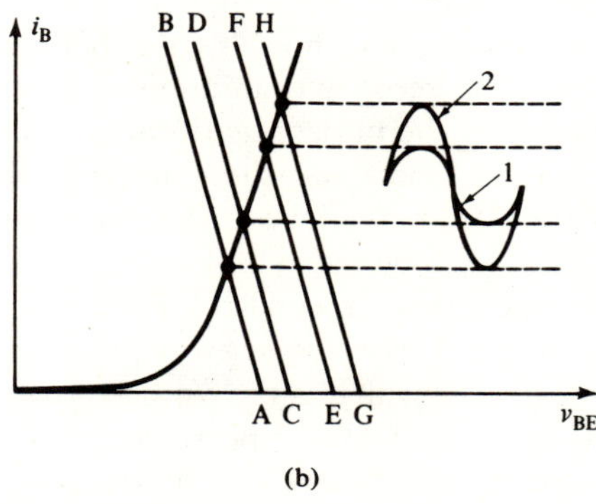

(b)

Figure 6.7

values of the resistors, and the value of V_{CC}, and is *unaffected* by the source. Second, the Q point is at the steep portion of the input characteristic where the value of r_I is small; thus a reasonable amount of ac base current will flow (waveform 2 in Fig. 6.7(a)). The 10 μF capacitor in Fig. 6.4 is called the *coupling capacitor, C_C*; the coupling capacitor acts like an *open circuit* to dc current, but has a *small* reactance (15.9 Ω, in this case) to ac currents.

We could proceed with the ac analysis of the amplifier using the graphical technique illustrated in Figs. 6.6(c) and 6.7; this, however, is tedious and involves possession of the transistor input characteristic, as well as construction of both dc and ac load lines. These graphical techniques were presented to give the reader an insight into the mechanism which governs capacitor-coupled circuits; we have more to say about ac load lines later. In the following section, we turn to a method of analyzing the ac operation of the circuit of Fig. 6.4 using the small signal ac equivalent circuit.

6-2   ac ANALYSIS OF THE CAPACITOR-COUPLED AMPLIFIER CIRCUIT

To analyze the capacitor-coupled amplifier circuit for ac operation, we must first draw the ac equivalent circuit of the amplifier shown in Fig. 6.4. The first step in constructing the ac circuit is to draw the ac source consisting of the 1.0 mV voltage in series with the 1 kΩ internal resistance. This source is connected through the 10 μF coupling capacitor to the base of the transistor; i.e., the coupling capacitor is effectively in *series* with the internal resistance of the source, as shown in Fig. 6.8(a). This brings us to the base of the transistor. We know from chapter 4 that, between the base and emitter terminals, the transistor can be represented by a resistance, r_I; this is drawn between terminals B and E in Fig. 6.8(a). Between

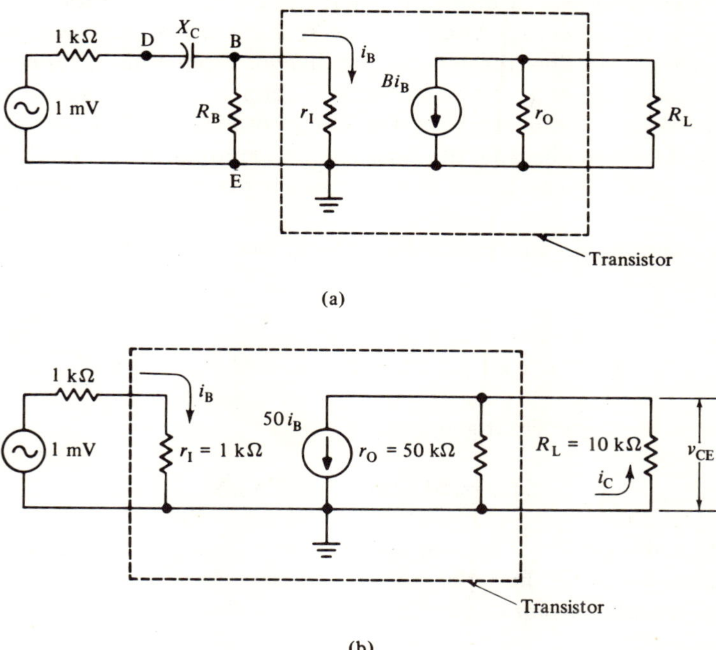

(a)

(b)

Figure 6.8

the collector and emitter terminals, the transistor is represented by a current generator, βi_B (where i_B is the ac base current—the current flowing in the resistance, r_I), in parallel with a resistance, r_O; see Fig. 6.8(a). Now, in the original circuit of Fig. 6.4, there is a bias resistor (1 MΩ, in this case) connected between the base and the positive terminal of V_{CC}. We also know from chapter 4 that all dc sources are excluded from (shorted out of) the ac circuit. If V_{CC} is shorted, the positive terminal can be considered at ground potential; we say that the positive terminal of the battery *is at ac ground potential*. Therefore, the bias resistor is connected between the *base and ground* (emitter) in the ac equivalent circuit of Fig. 6.8(a);

this puts R_B effectively in *parallel* with r_I. Similarly, the load resistor, R_L, is connected between the collector and the positive terminal of V_{CC}. Since V_{CC} is shorted in the ac circuit, R_L is connected between *collector and ground* in Fig. 6.8(a); it can be seen that R_L is now effectively in *parallel* with r_O.

The circuit of Fig. 6.8(a) now represents the capacitor-coupled amplifier circuit of Fig. 6.4 as far as ac operation is concerned; it should be emphasized again that *this circuit is valid for only the ac analysis* of the amplifier. The portion of the circuit *within* the dotted box of Fig. 6.8(a) (r_I, βi_B, and r_O) is part of the transistor itself; everything *outside* the box (the source, R_B, and R_L) is part of the amplifier, but *not* part of the transistor.

Now let us analyze the circuit of Fig. 6.8(a). First, it was found in the previous section that the reactance of the 10 μF coupling capacitor at 1.0 kHz was 15.9 Ω; we saw that we can neglect this in comparison with the 1 kΩ internal source resistance. Second, notice in Fig. 6.8(a) that the 1 MΩ bias resistor is in parallel with the input resistance, r_I. Assuming that r_I is 1 kΩ, it can be seen that most of the ac current will flow through the small 1 kΩ input resistance, rather than through the much larger 1 MΩ bias resistor. Therefore, we can neglect the effect of the bias resistor in the ac analysis. The circuit of Fig. 6.8(a) is re-drawn in Fig. 6.8(b) with the capacitor and the bias resistor omitted. Using small letters to represent the peak-to-peak values of the ac quantities, we now proceed with the analysis of the circuit.

The base current, i_B, can be found by Ohm's and Kirchhoff's laws as follows, from Fig. 6.8(b):

$$i_B = \frac{1 \text{ mV}}{1 \text{ k}\Omega + 1 \text{ k}\Omega} = \frac{1 \times 10^{-3}}{2 \times 10^3} = 0.5 \times 10^{-6} = 0.5 \ \mu\text{A}$$

The current in the βi_B current generator is:

$$\beta i_B = 50 \ (0.5 \ \mu\text{A}) = 25 \ \mu\text{A}$$

We know that not all of this current flows in the collector; some of it recirculates in the resistance, r_O. To find the collector current (current in the 10 kΩ load resistor), we use the current divider expression as follows:

$$i_C = \frac{(\beta i_B)r_O}{r_O + R_L} = \frac{(25 \ \mu\text{A})(50 \text{ k}\Omega)}{50 \text{ k}\Omega + 10 \text{ k}\Omega} = 20.8 \ \mu\text{A}$$

This means that 25 μA $-$ 20.8 μA, or 4.2 μA are lost as re-circulation current through r_O (50 kΩ). Finally, the output voltage (voltage across the 10 kΩ load resistor, or voltage across the collector and emitter terminals) is found by Ohm's law as follows:

$$v_{CE} = i_C R_L = (20.8 \ \mu\text{A})(10 \text{ k}\Omega) = 0.208 \text{ V} = 208 \text{ mV}$$

Notice how simple it is to analyze the ac behavior of the circuit once the transistor parameters (β, r_I, and r_O) are known. The results indicate that both the voltage and the current have been amplified; the voltage has been increased from 1.0 mV to 208 mV, and the current from 0.5 μA to 20.8 μA.

This analysis can proceed in a slightly different manner; the ac *voltage* source consisting of the 1.0 mV voltage in series with the 1 kΩ resistance can be converted to a *current* source. Following the conversion rules of chapter 4, the voltage source of Fig. 6.8 (1.0 mV in series with 1 kΩ) is converted to the current source shown in Fig. 6.9 (1.0 μA in parallel with 1 kΩ). Now the analysis of the circuit of Fig. 6.9

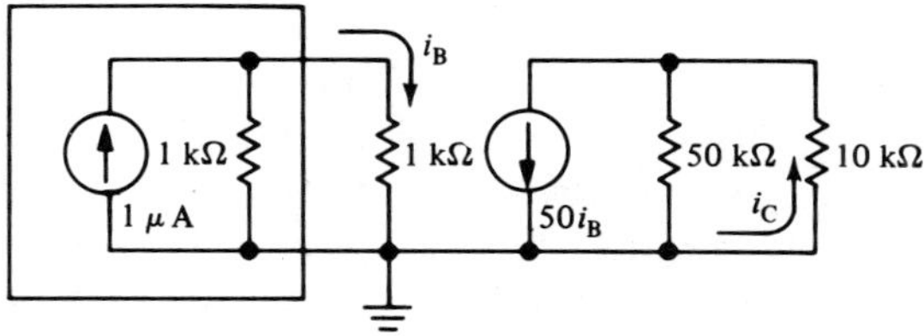

Figure 6.9

proceeds as follows: The current source is located within the box shown in Fig. 6.9. The 1.0 μA ac current generator is in parallel with two 1 kΩ resistances; one of these resistances is the internal resistance of the source and the other is the input resistance of the transistor. The ac base current, i_B, can be computed by the current divider. However, in this case it is obvious that the 1.0 μA *splits evenly* between the two 1 kΩ resistances, so that the ac base current is 0.5 μA; this is the same value found using the voltage source configuration of Fig. 6.8(b). From this point, the analysis proceeds as before, and the output voltage is again found to be 208 mV. The above discussion illustrates that either voltage *or* current source representation can be used in analyzing the ac behavior of the amplifier. We see in future sections that most amplifier analyses lend themselves more readily to current source representations.

It is important to bear in mind that the currents and voltages associated with the amplifier are *complex;* i.e., they consist of an ac *and* a dc level. We now show the complete waveforms of all of the currents and voltages. The amplifier circuit of Fig. 6.4 is re-drawn in Fig. 6.10 for convenience. Let us consider first the waveform of the voltage, v_{DE}, at the output terminals of the ac source; note that

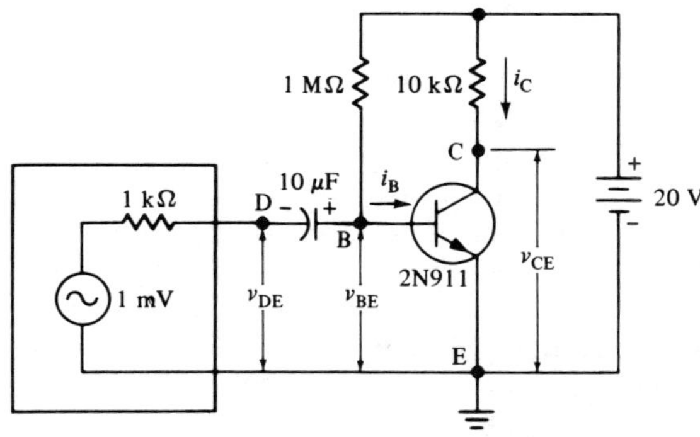

Figure 6.10

terminal D is just to the *left* of the 10 μF coupling capacitor. We saw in the previous ac analysis that the ac current flowing out of the source was 0.5 μA. This current must flow through the 1 kΩ internal source resistance; therefore the voltage drop across this internal resistance is, by Ohm's law, equal to:

$$(0.5 \ \mu A)(1 \ k\Omega) = (0.5 \times 10^{-6} \ A)(1 \times 10^{3} \ \Omega) = 0.5 \times 10^{-3} \ V$$
$$= 0.5 \ mV$$

This 0.5 mV drop across the internal resistance *subtracts* from the original 1.0 mV voltage leaving 1.0 mV $-$ 0.5 mV, or 0.5 mV at the output terminals of the source, terminals D and E. Recall from basic circuits theory that v_{DE} is the only voltage associated with the source *which can actually be measured* since the portion of the circuit within the box in Fig. 6.10 is *inside* the source; v_{DE} will be equal to 1.0 mV *only when the output terminals of the source are open-circuited* (no voltage drop across the internal resistance). We know that the coupling capacitor blocks any dc current which tries to flow through the source from the amplifier; thus there are *no dc currents or voltages* in the source. The waveform of the voltage, v_{DE}, appears in Fig. 6.11(a); note that it is a 0.5 mV peak-to-peak ac riding on a *zero-volt dc level* (a pure ac sinusoid).

Next consider the waveform of v_{BE} (terminal B is the base of the transistor, to the *right* of the capacitor). The ac portion of this voltage at terminals B and E is equal, by Kirchhoff's law, to v_{DE} *minus the ac drop* across the capacitor. We saw that the capacitor's reactance at 1.0 kHz was 15.9 Ω; this was small enough to be neglected in comparison with the other impedances in the circuit. Therefore, if we assume that there is practically *no ac voltage* across the coupling capacitor, we can say that the ac portion of v_{BE} is equal to the ac portion of v_{DE}, i.e., 0.5 mV. To check our assumption that there is a negligible ac drop across the capacitor, we can compute this voltage drop since we know the value of the ac current in the capacitor (0.5 μA) and the reactance of the capacitor (15.9 Ω); this ac drop is:

$$(0.5 \ \mu A)(15.9 \ \Omega) = (0.5 \times 10^{-6} \ A)(15.9 \ \Omega) \simeq 8 \times 10^{-6} \ V$$
$$= 0.008 \ mV$$

Since the 0.008 mV ac drop across the capacitor is negligible in comparison with 0.5 mV, our assumption was correct and we shall assume that *the ac drop* across the capacitor is approximately *zero;* therefore *the ac portion* of v_{BE} is approximately equal to *the ac portion* of v_{DE}. However, there is also *a dc drop* across the base and emitter terminals of the transistor; this is the 0.6 V forward drop. The complete waveform of v_{BE} is shown in Fig. 6.11(b); note that this is composed of a 0.5 mV peak-to-peak ac sine wave (same as v_{DE}) riding on a *0.6 V dc level.*

Figure 6.11(c) shows the waveform of v_{c_0}, the voltage across the coupling capacitor. As we have just shown, there is almost no ac voltage across the capacitor; however, there is a 0.6 V dc level. The dc voltage with respect to ground at the left-hand plate of the capacitor (terminal D in Fig. 6.10) is zero volts while that at the right-hand plate (terminal B) is 0.6 V; therefore the voltage at terminal B with

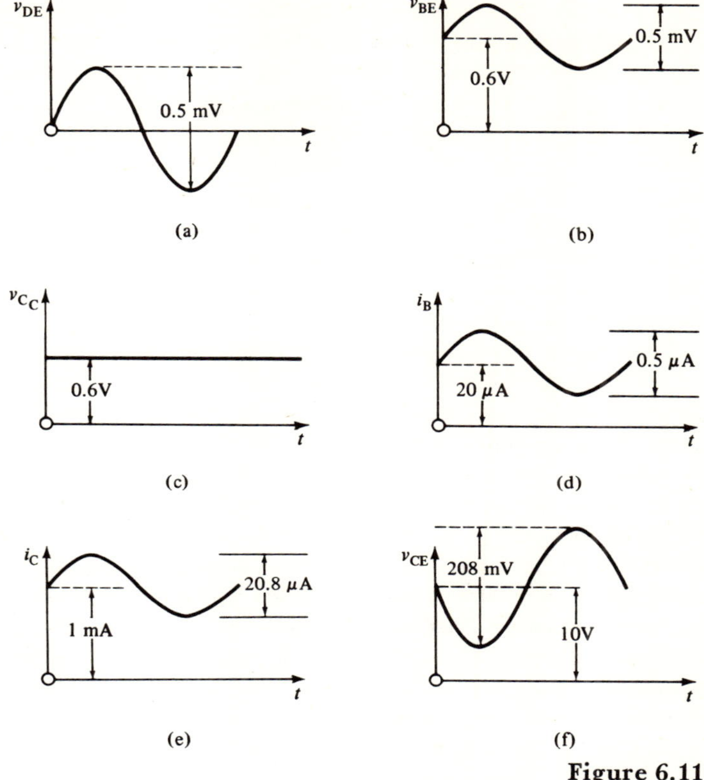

Figure 6.11

respect to D (v_{BD}) is 0.6 V dc. In other words, terminal B is at a *higher dc potential* than terminal D; this explains the polarity of the 10 μF polarized capacitor.

Figures 6.11(a) through (c) illustrate further the action of the coupling capacitor. It takes the pure ac voltage from the source (v_{DE}) and *raises it up by the 0.6 V dc level*, so that the voltage which appears at the base of the transistor is the complex waveform shown in Fig. 6.11(b).

Throughout this entire analysis we have assumed that the capacitive reactance of the coupling capacitor is very small. We know that X_C is given as follows:

$$X_C = \frac{1}{2\pi f C}$$

We can see from the equation that, if we desire that X_C be very *small*, we should like both the frequency and the value of the capacitance to be *large*. We see in chapter 8 that the frequency of the input source voltage is beyond the control of the designer; thus all we can do to make X_C small is to make the capacitance large. A large value of capacitance, however, usually means a large physical size. The primary advantage of using transistors in any electronic circuit (as opposed to vacuum tubes) is their inherently small size and weight; therefore, the inclusion of a physically large component (such as large capacitor) in the circuit will partly counteract the advantage gained by using transistors. Electrolytic and tantalytic capacitors are polarized units which have relatively large values of capacitance, yet still can be manufactured in small packages. Thus, care must be taken when connecting these polarized capacitors into the amplifier circuit. The terminal marked "plus" is to be connected so that the dc voltage at this terminal is *positive* with respect to the voltage at the other terminal. Note in Fig. 6.10 that the *positive* terminal of the 10 μF capacitor is connected to the *base* of the transistor (0.6 V dc more positive than ground) and the negative terminal is connected to terminal D (at dc ground potential); if the transistor were *pnp* and using a negative V_{CC}, the capacitor would have to be reversed.

Figure 6.11(d) shows the complete waveform of the base current; note that it consists of a 0.5 μA peak-to-peak ac riding on a 20 μA dc level. Observation of Fig. 6.10 reveals some information concerning the base current waveform. The dc portion of the current (20 μA) is supplied to the base through the 1 MΩ bias resistor, while the 0.5 μA ac enters from the source through the coupling capacitor; these two portions combine to produce the waveform of Fig. 6.11(d). The waveforms of i_C and v_{CE} are shown in Figs. 6.11(e) and (f), respectively. It should be obvious from observation of the waveforms in Fig. 6.11 that they are *not* drawn to scale. Knowing how to construct and interpret these waveforms is important for *two* reasons: First, as we shall soon see, a knowledge of the dc values gives us an indication of how close we are either to saturation or to cut-off; this in turn tells us how large an ac sinusoid the amplifier can accommodate without distortion. The second reason concerns the waveforms shown in Fig. 6.11. The *voltage* waveforms (Figs. 6.11(a), (b), (c), and (f)) are precisely those waveforms which would be observed if an *oscilloscope* were connected across the appropriate terminals. Since the oscilloscope is the single most important tool used in the design of electronic circuits, it is important that the reader be able to recognize as well as predict the appearance of the various waveforms.

6-3   THE CASCADE CONNECTION

We saw in the last section that the output voltage of the amplifier was 208 mV and the current was 20.8 μA; the ac output power is:

$$P_O = \frac{v_{OP\text{-}P}\, i_{OP\text{-}P}}{8} = \frac{(208\text{ mV})(20.8\ \mu\text{A})}{8} = 0.54 \times 10^{-6}\text{ W}$$

$$= 0.54\ \mu\text{W}$$

As we see in later chapters, this amplified power is usually delivered to some type of load, such as a loudspeaker. Now suppose that the power output computed above is not large enough for the particular application at hand; therefore, we should increase the ac power to some value greater than 0.54 μW, *using the same source as the input* (1.0 mV voltage with a 1 kΩ internal resistance). One solution is to obtain a transistor with a larger β, since (as we saw in chapter 4) the gain of the amplifier is dependent upon the β of the transistor. This poses a number of problems: First, there is a *definite limit* to the size of the β of a transistor which can be obtained with present manufacturing techniques. Second (as we see in chapter 7), the variation in the β due to temperature and transistor replacement causes a corresponding variation in the *gain* of the amplifier, which is undesirable; e.g., even if we use a transistor with a large β, the gain may drop at the lower temperatures due to the decrease in β at these temperatures.

The method used to increase further the ac power consists of connecting the *output* terminals (collector and emitter) of the amplifier of Figs. 6.4 and 6.10 to the *input* terminals (base and emitter) of a *second* amplifier. This situation is pictured in Fig. 6.12. Note that for simplicity, we have replaced the usual symbol

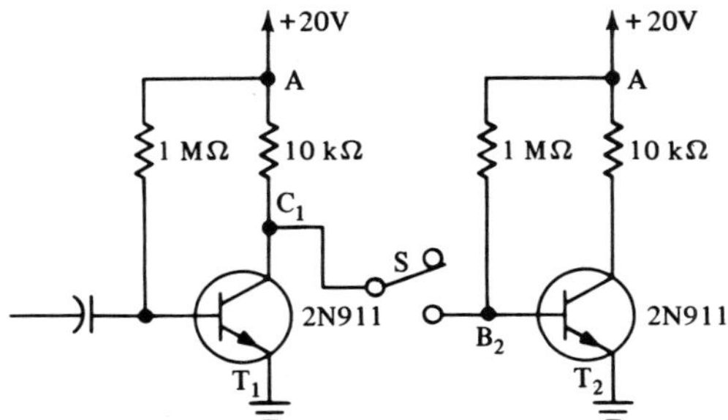

Figure 6.12

for the V_{CC} battery by an arrow with a "$+20$ V" sign next to it. This indicates that the terminals common to both the 1 MΩ bias resistor and the 10 kΩ load resistor (terminals A in Fig. 6.12) are connected to the *positive* terminal of the 20 V battery, and it is assumed that the *negative* terminal of the battery is connected to ground; we use this representation for the duration of the text. The reader should *not* get the impression that there are two separate 20 V batteries in Fig. 6.12; terminals A in the figure are connected *to each other* and also to the positive terminal of a *single* 20 V battery. Similarly, the two ground terminals (connected to the emitters) in Fig. 6.12 are connected together.

The object in Fig. 6.12 is to *transfer* the ac power (voltage and current) from the *output* of the first amplifier (at the left) to the *input* of the second. Since the output is obtained from the collector of the transistor, and the input is applied to the base, the logical approach is to connect the collector of the first transistor (C_1) to the base of the second (B_2). Let us first try to couple the output of the first transistor (T_1) to the input of the second (T_2), *directly;* i.e., we again attempt the direct coupling approach which was discussed earlier in this chapter when we coupled the source to the amplifier. The two amplifiers shown in Fig. 6.12 we assume to be identical, both transistors having a β of 50. Let us now analyze the circuit to see what happens to *the dc conditions* when the switch, S, in Fig. 6.12 is closed.

When the switch is closed, the circuit of Fig. 6.12 can be re-drawn as shown in Fig. 6.13(a). Note first that, once the switch is closed, C_2 is connected directly to

B_1; in other words, the collector of T_1 is at the *same potential* as the base of T_2. Second, it should be obvious from Fig. 6.13(a) that the 10 kΩ load resistor connected to C_1 is in *parallel* with the 1 MΩ bias resistor connected to B_2; since the 1 MΩ resistor is so much larger than 10 kΩ, we can neglect its effect on the circuit.

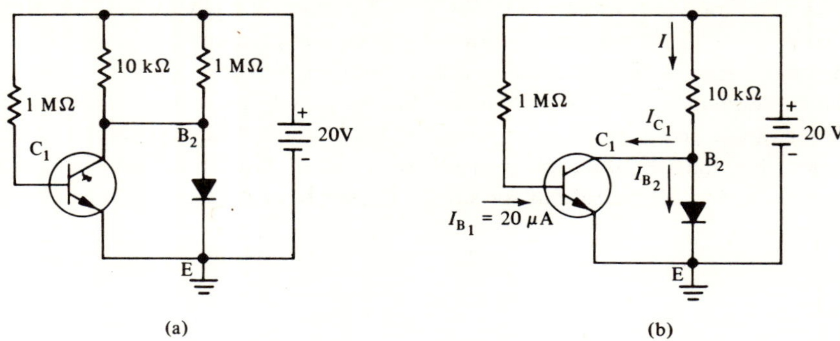

Figure 6.13

The third point to note is that the base-emitter junction of T_2 is represented by the diode shown in Fig. 6.13(a).

The circuit of Fig. 6.13(a) is again re-drawn in Fig. 6.13(b); note that the 1 MΩ resistor has been omitted. It can be seen from the circuit of Fig. 6.13(b) that the current, I (current in the 10 kΩ resistor), is equal to the sum of I_{C_1} (current in the collector of the first transistor) and I_{B_2} (current in the base of the second transistor). Now let us find the values of these various currents. We can begin by applying Kirchhoff's voltage law to the extreme right-hand loop of the circuit of Fig. 6.13(b). Summing the voltages around the loop, and setting equal to zero we get:

$$20 = (10 \text{ k}\Omega)(I) + V_{\text{B}_2\text{E}}$$

It can be seen from Fig. 6.13(b) that the base-emitter junction of T_2 is forward biased, and thus $V_{\text{BE}} = 0.6$ V. Neglecting the 0.6 V drop in comparison with the 20 V battery voltage, we say:

$$20 \text{ V} \simeq 10 \, I$$

The current in the 10 kΩ resistor is therefore found as follows:

$$I = \frac{20 \text{ V}}{10 \text{ k}\Omega} = 2 \text{ mA}$$

The base current in T_1 (I_{B_1}) is still 20 μA since the interconnection of the two amplifiers has no effect on the base circuit of T_1 (the voltage across the 1 MΩ bias resistor of T_1 is approximately 20 V, neglecting the 0.6 V base-to-emitter drop, and therefore the base current is approximately 20 μA). Now if the β of T_1 is 50, the collector current of T_1 (I_{C_1}) is found as follows:

$$I_{\text{C}_1} = \beta I_{\text{B}_1} = 50 \, (20 \, \mu\text{A}) = 1 \text{ mA}$$

Applying Kirchhoff's current law to the circuit, we see that:

$$I = I_{C_1} + I_{B_2}$$
$$I_{B_2} = I - I_{C_1}$$

Solving for I_{B_2}, we get:

$$I_{B_2} = I - I_{C_1} = 2 \text{ mA} - 1 \text{ mA} = 1 \text{ mA}$$

Now if the β of T_2 is also 50, the collector current of T_2 (I_{C_2}) is:

$$I_{C_2} = \beta I_{B_2} = 50 \ (1 \text{ mA}) = 50 \text{ mA}$$

However, the collector saturation current of the second amplifier circuit is:

$$I_{Cs} = \frac{V_{CC}}{R_L} = \frac{20 \text{ V}}{10 \text{ k}\Omega} = 2 \text{ mA}$$

Therefore, the collector current of T_2 cannot possibly be 50 mA, since the current cannot rise above 2 mA; in other words T_2 *is very far into saturation.* The collector-to-emitter voltage of T_2 (V_{CE_2}) is approximately 0.2 V; the collector-to-emitter voltage of T_1 (V_{CE_1}) must be the *same voltage as that across the base-emitter junction of T_2,* 0.6 V.

The situation pictured in Fig. 6.12, that of direct coupling the two fixed bias circuits together, is thus totally unworkable. The problem involved here is similar to that encountered when we attempted to direct couple the source to the amplifier in Sec. 6-1. In Fig. 6.12, *the direct connection of C_1 to B_2 caused major changes in the dc conditions of both amplifiers;* V_{CE_1} dropped from 10 V to 0.6 V, and V_{CE_2} dropped from 10 V to 0.2 V. In other words, the interconnection of the two amplifiers (closing of switch S in Fig. 6.12) resulted in an *interaction* between them such as to change the dc conditions. In particular, this interaction has caused T_2 to be biased far into saturation; recall from chapter 4 that an amplifier biased at saturation is *unable* to produce a sinusoidal output. Actually, there is an even more serious problem resulting from the direct coupling of amplifiers, namely, the problem of *drift.* Although drift causes major additional problems in the circuit of Fig. 6.12, we wait until later in this chapter to consider it in some detail. However, the fact that T_2 moves into saturation is reason enough for looking for an alternate method of coupling the amplifiers.

We return again to the capacitor, or resistor-capacitor (R-C) method of coupling shown in Fig. 6.14. A 10 μF capacitor is connected between C_1 and B_2; since the capacitor draws *no dc current,* there is no interaction between the two amplifiers *as far as dc conditions* are concerned. In other words, the capacitor effectively *separates* the dc operation of the two amplifiers such that the dc voltages and currents in the two circuits are *completely independent* of each other. The collector-to-emitter voltage of T_1 is 10 V and the base-to-emitter voltage of T_2 is 0.6 V; thus the voltage drop across the capacitor is 9.4 V (note the polarity). This would seem to indicate that the voltage rating of the capacitor should be at least 10 V.

However, if T_1 failed in some way, it is possible for I_{C_1} to drop to zero. If this happens, the drop across the 10 kΩ resistor becomes zero, and V_{CE_1} increases to 20 V (V_{CC}). The drop across the capacitor therefore becomes equal to 19.4 V; if the coupling capacitor is rated for only 10 or 15 V, the failure of transistor T_1 will cause a consequent failure of the 10 μF coupling capacitor. For this reason, the voltage rating of the capacitor in this case would be about 25 V, for safety.

When the output of one amplifier is connected to the input of a second, the two amplifiers are said to be connected in *cascade;* each amplifier is referred to as a *stage* of amplification. The circuit of Fig. 6.14 is that of a *two-stage amplifier;* the

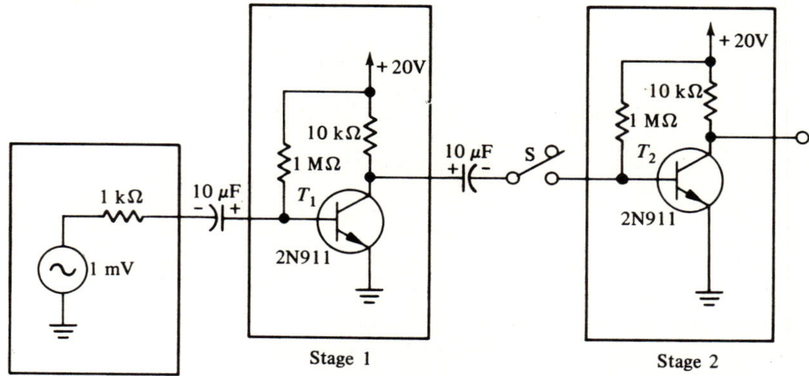

Figure 6.14

amplifier stage connected to the source is called the *first* stage, the amplifier whose input is fed from the output of the first stage is called the *second* stage, and so on.

Let us now study the ac operation of this two-stage amplifier. Figure 6.15(a) shows the ac equivalent circuit of the entire two-stage amplifier of Fig. 6.14; the capacitive reactances of the two coupling capacitors have been omitted from the circuit since we have seen that they present negligible impedance at a frequency of 1.0 kHz, namely, 15.9 Ω. The equivalent circuits of the two stages are indicated within the dotted boxes. Note that the arrow directions of the βi_B current sources *reverse* on alternate stages to account for the 180° phase shift between input and output voltages caused by each amplifier stage. Each stage consists of not only the transistor (r_I, r_O, and the βi_B current source), but also the components associated with the transistor (R_B and R_L). With the switch *open*, as shown in Fig. 6.15(a), the two stages are *not* connected; the current in the $\beta_1 i_{B_1}$ generator (which was found to be 25 μA in Sec. 6-2) splits between the 50 kΩ output resistance and the 10 kΩ load resistance, *most* of it (20.8 μA) going to the 10 kΩ load resistance (R_{L_1}). Of course when the switch is open, *no* current gets to the base of the *second* transistor; i.e., $i_{B_2} = 0$.

Now let us close the switch S in Fig. 6.15(a). The $\beta_1 i_{B_1}$ current generator is now in parallel with *four* resistances: r_{O_1}, R_L, R_{B_2}, and r_{I_2}. We can immediately neglect the effect of the bias resistor of stage 2 (1 MΩ) since it is much larger than the other three resistances in parallel with it. A portion of the circuit of Fig. 6.15(a) is

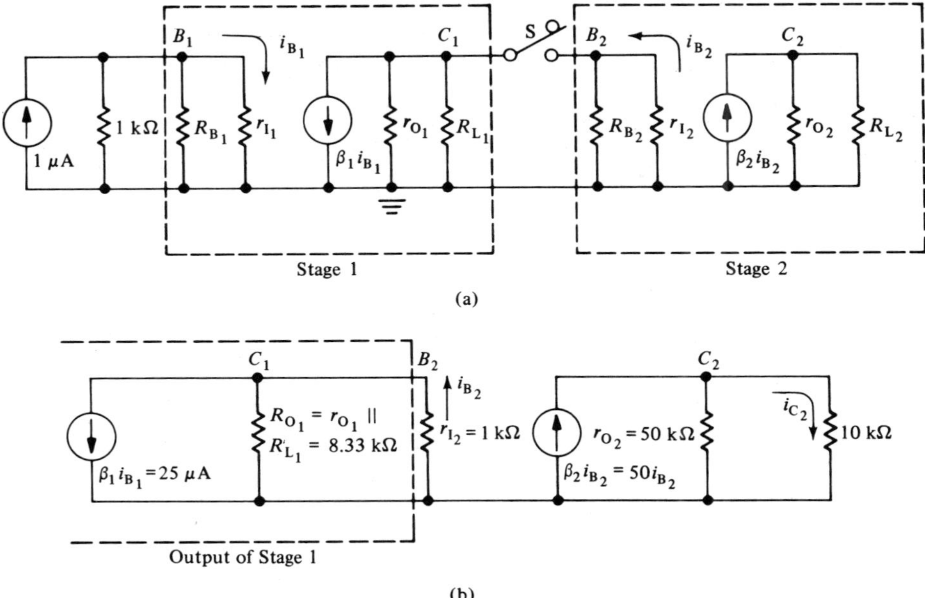

Figure 6.15

re-drawn in Fig. 6.15(b). Note that when the switch is closed, the collector of T_1 is effectively connected to the base of T_2, *as far as ac conditions are concerned;* we know that *the dc voltages* at C_1 and B_2 are *not* the same because the two terminals are separated by the coupling capacitor. C_1 and B_2 are at the *same ac potential* because the coupling capacitor is considered to be a *short circuit* (small impedance) at ac.

We now define a new resistance, R_O, which is equal to the parallel combination of r_O and R_L:

$$R_O = \frac{r_O R_L}{r_O + R_L}$$

Computing this value for the circuit of Fig. 6.15(b), we get:

$$R_O = \frac{r_O R_L}{r_O + R_L} = \frac{50(10)}{50 + 10} = 8.33 \text{ k}\Omega$$

The term r_O is the output resistance of the *transistor* in the common emitter configuration and is approximately equal to the parameter $1/h_{oe}$; it depends primarily upon the transistor. The term R_O is the output resistance of the *amplifier stage,* and depends upon the transistor *and* the load resistor. Thus the output circuit (collector circuit) of any amplifier stage can always be represented by a *current* generator in *parallel* with some output resistance, R_O, as shown in Fig. 6.15(b). Alternatively, the circuit can be represented by a *voltage* generator in *series* with an output resistance. We see, in later chapters, that the method of determining the value of R_O differs for the different amplifier circuits; this is particularly true of feedback circuits which are discussed in the next chapter. However, regardless of the type of circuit, the output circuit of the amplifier stage can be represented by a current generator in parallel with a resistance, R_O. Thus, the output circuit of stage 1 is represented by a 25 μA current generator in parallel with an 8.33 kΩ resistance, as shown within the dotted lines of Fig. 6.15(b).

It can be seen in Fig. 6.15(b) that the 25 μA from the current generator splits between the 8.33 kΩ output resistance of stage 1 and the 1 kΩ input resistance of stage 2. Using the current divider, we find that the base current flowing to the second stage is:

$$i_{B_2} = \frac{(\beta_1 i_{B_1})R_{O_1}}{R_{O_1} + r_{I_2}} = \frac{(25\ \mu A)(8.33)}{8.33 + 1} = 22.3\ \mu A$$

Note that this indicates that a large percentage of the 25 μA is transferred on to the second stage. This base current in the second stage is amplified further as follows: The value of the current generator in the second stage, $\beta_2 i_{B_2}$, is:

$$\beta_2 i_{B_2} = 50\ (22.3\ \mu A) = 1115\ \mu A$$

The collector current in the second stage, i_{C_2}, is:

$$i_{C_2} = \frac{(\beta_2 i_{B_2})r_{O_2}}{r_{O_2} + R_{L_2}} = \frac{(1115\ \mu A)(50)}{50 + 10} = 930\ \mu A$$

The collector-to-emitter voltage at the output of the second stage, v_{CE_2}, is:

$$v_{CE_2} = i_{C_2}R_{L_2} = (0.93\ mA)(10\ k\Omega) = 9.3\ V$$

The ac power output from the second stage is:

$$P_O = \frac{v_{OP\_P}i_{OP\_P}}{8} = \frac{(9.3\ V)(0.93\ mA)}{8} = 1.08 \times 10^{-3}\ W$$

$$= 1.08\ mW$$

The results of the above computations indicate that, by cascading amplifier stages, we can obtain larger amounts of current, voltage, and power; compare the current, voltage, and power outputs of stage 1 *before* cascading (20.8 μA, 208 mV, and 54 μW), with the same outputs from stage 2 when the two stages are *connected in cascade* (930 μA, 9.3 V, and 1.08 mW). A *portion* of the amplified current of stage 1 is fed to the base of stage 2 where it is amplified further.

Let us now study this problem of power transfer in more detail. Figure 6.16(a) shows a portion of the ac equivalent circuit of the two-stage amplifier; the switch is *open*. Of course, with the switch open *no* current and *no* voltage are transferred to the input of the second stage; there are zero microamperes through, and zero millivolts across, the 1 kΩ input resistance of the second stage in Fig. 6.16(a). The output current (current in the 10 kΩ load resistor of the first stage) is 20.8 μA and the output voltage (voltage across this load resistor) is 208 mV.

Figure 6.16(b) shows the same circuit when the switch is *closed*. As far as *ac*

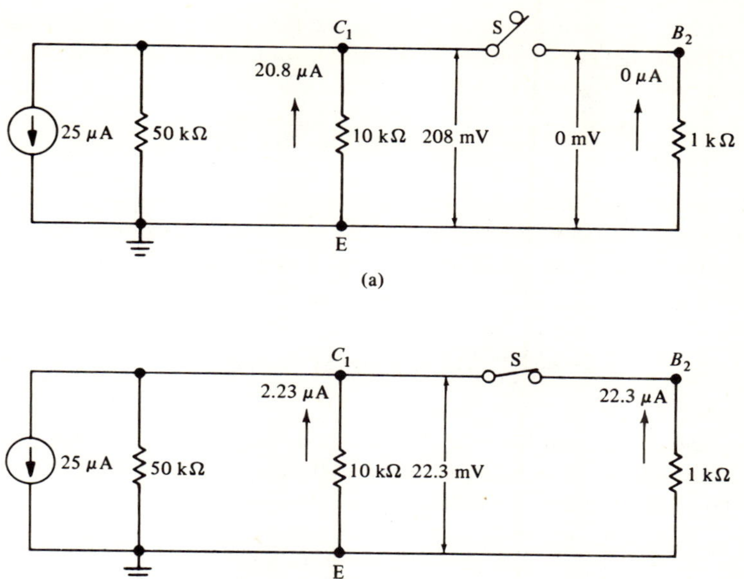

(a)

(b)

Figure 6.16

conditions are concerned, C_1 and B_2 are at the same potential, and, therefore, the three resistances (50 kΩ, 10 kΩ, and 1 kΩ) are effectively connected in *parallel* as shown in the figure. We saw before that most of the current (22.3 µA) from the 25 µA current generator flows through the 1 kΩ input resistance of the second stage; this is reasonable since the *smallest* resistance (1 kΩ) receives the *largest* current (22.3 µA). It can be seen from Fig. 6.16(b) that the input resistance of the second stage is in parallel with the output resistance of the first stage, and thus both resistances have the *same* voltage drop across them. Therefore, we say that the *output* voltage of the *first* stage (v_{CE_1}) is the *same* as the *input* voltage of the *second* stage (v_{BE_2}).

We know that v_{CE_1} was 208 mV *before* the switch was closed. To find the value of this quantity *after* the switch is closed, we need only apply Ohm's law to any one of the three parallel resistances. Since we have already found that the current in the 1 kΩ input resistance is 22.3 µA, the voltage is found as follows:

$$v_{CE_1} = v_{BE_2} = i_{B_2} r_{I_2} = (22.3 \ \mu A)(1 \ k\Omega)$$
$$v_{CE_1} = (22.3 \times 10^{-6})(10^3) = 22.3 \ mV$$

The above computation illustrates a very important concept. While the output voltage of the first stage was 208 mV *before* the switch was closed, it *dropped* to 22.3 mV *after* the switch was closed. In other words, *the connection of the input of stage 2 to the output of stage 1 results in a drastic reduction in the output voltage of stage 1.*

This phenomenon is explained as follows: With the switch open, as in Fig. 6.16(a), the 25 µA generator current circulates in only *two* resistances, the 50 kΩ and the 10 kΩ. We can say that this 25 μA flows through one resistance, R_o, equal to the parallel combination of the 10 kΩ and the 50 kΩ, or 8.33 kΩ. Therefore, the voltage across this 8.33 kΩ resistance is:

$$(25 \ \mu A)(8.33 \ k\Omega) = 208 \ mV$$

This checks with our previous computations. Now when the switch is closed, as in Fig. 6.16(b), *this same 25 μA* must circulate in *three* resistances, the 50 kΩ, the 10 kΩ, and the 1 kΩ. These three resistances can be replaced by a single resistance whose value is:

$$\frac{(8.33)(1)}{8.33 + 1} = 0.892 \ k\Omega = 892 \ \Omega$$

Since the *same 25 μA* must flow through a much *smaller* resistance when the switch is closed, the *voltage* across this resistance will also be *smaller*. The voltage across this new equivalent resistance (with the switch closed) is:

$$(25 \ \mu A)(892 \ \Omega) = 22.3 \ mV$$

This checks again with our previous computations.

In general, whenever a resistance is placed across two terminals of a circuit (as when the input resistance of the second stage is connected across the output terminals of the first), the voltage across those two terminals will *decrease* as soon as the resistance is connected. We say that the input circuit (1 kΩ input resistance) of the second stage *loads down* the output of the first. This is a common phenomenon which is encountered in all types of electrical and electronic circuits.

One might ask the questions: Why does the output voltage of the first stage drop from 208 mV to 22.3 mV when the second stage is connected? Why doesn't the voltage remain equal to 208 mV? We have already given a mathematical explanation for this phenomenon (see above computations). Following is an alternate means of viewing the situation: When the switch is closed in Fig. 6.16(b) (placing r_{I_2} across the collector and emitter terminals of the first stage), a *new circuit* results; the circuit pictured in Fig. 6.16(a) is obviously *not* the same as that shown in Fig. 6.16(b). If the two circuits are not the same, there is no reason to believe that the currents and voltages in the two circuits will be the same; therefore it should come as no surprise that the output voltage of the first stage changes when the switch is closed.

This is the same situation encountered by the reader when he studied dc circuits theory. You may recall that when a voltmeter is connected across two terminals in a circuit, the meter always reads less than the value predicted on the basis of theory. The reason is that the voltmeter itself has some finite resistance; when the meter is connected into the circuit, the effect is as if a resistor were inserted. Therefore, the resistance of the voltmeter loads down the original circuit in much the same way that the input resistance of the second stage loads down the output of the first.

Whenever two or more electronic circuits are connected together in some configuration, there will always be an *interaction* between them; this interaction causes the currents and voltages established in the circuit *after* the connection has been made to be *different* from those present *before* connecting. The interaction resulting from the interconnection of electronic circuits is an extremely important process which occurs in all types of circuitry; the designer must always take these factors into account when dealing with a particular problem. We have more to say about loading in the following chapters.

To illustrate the interaction of the two amplifier circuits, we shall summarize the various voltages and currents in the circuits both before and after the switch is closed:

<table>
<tr><td align="center">S OPEN:</td><td align="center">S CLOSED:</td></tr>
<tr><td align="center">$v_{CE_1} = 208$ mV</td><td align="center">$v_{CE_1} = 22.3$ mV</td></tr>
<tr><td align="center">$v_{BE_2} = 0$ mV</td><td align="center">$v_{BE_2} = 22.3$ mV</td></tr>
<tr><td align="center">$i_{B_2} = 0$ μA</td><td align="center">$i_{B_2} = 22.3$ μA</td></tr>
<tr><td align="center">$i_{R_{L_1}} = 20.8$ μA</td><td align="center">$i_{R_{L_1}} = \dfrac{v_{CE_1}}{R_{L_1}} = \dfrac{22.3 \text{ mV}}{10 \text{ k}\Omega} = 2.23$ μA</td></tr>
</table>

Summarizing these results briefly, we see that the output voltage of the first stage drops from 208 mV to 22.3 mV when the switch is closed; in other words, only a *very small portion* of the amplified voltage from the first stage is transferred to the second. The reason for this action is that *as soon as the connection is made, there are no longer 208 mV available from the output of the first stage.* This effect can be verified quite easily in the lab by connecting an oscilloscope across the output terminals of the first stage with the switch open, and observing the waveform. Upon closing the switch, it will be observed that the amplitude of the waveform decreases. Although only a *small* portion of the *voltage* is transferred, observation of the chart indicates that a *large* portion of the 25 μA from the *current generator* is transferred to the base of the second stage (22.3 μA); we shall discuss some of the reasons for these effects in the following section. Similarly, the current in the 10 kΩ load resistor of the *first* stage *drops* from 20.8 μA to 2.23 μA; this again indicates the efficiency of the current transfer from stage 1 to stage 2. These results are summarized in the chart as well as in Figs. 6.16(a) and (b).

At this point we should not be overly concerned about the fact that the second stage loads down the output of the first. The single most important transistor parameter is its β; it is this number by which the base current is multiplied to obtain the value of the current generator in the collector circuit, as shown in Fig. 6.17. The larger the value of this current generator, the more collector current will flow,

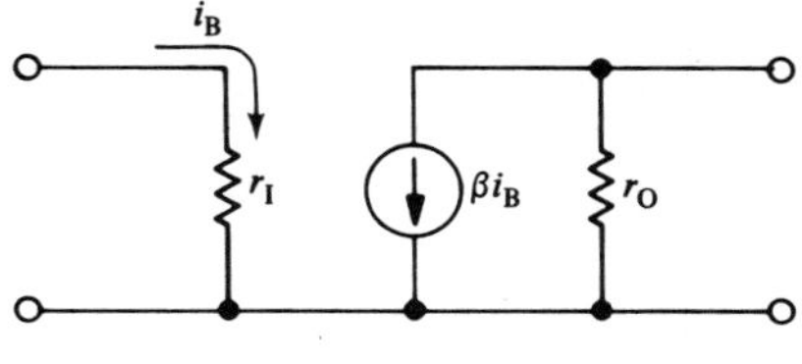

Figure 6.17

and consequently, the larger will be the collector-to-emitter voltage. To obtain a large value for the current generator, we should like both β and i_B to be *large;* it is important that we supply a lot of *current* to the base of the transistor. Therefore, although we have not transferred much *voltage* to the second stage, this should not concern us since we *have* transferred a relatively large amount of *current* (22.3 μA out of a possible 25 μA). The base current in the first stage (0.5 μA) was multiplied by β (50) to yield a 25 μA current generator in the collector circuit of the first stage; out of this 25 μA, 22.3 μA was transferred to the base of the second stage, and was also multiplied by 50 to yield a 1115 μA current generator in the collector circuit of the second stage. In other words, the transistor amplifier should be viewed as amplifying *current;* the *voltage* amplification is a consequence of the current amplification. As more and more current is transferred from stage to stage, more and more voltage is developed across the various resistors.

It is important to bear in mind that, although the second stage loads down the first, *the amplifier stage still provides voltage gain* (output voltage larger than input). From the preceding section you recall that the base current in the first stage was 0.5 μA and the input resistance of the transistor was 1 kΩ; thus the input voltage to the first stage was:

$$v_{I_1} = i_{B_1} r_{I_1} = (0.5 \ \mu\text{A})(1 \ \text{k}\Omega) = 0.5 \ \text{mV}$$

We have just seen that the output voltage of the first stage was 22.3 mV; thus the voltage gain of stage 1 is:

$$A_{V_1} = \frac{v_{O_1}}{v_{I_1}} = \frac{22.3 \ \text{mV}}{0.5 \ \text{mV}} = 44.6$$

We see in later chapters that R-C coupling is just one method of interconnecting amplifier stages. There is a more efficient transfer of power between stages if a transformer is used as the vehicle of interstage coupling. However, for a number of reasons, transformer coupling is avoided where possible.

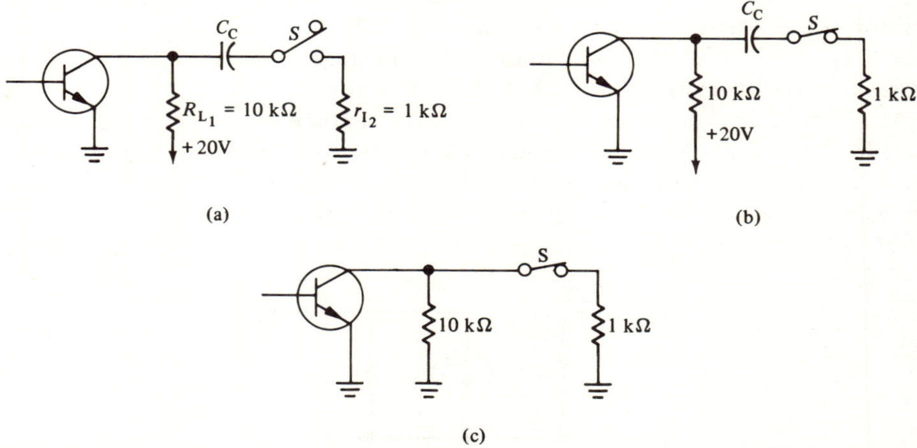

Figure 6.18

Before leaving the topic of loading, we illustrate the concept from a different perspective. Figure 6.18(a) shows the output circuit of the first stage connected through the coupling capacitor and switch to the input circuit of the second stage. The load line is drawn on the characteristics in the usual fashion as shown in Fig. 6.19. This load line is designated *the dc load line* because it is used to establish

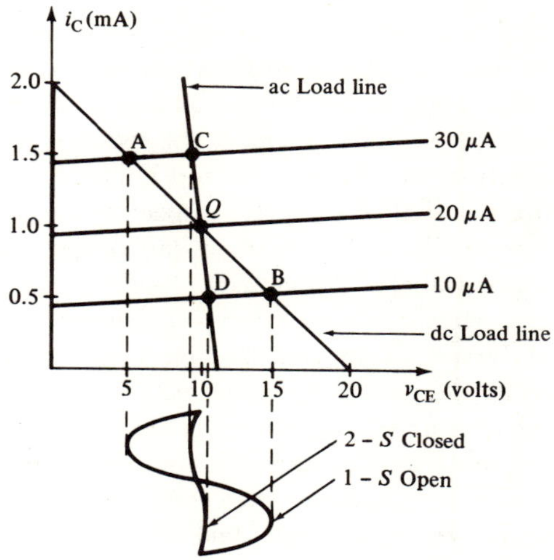

Figure 6.19

the dc operating point, or Q point. Now assume for a moment that the switch remains open as in Fig. 6.18(a); the two stages are disconnected. Assume also that the ac base current of the first stage is 20 μA peak-to-peak. If the transistor is biased with 20 μA (dc) of base current as shown in Fig. 6.19, then the base current will increase to 10 μA *above* the dc value of 20 μA (to 30 μA), and decrease to 10 μA *below* the dc value (to 10 μA). This movement will take place along the dc load line between points A and B (intersection of the dc load line with the 30 μA and 10 μA curves, respectively). If points A and B (in Fig. 6.19) are projected vertically downward below the horizontal axis, it can be seen that the collector-to-emitter voltage developed (labeled 1-*S* open) is approximately 10 V peak-to-peak.

Now let us close the switch so that the circuit appears as shown in Fig. 6.18(b). Because of the coupling capacitor, no dc current can flow through to the second stage, and thus the Q point is again established by the dc load line. In other words, *as far as dc conditions* are concerned, the portion of the circuit to the *right* of the dotted line in Fig. 6.18(b) does not exist. We know, however, that the coupling capacitor does not block the ac current. If the reactance of the capacitor is small compared with the resistances, we can neglect X_C as before and consider the 1 kΩ input resistance of stage 2 to be in parallel with the 10 kΩ load resistance of stage 1, as shown in Fig. 6.18(c). Therefore, *as far as ac conditions* are concerned, the transistor "sees" a load resistance equal to the parallel combination of R_L and r_{I_2}, or R_L'. In this case, R_L' is equal to:

$$R_L' = \frac{R_{L_1} r_{I_2}}{R_{L_1} + r_{I_2}} = \frac{10(1)}{10 + 1} = 0.909 \text{ k}\Omega = 909 \ \Omega$$

From chapter 1 you may recall that a small load resistance indicates a steep load line. *The ac load line* is drawn through the Q point in Fig. 6.19; its slope is equal to the reciprocal of 909 Ω. We discuss the construction of this load line in a later chapter.

Now if the same 20 μA peak-to-peak current flows through the base, the base current will again vary between 30 μA and 10 μA. However, with the switch closed, the movement now takes place along *the ac load line;* instead of moving between points A and B as when the switch was open, the swing is now between points C and D because the switch is closed (points C and D are the intersections of the ac load line with the 30 μA and 10 μA curves, respectively). If points C and D are projected downward vertically in Fig. 6.19, it can be seen that the collector-to-emitter voltage developed (labeled 2-*S* closed) *is much smaller* than the 10 V peak-to-peak voltage present when the switch is open. This is a graphical illustration of the loading effect of the second stage on the first; compare waveforms 1 (*S*-open) and 2 (*S*-closed). This action comes about because for dc currents, the transistor sees only its own load resistance, R_L, while for ac currents, it sees the parallel combination of R_L and the input resistance of the next stage. The dc load line is dependent upon R_L and V_{CC}; this, along with the characteristics and the bias current, sets the Q point. However, the path taken on the curves is determined by both R_L of the stage and r_I of the following stage. We could solve problems by this graphical technique, but it is faster and easier to use the analytical technique developed earlier in this section. The above presentation was included to give the reader an insight into the loading problem. We return in later chapters to a more detailed study of the ac load line.

6-4   DESIGN FACTORS

We now study some of the factors which determine the amount of current transferred between stages. The output circuit of the first stage and the input of the second are shown re-drawn in Fig. 6.20. The object is to transfer as much current as possible from the current generator $(\beta_1 i_{B_1})$ into the base of the next stage (i_{B_2}).

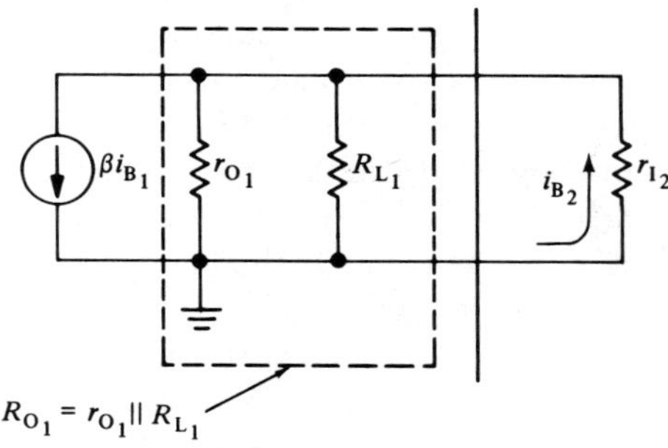

$$R_{O_1} = r_{O_1} \| R_{L_1}$$

Figure 6.20

Since the largest amount of current goes to the smallest resistor, we should like r_{I_2} to be *small* and R_{O_1} to be *large*. Since R_{O_1} is equal to the parallel combination of r_{O_1} and R_{L_1}, we should like both r_{O_1} *and* R_{L_1} to be as large as possible. Summarizing, for the most efficient transfer of current from stage 1 to stage 2, we should like r_{I_2} to be small, and r_{O_1} and R_{L_1} to be large. We know that the quantities r_I and r_O depend primarily upon the transistor itself; thus once the transistor and the Q point are selected, the values of r_I and r_O are fixed. The transistor to be used in a particular application is usually selected on the basis of its β, breakdown voltage, leakage current, and other factors as we see in later chapters. Once the transistor type is selected for the particular circuit, the designer has little control over the values of r_I and r_O. In addition, there is a relationship between the β of a transistor and its r_I, as we see in chapter 7; in other words, once the designer selects a transistor for a particular β and establishes the Q point, the value of r_I is fixed.

Since we have little control over the values of r_I and r_O, the only action we can take is to make R_L as large as possible. Let us first see what happens if the value of R_L is *reduced* from 10 kΩ to 2 kΩ in the amplifier of the previous section. It is true that the Q point will change, but we are now interested in only the ac operation and thus we do not concern ourselves with the dc quantities. Assuming that all of the other quantities remain the same (r_{O_1} equals 50 kΩ, $\beta_1 i_{B_1}$ equals 25 μA, and r_{I_2} equals 1 kΩ), we proceed in the same manner as in the previous section. We find R_{O_1} as follows:

$$R_{O_1} = \frac{r_{O_1} R_{L_1}}{r_{O_1} + R_{L_1}} = \frac{50(2)}{50 + 2} \simeq 2 \text{ k}\Omega$$

Next we find the current in the base of the second stage, i_{B_2}:

$$i_{B_2} = \frac{(\beta_1 i_{B_1}) R_{O_1}}{R_{O_1} + r_{I_2}} = \frac{(25 \ \mu\text{A})2}{2 + 1} = 16.67 \ \mu\text{A}$$

This indicates that with R_L equal to 2 kΩ, *only 16.67 μA* is transferred to the second stage; compare this with *22.3 μA* when R_L equals 10 kΩ. There is also less voltage transferred since:

$$v_{CE_1} = v_{BE_2} = i_{B_2} r_{I_2} = (16.67 \ \mu\text{A})(1 \text{ k}\Omega) = 16.67 \text{ mV}$$

Compare this with 22.3 mV when R_L equals 10 kΩ. These computations indicate clearly that the efficiency of current transfer from stage to stage is poorer if the value of the load resistance is reduced.

This concept can be explained quite readily with the aid of Fig. 6.20. The portion of the circuit to the *left* of the *solid* line can be considered to be a *current source* consisting of a current generator, $\beta_1 i_{B_1}$, and an internal resistance, R_{O_1}; this current source feeds a resistance, r_{I_2}. If the value of R_{O_1} were *infinite*, the current source would be *ideal*, and the *full value* of the generator current ($\beta_1 i_{B_1}$) would be transferred to r_{I_2}. However, R_{O_1} cannot be made infinite because it can never be larger than r_O, and r_O is usually between 20 kΩ and 100 kΩ for a small signal transistor in the common emitter configuration. We saw that if R_L is made too small, the current flowing to r_{I_2} is reduced; this is because *more* current now flows through

R_{L_1}. Any ac current which flows through R_{L_1} is "wasted"; i.e., it accomplishes no useful purpose since it is not transferred on to the next stage for further amplification.

Now that we know that it is desirable to have a large value of R_L, the next question to ask is: How large can we make it? Let us change the value of R_L from 10 kΩ to 100 kΩ. Refer to Fig. 6.21. The original dc load line (R_L equal to 10 kΩ) is shown drawn on the axes with the Q point in the center at point A. If R_L is changed to 100 kΩ, the *new* dc load line will have a much *smaller* slope, as shown in Fig. 6.21. If it is desired that the new Q point be in the center of this new load

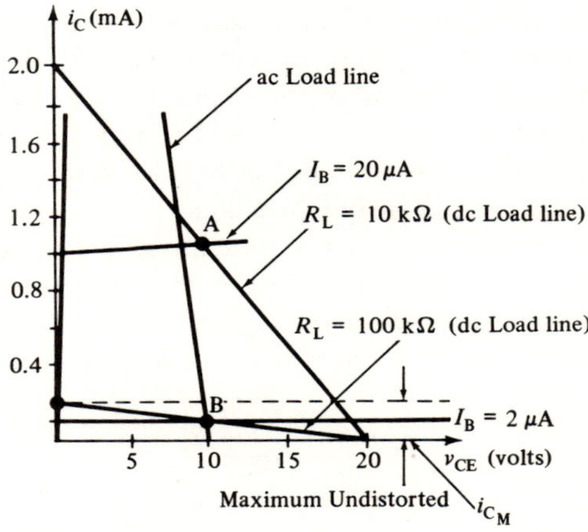

Figure 6.21

line (at point B in Fig. 6.21), the transistor bias will have to be changed. The quiescent collector current must be decreased to *one-tenth* of its original value, from 1 mA to 0.1 mA. This means that the quiescent base current must also be reduced, from 20 μA to 2 μA, assuming that the β of the transistor remains equal to 50 at values of collector current as small as 0.1 mA; i.e., we are assuming uniform spacing of the characteristic curves. Finally, in order to *reduce* the base current to 2 μA, we must *increase* the value of the bias resistor from 1 to 10 MΩ (assuming we are still using a 20 V V_{CC}). If this change in the value of R_B were *not* made, the base current would remain at 20 μA, and the transistor would saturate (Q point located at point C in Fig. 6.21).

Now let us examine the behavior of the amplifier when R_{L_1} is set equal to 100 kΩ and the Q point is at point B in Fig. 6.21. The ac load line is shown in the figure; note that it passes through point B. The first point to consider is that the change in the Q point (from A to B) necessitated by the change in R_{L_1} (from 10 kΩ to 100 kΩ) *has reduced the maximum undistorted peak-to-peak collector current available from the amplifier stage.* Because the quiescent current has been reduced to such a small value (0.1 mA) with the consequence that the Q point (point B) is very close to the horizontal axis, the maximum undistorted peak-to-peak collector current is now only 0.2 mA, as indicated in Fig. 6.21. Compare this with 2 mA, when the transistor is biased at point A. This is not too serious if the ac currents are quite small.

For example, the value of the current generator in the circuit being discussed is 25 μA, or 0.025 mA. Since this is much smaller than the maximum swing of 0.2 mA, there is no chance that we might over-drive the *first* stage. However, we saw that the current generator for the *second* stage ($\beta_2 i_{B_2}$) was *1115* μA; thus we very definitely *would over-drive this stage* if R_{L_2} were also set equal to 100 kΩ, since we anticipate a 1 mA swing and we can only accommodate a 0.2 mA swing. Therefore, the reduction in the maximum allowable swing is not serious in the *early* stages of amplification, but may become troublesome in the later stages.

There is a second, and far more serious problem associated with the change in Q point resulting from the use of the larger load resistance. From chapter 3 you may recall that the β of the transistor is not uniform at every point on the characteristics; the spacing varies somewhat, reaching a maximum at some value of collector current. At low values of collector current (such as 0.1 mA), the β is quite often considerably less than its value at larger values of I_C. Let us say that the β of the 2N911 drops to 25 at 0.1 mA. Figure 6.22 shows the characteristics of the transistor as they would actually appear. Note that since the β at 0.1 mA is *25* and not *50*, the curve at that point is labeled 4 μA (0.1 mA/25 equals 4 μA) and not 2 μA (the curve labeled 2 μA in Fig. 6.21 was drawn on the assumption of uniform spacing, and is, strictly speaking, incorrectly labeled). Note that the three curves drawn at the lower currents (0, 4, and 8 μA) and those three at the higher currents (16, 20, and 24 μA) are all drawn at 4 μA intervals of base current. The larger spacing at the higher currents, however, indicates a larger value of β at these currents.

Now let us assume that the ac input base current to the amplifier stage is 8 μA peak-to-peak. If the transistor is biased at point A in Fig. 6.22, the base current

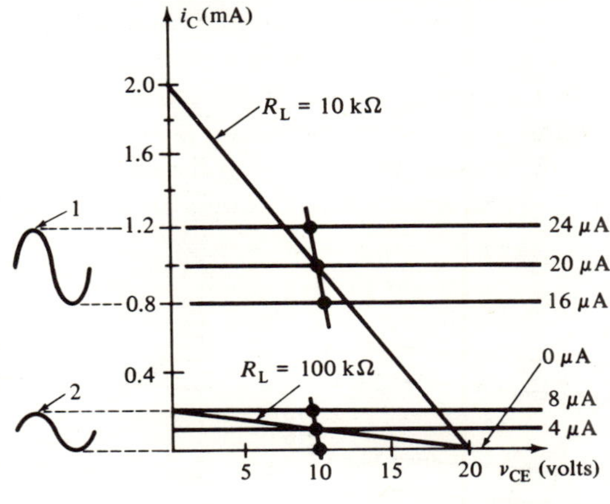

Figure 6.22

will vary between 16 and 24 μA, yielding a peak-to-peak collector current of approximately 0.4 mA (waveform 1). On the other hand, if the transistor is biased at point B, the same 8 μA peak-to-peak ac input will cause a variation in base current between 0 and 8 μA, yielding a peak-to-peak collector current of approximately 0.2 mA (waveform 2). The smaller output is caused by the reduction in β at the lower collector currents. In other words, *by increasing the value of R_L, we have actually reduced the ac current available because the β of the transistor is smaller at low values of collector current.*

We can illustrate this point analytically. We saw earlier in the chapter that at a collector current of 1 mA, β was 50, $\beta_1 i_{B_1}$ was 25 μA, R_{O_1} was 8.33 kΩ (10 kΩ in parallel with 50 kΩ), and r_{I_2} was 1 kΩ; i_{B_2} was found to be 22.3 μA, or 89% of the 25 μA current generator. Now, if the transistor is biased at 0.1 mA, β is 25, $\beta_1 i_{B_1}$ is 12.5 μA, R_{O_1} is:

$$R_{O_1} = \frac{r_{O_1} R_{L_1}}{r_{O_1} + R_{L_1}} = \frac{50(100)}{50 + 100} = 33.3 \text{ k}\Omega$$

The current delivered to the next base is:

$$i_{B_2} = \frac{(\beta_1 i_{B_1}) R_{O_1}}{R_{O_1} + r_{I_2}} = \frac{12.5 \ \mu\text{A}(33.3 \text{ k}\Omega)}{33.3 \text{ k}\Omega + 1 \text{ k}\Omega} = 12.1 \ \mu\text{A}$$

This illustrates that there is *less* current delivered to the next stage (12.1 μA) when R_L is 100 kΩ than when R_L is 10 kΩ. In other words, although a greater *percentage* of the current generator is delivered to the next stage when R_L is 100 kΩ (12.1 μA out of a possible 12.5 μA, or 97%), there is less *total* current delivered (12.1 μA is less than 22.3 μA) because the decrease in the β at the lower collector currents causes a consequent decrease in the size of the current generator (12.5 μA as compared with 25 μA).

We can see, therefore, that the use of a very *large* value of R_L usually results in a very *small* value of quiescent collector current. This small I_{C_Q} is in turn responsible for *two* problems: reduction in the maximum allowable swing, and a decrease in the gain due to a reduction in β. There is a definite limit to the size which we can make R_L without the process becoming self-defeating. The value of R_L is usually selected so that the Q point is in the region of *maximum β*.

We can now see why we use the expression "small signal" equivalent circuit. The β (spacing) of the transistor remains uniform in *only a small region* on the

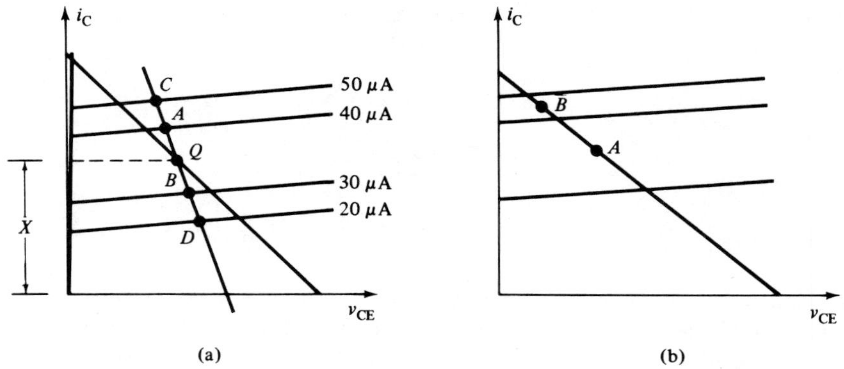

Figure 6.23

characteristics. The characteristics shown in Fig. 6.23(a) are exaggerated slightly to illustrate a point. Note first that the Q point has been established in the region of maximum β; the *spacing is largest* at collector current "X" (between the 30 and 40 μA curves). If the input signal swing is relatively small (less than or equal to the length A–B in Fig. 6.23(a)), the small signal equivalent circuit can be used with a high degree of accuracy because the β of the transistor is relatively uniform in this region. However, if the signal swing becomes much larger, say 30 μA peak-to-peak (equal to the length C–D in Fig. 6.23(a)), the small signal equivalent circuit analysis gives only an approximate result because the β varies as the input ac current swings between points C and D. The small signal equivalent circuit yields accurate results as long as the β remains constant; this is generally true for only *small* swings (equal to or less than A–B in the figure). As the signal swing becomes larger and larger, the results obtained from use of the equivalent circuit become more *approximate* because the transistor parameters (particularly β) vary from point-to-point on the curves as explained in chapter 3. However, because of the tremendous time savings resulting from the use of the equivalent circuit, it is still preferred to the point-by-point graphical technique described in chapter 4; the results obtained, even for large swings, give a reasonably good indication of the ac operation of the circuit.

From chapter 5 you will recall that it was desirable to keep the Q point shift as small as possible; it was mentioned that if the Q point shifts too far, the transistor may saturate. It can be seen from Fig. 6.23(b) that there is another reason why we do not want a large Q point shift. Suppose that the Q point shifts from A to B in Fig. 6.23(b). The β is obviously *smaller* at point B, and this will result in a reduction in the gain. In other words, we should like to prevent the Q point from shifting into a region of smaller β.

We can make R_L equal to 100 kΩ and still maintain I_{CQ} at 1 mA (region of maximum β) if we *increase the value of the supply voltage, V_{CC}, as shown in Fig. 6.24 (a).* Note that if V_{CC} is increased to 110 V, and R_L is made equal to 100 kΩ, it is possible to bias the transistor at a collector current of 1 mA. However, this would involve changing V_{CC} from 20 V to 110 V. If the electronic system in which this amplifier is to be used operates from a 20 V supply, it would be inefficient to include a 110 V dc supply just for this one amplifier stage.

If the ac output swing is small (such as the 25 μA swing encountered in the amplifier discussed earlier in the chapter), *it is not necessary that the Q point be in the center of the load line.* Figure 6.24(b) illustrates that we can use a 16 kΩ load resistor with the same 20 V battery and still maintain the quiescent collector current at 1 mA; if R_L is changed from 10 kΩ to 16 kΩ, the Q point is changed from point A to point B. In this case we must be sure that the amplifier has good stability because only a small change in I_{CQ} is necessary to shift the Q point from point B to saturation at point C.

There is one final point to be made concerning the maximum undistorted peak-to-peak sine wave available from the amplifier, which we shall designate i_{CM}. From chapter 4 you will recall that we made the following statements concerning i_{CM} *for a one-stage amplifier:*

$$(1) \qquad \left.\begin{array}{l} I_{CQ} < \dfrac{1}{2} I_{Cs} \\[2ex] I_{CQ} = \dfrac{1}{2} I_{Cs} \end{array}\right\} : \qquad i_{CM} = 2I_{CQ}$$

$$(2) \qquad I_{CQ} > \dfrac{1}{2} I_{Cs}: \qquad i_{CM} = 2(I_{Cs} - I_{CQ})$$

These statements assume that we are dealing with only a one-stage amplifier; i.e., *the dc and ac load lines are one and the same*. When two stages are connected in cascade, the dc load line establishes the Q point, while the ac load line (which intersects the Q point) determines the path of motion on the characteristics. When the dc and ac

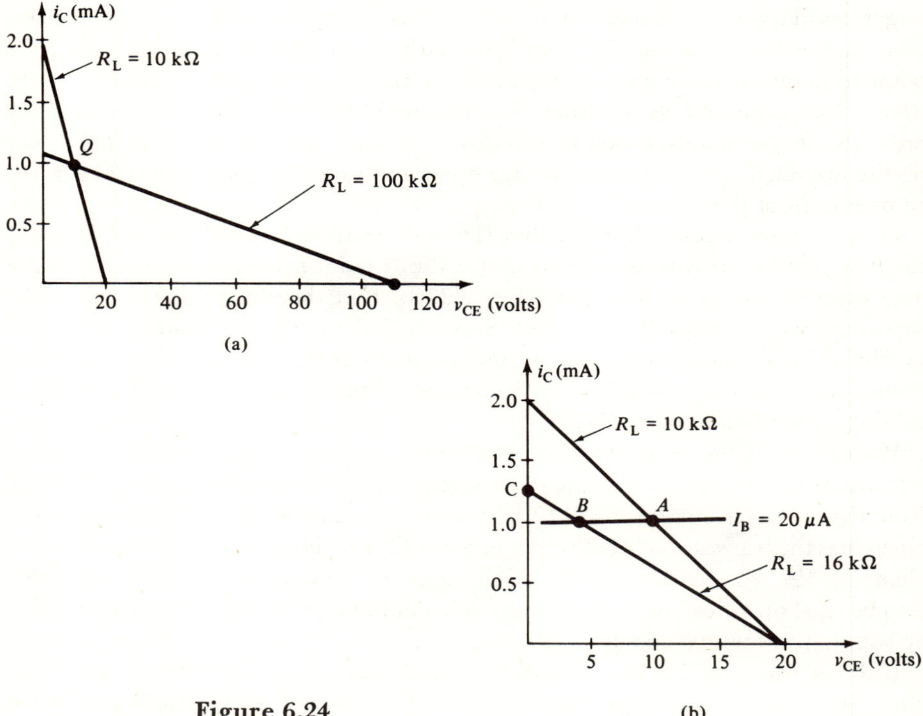

Figure 6.24

load lines are *not* the same, we have to make a slight correction in the previous statements.

Consider the following amplifier: V_{CC} equals 10 V, R_L equals 1 kΩ, and r_I (of the next stage) equals 1 kΩ. Figure 6.25(a) shows the dc load line and two ac load lines drawn for two different Q points, A and B. The quiescent collector current at Q point A is *less* than one-half of the collector saturation current (2.5 mA is less than 10 mA). It can be seen from the figure that i_{CM} for this Q point is equal to twice I_{CQ}, this quantity being labeled "X" in the figure. Similarly, I_{CQ} at Q point B is equal to *exactly one-half* I_{Cs}; i_{CM} for this Q point is again equal to twice I_{CQ}, labeled "Y" in Fig. 6.25(a). Thus, if I_{CQ} is equal to or less than one-half I_{Cs}, i_{CM} is equal to

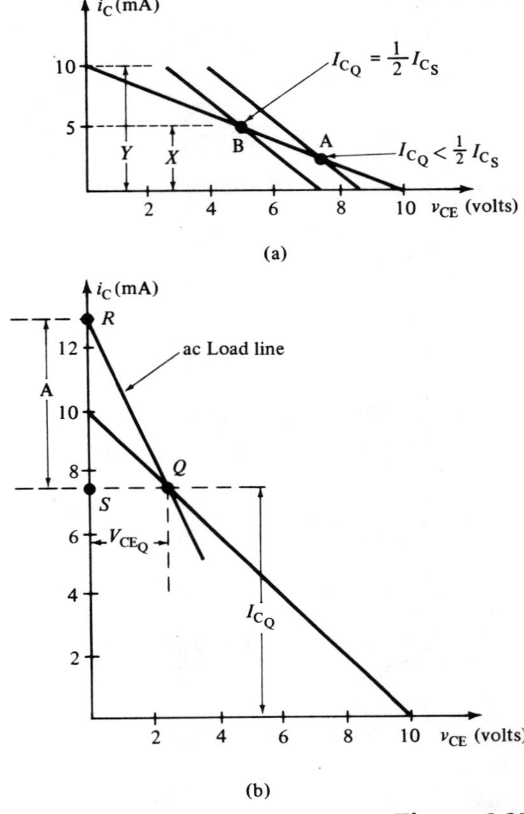

Figure 6.25

twice I_{C_Q}; in other words, the statements for the above conditions are true whether or not the dc and ac load lines are the same.

The situation changes slightly when the value of I_{C_Q} is *greater* than one-half of I_{C_S}; the diagram for this phenomenon is pictured in Fig. 6.25(b). Note from the figure that the collector current can swing *above* the quiescent value of 7.5 mA to point R (a distance equal to A), and *below* 7.5 mA down to the horizontal axis (a distance equal to I_{C_Q}). The maximum allowable sinusoidal output will be limited by whichever is *smaller*, A or I_{C_Q}; if A is *smaller* than I_{C_Q}, i_{C_M} is equal to 2A, and if the reverse is true, i_{C_M} is equal to $2I_{C_Q}$. The problem, then, is to compute the value "A" and then compare it with I_{C_Q}. This can be done graphically directly from Fig. 6.25(b), since it can be seen that A is equal to 12.5 mA minus 7.5 mA, or 5 mA, while I_{C_Q} is equal to 7.5 mA. In this case, since A (5 mA) is *smaller* than I_{C_Q} (7.5 mA), i_{C_M} is equal to 2A, or 10 mA (current can increase to 5 mA above 7.5 mA and can decrease to 5 mA below 7.5 mA).

To save time, we should like to have an analytic method of computing the value of A. Referring to Fig. 6.25(a), we can say that the slope of the ac load line (line R–Q) is equal to the ratio of A (line R–S) to V_{CEQ} (line S–Q) in triangle *RSQ*. However, we know that the slope of the ac load line is also equal to the reciprocal of the ac load resistance, or $1/R_L'$. Setting these two quantities equal to each other, we get:

$$\frac{A}{V_{\text{CEQ}}} = \frac{1}{R_L'}$$

Re-arranging and substituting, we get the following expression:

$$A = \frac{V_{\text{CEQ}}}{R_L'} = \frac{V_{\text{CC}} - I_{\text{CQ}}R_L}{R_L'}$$

where R_L' is:

$$R_L' = R_{L_1} \parallel r_{I_2}$$

For the situation pictured in Fig. 6.25, it was mentioned that V_{CC} equals 10 V, R_{L_1} equals 1 kΩ, r_{I_2} equals 1 kΩ, and I_{CQ} equals 7.5 mA; R_L' is thus equal to:

$$R_L' = \frac{R_{L_1}r_{I_2}}{R_{L_1} + r_{I_2}} = \frac{1(1)}{1 + 1} = 0.5 \text{ k}\Omega$$

Substituting these values into the equation we find that A is:

$$A = \frac{V_{\text{CC}} - I_{\text{CQ}}R_L}{R_L'} = \frac{10 - 7.5(1)}{0.5 \text{ k}\Omega} = 5 \text{ mA}$$

This checks with the result of the graphical solution; since A is smaller than I_{CQ}, i_{CM} is equal to 2A, or 10 mA.

Suppose that the bias is changed so that I_{CQ} is now equal to 6 mA. Substituting into the equation to find A, we get:

$$A = \frac{V_{\text{CC}} - I_{\text{CQ}}R_L}{R_L'} = \frac{10 - 6(1)}{0.5} = 8 \text{ mA}$$

Note that A (8 mA) is now *greater* than I_{CQ} (6 mA); therefore, i_{CM} is now equal to $2I_{\text{CQ}}$, or 12 mA.

Summarizing, we can make the following statements:

1. For $0 < I_{\text{CQ}} \leq \frac{1}{2}I_{\text{CS}}$: $i_{\text{CM}} = 2I_{\text{CQ}}$
2. For $\frac{1}{2}I_{\text{CS}} < I_{\text{CQ}} < I_{\text{CS}}$:

    Compute $A = \dfrac{V_{\text{CC}} - I_{\text{CQ}}R_L}{R_L'}$

    a. If $A < I_{\text{CQ}}$: $i_{\text{CM}} = 2A$
    b. If $A > I_{\text{CQ}}$: $i_{\text{CM}} = 2I_{\text{CQ}}$

If there is no second stage, r_{I_2} equals infinity and R_L equals R_L' (dc and ac load lines are the same). Thus, if R_L equals R_L', we can write the expression for A as follows:

$$A = \frac{V_{CC} - I_{CQ}R_L}{R_L} = \frac{V_{CC}}{R_L} - I_{CQ}$$

But we know that:

$$I_{Cs} = \frac{V_{CC}}{R_L}$$

Therefore, A is:

$$A = I_{Cs} - I_{CQ}$$

The quantity i_{CM} is thus:

$$i_{CM} = 2A = 2(I_{Cs} - I_{CQ})$$

Note that this is the same expression developed for one-stage amplifiers in chapter 4, and repeated again a few paragraphs back; this illustrates that the expression for the one-stage amplifier is a special case of the more general expression for multi-stage amplifiers derived above.

Most of the discussions in this section have neglected an important relationship between the β of a transistor and its r_I; we shall see in the next chapter that such a relationship exists. The relationship was not introduced at this time because the authors felt that it would complicate the major concepts discussed.

The concepts developed in this section concerning the various design factors were not discussed primarily for the purpose of developing the reader into a circuit designer. The object here is to illustrate the fact that the design of any electronic circuit or system involves the consideration of many factors, and that the completed design, as we shall see in chapter 10, is usually the result of a compromise between a number of conflicting requirements.

6-5 GAIN OF THE CASCADED AMPLIFIER

Figure 6.26 shows the ac equivalent circuit of a two-stage amplifier using fixed bias; in this diagram i_S is the source (or input) current and R_S is the internal

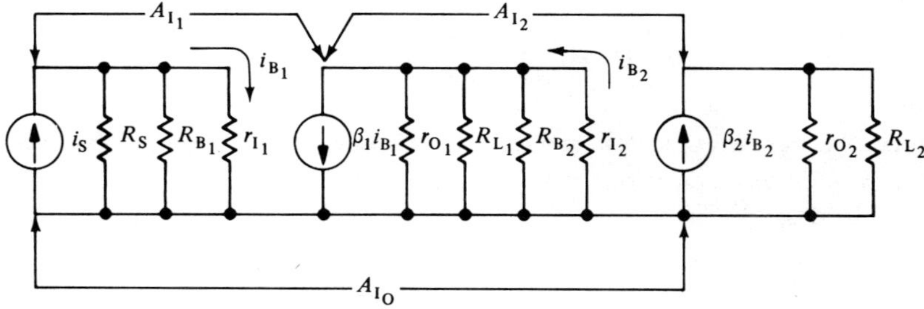

Figure 6.26

resistance of the source. We should like to develop a convenient means of designating the gain of this amplifier, one which will be used throughout the remainder of this text. There are many ways in which we can express the gain of an amplifier; we can refer to its current gain, voltage gain, or power gain. We use the *current gain* as an indication of the effectiveness of the amplifier. Returning to the two-stage amplifier discussed in Secs. 6-3 and 6-4, we repeat the values of the various peak-to-peak ac currents computed for that amplifier as follows:

$$i_S = 1 \ \mu A \qquad\qquad R_S = 1 \ k\Omega$$

$$i_{B_1} = 0.5 \ \mu A \qquad \beta_1 i_{B_1} = 50(0.5 \ \mu A) = 25 \ \mu A$$

$$i_{B_2} = 22.3 \ \mu A \qquad \beta_2 i_{B_2} = 50(22.3 \ \mu A) = 1115 \ \mu A$$

Before we deal with the current gain of the amplifier, we must determine which is the input current and which is the output. Is the *input* current to the first stage in Fig. 6.26 i_S, or is it i_{B_1}? Is the *output* current of that stage $\beta_1 i_{B_1}$, or the current through R_{L_1}, or the current through r_{I_2} (i_{B_2})? By convention, we define the *input* current to stage 1 as i_S and the *ouput* current as $\beta_1 i_{B_1}$; similarly, we define the input current to stage 2 as $\beta_1 i_{B_1}$ and the output current as $\beta_2 i_{B_2}$. In other words, the current gains of the various stages are equal to the ratios of the *appropriate current generators.*

We compute the current gains of stages 1 and 2 as follows:

$$\text{STAGE 1:} \quad A_{I_1} = \frac{\beta_1 i_{B_1}}{i_S} = \frac{25 \ \mu A}{1 \ \mu A} = 25$$

$$\text{STAGE 2:} \quad A_{I_2} = \frac{\beta_2 i_{B_2}}{\beta_1 i_{B_1}} = \frac{1115 \ \mu A}{25 \ \mu A} = 44.6$$

We define the *over-all gain* of the entire two-stage amplifier as the value of the *final* current generator (in this case, $\beta_2 i_{B_2}$) divided by the value of the *initial* current generator (i_S); thus:

$$\text{OVER-ALL GAIN} = A_{I_0} = \frac{\beta_2 i_{B_2}}{i_S} = \frac{1115 \ \mu A}{1 \ \mu A} = 1115$$

If we multiply the gains of the individual stages together, we arrive at the following results:

$$A_{I_1} \times A_{I_2} = \frac{\beta_1 i_{B_1}}{i_S} \times \frac{\beta_2 i_{B_2}}{\beta_1 i_{B_1}} = \frac{\beta_2 i_{B_2}}{i_S} = A_{I_0}$$

Therefore, we can say that *the over-all gain of the two-stage amplifier is equal to the product of the individual gains;* this can be shown to be true for any number of stages connected in cascade.

$$A_{I_0} = A_{I_1} \times A_{I_2}$$

Substituting into the above equation, we see that:

$$A_{I_0} = A_{I_1} \times A_{I_2} = 25 \times 44.6 = 1115$$

This checks with the above computations.

In the circuit of Fig. 6.26, the source current, i_S, can be viewed as the βi_B *current generator* of a *preceding* stage, and the internal source resistance, R_S, as the *output resistance* (R_O) of that stage. In other words, as far as our definition of gain is concerned, we actually do not care whether the current source composed of i_S and R_S represents an electromagnetic energy conversion device (such as a magnetic cartridge), or whether it represents the output circuit of some preceding (earlier) stage of amplification; you will recall that the output circuit of an amplifier can always be represented by a current generator in parallel with some output resistance.

The equations derived above concerning the gain of cascaded amplifier stages prove useful in predicting the number of stages of amplification required for a particular application. Actually, as we shall see in later chapters, the design of a multi-stage amplifier is a complex process; the number of stages required to meet the particular specifications depends upon many factors, some of which have not as yet been discussed. Let us assume for a moment that all of the stages in a multi-stage amplifier are to have the identical gain. We can then write the expression for the over-all gain as follows:

$$A_{I_O} = A_{I_1} \times A_{I_2} \times A_{I_3} \dots$$

$$\text{If:} \quad A_{I_1} = A_{I_2} = A_{I_3} \dots A_I$$

$$\text{Then:} \quad A_{I_O} = A_I \times A_I \times \dots$$

$$A_{I_O} = A_I{}^n$$

In this case, n represents the number of stages used. Suppose it is required that the over-all current gain of a particular multi-stage amplifier be equal to 1000; assume as a first approximation that the gains of the individual stages are identical. We can then write the following:

$$A_{I_O} = A_I{}^n = 1000$$

Let us see if it is possible for *two* cascaded stages to supply an over-all gain of 1000; i.e., n is equal to 2. Solving for A_I (the gain of each individual stage), we get:

$$1000 = A_I{}^2$$

$$A_I = \sqrt{1000} = 31.6$$

This means that each stage should have a gain of approximately 30. This can be achieved if the β of the transistors used in each of the two stages is *greater than 30;* the β of the transistors must be larger than the required gain to account for the current "lost" or "wasted" in R_L and the bias resistors (we see later on in the chapter that the values of the bias resistors, particularly in the emitter bias circuit, have an important effect upon the gain). It is *not* necessary, of course, that the gains of the two stages each be equal to 31.6; the gain of the first stage might be 25 and that of the second, 40. The gain required for *each individual stage* depends upon a number of factors, including the value of the source resistance, R_S, as well as the value of the final load to which the amplified power is being delivered. These concepts are discussed in greater detail later in this text; they are introduced at this point to indicate a very rough method for determining the number of stages required in a particular multi-stage amplifier. We could, of course, use *three* stages in the above amplifier, each with a gain of 10. This has the advantage that transistors with smaller β's can be used, but has the disadvantage that the extra stage contributes additional space and weight.

6-6 LIMITATIONS ON CASCADING

Suppose now that we add a *third identical stage* to the cascaded amplifier chain discussed earlier in this chapter. This third stage is shown connected to the output (collector) circuit of the second stage in Fig. 6.27(a). You may recall that the βi_B

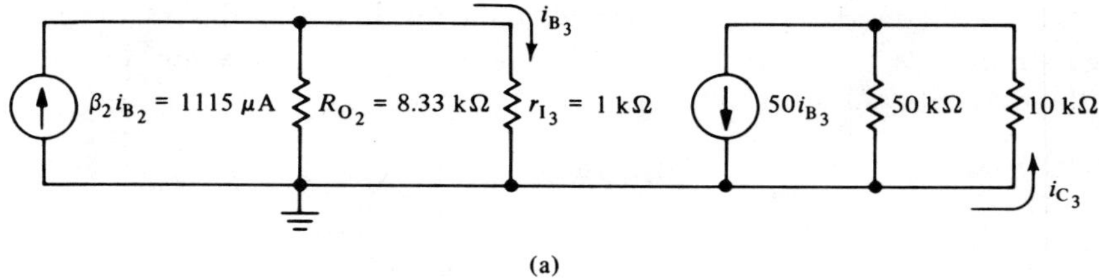

(a)

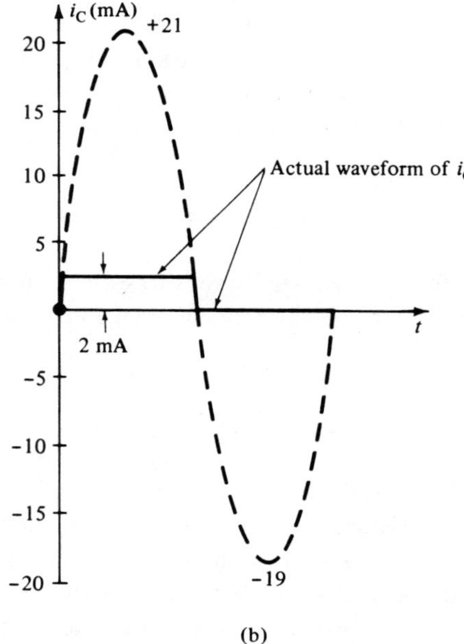

(b)

Figure 6.27

current generator for the second stage was 1115 μA and the output resistance of the stage was 8.33 kΩ; we again neglect the effect of the relatively large 1 MΩ bias resistor, which is not shown in the figure. Let us analyze the circuit of Fig. 6.27 (a). The first step is to compute the value of the ac base current in stage 3:

$$i_{B_3} = \frac{(\beta_2 i_{B_2})R_{O_2}}{R_{O_2} + r_{I_3}} = \frac{(1115 \ \mu A)(8.33)}{8.33 + 1 \ k\Omega} \simeq 1000 \ \mu A = 1.0 \ mA$$

The value of the current generator for stage 3 is:

$$\beta_3 i_{B_3} = 50(1) = 50 \ mA$$

Finally, the collector current in stage 3 is:

$$i_{C_3} = \frac{(\beta_3 i_{B_3})r_{O_3}}{r_{O_3} + R_{L_3}} = \frac{(50 \ mA)(50)}{50 + 10} = 41.6 \ mA$$

However, we know that the value of the collector saturation current is:

$$I_{Cs} = \frac{V_{CC}}{R_L} = \frac{20 \ V}{10 \ k\Omega} = 2 \ mA$$

Since the Q point is in the center of the load line (I_{CQ} equals 1 mA), i_{CM} is $2I_{CQ}$, or 2 mA. In other words, although the equations "predict" a peak-to-peak collector current of 41.6 mA, this waveform will *not* be generated because the maximum undistorted sinusoidal waveform available from the amplifier stage is 2 mA. Therefore, this amplifier stage is considerably *over-driven*, resulting in a severely distorted collector current waveform, as shown in Fig. 6.27(b).

Let us assume for a moment that the peak-to-peak collector current predicted by the equivalent circuit analysis is an even 40 mA. This means that the collector current will "try" to increase to 20 mA *above* the quiescent value of 1 mA (to 21 mA), and decrease to 20 mA *below* I_{CQ} (to *minus* 19 mA), as shown by the dotted waveform in Fig. 6.27(b). However, since the collector current cannot increase above 2 mA, nor decrease below 0 mA, the *actual* waveform of i_C is the 2 mA peak-to-peak waveform with the flattened top and bottom shown in the figure.

Thus the output of the third stage *is not a pure sine wave* like the outputs of stages 1 and 2. The reason is that the current has been amplified to such a large value that it can no longer be accommodated by the amplifier. This "squared" waveform shown in Fig. 6.27 is desirable in certain types of pulse circuits, but cannot be tolerated in an amplifier; it will result in distorted sound in an audio system, and a distorted picture in a video system. The third stage can be re-designed to accommodate a 40 mA peak-to-peak waveform by increasing V_{CC} and/or decreasing R_L, and changing the bias; we return to this particular concept in later chapters. The collector-to-emitter voltage will try to increase to:

$$v_{CE_3} = i_{C_3}R_L = (40 \ mA)(10 \ k\Omega) = 400 \ V$$

We know, however, that the maximum undistorted peak-to-peak collector-to-emitter voltage available from the amplifier is equal to V_{CC} if the Q point is in the center of the load line; i.e., $v_{CEM} = V_{CC}$. Therefore, instead of the *predicted* 400 V$_{P-P}$

sine wave, the *actual* collector-to-emitter waveform will be a 20 V_{P-P} voltage (from zero volts at saturation, to 20 V at cut-off) *with flattened top and bottom*, similar to the waveform of i_C shown in Fig. 6.27(b). The greater the ratio of i_C (the predicted peak-to-peak collector current, 40 mA in this case) to i_{CM} (the maximum undistorted peak-to-peak collector current available from the amplifier, 2 mA in this case), the closer the actual waveform (waveform of i_C shown in Fig. 6.27(b)) will resemble a square wave; this is an important consideration in the design of pulse circuitry.

The purpose of the above discussion has been to illustrate the fact that we cannot connect stages in cascade indefinitely and still expect the output to be sinusoidal; the maximum output of each amplifier stage is limited by the size of V_{CC} and R_L, by the transistor characteristics, and by the location of the Q point. These factors are discussed in greater detail in a later chapter.

EXAMPLE 1

Analyze the circuit shown in Fig. 6.28(a); find both the dc levels and the ac quantities for the one-stage amplifier shown.

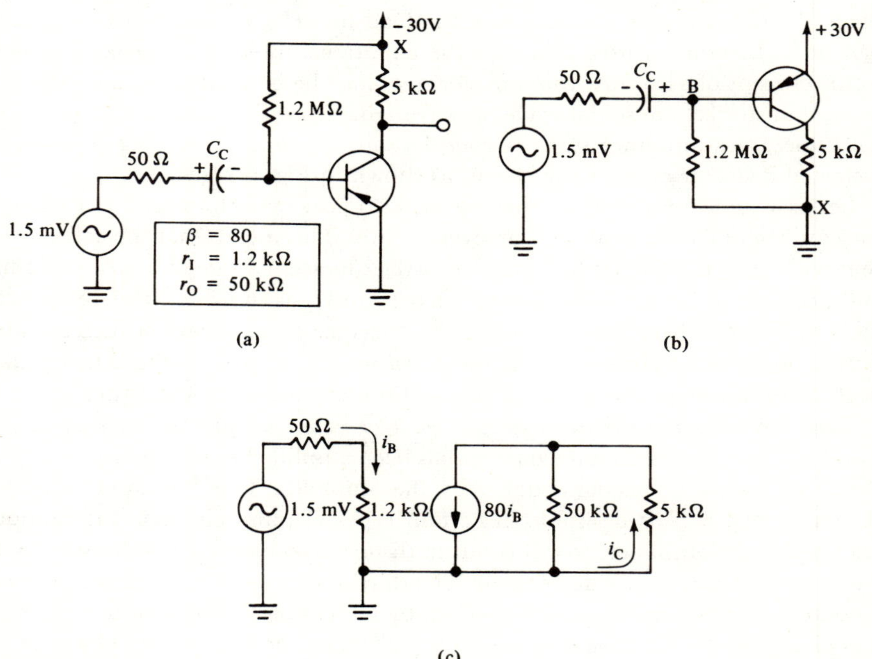

Figure 6.28

SOLUTION

Note first that the transistor used in the amplifier is a *pnp* type; this explains both the polarity on the coupling capacitor (the base of the transistor is a few tenths of a volt *negative* with respect to ground) and the *minus* 30 V dc supply. It should also be borne in mind that the dc currents in a *pnp* transistor flow in directions *opposite* to those in an *npn* transistor. The voltage rating for the coupling capacitor in the circuit of Fig. 6.28(a) need only be approximately 1.0 V, since the dc level of the input voltage (1.5 mV in series with 50 Ω) is 0 V and the dc base-to-emitter drop (V_{BE}) for the transistor is approximately 0.5 V. It should be obvious to the reader at this point that the *negative* terminal of the 30 V supply is connected to the common terminal of the 1.2 MΩ bias resistor and the 5 kΩ load resistor, and the *positive* terminal is connected to ground (not shown).

Sometimes it is desired that a *pnp* transistor be used in a system where the negative terminal of V_{CC} is connected to the circuit ground. In this case, the same transistor can be connected as shown in Fig. 6.28(b). The dc behavior of the circuit of Fig. 6.28(b) is the *same* as that of Fig. 6.28(a), since, in both cases, the emitter of the transistor is 30 V more positive (or 30 V less negative) than the common terminal of the bias and load resistors (terminal "X" in Figs. 6.28(a) and (b)). In other words, it makes no difference which terminal of V_{CC} is grounded as long as the emitter is 30 V *more positive* than terminal X. Neglecting the base-to-emitter drop in Fig. 6.28(b), we can say that the voltage drop across the 1.2 MΩ bias resistor, V_{BX}, is approximately $+30$ V; i.e., the base of the transistor is 30 V more positive than ground. Therefore, not only must the polarity of the coupling capacitor be *reversed* when moving from the circuit of Fig. 6.28(a) to that of 6.28(b), but the voltage rating of the capacitor used must be in excess of 30 V. These points were discussed to illustrate to the reader the many intricacies involved in circuit design.

Now let us proceed with the dc analysis of the circuit of Fig. 6.28(a). We first compute the quiescent base current:

$$I_{BQ} = \frac{V_{CC}}{R_B} = \frac{30 \text{ V}}{1.2 \text{ M}\Omega} = 25 \text{ } \mu A$$

Neglecting the leakage effects, the quiescent collector current at room temperature is:

$$I_{CQ} = \beta I_{BQ} = 80(25 \text{ } \mu A) = 2 \text{ mA}$$

The quiescent collector-to-emitter voltage is:

$$V_{CEQ} = V_{CC} - I_{CQ}R_L = 30 - (2)(5) = 20 \text{ V}$$

Observation of the circuit of Fig. 6.28(a) reveals that the collector of the transistor is at a potential 20 V *more negative* than the emitter (ground).

To analyze the ac behavior of the circuit, we first draw the ac equivalent circuit shown in Fig. 6.28(c); note that the capacitive reactance of C_C has been omitted since we again assume that the capacitor is a short circuit at ac. The equivalent circuit representation for a *pnp* transistor is the same as for an *npn* since dc voltages, currents, and polarities are not shown in the ac equivalent circuit. The ac base current, i_B, is found as follows:

$$i_B = \frac{1.5 \text{ mV}}{50 \text{ } \Omega + 1.2 \text{ k}\Omega} = \frac{1.5 \times 10^{-3}}{(0.05 + 1.2) \times 10^3} = 1.2 \times 10^{-6} \text{ A}$$

$$= 1.2 \text{ } \mu A$$

The value of the current generator is:

$$\beta i_B = 80(1.2 \ \mu A) = 96 \ \mu A$$

The ac collector current is:

$$i_C = \frac{(\beta i_B)r_O}{r_O + R_L} = \frac{(96 \ \mu A)50}{50 + 5} = 87 \ \mu A = 0.087 \ mA$$

Finally, v_{CE} is:

$$v_{CE} = i_C R_L = (0.087 \ mA)(5 \ k\Omega) = 0.436 \ V$$

Let us check to see if the amplifier can accommodate this ac output. The collector saturation current is:

$$I_{Cs} = \frac{V_{CC}}{R_L} = \frac{30 \ V}{5 \ k\Omega} = 6 \ mA$$

Since I_{CQ} (2 mA) is less than one-half I_{Cs} (6 mA), $i_{CM} = 2I_{CQ}$, or:

$$i_{CM} = 2I_{CQ} = 2(2 \ mA) = 4 \ mA$$

Since i_{CM} (4 mA) is *greater* than the computed value of i_C (0.087 mA), the amplifier stage *can* accommodate the ac output, and, therefore, the transistor will neither saturate nor cut-off.

Many times, the designer wishes to analyze some particular circuit to obtain an approximate idea of its behavior. The following expressions yield an excellent indication of the dc behavior for the fixed bias circuit:

1. $I_{BQ} = \dfrac{V_{CC}}{R_B}$ 3. $V_{CEQ} = V_{CC} - I_{CQ}R_L$

2. $I_{CQ} = \beta I_{BQ}$ 4. $I_{Cs} = \dfrac{V_{CC}}{R_L}$

To find the ac quantities, we must first compute i_B. Then, if r_O is greater than 10 times R_L, we can neglect the effect of r_O and assume that the collector current, i_C, is approximately equal to the value of the βi_B current generator.

1. Compute i_B

2. $i_C \simeq \beta i_B$ for $r_O \gg R_L$

Even if r_O is *not* much greater than R_L, the above procedure yields a reasonably good approximation if a rapid estimate of the circuit behavior is required.

EXAMPLE 2

A second identical stage is connected in cascade with the amplifier stage of example 1. The complete two-stage amplifier along with the input source is shown in Fig. 6.29(a). Analyze the behavior of the second stage.

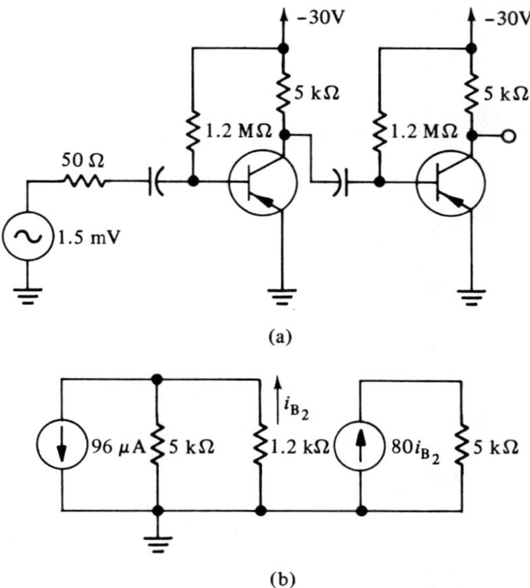

(a)

(b)

Figure 6.29

SOLUTION

Since the second stage is identical to the first, the dc values will be the same as those found in example 1:

$$I_{\text{BQ}} = 25 \ \mu\text{A} \qquad V_{\text{CEQ}} = 20 \text{ V}$$

$$I_{\text{CQ}} = 2 \text{ mA} \qquad I_{\text{Cs}} = 6 \text{ mA}$$

Figure 6.29(b) shows the ac equivalent circuit of Fig. 6.29(a); both r_{O} and R_{B} have been omitted from the diagram. The base current, i_{B}, is found as:

$$i_{\text{B}_2} = \frac{(96 \ \mu\text{A})(5)}{5 + 1.2} = 77.5 \ \mu\text{A}$$

The collector current, i_{C}, is approximated as:

$$i_{\text{C}_2} \simeq \beta i_{\text{B}_2} = 80(77.5 \ \mu\text{A}) = 6200 \ \mu\text{A} = 6.2 \text{ mA}$$

However, we saw before that $i_{\text{CM}} = 4$ mA if the transistor is biased at a collector current of 2 mA; therefore, this second stage *cannot* accommodate a 6.2 mA$_{\text{P-P}}$ ac collector current and the output waveforms will be distorted.

It is important for the reader to bear in mind that the above example was solved using a number of approximations: We neglected the effects of both R_{B} and r_{O}. In addition, we have been using the *small* signal ac equivalent circuit in the solution of problems involving relatively *large* signals; recall from Sec. 6-4 that the *accuracy* of the results of an analysis using the equivalent circuit *decreases* as the *size* of the signal *increases*. However, the methods of analysis used in examples 1 and 2 are extremely important to the designer in his estimation of the *approximate* operation of the amplifier circuit.

6-7 DIRECT COUPLING THE EMITTER BIAS CIRCUIT

Up to this point in the chapter we have been considering the ac operation of the fixed bias amplifier circuit; we now turn our attention to the emitter bias circuit. The first problem to be discussed is the method of coupling the amplifier stages. We saw earlier in this chapter that when we attempted to direct couple two fixed bias stages, one of the transistors was driven very far into saturation (Sec. 6-3). Let us now see what happens when we dc couple into an amplifier stage using emitter bias. Figure 6.30 shows a fixed bias stage (using transistor T_1) directly coupled to

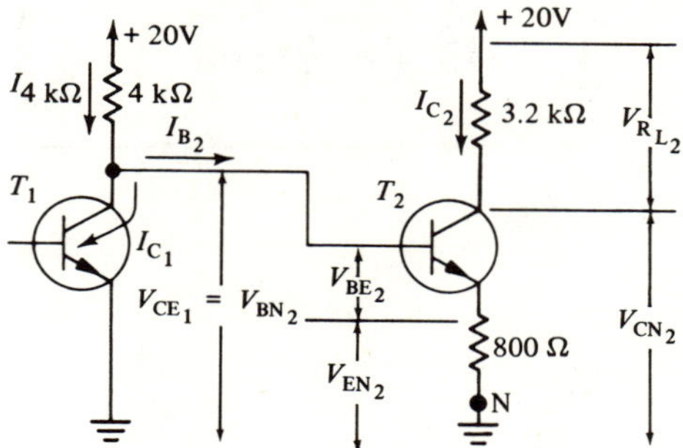

Figure 6.30

an emitter biased stage (using T_2). Note that the current flowing through the 4 kΩ load resistor in stage 1 ($I_{4 k\Omega}$) supplies both the *collector* current for T_1 (I_{C_1}) and the *base* current for T_2 (I_{B_2}). Since the base current for the transistor of stage 2 is supplied from the load resistor of stage 1, the two bias resistors (R_1 and R_2, which are usually used to supply dc base current to the transistor in the emitter bias circuit as shown in chapter 5) are not necessary and are thus omitted from the circuit in Fig. 6.30.

The actual analysis of the dc coupled amplifier circuit of Fig. 6.30 is quite complex, involving a liberal use of Thevenin's theorem. To avoid long, complicated mathematical analyses, we shall make some approximations which will serve adequately in illustrating the major points to be made in this section. It can be shown that the base current of T_2 is *much smaller* than the collector current of T_1. Therefore, we shall assume that *almost all* of the current in the 4 kΩ load resistor of stage 2 flows into the collector of T_1; i.e.:

$$I_{4 k\Omega} \simeq I_{C_1}$$

We are now ready to analyze the circuit of Fig. 6.30. Suppose that $I_{C_1} = 4.4$ mA; this means that $I_{4 k\Omega}$ is *also* equal to 4.4 mA. According to Ohm's law, the voltage across the 4 kΩ resistor is:

$$V_{4 k\Omega} = (I_{4 k\Omega})(4 k\Omega) = (4.4 \text{ mA})(4 k\Omega) = 17.6 \text{ V}$$

V_{CE_1} is thus found as follows:

$$V_{CE_1} = V_{CC} - V_{4\ k\Omega} = 20 - 17.6 = 2.4 \text{ V}$$

Now it can be seen from Fig. 6.30 that $V_{CE_1} = V_{BN_2}$; this is because the collector of T_1 is connected directly to the base of T_2, and the emitter of T_1 is connected to ground (terminal N). It can also be seen from the figure that the following relationship is true, according to Kirchhoff's voltage law:

$$V_{BN_2} = V_{BE_2} + V_{EN_2}$$
$$V_{EN_2} = V_{BN_2} - V_{BE_2}$$

Let us assume that $V_{BE_2} = 0.4$ V; then:

$$V_{EN_2} = V_{BN_2} - V_{BE_2} = 2.4 \text{ V} - 0.4 \text{ V} = 2.0 \text{ V}$$

Assuming the collector current of T_2 is approximately equal to the emitter current, we find I_{C_2} as:

$$I_{C_2} \simeq I_{E_2} = \frac{V_{EN_2}}{R_E} = \frac{2.0 \text{ V}}{0.8 \text{ k}\Omega} = 2.5 \text{ mA}$$

The voltage across the 3.2 kΩ load resistor of stage 2 is:

$$V_{RL_2} = (I_{C_2})(R_{L_2}) = (2.5 \text{ mA})(3.2 \text{ k}\Omega) = 8.0 \text{ V}$$

Finally, the collector-to-ground voltage of stage 2 (V_{CN_2}) is:

$$V_{CN_2} = V_{CC} - V_{RL_2} = 20 \text{ V} - 8 \text{ V} = 12.0 \text{ V}$$

We have completely analyzed the dc behavior of the circuit of Fig. 6.30.

Now suppose that, due to a temperature increase or replacement of T_1 by another unit, the collector current of stage 1 *increases* by 0.1 mA; i.e.:

$$I_{C_1} = 4.5 \text{ mA}$$

Let us now repeat the analysis for these new conditions.

$$V_{4\ k\Omega} = (4.5 \text{ mA})(4 \text{ k}\Omega) = 18 \text{ V}$$
$$V_{CE_1} = 20 - 18 = 2.0 \text{ V}$$

To be realistic, we should make provision for the fact that the base-to-emitter voltage of T_2 *decreases* with an *increase* in temperature; let us assume that V_{BE_2} *drops* from 0.4 V to 0.3 V. Continuing with the analysis, we get:

$$V_{EN_2} = V_{BN_2} - V_{BE_2} = 2 \text{ V} - 0.3 \text{ V} = 1.7 \text{ V}$$
$$I_{C_2} = \frac{V_{EN_2}}{R_E} = \frac{1.7 \text{ V}}{0.8 \text{ k}\Omega} = 2.125 \text{ mA}$$
$$V_{RL_2} = (2.125 \text{ mA})(3.2 \text{ k}\Omega) = 6.8 \text{ V}$$
$$V_{CN_2} = 20 - 6.8 = 13.2 \text{ V}$$

The results of this analysis illustrate a number of significant points. Note first that, although the total change in collector current in stage 1 was *0.1 mA* (from 4.4 mA to 4.5 mA), the total change for stage 2 was *0.375 mA* (from 2.5 mA to 2.125 mA). Similarly, the total change in V_{CE_1} was *0.4 V* (from 2.4 V to 2.0 V), while the total change in V_{CN_2} was *1.2 V* (from 12 V to 13.2 V). In other words, when direct coupling is used, *the Q point shift in the first stage is transmitted and amplified so that the Q point shift in the second stage is greater than in the first*. This phenomenon is known as *thermal drift*. Thermal drift is a consequence of the fact that the amplifier provides both current and voltage gain. We saw in chapter 4 that a small signal present at the base of the transistor (input of the amplifier) will be amplified at the collector (output); we can thus amplify small ac currents and voltages. The amplifier, however, *cannot* distinguish between a change caused by a sinusoidal variation, and a small shift due to the change in the Q point of the previous stage. Since the collector of stage 1 is connected directly to the base of stage 2, any Q point shift at the collector of stage 1 is transmitted directly to the base of stage 2, and then amplified in the collector circuit of stage 2. In other words, while the gain of the amplifier is important in amplifying small ac signals, it creates a problem by *also* amplifying any Q point shifts.

Thermal drift is an undesirable feature of the direct coupled amplifier. A 0.4 V shift in the collector-to-emitter voltage of stage 1 was followed by a 1.2 V shift in the collector-to-ground voltage of stage 2. If the collector of stage 2 is now direct coupled to the base of yet a *third* stage, the gain of this third stage will produce an *even larger* Q point shift in *its* collector circuit. As the Q point shift becomes larger with each successive amplifier stage, some later stage will eventually be driven either to cut-off or saturation. It is for this reason that the designer must use great care in working with direct coupled amplifiers.

The reader should *not* get the impression that because of thermal drift we never use direct coupled amplifiers. Direct coupled amplifiers *are* used in practice in many applications. However, we usually provide elaborate feedback circuitry to stabilize the Q points, and thus minimize the problems caused by thermal drift. Because of the complexity of this feedback circuitry, we do not discuss direct coupled amplifiers at this point; for the remainder of this chapter we concern ourselves *only* with R-C coupled amplifier circuits. The major advantage of R-C coupled circuits is that *the dc conditions of each stage are completely independent of those of the other stages;* the coupling capacitor prevents amplification of the Q point shift of one stage by the following stage, and thus there is no problem of thermal drift.

6-8   EMITTER BIAS CIRCUIT: INPUT RESISTANCE

Before undertaking a detailed study of the ac behavior of the emitter biased amplifier, we must discuss some very important points concerning this circuit which did not exist in the case of the fixed bias amplifier circuit. Up to this point in the text, we have been referring to the input resistance of the amplifier as r_I; you may recall from chapters 3 and 4 that r_I is approximately equal to h_{ie}, which in turn is the dynamic resistance (reciprocal of the slope) of the input characteristic.

The input signal to an amplifier is usually applied directly between the *base* and *ground* terminals. *Since the emitter is grounded in the case of the fixed bias circuit, the input signal is effectively connected directly across the terminals of the base-emitter diode (base and emitter terminals); the resistance between these two terminals is r_I. Therefore, since the input is connected across the terminals of a device whose resistance is r_I, we can say that the input resistance to the amplifier is equal to r_I in the case of the fixed bias circuit.*

The situation is entirely different in the case of the emitter biased amplifier. Here, the emitter is *not* connected to ground; there is an emitter resistor, R_E, connected between the emitter terminal and ground. Therefore, the input resistance to the amplifier is *not* equal to r_I. At this point it is important to make a distinction between R_I, *the general term for the input resistance to an amplifier*, and r_I, the dynamic resistance of the base-emitter diode; in general, these two terms are *not* the same, as we shall see.

Let us now compute R_I for the emitter-biased amplifier circuit. To find this quantity for any type of circuit, an input voltage, v_I, is applied to the two input terminals in question, terminals X and Y, as shown in Fig. 6.31(a). The input current, i_I, is measured. The input resistance to the two terminals, R_I, is equal to v_I divided by i_I.

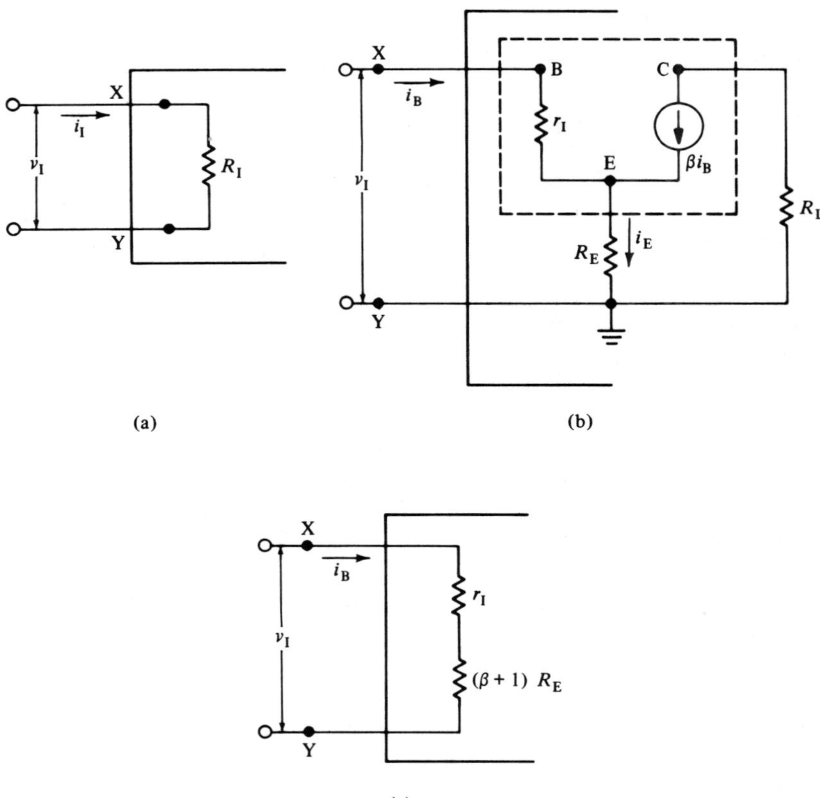

(a)   (b)

(c)

Figure 6.31

$$R_I = \frac{v_I}{i_I}$$

Figure 6.31(b) shows the ac equivalent circuit of the emitter-biased amplifier; the two bias resistors, R_1 and R_2, have been omitted for simplicity. The transistor itself is located within the dotted box in the figure. Note that the output resistance of the transistor, r_0, has been omitted from the circuit. This has been done to simplify the following manipulations; the result still possesses a high degree of accuracy. It can also be seen from the figure that the emitter terminal is no longer at ground potential; thus R_L is connected between collector and *ground*, *not* between collector and *emitter* as in the case of the fixed bias circuit.

We compute the input resistance of the circuit of Fig. 6.31(b) (between terminals X and Y) by an algebraic technique. An input voltage, v_I, is applied to terminals X and Y; the input current, i_I, is equal to the base current, i_B, in this case. Note that i_B flows through r_I, but it is the emitter current, i_E, which flows through R_E. Applying Kirchhoff's voltage law to the input loop, we obtain:

$$v_I = r_I i_B + R_E i_E$$

It can be seen from the circuit of Fig. 6.31(b) that i_E can be expressed as follows:

$$i_E = i_B + \beta i_B = i_B(\beta + 1)$$

Substituting and manipulating algebraically, we get:

$$v_I = r_I i_B + R_E[i_B(\beta + 1)]$$
$$v_I = r_I i_B + R_E(\beta + 1)i_B$$
$$v_I = i_B[r_I + (\beta + 1)R_E]$$

Now we know that the input resistance is equal to the input voltage divided by the input current; therefore:

$$R_I = \frac{v_I}{i_B} = \frac{i_B[r_I + (\beta + 1)R_E]}{i_B}$$
$$R_I = r_I + (\beta + 1)R_E$$

This is the expression for the input resistance *to this particular amplifier* circuit; in general, R_I is different for each circuit.

Now let us examine this expression more closely. Note first that it is composed of two parts; i.e., the effect is as if there are *two separate* resistors connected between terminals X and Y, as shown in Fig. 6.31(c). The first resistor, r_I, is the dynamic resistance of the base-emitter diode; this depends primarily upon the transistor. The second resistor, $(\beta + 1)R_E$, is known as the *reflected emitter resistance;* i.e., the emitter resistor, R_E, is reflected, or *multiplied* by the β of the transistor plus one. This reflected resistance depends upon the β of the transistor and the value of R_E. Usually, the β of the transistor is much greater than 1.0; therefore, we can make the following approximation:

$$R_\mathrm{I} \simeq r_\mathrm{I} + \beta R_\mathrm{E} \qquad \text{for } \beta \gg 1$$

EXAMPLE 3

A transistor amplifier using emitter bias has the following resistors and parameters associated with it: $r_\mathrm{I} = 1$ kΩ, $\beta = 50$, and $R_\mathrm{E} = 500$ Ω. Compute the input resistance to the amplifier.

SOLUTION

Substituting into the equation we get:

$$R_\mathrm{I} = r_\mathrm{I} + \beta R_\mathrm{E}$$
$$= 1 \text{ k}\Omega + 50(0.5 \text{ k}\Omega) = 1 \text{ k}\Omega + 25 \text{ k}\Omega = 26 \text{ k}\Omega$$

Note that the term due to the reflected resistance (25 kΩ) is *much larger* than that due to r_I (1 kΩ); this is generally the case. In other words, the reflected resistance due to the presence of the emitter resistor in the circuit is responsible for a *large increase* in the input resistance of the amplifier; this large input resistance has some very important effects upon the operation of the emitter biased amplifier, as we soon see.

EXAMPLE 4

The same transistor of example 3 is used in an amplifier where R_E is set equal to zero; i.e., *no emitter resistor* is included in the circuit. Compute the input resistance.

SOLUTION

If there is no emitter resistor in the circuit, the emitter terminal is connected to ground, and the amplifier is no longer emitter biased. Computing the input resistance we get:

$$R_\mathrm{I} = r_\mathrm{I} + \beta R_\mathrm{E} = 1 \text{ k}\Omega + 50(0) = 1 \text{ k}\Omega$$
$$R_\mathrm{I} = 1 \text{ k}\Omega = r_\mathrm{I} \qquad (R_\mathrm{E} = 0)$$

This shows that the input resistance of the amplifier is equal to r_I *only* when there is *no emitter resistor* present; i.e., when the amplifier is *fixed biased*.

6-9 EMITTER BIAS CIRCUIT: OUTPUT RESISTANCE

We have seen in the preceding section that the inclusion of the emitter resistor in the amplifier (for purposes of biasing and stability) results in a large increase in the *input* resistance of the amplifier. The *output* resistance of the amplifier is also affected by the presence of the emitter resistor. The methods used for computing the output resistance of the emitter biased amplifier circuit involve liberal use of the superposition theorem as well as a considerable amount of algebraic manipulation. This derivation will not be presented here. We saw earlier in the chapter that R_O for the *fixed bias* circuit was equal to the parallel combination of r_O and R_L. The results of the algebraic derivation reveal that the output resistance of the *emitter bias* circuit is:

$$R_L \left\| r_O \left[1 + \frac{\beta R_E}{R_E + r_I + R_{O_1}} \right] = R_O \right.$$

In this expression, R_O is the output resistance of the entire amplifier stage, R_L is the load resistor of the amplifier stage, $r_O = 1/h_{ce}$, and R_{O_1} is the output resistance of the *previous stage*.

EXAMPLE 5

The following are the component and parameter values for a particular amplifier: $R_L = 4$ kΩ, $r_O = 30$ kΩ, $\beta = 50$, $R_E = 500$ Ω, $r_I = 1$ kΩ, and $R_{O_1} = 4$ kΩ. Compute the output resistance.

SOLUTION

It can be seen from the expression that the inclusion of R_E in the circuit has the effect of multiplying r_O by the quantity within the brackets, which we call X; i.e.:

$$X = 1 + \frac{\beta R_E}{R_E + r_I + R_{O_1}}$$

Substituting into the equation we find that X is:

$$X = 1 + \frac{\beta R_E}{R_E + r_I + R_{O_1}} = 1 + \frac{50(0.5)}{0.5 + 1 + 4} = 5.54$$

Therefore, r_O (30 kΩ) is multiplied by 5.54. The original expression states that the output resistance is equal to the parallel combination of R_L (4 kΩ) and $r_O X$; $r_O X$ is:

$$r_O X = (30 \text{ k}\Omega)(5.54) = 166 \text{ k}\Omega$$

The output resistance is the parallel combination of 4 kΩ and 166 kΩ; since 4 kΩ is *much smaller* than 166 kΩ, we can say that R_O for this circuit *is approximately equal to 4 kΩ, or R_L*.

The above example illustrates the fact that the inclusion of R_E in the circuit is responsible for the multiplication of r_O by the factor X; since, as we have seen, $r_O X$ (166 kΩ) *is much larger than R_L (4 kΩ), we can say that *the output resistance of the emitter-biased amplifier circuit is approximately equal to R_L.*

$$R_O \simeq R_L \qquad \textit{Emitter Bias}$$

EXAMPLE 6

Repeat example 5 if $R_E = 0$ Ω.

SOLUTION

Solving first for X, we see that:

$$X = 1 + \frac{\beta R_\text{E}}{R_\text{E} + r_\text{I} + R_{\text{O}_1}} = 1 + \frac{\beta(0)}{0 + r_\text{I} + R_{\text{O}_1}} = 1$$

Therefore, since $X = 1$, $r_\text{O}X = r_\text{O}$; R_O is the parallel combination of r_O and R_L, or:

$$R_\text{O} = \frac{r_\text{O}R_\text{L}}{r_\text{O} + R_\text{L}} = \frac{30(4)}{30 + 4} = 3.53 \text{ k}\Omega$$

This example illustrates that *when R_E is removed from the circuit* (set equal to zero ohms), *the output resistance of the amplifier is the same as for the fixed bias circuit* (see Sec. 6-3), namely:

$$R_\text{O} = \frac{r_\text{O}R_\text{L}}{r_\text{O} + R_\text{L}} \qquad \textit{Fixed Bias}$$

In other words, since it is the emitter resistor which is responsible for the multiplication of r_O by the relatively large factor of X (5.54, in example 5), when $R_\text{E} = 0$, as in example 6, X reduces to 1 (r_O is multiplied by 1).

We just saw that when $R_\text{E} = 0$, R_O is equal to the parallel combination of r_O and R_L (3.53 kΩ). Since r_O (30 kΩ) is generally much larger than R_L (4 kΩ), little error results from saying that R_O is approximately equal to R_L (4 kΩ), *even when* $R_\text{E} = 0$; compare the value of R_O computed taking r_O into account (3.53 kΩ) with that value computed neglecting r_O (4 kΩ). Throughout the remainder of this text, *we generally neglect r_O in comparison with R_L.*

The ac equivalent circuit of the emitter-biased amplifier is shown in Fig. 6.32(a).

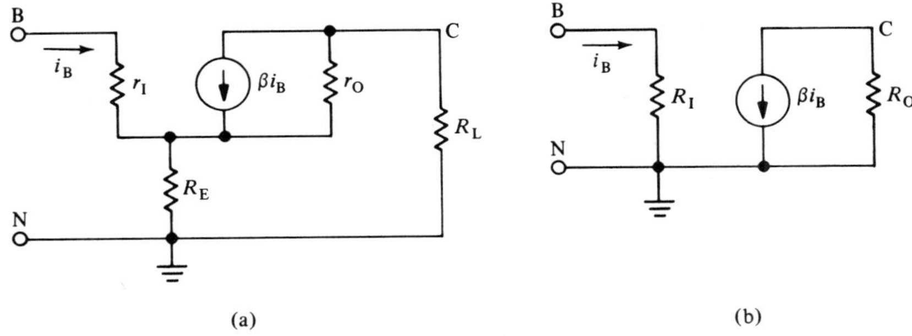

Figure 6.32

Because of the interconnection of the various resistors and the current generator, the analysis of this circuit can be quite complex. Therefore, we use instead the equivalent circuit of Fig. 6.32(b); obviously, this circuit is much simpler to work with since it consists of only *two resistors and one current generator*. As previously mentioned, most amplifier circuits can be reduced to the form shown in Fig. 6.32(b). For the emitter-biased circuit, $R_\text{I} = r_\text{I} + \beta R_\text{E}$, as shown in Sec. 6-8, and

R_O is approximately equal to R_L. It is important to mention again that the statements made in the preceding sentence *hold only for the emitter-biased circuit;* the expressions for both R_I and R_O are, in general, different for other circuit configurations.

6-10 ANALYSIS OF THE EMITTER-BIASED AMPLIFIER

Figure 6.33(a) shows a fixed bias amplifier stage capacitor-coupled to an emitter-biased stage. Note that the bias resistors (47 kΩ and 10 kΩ) are needed to supply dc

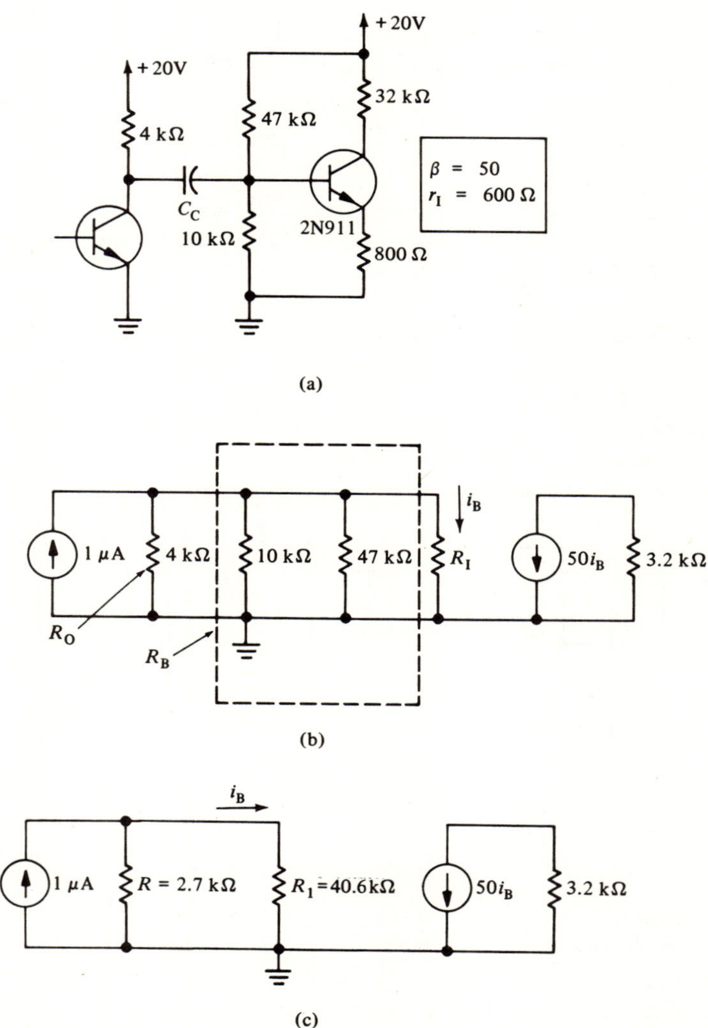

Figure 6.33

current to the base of the transistor; in addition, along with the 800 Ω emitter resistor they help to stabilize the Q point, as was shown in chapter 5. Figure 6.33(b) shows the ac equivalent circuit of the configuration of Fig. 6.33(a). Note first that the current generator for the first stage is 1.0 μA. Second, we have neglected the r_O of the first stage; therefore, R_O for that stage is approximately equal to the 4 kΩ load resistor, as shown in Fig. 6.33(b). The output circuit of the first stage thus consists of a 1.0 μA current generator in parallel with a 4 kΩ resistor. The third important point concerns the two bias resistors, R_1 (47 kΩ) and R_2 (10 kΩ). You may recall that the 1 MΩ bias resistor used in the fixed bias circuit discussed earlier in this chapter was large enough to be neglected in comparison with the resistors in parallel with it; it is generally true that R_B is large enough to be neg-lected in the fixed bias circuit. This is *not* true, however, in the case of the emitter bias circuit *because R_1 and R_2 are not always very large;* therefore, they must be included in the equivalent circuit as shown in Fig. 6.33(b). R_2 is connected between base and ground in Fig. 6.33(a) and thus appears the same way in Fig. 6.33(b). R_1 is connected between the base and positive terminal of the 20 V dc supply; since dc supplies are shorted in the ac circuit, R_1 is *also* connected between base and ground in the ac circuit of Fig. 6.33(b). Thus, R_1 and R_2 (47 kΩ and 10 kΩ, in this case) are connected in *parallel* in the ac equivalent circuit, as shown within the dotted box of Fig. 6.33(b); the parallel combination of R_1 and R_2 is referred to as R_B (see chapter 5). Finally, the remainder of the circuit is represented as described in the preceding two sections: an input resistance (R_I), a current generator (βi_B), and an output resistance ($R_O = R_L = 3.2$ kΩ).

The next step is to analyze the ac behavior of the circuit of Fig. 6.33(a). We must first know the value of the input resistance, R_I. Substituting into the equation derived in Sec. 6-8, we get:

$$R_I = r_I + \beta R_E = 0.6 \text{ k}\Omega + 50(0.8 \text{ k}\Omega)$$
$$R_I = 40.6 \text{ k}\Omega$$

Now we are interested in finding the value of the base current, i_B. The 1.0 μA from the current generator will split between *four* resistances: $R_O = 4$ kΩ, $R_2 = 10$ kΩ, $R_1 = 47$ kΩ, and $R_I = 40.6$ kΩ. Since we wish to know i_B, the current flowing in R_I, we must combine the other three resistances in parallel. We already know that R_B is equal to the parallel combination of R_1 and R_2; thus:

$$R_B = \frac{R_1 R_2}{R_1 + R_2} = \frac{(47)(10)}{47 + 10} = 8.25 \text{ k}\Omega$$

We must now combine R_B in parallel with R_O. Let us call this combination "R"; therefore, R represents the parallel combination of R_O, R_1, and R_2.

$$\boxed{R = \frac{R_O R_B}{R_O + R_B}} = \frac{4(8.25)}{4 + 8.25} = 2.7 \text{ k}\Omega$$

The ac circuit can now be represented as shown in Fig. 6.33(c). Using the current divider, we solve for the base current as follows:

$$i_B = \frac{(1\ \mu A)R}{R + R_I} = \frac{(1)(2.7)}{2.7 + 40.6} = 0.062\ \mu A$$

The βi_B current generator in the collector circuit is:

$$(50)(0.062\ \mu A) = 3.1\ \mu A$$

Finally, the gain of the amplifier is:

$$A_I = \frac{3.1\ \mu A}{1\ \mu A} = 3.1$$

This analysis points up a number of very important ideas. The most important point concerns the fact that *the gain of the amplifier is very small*, only 3.1; i.e., the output current is only 3.1 times as large as the input current. Let us examine the reasons for this unusually small value of gain; refer again to Figs. 6.33(b) and (c). From earlier sections of this chapter you may recall that, in order to have a large gain, we must supply a large percentage of the current from the current generator of the previous stage (in this case, 1.0 μA) to the base of the following stage. This means that we should like the *internal resistance* of the current source to be *large*, and the *input resistance* of the next stage to be *small*. It can be seen in the circuit of Fig. 6.33(c) that the *opposite* is true; the 2.7 kΩ resistance (R) can be considered as the *internal resistance* of the current source whose *current generator* is 1.0 μA. This internal resistance $(R = 2.7\ k\Omega)$ is *much smaller* than the input resistance $(R_I = 40.6\ k\Omega)$ of the stage; therefore, *only a small portion* (0.062 μA) *of the 1.0 μA current generator flows into* R_I. The remainder of the 1.0 μA (0.938 μA) flows through the 2.7 kΩ resistance. Since only 0.062 μA flows into the transistor itself as i_B, only this amount is multiplied by β, which accounts for the small value of gain.

Thus, there are actually *two* reasons for the small current gain: R_I is *too large* and R is *too small*. We have already studied in some detail the reason for the large size of R_I; we saw in Sec. 6-8 that the reflected emitter resistance, βR_E (in this case, 40 kΩ), adds to r_I and therefore has the effect of *increasing* the total input resistance to the amplifier. The results of the previous computations indicate that this increase in the input resistance is responsible for the small value of gain. The next question to ask is: Is there any way of increasing the gain of the amplifier while still using emitter bias to stabilize the Q point? One method would involve using a transistor with a *smaller β* to *decrease* the reflected resistance; however, although this *will* decrease the input resistance and thus allow a *greater* percentage of the 1.0 μA current to flow into the transistor, it will *also decrease* the value of the βi_B current generator (since the value of the current generator is directly proportional to β), with the consequence that the total gain will actually be smaller (less than 3.1). Thus this method will not work because it is self-defeating. Another solution might be to decrease the value of R_E; this also will result in a decrease in the value of R_I, and will, in fact, increase the gain. However, we saw in chapter 5 that *decreasing* the value of R_E *increases* the stability factor; thus, although the gain is improved, the stability becomes worse.

The above discussion points up the fact that we should like R_E to be *large* to provide good *dc* stability, but we also want it to be *small* for *proper ac operation* (large gain). Both of these objectives can be achieved by connecting a large capacitor, $C_E = 100\ \mu F$, in *parallel* with R_E as shown in Fig. 6.34(a). Let us now re-analyze the ac behavior of the amplifier to see how the presence of this capacitor improves

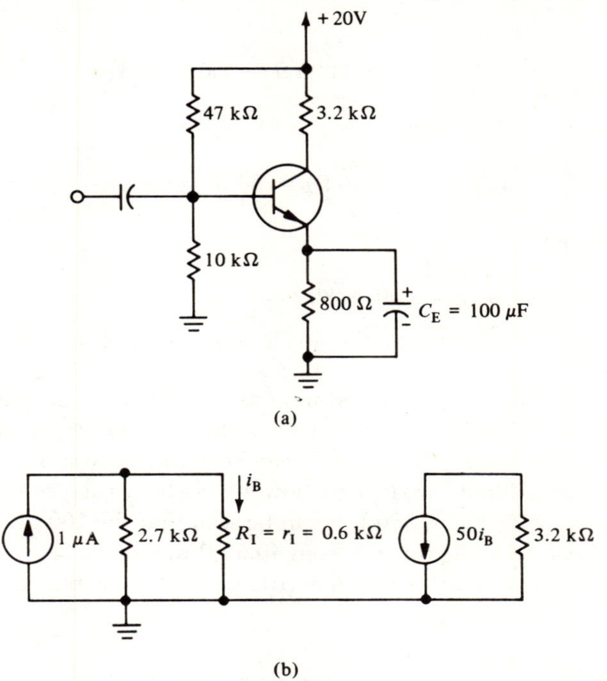

Figure 6.34

the gain. We assume that the frequency of the ac signal is 1.0 kHz; then the reactance of the 100 μF capacitor is:

$$X_{C_E} = \frac{1}{2\pi f C_E} = \frac{0.159}{10^3 \times 100 \times 10^{-6}} = 1.59\ \Omega \simeq 1.6\ \Omega$$

In deriving the expression for the input resistance in Sec. 6-8, we assumed that the only component connected between the emitter and ground was the emitter resistor, R_E. Now, however, there is both a resistor (R_E) *and* a capacitor (C_E). Therefore, to be more general, we should derive an expression for the input *impedance* instead of the input *resistance*. The expression for obtaining the input impedance is almost the same as for the input resistance, except for some slight modifications which are shown as follows:

$$Z_I = r_I + \beta Z_E$$

In this expression, Z_I is the input impedance, and Z_E is the total impedance between the emitter and ground terminals; the arrow in the expression indicates that a *vector* addition must be made.

In the circuit of Fig. 6.34(a), Z_E is the equivalent impedance of the reactance of the capacitor (X_{C_E}) taken in parallel with the emitter resistance (R_E). Since, as we just saw, $X_{C_E} = 1.6\ \Omega$ and $R_E = 800\ \Omega$, we can neglect the 800 Ω resistance because it is very large in comparison with the 1.6 Ω reactance; thus Z_E is expressed as follows:

$$Z_E = X_{C_E} \parallel R_E = 1.6\ \Omega \parallel 800\ \Omega \simeq 1.6\ \Omega = X_{C_E}$$

The reflected impedance, βZ_E, is now:

$$\beta Z_E = (50)(1.6\ \Omega) = 80\ \Omega$$

The input impedance is:

$$Z_I = r_I \oplus \beta Z_E = 600 \oplus 80$$

$$Z_I = \sqrt{(600)^2 + (80)^2} = 605\ \Omega \simeq 600\ \Omega = r_I$$

It can be seen above that the 80 Ω reflected impedance can easily be neglected in comparison with the 600 Ω value of r_I; *thus the input impedance to the amplifier is approximately equal to r_I, the same value obtained when no emitter resistor is in the circuit* (fixed bias). The ac equivalent circuit can now be re-drawn as shown in Fig. 6.34(b); comparing this with Fig. 6.33(c), it can be seen that the 100 μF capacitor has had the effect of dropping R_I (or Z_I) from 40.6 kΩ to 0.6 kΩ.

Let us now analyze the circuit of Fig. 6.34(b). Using the current divider, we find i_B to be:

$$i_B = \frac{(1\ \mu A)R}{R + r_I} = \frac{(1\ \mu A)(2.7\ k\Omega)}{2.7\ k\Omega + 0.6\ k\Omega} = 0.82\ \mu A$$

Continuing with the analysis we get:

$$\beta i_B = 50(0.82\ \mu A) = 41\ \mu A$$

$$A_I = \frac{41\ \mu A}{1\ \mu A} = 41$$

Notice that *the gain of the amplifier is now equal to 41, a significant improvement over 3.1.* The larger the gain of an amplifier stage, the fewer the number of stages needed for a particular application; a smaller number of stages means less components, with a corresponding reduction in the size, weight, and cost of the unit. For example, suppose that we connect three stages in cascade, each with a gain of 3.1; then the overall gain will be:

$$A_{Io} = A_I{}^n = (3.1)^3 = 29.8$$

Thus, *one stage with a gain of 41 provides more gain than three cascaded stages, each with a gain of 3.1.* This shows how important it is to try to obtain as much gain as possible from each amplifier stage.

We now study in more detail the operation of the capacitor, C_E. By placing this capacitor in parallel with R_E, we have reduced the value of the reflected impedance; this reflected impedance was equal to 40 kΩ *before* the capacitor was connected, and dropped markedly to 80 Ω *after* the connection was made. This *reduction* in the input impedance (from 40.6 kΩ to 0.6 kΩ) resulted in an *increase* in the base current (from 0.062 μA to 0.82 μA) with a corresponding *increase* in the gain (from 3.1 to 41). It can be seen now why the value of C_E must be made quite large; a *large* value of C_E means a *small* value of X_{CE}, and hence a *small* value of reflected impedance because X_{CE} shunts R_E.

Although C_E shunts R_E and thus provides a small input impedance for *ac* signals, *it has absolutely no effect on the dc conditions,* since a capacitor acts like an open circuit to dc currents. Therefore, this capacitor fulfills the requirements: It reduces the input impedance of the amplifier, but leaves the dc conditions unaffected; i.e., the stability factor of the amplifier stage remains unchanged when C_E is connected.

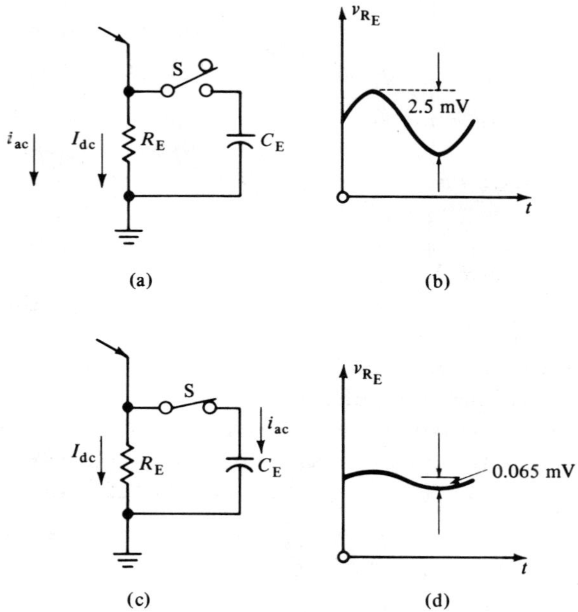

Figure 6.35

Figure 6.35(a) shows that when the switch is open (no C_E connected in circuit), both the dc *and* the ac portions of the emitter current flow through R_E. The waveform of the voltage across the emitter resistor, v_{RE}, is shown in Fig. 6.35(b). Note that the waveform consists of both an ac and a dc level. To find the value of the peak-to-peak ac voltage, we must multiply the emitter current by the emitter resistance. We saw that when no C_E was present, i_C was equal to 3.1 μA. Assuming that i_C is approximately equal to i_E, we can find the ac voltage to be:

$$v = i_E R_E = (3.1 \times 10^{-6})(800) \simeq 2.5 \text{ mV}$$

This is shown in Fig. 6.35(b); it should be mentioned that the waveforms shown in

Fig. 6.35 are *not* drawn to scale. Figure 6.35(c) shows that when the switch is closed (C_E connected into circuit), *all of the dc current* flows through R_E (since none can flow through the capacitor), while *almost all* of *the ac current* flows through the capacitor. This is obvious since the capacitor presents only a 1.6 Ω reactance to the ac current, while the resistance of R_E is 800 Ω. To find the new value of the ac portion of v_{R_E}, we must multiply i_C by Z_E. We saw that when the capacitor was connected, i_C was equal to 41 μA; we also know that Z_E is approximately equal to 1.6 Ω. Therefore, we find the ac voltage to be:

$$v = i_E Z_E \simeq i_C Z_E = (41 \times 10^{-6})(1.6) \simeq 0.065 \text{ mV}$$

The waveform is shown in Fig. 6.35(d); note the large reduction in the value of the ac voltage (from 2.5 mV to 0.065 mV). Note also that *the dc level remains unchanged;* this is because neither the dc emitter current nor the value of R_E has been changed by the addition of C_E to the circuit.

Because most of the ac current flows through C_E instead of R_E, we say that the ac current *by-passes* the emitter resistor to flow through the capacitor, and thus we call C_E the *emitter by-pass capacitor.*

From Fig. 6.33(c) you may recall that the reasons given for the small value of gain (3.1, before C_E was added) was that R (2.7 kΩ) was too small and R_I (40.6 kΩ) was too large. We have just solved the problem connected with R_I by adding a 100 μF by-pass capacitor; R_I was reduced from 40.6 kΩ to 0.6 kΩ and the gain was increased from 3.1 to 41. Let us now see if we can improve the gain further by increasing the value of R. You may recall that R is equal to the parallel combination of R_B (parallel combination of R_1 and R_2) and R_O of the previous stage. There is little we can do about R_O, since this is fixed by the dc conditions of the previous stage. We can increase R_B by increasing the values of both R_1 and R_2. It can be seen from Fig. 6.33(b) that the *larger* the values of both R_1 and R_2, the *less* ac current is diverted through them, and the *more* ac current flows through R_I with a consequent increase in the gain. Let us see the effect on the gain if both R_1 and R_2 are *increased ten times;* i.e., $R_1 = 470$ kΩ and $R_2 = 100$ kΩ. We must first find R_B:

$$R_B = \frac{R_1 R_2}{R_1 + R_2} = \frac{(470)(100)}{470 + 100} = 82.5 \text{ k}\Omega$$

R is equal to:

$$R = \frac{R_B R_O}{R_B + R_O} = \frac{(82.5)(4)}{82.5 + 4} = 3.8 \text{ k}\Omega$$

Proceeding as before, we find the gain to be:

$$i_B = \frac{(1 \ \mu A) R}{R + R_I} = \frac{(1 \ \mu A)(3.8)}{3.8 + 0.6} = 0.86 \ \mu A$$

$$\beta i_B = 50(0.86 \ \mu A) = 43 \ \mu A$$

$$A_I = \frac{43 \ \mu A}{1 \ \mu A} = 43$$

As expected, increasing the values of both R_1 and R_2 has resulted in an increase in the gain (from 41 to 43). From chapter 5 you may recall, however, that *increasing R_1 and R_2 causes a corresponding increase* in the stability factor (poorer stability). Computing S first for the original case ($R_1 = 47$ kΩ and $R_2 = 10$ kΩ), we get:

$$S = \frac{1 + \dfrac{R_E}{R_B}}{\dfrac{1}{\beta} + \dfrac{R_E}{R_B}} = \frac{1 + \dfrac{0.8}{8.25}}{\dfrac{1}{50} + \dfrac{0.8}{8.25}} = 9.36$$

Next we find S for $R_1 = 470$ kΩ and $R_2 = 100$ kΩ:

$$S = \frac{1 + \dfrac{R_E}{R_B}}{\dfrac{1}{\beta} + \dfrac{R_E}{R_B}} = \frac{1 + \dfrac{0.8}{82.5}}{\dfrac{1}{50} + \dfrac{0.8}{82.5}} = 33.9$$

This indicates that increasing R_1 and R_2 has improved the gain (from 41 to 43) *but has made the stability worse* (S increased from 9.36 to 33.9). In other words, by increasing R_1 and R_2 we have bought gain at the expense of stability. Therefore, we must *compromise* when it comes to selecting the values of R_1 and R_2. Generally speaking, we can tolerate more Q point shift in the *earlier* stages where the total signal swing is still *small*. In Fig. 6.36(a) the peak-to-peak swing is designated by the letter A.

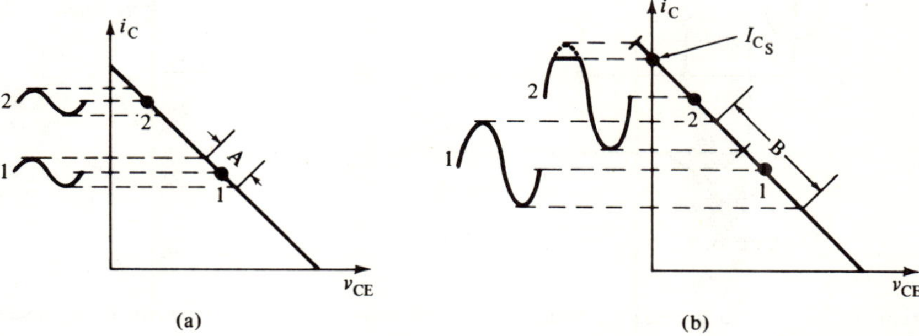

Figure 6.36

Note that if the Q point shifts from point 1 to point 2, the collector current waveform remains unchanged (assuming equal spacing of the curves); collector current waveforms 1 and 2 appear at the left of the figure. Therefore, we can tolerate a considerable Q point shift, S need not be very small, and we can make the values of R_1 and R_2 large to increase the gain. In Fig. 6.36(b), the swing is *much larger* (designated by the letter B). It can be seen from the diagram that if the Q point shifts from point 1 to point 2 (same shift as in Fig. 6.36(a)), the transistor will *saturate* with the result that the top of the waveform will *flatten;* compare collector current waveforms 1 and 2 in Fig. 6.36(b). Thus, a Q point shift from point 1 to point 2 *cannot* be tolerated with the larger swing (B) shown in Fig. 6.36(b); the values of R_1 and R_2 must be reduced to give a better (smaller) stability factor, and consequently the gain of that stage will decrease.

Another alternative lies in making R_1 and R_2 large so as to achieve large gain, *and then making R_E large*, since this is another method of decreasing the value of the stability factor. But there is a limit to the size which we can make R_E. You may recall from earlier chapters that the dc load line depends upon both R_L *and* R_E when R_E is connected in the circuit. Therefore, if R_E is made larger and larger, the slope of the load line will become smaller, with a consequent reduction in the value of I_{Cs}. This means that the Q point will be close to the horizontal axis; we saw in Sec. 6-4 that this situation entails many problems. We return again to the fact that the design of an amplifier such as this involves many compromises; these will be illustrated in more detail in chapter 10.

6-11   GENERAL GAIN EQUATION

It is useful at this point to derive an equation for the gain of the amplifiers studied in this chapter. With this equation, the designer can make a rapid estimate of the circuit behavior; i.e., it will not be necessary to perform a complete step-by-step analysis. The basic circuit is shown in Fig. 6.37. This is the circuit for the emitter-

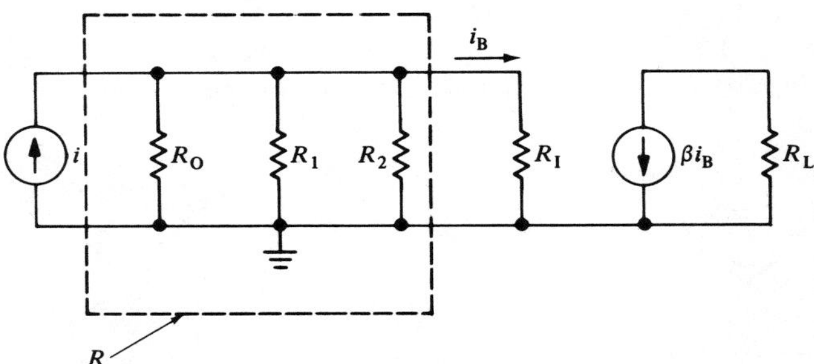

Figure 6.37

biased amplifier; when dealing with the fixed bias amplifier, just replace resistors R_1 and R_2 by R_B (the values used for R_B in the fixed bias amplifier are generally large enough to be neglected). It can be seen from Fig. 6.37 that R_1 and R_2 are both in parallel with R_I; therefore, the *smaller* the values of these resistors, the *more* current will be drawn by them, with correspondingly *less* current drawn by R_I (the current in R_I is i_B). This illustrates that the designer should always try, if possible, to avoid placing small resistors in parallel with the input terminals of the amplifier; you may recall from the last section of chapter 5 that it was necessary to use two supplies (positive and negative) in the collector-to-base bias circuit to avoid placing a small resistor in parallel with the amplifier input terminals.

Now, proceeding with the derivation of the gain equation, we define R as being the parallel equivalent of R_0, R_1, and R_2.

$$R = R_0 \,\|\, R_1 \,\|\, R_2$$

To find i_B, we use the current divider, as follows:

$$i_B = \frac{(i)R}{R + R_I}$$

In this expression, i is the current generator of the previous stage, and R_I is the input resistance of the stage in question. The value of R_I depends upon whether or not the emitter resistor is by-passed by a capacitor; we see in the next chapter that we sometimes intentionally omit the by-pass capacitor from the circuit for very important reasons. The gain of the amplifier of Fig. 6.37 can now be computed as follows:

$$A_I = \frac{\beta i_B}{i} = \frac{\beta \dfrac{iR}{R + R_I}}{i} = \frac{\beta R}{R + R_I}$$

$$\boxed{A_I = \frac{\beta R}{R + R_I}}$$

The equation again illustrates the fact that, for large gain, we should like the transistor to have a *large* β, we should like to have a *large* value of R (this means large values of R_O, R_1, and R_2), and a *small* value of R_I. This equation will be of great use when we study feedback amplifiers in the following chapter. It should be mentioned that this equation is valid only for the fixed bias and emitter-biased circuits.

EXAMPLE 7

For the amplifier circuit of Fig. 6.38(a):
    (a)   Compute the gain.
    (b)   Compute I_{C_Q} and I_{C_S}.
    (c)   Compute i_C (the ac collector current in the 2 kΩ resistor) if the ac collector current in the previous stage is 100 μA. In this example, neglect r_O and assume that both C_C and C_E are large enough so that they act like ac shorts.

SOLUTION

    (a)   The ac equivalent circuit of the amplifier of Fig. 6.38(a) is drawn in Fig. 6.38(b). To find the gain, we must first find the value of R:

$$R = R_O \parallel R_B$$

$$R_B = \frac{R_1 R_2}{R_1 + R_2} = \frac{(25)(2.5)}{25 + 2.5} = 2.27 \text{ k}\Omega$$

$$R = \frac{R_O R_B}{R_O + R_B} = \frac{5(2.27)}{5 + 2.27} = 1.56 \text{ k}\Omega$$

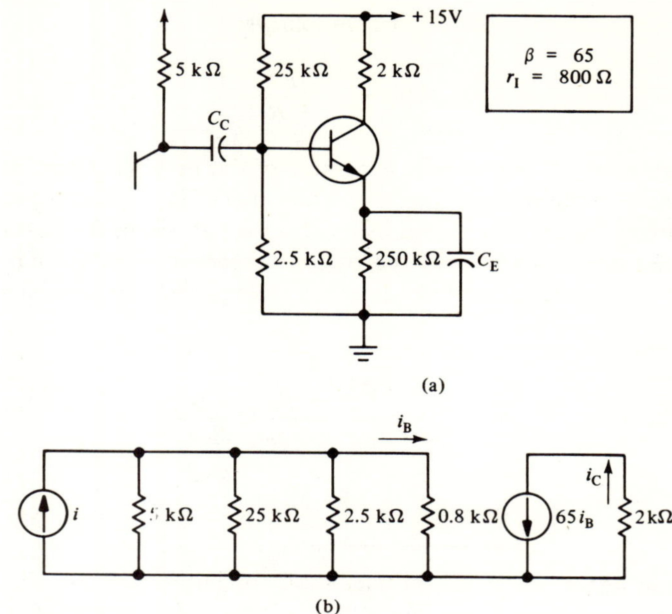

Figure 6.38

As can be seen from Fig. 6.38(b), $R_I = r_I = 800\ \Omega$, since we assumed that C_E effectively by-passes the 250 Ω emitter resistor. Thus, the gain is:

$$A_I = \frac{\beta R}{R + R_I} = \frac{65(1.56)}{1.56 + 0.8} = 43$$

(b)   Assume that $V_{BE} = 0.6$ V. First we must find V_B:

$$V_B = \frac{R_2 V_{CC}}{R_1 + R_2} = \frac{2.5(15)}{25 + 2.5} = 1.36\ \text{V}$$

We already know that $R_B = 2.27$ kΩ. Therefore, I_{CQ} is:

$$I_{CQ} = \frac{\beta(V_B - V_{BE})}{R_B + \beta R_E} = \frac{65(1.36 - 0.6)}{2.27 + 65(0.25)} = 2.66\ \text{mA}$$

I_{Cs} is found as follows:

$$I_{Cs} = \frac{V_{CC}}{R_L + R_E} = \frac{15}{2 + 0.25} = 6.66\ \text{mA}$$

(c)   Since we have neglected the effect of r_O, i_C, the ac collector current in the 2 kΩ resistor (see Fig. 6.38(b)) is equal to the current generator ($65\ i_B$). The current generator in the previous stage (i, in Fig. 6.38(b)) is given as $100\ \mu$A. Since the gain is defined as the ratio of βi_B to i, we can write the following:

$$A_I = \frac{\beta i_B}{i}$$

$$i_C = \beta i_B = (A_I)i = 43(100 \ \mu A) = 4300 \ \mu A = 4.3 \ \text{mA}$$

Let us now see if this amplifier can accommodate a 4.3 mA peak-to-peak collector current. We saw that $I_{CQ} = 2.66$ mA and $I_{Cs} = 6.66$ mA; thus I_{CQ} is less than one-half I_{Cs}. Therefore, $i_{CM} = 2I_{CQ}$, or:

$$i_{CM} = 2I_{CQ} = 2(2.66 \ \text{mA}) = 5.32 \ \text{mA}$$

Since i_{CM} (5.32 mA) is *greater* than i_C (4.3 mA), the amplifier *can* accommodate the signal, and there will be no distortion.

One final point should be mentioned. The resistance used for drawing the ac load line with the fixed bias circuit was defined as:

$$R'_L = r_I \parallel R_L$$

This was true because under ac conditions, r_I was effectively in parallel with R_L. With the emitter bias circuit, the equation is changed slightly to the following form, because R_B (R_1 and R_2) is also in parallel with R_L:

$$R'_L = R_I \parallel R_L \parallel R_B$$

Note that R_E does *not* appear in the expression for R'_L. This is because, *under ac conditions*, R_E is effectively "removed from the circuit" by the action of C_E. *Under dc conditions*, however, R_E must be added to R_L when drawing the dc load line.

6-12   ac ANALYSIS OF THE COLLECTOR-TO-BASE BIAS CIRCUIT

Now that we have dealt with both the fixed bias and emitter bias circuits, the final step is to perform an ac analysis upon the collector-to-base bias circuit. An amplifier using collector-to-base bias is shown in Fig. 6.39(a); note that this amplifier stage is fed from a source consisting of a 1 μA current generator and an internal resistance of 10 kΩ. The ac equivalent circuit is shown in Fig. 6.39(b); the transistor itself is located within the dotted box. We could analyze the equivalent circuit of Fig. 6.39(b) directly. However, the fact that the bias resistor ($R_B = 50$ kΩ) is connected between the base and collector terminals makes the analysis of this circuit quite complex; i.e., it involves repeated use of Kirchhoff's laws as well as a considerable amount of algebraic manipulation. Fortunately, there is a method of analyzing this circuit which is similar to those methods used previously in this chapter; in addition, it yields some insight into the operation of the circuit.

Figure 6.40(a) shows the same ac equivalent circuit with various voltages and currents indicated. Notice that, because of the assumed directions of the currents (i_B, i_{RB}, and i_C), the polarities of the voltages across the various resistors (v_I, v_{RB},

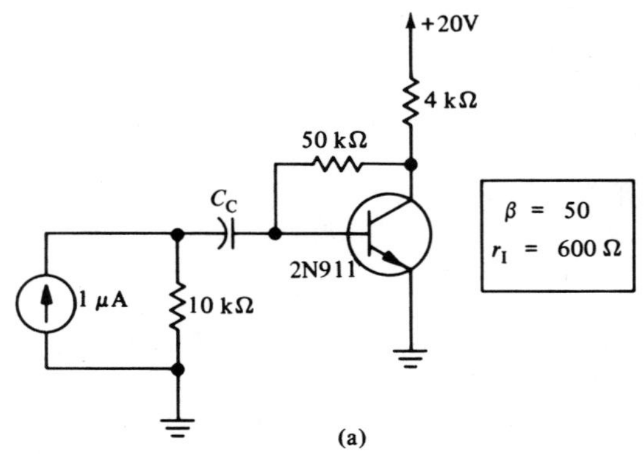

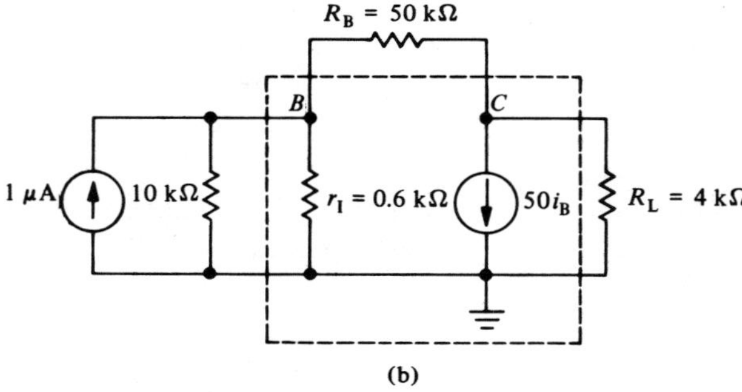

Figure 6.39

and v_O) are indicated as shown. Writing Kirchhoff's voltage law for the center loop of the circuit of Fig. 6.40(a), we get the following equation:

$$v_I - v_{R_B} + v_O = 0$$

$$v_{R_B} = v_O + v_I$$

Now we know that the ratio of the output voltage (v_O) to the input voltage (v_I) is equal to the voltage gain of the amplifier, A_V.

$$A_V = \frac{v_O}{v_I}; \qquad v_O = A_V v_I$$

Therefore, if we substitute into the original equation, we get:

$$v_{R_B} = v_O + v_I = A_V v_I + v_I = v_I(A_V + 1)$$

The ac current through the bias resistor, R_B, is found by Ohm's law as follows:

$$i_{R_B} = \frac{v_{R_B}}{R_B} = \frac{v_I(A_V + 1)}{R_B} = \frac{v_I}{\left(\dfrac{R_B}{A_V + 1}\right)} = \frac{v_I}{(R_B')}$$

$$R_B' = \frac{R_B}{A_V + 1}$$

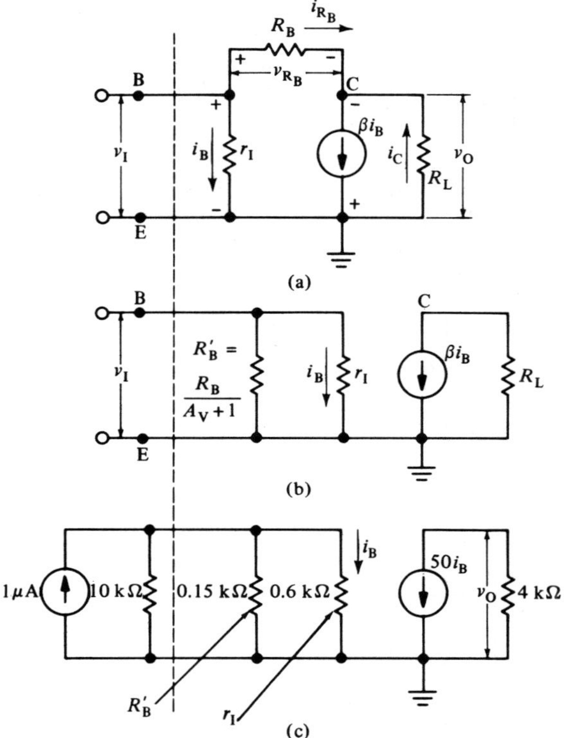

Figure 6.40

Now let us examine the above equation for a moment. We know that, in fact, R_B is connected between the base and collector terminals of the transistor. We can, however, view the above equation in a different manner. Note that the current through the bias resistor, i_{R_B}, is equal to the ratio of the input voltage, v_I, to a *new resistance*, R_B'. *If, as the equation indicates, v_I is the voltage across this new resistance, R_B', then R_B' must be connected across the input terminals of the amplifier,* as shown in Fig. 6.40(b). In other words, instead of representing the circuit as shown in Fig. 6.40(a) (with R_B connected between the base and collector terminals), *we may disconnect R_B from between the base and collector terminals and connect a new resistance, R_B', across the input* (base and emitter) *terminals of the amplifier,* as shown in Fig. 6.40(b).

Two points must now be mentioned in connection with the above statements: First, the bias resistor is *not actually disconnected* from the base and collector terminals; we merely make use of the circuit of Fig. 6.40(b) (rather than that of Fig. 6.40(a)) because it facilitates the analysis of the amplifier circuit, as we soon see. Second, as far as the input voltage (v_I) is concerned, the portions of the circuits of Figs. 6.40(a) and (b) are identical to the *right* of the dotted line. We are going to see shortly that the output resistance of the amplifier is *also* affected by the presence of R_B, and we must therefore view the circuit in a slightly different manner when dealing with cascade connections.

The phenomenon illustrated by Figs. 6.40(a) and (b) is known as the *Miller*

Effect. This states that any impedance connected between the input and output terminals (base and collector, in this case) of an amplifier may be shunted across the input terminals *if it is divided by the voltage gain of the amplifier plus one;* we shall be using this concept again in chapter 8 when we consider frequency response.

The next step in this analysis is to compute the voltage gain of the amplifier. To do this we shall again make use of Fig. 6.40(a). We assume, in deriving an expression for the voltage gain, that R_B *is much larger than* R_L; this is usually the case. If this is true, most of the current from the βi_B current generator will flow through R_L rather than R_B; i.e., the *collector current* (i_C) *will be much larger than the current in* R_B (i_{R_B}). Therefore, we can assume that *all* of the current from the βi_B current generator will flow through R_L; this may be expressed as follows:

$$\text{For } R_B \gg R_L: \qquad i_C \gg i_{R_B}$$

$$i_C = \beta i_B - i_{R_B} \simeq \beta i_B$$

The voltage gain of the amplifier of Fig. 6.40(a) is expressed as follows:

$$A_V = \frac{v_O}{v_I} = \frac{i_C R_L}{i_B r_I} \simeq \frac{\beta i_B R_L}{i_B r_I} = \frac{\beta R_L}{r_I}$$

$$\boxed{A_V = \frac{\beta R_L}{r_I}}$$

This is a good approximation if R_B is much greater than R_L. The exact expression, which can be derived using a considerable amount of algebraic manipulation, is:

$$A_V = \frac{\beta R_L}{r_I\left(1 + \dfrac{R_L}{R_B}\right)}$$

We can assume, however, that R_B is much greater than R_L, and therefore we can use the simpler expression for the voltage gain.

The voltage gain for amplifiers in the common emitter configuration is usually much greater than one; thus the expression for R_B' can be written:

$$R_B' = \frac{R_B}{A_V + 1} \simeq \frac{R_B}{A_V} = \frac{R_B}{\dfrac{\beta R_L}{r_I}} = \frac{R_B r_I}{\beta R_L}$$

Let us now analyze the circuit of Fig. 6.39(a). In order to do this we must first compute the value of R_B' for the circuit; this is found to be:

$$R_B' = \frac{R_B r_I}{\beta R_L} = \frac{50(0.6)}{50(4)} = 0.15 \text{ k}\Omega = 150 \text{ }\Omega$$

The equivalent circuit may now be represented as shown in Fig. 6.40(c). Note that R_B' is a 150 Ω resistor in shunt with the input terminals. A comparison of the circuits of Figs. 6.40(a) and (c) reveals that *it is much easier to analyze the circuit of Fig. 6.40(c) since it has the same form as the other circuits which we discussed in this chapter;* i.e., the equivalent circuit consists of a current generator feeding a number of resistors (R_O, R_B', r_I) connected in parallel.

Since the 10 kΩ output resistance of the previous stage is much larger than R_B' (0.15 kΩ) and r_I (0.6 kΩ), we neglect its effect in the following analysis. The base current is:

$$i_B = \frac{iR_B'}{R_B' + r_I} = \frac{(1\mu A)(0.15)}{0.15 + 0.6} = 0.2 \ \mu A$$

The current gain is therefore equal to:

$$A_I = \frac{\beta i_B}{i} = \frac{50(0.2 \ \mu A)}{1 \ \mu A} = \frac{10 \ \mu A}{1 \ \mu A} = 10$$

While 10 is not a particularly small value of current gain, it seems as though we should be able to obtain a larger value using a transistor whose β is 50. The reason for this value of gain can be understood by examining the circuit of Fig. 6.40(c). Notice that a *small resistance* ($R_B' = 150 \ \Omega$) is in *shunt* with the input terminals. Since this resistance is *smaller* than r_I (600 Ω), it will divert *most* of the 1 μA current away from r_I; in this case, the base current is 0.2 μA while the current diverted through R_B' is 0.8 μA (1.0 μA minus 0.2 μA). This situation is similar in many respects to that encountered in the emitter bias circuit. In the emitter bias circuit, the presence of R_E *raised the input resistance* of the amplifier, thus *decreasing* the gain (reduced value of i_B). In the case of the collector-to-base bias circuit, R_B' is a *small resistance in parallel with the input terminals;* thus i_B (and the gain) is *reduced* by the amount of current diverted away from the amplifier through the relatively small R_B'.

Before proceeding with a description of the remedy for the above situation, it is instructive to examine another aspect of the collector-to-base bias circuit. Suppose that a 1 kΩ resistor is connected to the output of the amplifier through a coupling capacitor as shown in Fig. 6.41(a). There are two methods of analyzing this circuit. The first involves considering the load resistance as the parallel combination of the original 4 kΩ load resistance and the additional 1 kΩ resistor, since, under ac conditions, the two resistors are effectively connected in parallel. Thus:

$$R_L' = \frac{(4 \ k\Omega)(1 \ k\Omega)}{4 \ k\Omega + 1 \ k\Omega} = 0.8 \ k\Omega$$

Since R_B' depends upon the value of the load resistance, the new value of this resistance is:

$$R_B' = \frac{R_B r_I}{\beta R_L'} = \frac{50(0.6)}{50(0.8)} = 0.75 \ \Omega$$

The base current is:

$$i_B = \frac{iR_B'}{R_B' + r_I} = \frac{(1 \ \mu A)(0.75)}{0.75 + 0.6} = 0.555 \ \mu A$$

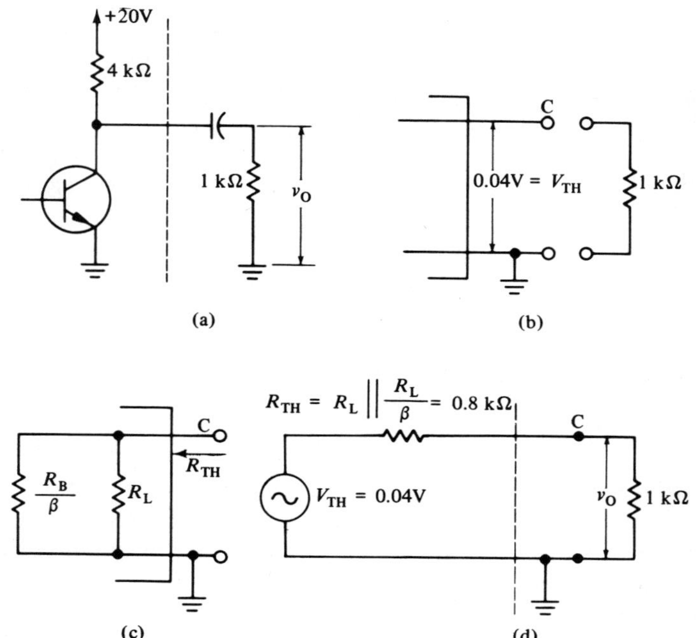

Figure 6.41

The βi_B current generator is:

$$\beta i_B = 50(0.555\ \mu A) = 27.75\ \mu A$$

The output voltage, v_O, is equal to the product of the βi_B current generator and the new load resistance, R_L'; thus, v_O is:

$$v_O = (\beta i_B)R_L' = (27.75\ \mu A)(0.8\ k\Omega) = 0.022\ V$$

As can be seen, this method involves repeating the entire analysis.

A second, more sophisticated method is now illustrated. The object here is to utilize Thevenin's theorem in replacing that portion of the circuit of Fig. 6.41 (a) to the *left* of the dotted line by a simpler equivalent. The first step is to find the Thevenin open circuit voltage, V_{TH}. We saw before (when no additional resistor was connected to the output terminals of the amplifier) that the current generator was:

$$\beta i_B = 50(0.2\ \mu A) = 10\ \mu A$$

The output voltage under these conditions is equal to the Thevenin open circuit voltage, since the additional 1 kΩ resistor is not connected to the output terminals (output terminals open-circuited). Thus, V_{TH} is:

$$V_{TH} = (\beta i_B)R_L = (10\ \mu A)(4\ k\Omega) = 0.04\ V$$

Therefore, the Thevenin open circuit voltage, V_{TH}, is 0.04 V, as shown in Fig.

6.41 (b). The next step is to find the Thevenin resistance (or impedance), R_{TH}. It can be shown algebraically that the output resistance of the collector-to-base biased amplifier is equal to the parallel combination of R_L and a resistance equal to R_B/β, as shown in Fig. 6.41 (c). Thus the value of R_{TH} for this particular circuit is:

$$R_O = R_{\text{TH}} = \frac{R_L \dfrac{R_B}{\beta}}{R_L + \dfrac{R_B}{\beta}} = \frac{(4) \dfrac{50}{50}}{4 + \dfrac{50}{50}} = 0.8 \text{ k}\Omega$$

Now that we know the values of both V_{TH} and R_{TH}, we can re-draw the circuit as shown in Fig. 6.41 (d); the portion of the circuit to the left of the dotted line in Fig. 6.41 (d) represents the *same* portion of the circuit to the left of the dotted line in Fig. 6.41 (a). After the 1 kΩ resistor is added to the circuit of Fig. 6.41 (d), the output voltage can be found by the voltage divider as follows:

$$v_O = \frac{(0.04)(1)}{1 + 0.8} = 0.022 \text{ V}$$

Note that this agrees with the value found using the first method. This second method is simpler, but it involves knowing the output resistance of the amplifier.

Let us turn now to a description of a method for increasing the current gain of the amplifier of Fig. 6.39 (a). The 50 kΩ bias resistor is replaced by two 25 kΩ resistors, and the common terminal of these two resistors (terminal A) is connected through a capacitor to ground, as shown in Fig. 6.42 (a). Assuming the frequency of the input signal is 1 kHz, we can compute the reactance of the 10 μF capacitor as:

$$X_C = \frac{1}{2\pi fC} = \frac{0.159}{10^3 \times 10 \times 10^{-6}} = 15.9 \ \Omega \simeq 16 \ \Omega$$

A portion of the circuit of Fig. 6.42 (a) is re-drawn in Fig. 6.42 (b). The input voltage, v_I, is applied between base (terminal B) and ground as shown. Terminal B is connected through a 25 kΩ resistor to terminal A; terminal A is connected through the other 25 kΩ resistor to the collector of the transistor (terminal C). The capacitor is represented by a 16 Ω reactance between terminal A and ground. We neglect any ac current flowing through the *right*-hand 25 kΩ resistor, since *almost all* of the ac current in the *left*-hand 25 kΩ resistor will flow through the relatively *small* 16 ohm impedance rather than through the *right*-hand 25 kΩ resistor. Thus we can apply the voltage divider to that portion of the circuit of Fig. 6.42 (b) composed of the *left*-hand 25 kΩ resistor and the 16 Ω impedance. Assume that the input voltage, v_I, is equal to 0.1 V. Then the voltage across the 16 Ω impedance, v_A, can be found by the voltage-divider as follows:

$$v_A = \frac{(v_I) X_C}{25 \text{ k}\Omega + X_C} = \frac{(0.1 \text{ V})(16)}{25,000 + 16} \simeq \frac{(0.1 \text{ V})(16)}{25,000} = 0.000064 \text{ V}$$

$$v_A = 0.000064 \text{ V} = 64 \ \mu\text{V}$$

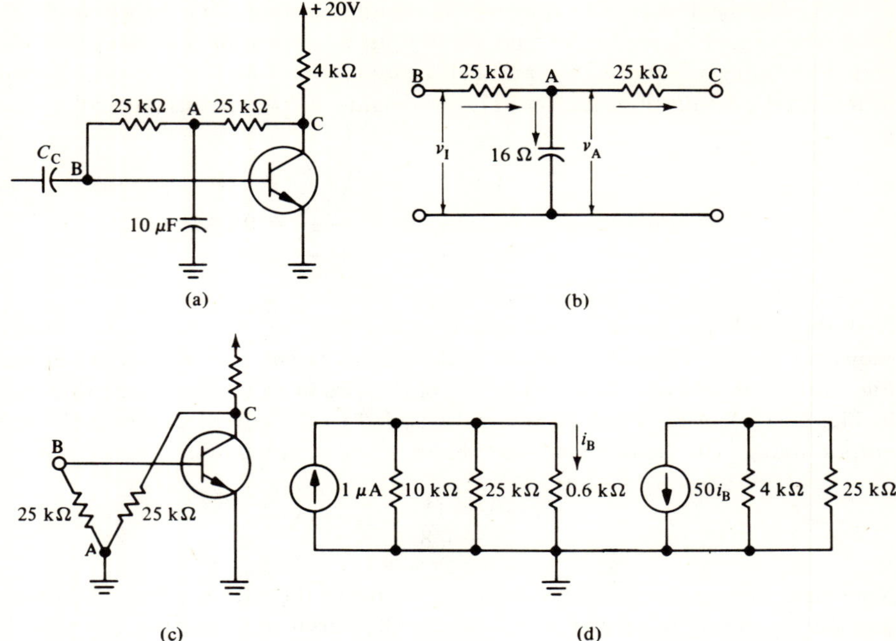

Figure 6.42

This very small voltage across the 10 μF capacitor (64 μV) is a result of the fact that the reactance of the capacitor (16 Ω) is very small in comparison with the 25 kΩ resistor. *Since this voltage is so much smaller than the input voltage (0.000064 V compared with 0.1 V), we can say that terminal A is approximately at ground potential.* If this is the case, the circuit can be re-drawn as shown in Fig. 6.42(c); terminal A is connected to ground. Thus, the *left*-hand 25 kΩ resistor shunts the *input* terminals of the amplifier, while the *right*-hand 25 kΩ resistor shunts the output terminals. Since 25 kΩ is *large* compared to R_B' (150 Ω), *less* current will be diverted through this resistor, and consequently, the gain will be larger than when no capacitor was used. Since the capacitor conducts *no dc current*, it has no effect on the dc conditions or the stability factor of the circuit; as far *as dc is* concerned, $R_B = 50$ kΩ (the two 25 kΩ resistors are *in dc series*).

The ac equivalent circuit can now be drawn as shown in Fig. 6.42(d); note that there are 25 kΩ resistors shunting both the input and the output. To find the new value of gain, we must first find R:

$$R = \frac{(10)(25)}{10 + 25} \simeq 7 \text{ k}\Omega$$

The gain is:

$$A_\text{I} = \frac{\beta R}{R + R_\text{I}} = \frac{50(7)}{7 + 0.6} = 46$$

Note that the gain is now equal to 46, a decided improvement over 10. The value of this by-pass capacitor, designated C_B, should be kept *large* in order that the impedance be *small*. It is not necessary that R_B be split evenly (25 kΩ and 25 kΩ); it is usually split so that the loading on both the input and output circuits is the same.

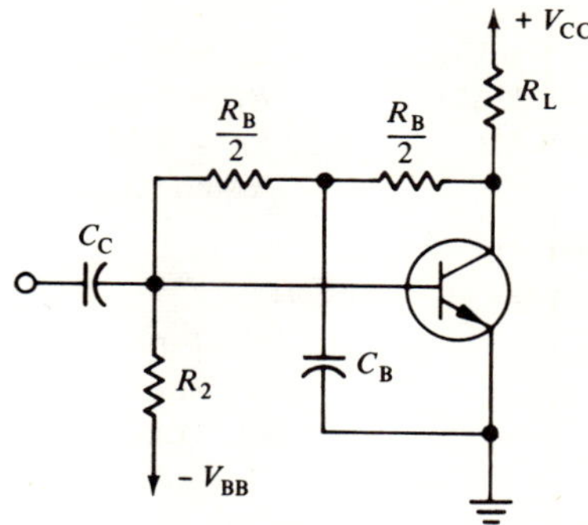

Figure 6.43

A complete circuit using collector-to-base feedback is shown in Fig. 6.43. From chapter 5 you may recall that it was sometimes necessary to use a second bias resistor, R_2, to maintain the Q point at a convenient position on the load line. We also saw that it was necessary to include a *second* supply (V_{BB}) in order that the value of R_2 not be too small. We can see now why it is *undesirable* for R_2 to be small; it shunts the input terminals, and if its value is small, it will cause a reduction in the gain. When computing the gain, the value of R_2 must be included in "R," since it shunts the input. The voltage rating of C_B should be made equal to V_{CC} for safety.

We saw in chapter 5 that negative dc feedback was necessary to stabilize the Q point; this necessitated use of either the collector-to-base or emitter bias circuit. However, we saw that use of both of these circuits *caused a considerable reduction in the gain;* therefore, it was necessary to include capacitors (C_E and C_B) in order to restore the gain.

PROBLEMS

1. The fixed bias circuit of Fig. 6.10 has the following values:

$$v_S = 1 \text{ mV}_{(P-P)} \qquad R_B = 1 \text{ M}\Omega$$
$$R_S \text{ (source resistance)} = 500 \text{ }\Omega \qquad r_I = 1 \text{ k}\Omega$$
$$V_{CC} = 30 \text{ V} \qquad \beta = 100$$
$$R_L = 5 \text{ k}\Omega \qquad r_O = 25 \text{ k}\Omega$$

$$\text{Assume } X_{C_C} = 0 \text{ }\Omega$$

Draw the ac equivalent circuit.
Sketch completely labeled waveforms of:

    (a) v_{DE} (d) i_B
    (b) v_{BE} (e) i_C
    (c) v_{C_C} (f) v_{CE}

2.   Repeat problem 1, except make $R_B = 2 \text{ M}\Omega$.

3.   Repeat problem 1, except make $v_S = 2 \text{ mV}$.

4.   The two-stage amplifier of Fig. 6.14 uses two identical stages each having the following values:

$$
\begin{aligned}
v_S &= 1 \text{ mV}_{(P-P)} & r_O &= 25 \text{ k}\Omega \\
R_S &= 500 \ \Omega & R_L &= 5 \text{ k}\Omega \\
r_I &= 1 \text{ k}\Omega & R_B &= 1 \text{ M}\Omega \\
\beta &= 75 & \text{Assume } X_{Cc} &= 0 \ \Omega
\end{aligned}
$$

Draw the ac equivalent circuit.
Compute the peak-to-peak ac values of:

(a) i_{B_1} (e) βi_{B_2}
(b) βi_{B_1} (f) i_{C_2}
(c) v_{CE_1} (g) v_{CE_2}
(d) i_{B_2}

5.   Repeat problem 4, except make $\beta = 150$. What effect does the value of β have upon the gain?

6.   Repeat problem 4, except make $r_I = 2 \text{ k}\Omega$. What effect does the value of r_I have upon the gain?

7.   Repeat problem 4, except make $r_O = 10 \text{ k}\Omega$. What effect does the value of r_O have upon the gain?

8.   Repeat problem 4, except make $R_L = 10 \text{ k}\Omega$. What effect does the value of R_L have upon the gain?

9.   A fixed bias circuit has the following values:

$$
\begin{aligned}
V_{CC} &= 20 \text{ V} & \beta &= 50 \\
R_L &= 5 \text{ k}\Omega & r_{I_2} \text{ (next stage)} &= 1 \text{ k}\Omega \\
R_B &= 1 \text{ M}\Omega & &
\end{aligned}
$$

Compute the value of i_{C_M}.

10.   Repeat problem 9, except make $\beta = 100$.

11.   Repeat problem 9, except make $\beta = 150$.

12.   A two-stage amplifier has $A_{I_1} = 23$ and $A_{I_2} = 40$.
Compute the value of the over-all gain, A_{I_O}.

13.   Repeat problem 12 if $A_{I_1} = 70$ and $A_{I_2} = 80$.

14.   Two identical stages are to be used in an amplifier whose over-all gain is 3000. What should be the gain of each stage?

15.   A fixed bias amplifier circuit has the following values:

$$
\begin{aligned}
v_S &= 10 \text{ mV}_{(P-P)} & R_B &= 1 \text{ M}\Omega \\
R_S &= 1 \text{ k}\Omega & r_{I_1} = r_{I_2} \text{ (next stage)} &= 1 \text{ k}\Omega \\
V_{CC} &= 30 \text{ V} & \beta &= 100 \\
R_L &= 5 \text{ k}\Omega & \text{Neglect } r_O. &
\end{aligned}
$$

Sketch the waveform of i_C. Is it distorted?

16.   Repeat problem 15, except make $v_S = 100 \text{ mV}_{(P-P)}$.

17.   An amplifier using emitter bias (with no by-pass capacitor) has the following values:

$$
r_I = 1 \text{ k}\Omega \qquad \beta = 100 \qquad R_E = 1 \text{ k}\Omega
$$

Compute the value of the input resistance, R_I, to the amplifier.

18. Repeat problem 17, except make $\beta = 200$.

19. The amplifier of Fig. 6.33(a) has the following values:

$$\begin{aligned}
R_{L_1} &= 5 \text{ k}\Omega & R_{E_2} &= 1 \text{ k}\Omega \\
R_1 &= 25 \text{ k}\Omega & \beta_2 &= 100 \\
R_2 &= 5 \text{ k}\Omega & r_{I_2} &= 1 \text{ k}\Omega \\
R_{L_2} &= 4 \text{ k}\Omega
\end{aligned}$$

Compute the gain of the amplifier.

20. Repeat problem 19, except make $R_E = 500 \ \Omega$.
What effect does the value of R_E have upon the gain?

21. Repeat problem 19 if a by-pass capacitor is connected across R_E as in Fig. 6.34(a).
Assume $X_{C_E} = 0 \ \Omega$.

22. Repeat problem 20 if a by-pass capacitor is connected across R_E as in Fig. 6.34(a).

23. The amplifier of Fig. 6.39(a) has the following values:

$$\begin{aligned}
R_B &= 25 \text{ k}\Omega & \beta &= 100 \\
r_I &= 1 \text{ k}\Omega & R_L &= 5 \text{ k}\Omega
\end{aligned}$$

Neglect the output resistance of the previous stage.

Compute the gain.

24. Repeat problem 23 if a capacitor is connected as in Fig. 6.42(a).

7

FEEDBACK AMPLIFIERS

In chapter 5 we saw that it was necessary to introduce some dc feedback into the amplifier circuit in order to stabilize the Q point; we discussed *two* such techniques, the collector-to-base and emitter-bias circuits. We learned in chapter 6 that the gains of the amplifiers are reduced considerably (compared to the case of the fixed-bias circuit) when these new bias methods are incorporated into the circuits; i.e., it appeared that the price to pay for better stability was reduced gain. It was then shown, however, that the placement of large capacitors at various points in the circuit contributed to a vast increase in the gain, *while still retaining the improved dc stability*. It may appear to the reader that we have now completed the conversion of the simple, elementary amplifier of chapter 4 to a practical, functioning unit. However, we have not yet discussed one very important concept in connection with amplifiers: *the stability of the gain.*

It was shown in the previous chapters that the gain of the amplifier is highly dependent upon the β of the transistor. Since β varies with temperature and transistor replacement, *the gain will also vary when the temperature is changed, or when a transistor is replaced;* this variation in the gain of an amplifier due to external factors is undesirable. In this chapter we study the extent of the variation in the gain of an amplifier due to changes in the transistor parameters. We also discuss changes in the amplifier circuit which will enable us to improve the stability of the gain. There are some simple changes which can be made in the emitter and collector-to-base bias circuits. In addition, there are other, slightly more sophisticated techniques which are to be described. By the conclusion of this chapter, the amplifier will appear in its final form; i.e., it will provide gain, and it will also exhibit both dc (stable Q point) and ac (stable gain) stability.

In this chapter, we also begin to study the problems involved in connecting the amplifier output to the final load. We see that the nature of the load itself frequently makes it necessary to include special circuits between the output of the amplifier and the load. These special circuits, although they appear to be simple, actually illustrate some extremely important concepts.

7-1   RELATIONSHIP BETWEEN β AND r_I

Before proceeding with a discussion of the major concepts of this chapter, it is instructive to examine a relationship between the β of a transistor and its r_I. We have waited until now to deal with this concept because it has a definite bearing upon the central theme of the chapter. The semiconductor physicists have shown us that r_I, the ac resistance between the base and emitter terminals of the transistor, can be split up into *two* components, r_B and r_BE, as shown in Fig. 7.1(a). This can be expressed algebraically as follows:

$$r_\mathrm{I} = r_\mathrm{B} + r_\mathrm{BE}$$

The *first* component, r_B, is known as the *base spreading resistance;* this is the ohmic

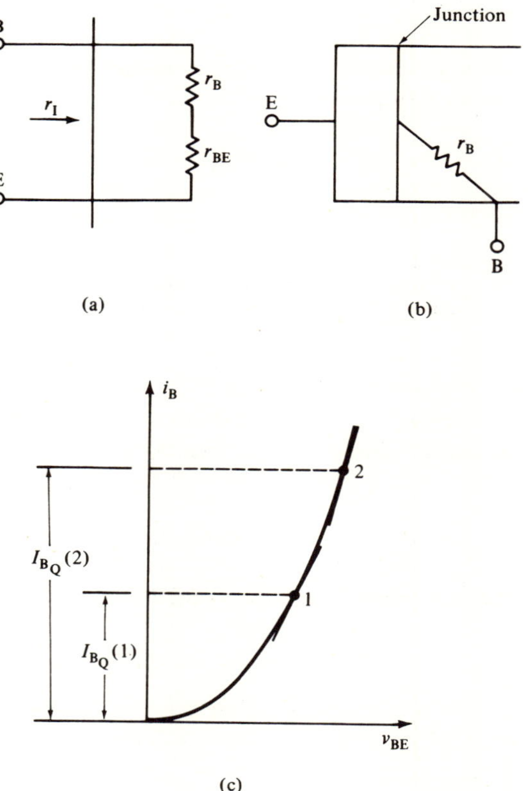

Figure 7.1

resistance of the base region (the actual resistance of the base material) between the base-emitter junction and the base terminal, as shown in Fig. 7.1(b). Strictly speaking, there is also a similar resistance in the *emitter* region between the junction and the *emitter* terminal; however, since the emitter is *much more heavily doped* than the base (see chapter 3), the resistance of the emitter region is *much smaller* than that of the base, and can thus be neglected. The value of r_B generally lies between 50 and 250 Ω. The *second* component, r_{BE}, is *the actual nonlinear resistance of the base-emitter junction;* i.e., it is the dynamic resistance of the input characteristic as described in chapters 3 and 4.

Let us now examine r_{BE} in more detail. We know that r_{BE} is equal to the reciprocal of the slope of the input characteristic; i.e., the *greater* the slope, the *smaller* the value of r_{BE}. We also know that the slope *increases* as the quiescent base current increases; e.g., the slope at point 2 in Fig. 7.1(c) is greater than that at point 1 (the quiescent base current is larger at point 2 than at point 1). From the results of the above discussion, we can say that r_{BE} *is inversely proportional to the quiescent base current, I_{BQ}*; this can be written as:

$$r_{BE} \sim \frac{1}{I_{BQ}}$$

This means that a larger value of I_{BQ} implies a smaller value of r_{BE}. We also know that the following relationship is true:

$$I_{BQ} = \frac{I_{CQ}}{\beta}$$

Therefore, we can write:

$$r_{BE} \sim \frac{1}{I_{BQ}} = \frac{1}{\dfrac{I_{CQ}}{\beta}}; \qquad r_{BE} \sim \frac{\beta}{I_{CQ}}$$

This shows that r_{BE} is *directly* proportional to β, and *inversely* proportional to the quiescent collector current, I_{CQ}. It can be shown with the aid of equations developed in atomic and semiconductor physics that r_{BE} can be expressed as follows:

$$r_{BE} = \frac{\beta}{40 I_{CQ}} \tag{7-1}$$

In this equation, r_{BE} is expressed in *kilohms* and I_{CQ} is in *milliamperes.*

EXAMPLE 1

A transistor amplifier uses a transistor whose β is 100. The amplifier is biased at a dc collector current of 10 mA ($I_{CQ} = 10$ mA), and the value of r_B for this particular transistor is 100 ohms. Compute r_I.

SOLUTION

Using the expression above, we first find r_{BE} to be:

$$r_{BE} = \frac{\beta}{40 I_{CQ}} = \frac{100}{40(10)} = 0.25 \text{ k}\Omega = 250 \ \Omega$$

Therefore, r_I is found as follows:

$$r_\mathrm{I} = r_\mathrm{B} + r_\mathrm{BE} = 100 + 250 = 350 \ \Omega$$

EXAMPLE 2

Repeat example 1 using the same transistor; the bias has been changed so that the quiescent collector current is now 1 mA.

SOLUTION

Solving as above, we find r_I to be:

$$r_\mathrm{I} = r_\mathrm{B} + \frac{\beta}{40 I_{\mathrm{C}_\mathrm{Q}}} = 100 + \frac{100}{40(1)} = 100 + 2500 = 2.6 \ \mathrm{k\Omega}$$

Note the large increase in the value of r_I when the value of $I_{\mathrm{C}_\mathrm{Q}}$ is decreased. This illustrates another reason why we cannot operate the amplifier at very small values of collector current. You will recall from Sec. 6-4 that two of the reasons given were reduction of allowable swing, and smaller gain due to reduced values of β at the lower values of I_C. Now we have a *third* reason. Operating the transistor at *small* values of $I_{\mathrm{C}_\mathrm{Q}}$ results in *large* values of r_I (examples 1 and 2 showed that r_I increased from 350 Ω to 2.6 kΩ when $I_{\mathrm{C}_\mathrm{Q}}$ decreased from 10 mA to 1 mA); an *increase* in the value of r_I results in a *reduction* in the gain, as we saw in chapter 6.

While we can now compute the value of r_BE once we know the values of both β and $I_{\mathrm{C}_\mathrm{Q}}$, we have no means of finding the value of r_B. This value is not always given on the transistor specification sheet, although there are some sophisticated methods for computing it. One approximate method for finding the value of r_I is to first find the value of r_BE using the equation, and then add 100 Ω (an average value for r_B) to the value of r_BE. For our purposes, it is not necessary that we know the exact value of r_I. Therefore, we shall neglect the effect of r_B, and assume that r_I *is approximately equal to* r_BE; i.e.,

$$r_\mathrm{I} \simeq r_\mathrm{BE} = \frac{\beta}{40 I_{\mathrm{C}_\mathrm{Q}}}$$

7-2   EFFECT OF β VARIATION ON AMPLIFIER GAIN

Figure 7.2 shows an emitter-biased amplifier stage using a 2N1574 transistor; the β of this particular unit at room temperature is 150. To find $I_{\mathrm{C}_\mathrm{Q}}$, we must first find V_B and R_B:

$$V_\mathrm{B} = \frac{R_2 V_\mathrm{CC}}{R_1 + R_2} = \frac{15(50)}{200 + 15} = 3.5 \ \mathrm{V}$$

$$R_\mathrm{B} = \frac{R_1 R_2}{R_1 + R_2} = \frac{(200)(15)}{200 + 15} = 14 \ \mathrm{k\Omega}$$

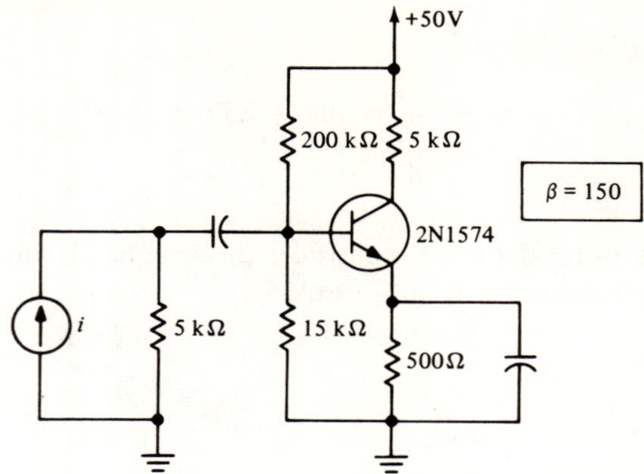

Figure 7.2

Assuming $V_{BE} = 0.5$ V, I_{C_Q} is:

$$I_{C_Q} = \frac{\beta(V_B - V_{BE})}{R_B + \beta R_E} = \frac{150(3.5 - 0.5)}{14 + 150(0.5)} = 5.05 \text{ mA} \simeq 5 \text{ mA}$$

Since the 500 Ω emitter resistor is by-passed, we can say that $R_I = r_I$. Using the expression developed in the previous section, r_I is:

$$r_I = \frac{\beta}{40 I_{C_Q}} = \frac{150}{40(5)} = 0.75 \text{ k}\Omega$$

Let us now find the gain of the amplifier stage shown in Fig. 7.2 at room temperature. R is found first as:

$$R = \frac{R_O R_B}{R_O + R_B} = \frac{(5)(14)}{5 + 14} = 3.7 \text{ k}\Omega$$

The gain at room temperature is now found to be:

$$A_I = \frac{\beta R}{R + r_I} = \frac{150(3.7)}{3.7 + 0.75} = 124.5$$

Now suppose that the temperature of the environment *increases* to 125°C. The β of the transistor will *also* increase; let us say that it *doubles*, so that the β of the transistor at 125°C is 300. We assume that the amplifier is biased very "tightly"; i.e., assume that the quiescent collector current remains relatively constant at 5 mA. Let us now find the gain at the higher temperature. First, it is important to realize that r_I *will also double* (neglecting the effect of r_B) since it is directly proportional to β. Therefore, the new value of r_I is:

$$r_I = \frac{\beta}{40 I_{C_Q}} = \frac{300}{40 \times 5} = 1.5 \text{ k}\Omega$$

The new value of gain is:

$$A_I = \frac{\beta R}{R + r_I} = \frac{300(3.7)}{3.7 + 1.5} = 213$$

Notice that there has been a *very marked increase in the gain* of the amplifier stage; as β *increased* from 150 to 300, the amplifier gain *increased* from 124.5 to 213. It should be noted that the same effect would occur if the original transistor (β = 150) were replaced with another unit (β = 300) at room temperature.

Generally speaking, this variation in the gain of the amplifier due to a variation in β is undesirable. The degree to which this instability disturbs the system depends upon the system itself; i.e., *just how important is it* that the ratio of the output to the input remain constant? For example, suppose that the gain of an amplifier in an audio system increases for some reason. If the gain *increases* while the input remains *constant*, then the output *increases*. Since the output of an audio system is the volume of sound produced in the speaker, this volume increases (louder sound) when the gain of the amplifier increases. If the sound is too loud for the listener's convenience, he merely reduces the gain of the amplifier by adjusting the volume (gain control) dial on his set. Thus the problem of gain variation is not critical in the case of some systems.

Consider next the following hypothetical situation pictured in Fig. 7.3: Suppose

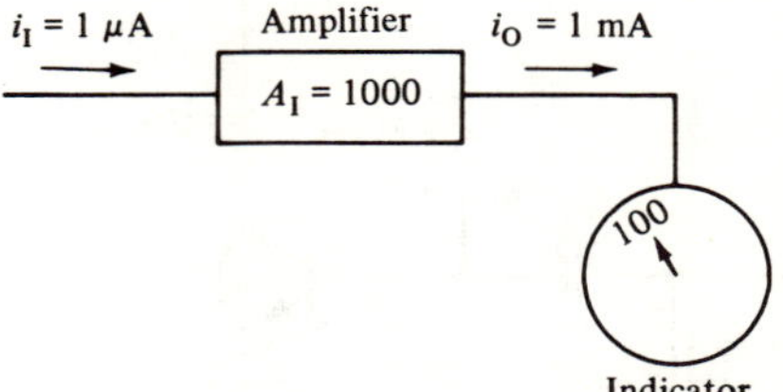

Figure 7.3

that we are designing a portion of a radar system. A 1 μA current coming from the "front end" (radio frequency amplifier section) of the radar set indicates that an airplane has been spotted *100 miles from the receiver*. This 1 μA current is too small to operate any type of indicator equipment; therefore, it is amplified 1000 times (as shown in Fig. 7.3). If the gain of the amplifier is exactly 1000, the output current will be 1 mA. The output circuit of the amplifier is then connected to some type of electro-mechanical indicator, and the indicator is adjusted so that when the output current of the amplifier is 1 mA, the indicator pointer will read 100 miles, as shown in Fig. 7.3. This way, the radar operator can easily read the distance between the detected airplane and the receiver. If the airplane is only 50 miles away, the input current will be *twice* as much, or 2 μA. The output current (output of the amplifier) will be 2 mA, and the pointer will read 50 miles.

Now suppose that the amplifier of Fig. 7.3 has recently been repaired; a transistor has failed and been replaced by another unit. Suppose also that the β of this new transistor is *larger* than that of the old (dead) one, and the gain of the amplifier *increases to 2000* as a result of the replacement. If an airplane is spotted 100 miles away from the receiver, the input current to the amplifier, i_I, will again be 1 μA. Since the gain is now 2000, the output current will be 2 mA, and the pointer will read *50 miles, even though the airplane is actually 100 miles away*. Therefore, the operator will receive an inaccurate reading, which could prove dangerous if he is directing traffic in the vicinity of a busy airport. The problem here is that the change in the gain of the amplifier due to a transistor replacement changed the calibration for the indicator. This is an example of a system in which it is desirable that the gain be as constant as possible; the more stable the gain, the more accurately the radar operator can judge the distance of approaching aircraft.

7-3   NEGATIVE ac FEEDBACK: EMITTER BIAS

The discussions of the previous section have shown that the gain of a transistor amplifier will vary because such gain depends very strongly upon the transistor parameters; it is the variation in these parameters, particularly β, which is responsible for the instability of the gain. We have also seen that *this variation in the gain is highly undesirable* in certain types of electronic systems. Obviously, our next task is to initiate some change in the amplifier circuit which will contribute toward a stabilization in the gain. Such a change is shown in the circuit of Fig. 7.4. This

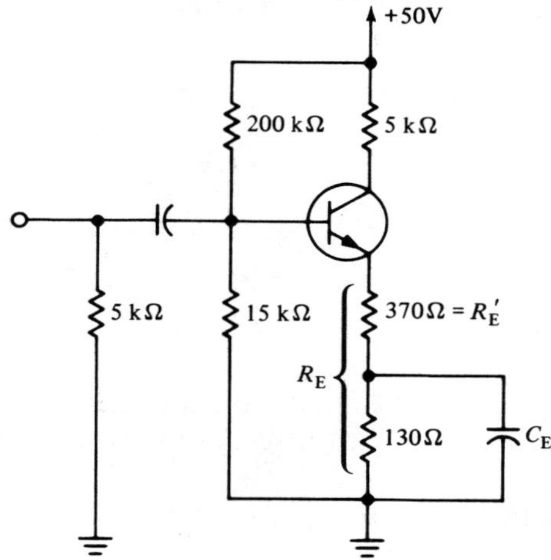

Figure 7.4

is the same basic amplifier circuit of Fig. 7.2; however, an important change has been made: The original 500 Ω emitter resistor has been replaced by *two new resistors:* a 370 Ω resistor (labeled R_E' in the figure) and a 130 Ω resistor. The by-pass capacitor, C_E, has been placed in parallel *with only the 130 Ω resistor*. Since C_E conducts no dc current, the 370 Ω and the 130 Ω resistors *are connected in series as far as dc conditions are concerned;* in other words, the stability factor remains unchanged because the total emitter resistance, R_E (as shown in Fig. 7.4), *is still equal to 500 Ω* (370 + 130 = 500). As mentioned, C_E by-passes only the 130 Ω resistor; we call R_E' (370 Ω, in this case) the *unby-passed* portion of the emitter resistance.

Let us now study the behavior of this circuit. We first compute the gain of the amplifier stage at room temperature ($\beta = 150$). The values of R (3.7 kΩ) and r_I (0.75 kΩ) are the same as for the circuit of Fig. 7.2. However, R_I is *not* equal to r_I for the circuit of Fig. 7.4 since a portion of the emitter resistance ($R_\mathrm{E}' = 370\ \Omega$) is *not* by-passed by the capacitor. For the circuit of Fig. 7.4, R_I is given as follows:

$$R_\mathrm{I} = r_\mathrm{I} + \beta R_\mathrm{E}'$$

In this expression, it is important to bear in mind that R_E' is the unby-passed portion of the emitter resistance. Writing the expression for the gain, and substituting into the equation, we get:

$$A_\mathrm{I} = \frac{\beta R}{R + R_\mathrm{I}} = \frac{\beta R}{R + r_\mathrm{I} + \beta R_\mathrm{E}'}$$

$$A_\mathrm{I} = \frac{150(3.7)}{3.7 + 0.75 + 150(0.37)} = 9.25$$

Note that the gain is quite small (9.25); it was equal to 124.5 (see previous section) when the *entire* emitter resistance was by-passed. This should come as no surprise since, as we saw in chapter 6, omission of the by-pass capacitor causes an *increase* in the input resistance (R_I), with a consequent *decrease* in the gain. Next we compute the gain of the amplifier at a temperature of 125°C ($\beta = 300$). R is still equal to 3.7 kΩ, while r_I increases to 1.5 kΩ (see previous section). The gain of the amplifier of Fig. 7.4 at 125°C is:

$$A_\mathrm{I} = \frac{\beta R}{R + r_\mathrm{I} + \beta R_\mathrm{E}'} = \frac{300(3.7)}{3.7 + 1.5 + 300(0.37)} = 9.53$$

In studying the results of the above computations, we notice some interesting phenomena. Although the temperature increased from *room temperature to 125°C,* and β correspondingly increased from *150 to 300,* the gain of the amplifier of Fig. 7.4 increased *only from 9.25 to 9.53, a very small change.* In contrast to this, under the *same* environmental conditions, the gain of the amplifier of Fig. 7.2 increased *from 124.5 to 213, a considerably larger change.* We conclude then, that the gain of the amplifier of Fig. 7.4 is more stable than that of the amplifier of Fig. 7.2. Note, however, that although the gain has been made much *more stable,* it has also been *markedly reduced;* the gain computed for the circuit of Fig. 7.4 (9.25, 9.53) *is stable, although small,* while that computed for the circuit of Fig. 7.2 (124.5, 213) *is large, although unstable.*

This improvement in the stability of the gain of the amplifier results directly from the change in the connection of the by-pass capacitor in Fig. 7.4. An *increase* in the β of the transistor (from 150 to 300) will cause an *increase* in the value of the βi_B *current generator* of the ac equivalent circuit (from $150i_\mathrm{B}$ to $300i_\mathrm{B}$). However, this *same increase* in β *also* causes an *increase* in the *input resistance* of the amplifier; the *increase* in the input resistance is computed as follows:

$$R_\mathrm{I} = r_\mathrm{I} + \beta R_\mathrm{E}' = \frac{\beta}{40 I_{C_\mathrm{Q}}} + \beta R_\mathrm{E}'$$

(a) $\beta = 150$

$$R_{\mathrm{I}} = \frac{150}{40 \times 5} + 150(0.37)$$

$$= 0.75 + 55.5 = 56.25 \text{ k}\Omega$$

(b) $\beta = 300$

$$R_{\mathrm{I}} = \frac{300}{40 \times 5} + 300(0.37)$$

$$= 1.5 + 111 = 112.5 \text{ k}\Omega$$

Note that the input resistance increased (from 56.25 kΩ to 112.5 kΩ) due to *two* separate factors: an increase in the value of r_{I} (from 0.75 kΩ to 1.5 kΩ), and an increase in the value of the reflected resistance, $\beta R_{\mathrm{E}}'$ (from 55.5 kΩ to 111 kΩ). Both r_{I} and βR_{E} are directly proportional to β, and thus they increase when β increases. The ac equivalent circuit at room temperature ($\beta = 150$) is shown in Fig. 7.5(a); the input resistance is 56.25 kΩ. Assuming that the ac current generated

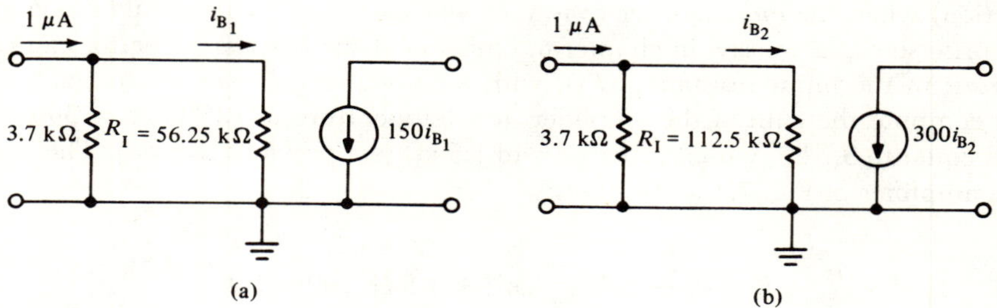

Figure 7.5

in the previous stage is 1 μA, we find the base current at room temperature, i_{B_1}, to be:

$$i_{B_1} = \frac{(1 \ \mu\text{A})R}{R + R_{\mathrm{I}}} = \frac{(1 \ \mu\text{A})(3.7)}{3.7 + 56.25} = 0.0617 \ \mu\text{A}$$

$$\beta i_{B_1} = 150(0.0617 \ \mu\text{A}) = 9.25 \ \mu\text{A}$$

Figure 7.5(b) shows the ac equivalent circuit at 125°C ($\beta = 300$); note that the input resistance has increased to 112.5 kΩ. Repeating the computations for this circuit, we get:

$$i_{B_2} = \frac{(1 \ \mu\text{A})R}{R + R_{\mathrm{I}}} = \frac{(1 \ \mu\text{A})(3.7)}{3.7 + 112.5} = 0.0318 \ \mu\text{A}$$

$$\beta i_{B_2} = 300(0.0318 \ \mu\text{A}) = 9.53 \ \mu\text{A}$$

The above computations illustrate the basic concepts involved in the operation of the circuit of Fig. 7.4. We can see from Fig. 7.5 that, as a result of the temperature increase, the current generator increased from $150 i_{B_1}$ to $300 i_{B_2}$; however, as a result of the *same* temperature increase, the input resistance of the amplifier *also* increased. This *increase* in the input resistance (from 56.25 kΩ to 112.5 kΩ) resulted in a *decrease* in the base current (from 0.0617 μA to 0.0318 μA), and *thereby compensated for the increase in the size of the βi_B current generator.* In other words, although the base current is multiplied by a *larger* factor in Fig. 7.5(b) (300), there is actually *less base current available to be multiplied* (0.0318 μA), because of the *increase* in the input resistance. Note that the final output current is approximately the same in both cases (9.25 μA as compared with 9.53 μA).

From chapter 5 you may recall that when a circuit stabilizes itself by compensating for some change in the system due to an external factor (such as a change in temperature), the phenomenon is known as negative feedback. The mechanism by which a circuit stabilizes its gain is similar to that which operates to maintain a stable Q point. We refer to the method of leaving a portion of the emitter resistance unby-passed as *negative ac feedback*, to distinguish it from *negative dc feedback* (which is used to stabilize the Q point). It is important to note that, although we *can* stabilize the gain by means of the method described, in addition to other methods, *the price to be paid for this stability is a reduced gain.* This means that more stages of amplification will have to be used to perform a particular function, with a consequent increase in the size, weight, and cost of the circuit.

7-4 GENERAL FEEDBACK THEORY

Now let us study the concept of feedback more closely. We saw in the previous section that when ac feedback was introduced (by *unby*-passing a portion of the emitter resistance), the expression for the gain can be written as follows:

$$A_I = \frac{\beta R}{R + R_I} = \frac{\beta R}{R + r_I + \beta R_E'}$$

If we divide both the numerator and denominator by the factor $(R + r_I)$, and manipulate the expression algebraically, we obtain the following results:

$$A_I = \frac{\dfrac{\beta R}{R + r_I}}{1 + \dfrac{\beta R_E'}{R + r_I}} = \frac{\dfrac{\beta R}{R + r_I}}{1 + \dfrac{\beta R_E'}{R + r_I} \times \dfrac{R}{R}} = \frac{\dfrac{\beta R}{R + r_I}}{1 + \dfrac{\beta R}{R + r_I} \times \dfrac{R_E'}{R}}$$

We can write this expression alternately as:

$$\frac{\left(\dfrac{\beta R}{R + r_I}\right)}{1 + \left(\dfrac{\beta R}{R + r_I}\right)\left(\dfrac{R_E'}{R}\right)} = A_{CL} = \frac{(A_{OL})}{1 + (A_{OL})(\beta_F)} \tag{7-2}$$

The equation is the *general feedback equation;* the gain of every amplifier which incorporates negative feedback can be written in this form. In this expression,

A_{OL} is known as the "open loop gain," or the gain *without* feedback, β_F is called the "feedback factor," and A_{CL} is the "closed loop gain," or the gain *with* feedback. The Greek letter "β" was used for the feedback factor before the advent of transistors; therefore, the subscript "F" has been included to distinguish the feedback factor (β_F) from the transistor parameter (β). It is important to note that these two quantities *bear no relation to one another;* an effort should be made not to confuse them.

It was mentioned above that the general feedback equation applies for any circuit which incorporates negative ac feedback. For the *particular* circuit studied in the previous sections (using the unby-passed emitter resistance), we can equate the following terms:

$$A_{OL} = \frac{\beta R}{R + r_I}$$

$$\beta_F = \frac{R_E'}{R}$$

Note that the expression for A_{OL} is indeed the gain without feedback, since this is the equation which would be used to compute the gain if *no feedback were used;* i.e., if the *entire* emitter resistance were by-passed by the capacitor, C_E, R_I would be equal to r_I. Suppose for a moment that the entire emitter resistance *is* by-passed by the capacitor; this means that R_E' is equal to zero. Then β_F is also equal to zero.

$$\beta_F = \frac{R_E'}{R} = \frac{0}{3.7 \text{ k}\Omega} = 0$$

The closed loop gain can now be written as follows:

$$A_{CL} = \frac{A_{OL}}{1 + A_{OL}\beta_F} = \frac{A_{OL}}{1 + (A_{OL})(0)} = A_{OL}$$

This indicates that if the entire emitter resistance is by-passed, $R_E' = 0$; thus $\beta_F = 0$, there is *no* feedback, and the closed loop gain, A_{CL}, reduces to the open loop gain, A_{OL}, as shown above.

We saw in the previous sections that the open loop gain changed from 124.5 to 213 when β changed from 150 to 300; i.e., the open loop gain is highly dependent upon the transistor parameters. Note that, even though β *doubled* (from 150 to 300), the open loop gain *did not double* (from 124.5 to 213). The reason for this phenomenon is that, since r_I is directly proportional to β, it increases when β increases. In other words, even if no feedback is *intentionally* included, the variation of r_I (due to its dependence on β) provides a very small amount of stabilization; instead of doubling from 124.5 to 249 (a 100% increase), the gain changes from 124.5 to 213 (a 71% increase). This is still, however, too large a variation for some applications, and thus we must include additional feedback in the circuit.

Ideally, we desire that the gain (closed loop) be *completely independent* of the transistor parameters, in much the same way that we should like the Q point to be independent of β, I_{CO}, and V_{BE}. Just as with the case of dc feedback, we can never achieve perfect correction; i.e., there will always be *some change* in the gain, even *with* feedback. Our task is to study the factors upon which this change in the gain depends.

Suppose that the term "$A_{OL}\beta_F$" is *much larger* than 1.0; then the general equation can be written as follows:

$$A_{CL} = \frac{\cancel{A_{OL}}}{1 + \cancel{A_{OL}}\beta_F} \simeq \frac{A_{OL}}{A_{OL}\beta_F} = \frac{1}{\beta_F}$$

$$\boxed{A_{CL} \simeq \frac{1}{\beta_F}} \quad \text{for } (A_{OL}\beta_F \gg 1) \tag{7-3}$$

This indicates that, *if "$A_{OL}\beta_F$" is a very large number, the closed loop gain is approximately equal to the reciprocal of the feedback factor.* Let us see if this situation applies to the amplifier of Fig. 7.4. Computing the factor $A_{OL}\beta_F$ at both the low and the high temperatures, we get:

$$\beta_F = \frac{R_E'}{R} = \frac{0.37}{3.7} = 0.1$$

(a) $\underline{\beta = 150}$

$$A_{OL}\beta_F = (124.5)(0.1) = 12.45 \gg 1$$

(b) $\underline{\beta = 300}$

$$A_{OL}\beta_F = (213)(0.1) = 21.3 \gg 1$$

These results indicate that $A_{OL}\beta_F$ is much greater than 1.0 for the amplifier of Fig. 7.4. In electronics, a "rule of thumb" states that, for one number to be considered *much larger* than another, it must be *at least ten times that number;* this requirement is met at *both* temperatures, as can be seen from the above computations (both 12.45 and 21.3 are much larger than 1.0). Actually, we only need check to see if $A_{OL}\beta_F$ is much larger than 1.0 at the *lower temperature* ($A_{OL}\beta_F = 12.45$ at room temperature), since, if the inequality is valid at the lower temperature, it will have *even greater validity* at the higher temperature because of the increase in the value of A_{OL} ($A_{OL}\beta_F = 21.3$ at 125°C).

Now that we have shown that the inequality is satisfied for the circuit of Fig. 7.4, we can approximate the closed loop gain as follows:

$$A_{CL} \simeq \frac{1}{\beta_F} = \frac{1}{\dfrac{R_E'}{R}} = \frac{1}{\dfrac{0.37}{3.7}} = 10$$

You may recall from the previous sections that, when the *exact* equations were used, the closed loop gains were computed to be 9.25 and 9.53; thus the approximation is a reasonably good one. The approximation may be used only when the inequality is satisfied ($A_{OL}\beta_F$ is greater than 10); the *larger* the value of $A_{OL}\beta_F$, the *closer* the closed loop gain approaches the above approximation. Therefore, if $A_{OL}\beta_F$ is greater than 10, we use the approximation to obtain a value for the closed loop gain.

The advantage of using feedback can be illustrated in another manner. If the inequality is valid, the expression for the gain can be written as follows:

$$A_{CL} \simeq \frac{1}{\beta_F} = \frac{1}{\dfrac{R_E'}{R}} = \frac{R}{R_E'}$$

Note that the gain now depends upon the ratio of the two resistances, R and R_E'. We know that R is equal to the parallel equivalent of the bias resistances and the output resistance of the previous stage, and R_E' is the *unby-passed* portion of the emitter resistance. Therefore, *the gain is relatively independent of the transistor parameters* since it depends only upon the values of R and R_E'; if we desire a very stable gain, we can pay a slightly greater price for these resistors to insure that their values do not change with temperature.

We saw above that in order for the approximation to be valid, $A_{OL}\beta_F$ must be as large as possible. This means that both A_{OL} and/or β_F must be *large*. To make A_{OL} large, we must have a transistor with a *large value of β*, and in addition, *R must also be large*, as was explained in chapter 6. To make β_F large, we must have a a *large value of R_E'*; this can be achieved by leaving a larger portion of the emitter resistance *unby-passed*. However, if the value of R_E' is increased, the value of β_F is *also* increased. *Since the gain can be approximated as the reciprocal of β_F, increasing β_F will decrease the gain.* Thus, we return again to the fact that, in order to have a reasonably large, but stable gain, we must use a transistor with a large β; the larger the β of the transistor, the larger will be the open loop gain, and the more stable will be the closed loop gain for a given value of R_E'. Once the transistor is selected, however, the range of variation of the open loop gain is fixed; then the designer has the choice of *increasing R_E'* (and β_F) and obtaining a *smaller*, though *more stable* gain, or *decreasing* it and having a *larger*, but *less stable* gain. The selection of the transistor and the determination of the necessary amount of feedback both depend upon the nature of the electronic system into which the amplifier is to be inserted. If the system requires an amplifier with a very stable gain (as in the case of the radar system described in Sec. 7-2), it will be necessary for the designer to reduce the gain of the stage in order to improve the stability, and thus more stages will be required to provide the necessary over-all gain. In chapter 10, we illustrate the design of a practical amplifier; we describe in some detail the manner in which the transistor is selected, and the determination of the required amount of feedback.

EXAMPLE 3

The amplifier shown in Fig. 7.6 is basically the same as that in Fig. 7.4; i.e., it uses the same transistor, same dc supply, and same bias and load resistors. The only difference is that the 500 Ω emitter resistor *has been split up differently;* R_E' is now equal to 50 Ω. Note that, since the bias resistors (15 kΩ and 200 kΩ) and the total emitter resistance (500 Ω) have *not* been changed, the dc conditions are the *same* as in those of the amplifier of Figs. 7.2 and 7.4. Compute the gain of the amplifier of Fig. 7.6 at room temperature and at 125°C.

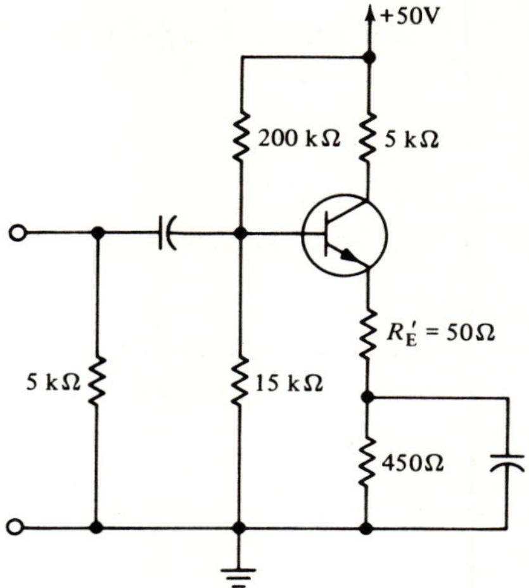

Figure 7.6

SOLUTION

We know that the open loop gain is given by the following expression:

$$A_{OL} = \frac{\beta R}{R + r_I}$$

Since neither β, nor R, nor r_I have been changed by the circuit modification, *the open loop gains will be the same* as computed for the circuit of Fig. 7.4, namely, 124.5 and 213. β_F *does* change because R_E' has been changed; the new value of β_F is:

$$\beta_F = \frac{R_E'}{R} = \frac{0.05}{3.7} = 0.0135$$

We can now compute the gains (closed loop) at room temperature and 125°C as follows:

(a) $\underline{\beta = 150}$

$$A_{CL} = \frac{A_{OL}}{1 + A_{OL}\beta_F} = \frac{124.5}{1 + (124.5)(0.0135)} = 46.5$$

(b) $\underline{\beta = 300}$

$$A_{CL} = \frac{A_{OL}}{1 + A_{OL}\beta_F} = \frac{213}{1 + (213)(0.0135)} = 54.9$$

Note that the gains are now *considerably larger* than those computed for the circuit of Fig. 7.4 (46.5 and 54.9 compared with 9.25 and 9.53); however, the gain of the amplifier of Fig. 7.6 is *less stable* than that of Fig. 7.4, since a change from 46.5 to 54.9 is a *greater percentage change* than that from 9.25 to 9.53. This action occurred because we have reduced the value of β_F (by reducing R_E'), thus increasing the gain, but sacrificing stability. The circuit of Fig. 7.6 uses *less feedback* than that of Fig. 7.4.

7-5   NEGATIVE ac FEEDBACK: COLLECTOR-TO-BASE BIAS

From chapter 6 you may recall that it was necessary to include a capacitor in the collector-to-base bias circuit to prevent loss of gain. Figure 7.7(a) shows a transistor

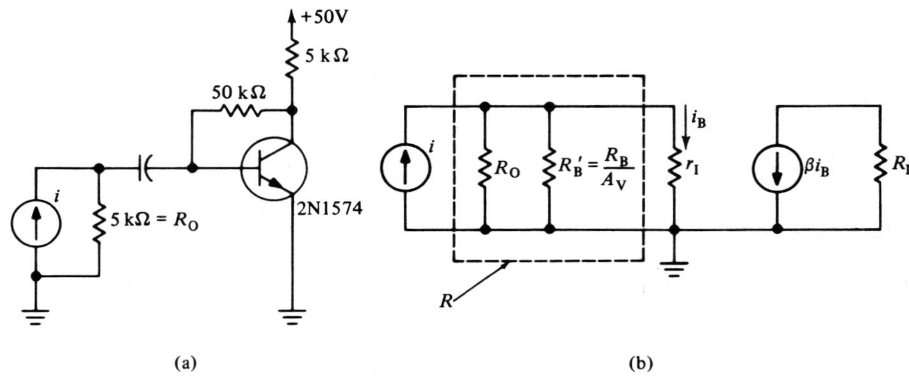

(a)   (b)

Figure 7.7

amplifier using collector-to-base bias; note that the capacitor has *not* been included in the circuit. The ac equivalent circuit appears in Fig. 7.7(b). We know that the expression for the gain of the amplifier in Fig. 7.7(b) can be written as follows:

$$A_\mathrm{I} = \frac{\beta R}{R + R_\mathrm{I}}$$

Since there is no emitter resistor in this circuit, $R_\mathrm{I} = r_\mathrm{I}$. The resistance, R, is equal to the parallel equivalent of R_O (the output resistance of the previous stage), and R_B' (which is equal to R_B divided by the voltage gain); this resistance is shown within the dotted box of Fig. 7.7(b). Substituting into the equation, we get:

$$A_\mathrm{I} = \frac{\beta \dfrac{R_\mathrm{O} R_\mathrm{B}'}{R_\mathrm{O} + R_\mathrm{B}'}}{\dfrac{R_\mathrm{O} R_\mathrm{B}'}{R_\mathrm{O} + R_\mathrm{B}'} + r_\mathrm{I}}$$

If we make the proper substitutions (such as $R_\mathrm{B}' = R_\mathrm{B}/A_\mathrm{V}$, and $A_\mathrm{V} = \beta R_\mathrm{L}/r_\mathrm{I}$), and manipulate the equations algebraically, we obtain the following result:

$$A_{\mathrm{CL}} = \frac{\left(\dfrac{\beta R_{\mathrm{O}}}{R_{\mathrm{O}} + r_{\mathrm{I}}}\right)}{1 + \left(\dfrac{\beta R_{\mathrm{O}}}{R_{\mathrm{O}} + r_i}\right)\left(\dfrac{R_{\mathrm{L}}}{R_{\mathrm{B}}}\right)} = \frac{A_{\mathrm{OL}}}{1 + A_{\mathrm{OL}}\beta_{\mathrm{F}}}$$

This equation indicates that the circuit of Fig. 7.7(a) provides negative feedback; in this case, the open loop gain and feedback factor are given as follows:

$$A_{\mathrm{OL}} = \frac{\beta R_{\mathrm{O}}}{R_{\mathrm{O}} + r_{\mathrm{I}}}$$

$$\beta_{\mathrm{F}} = \frac{R_{\mathrm{L}}}{R_{\mathrm{B}}}$$

Therefore, we can adjust β_{F} by varying the values of R_{L} and R_{B}, and thus provide the feedback necessary to stabilize the gain. From chapter 5 you may recall that the Q point is dependent upon the value of R_{B}. If R_{B} is made relatively small to yield a certain value of β_{F}, it may be necessary to include a second dc supply, as described in chapter 5, in order that the Q point be fixed in the appropriate part of the load line.

Let us assume that $I_{\mathrm{C_Q}} = 5$ mA, and the β of the transistor varies between 150 and 300. Computing first the open loop gains for the circuit of Fig. 7.7(a), we get:

(a) $\beta = 150$

$$A_{\mathrm{OL}} = \frac{\beta R_{\mathrm{O}}}{R_{\mathrm{O}} + r_{\mathrm{I}}} = \frac{150(5)}{5 + 0.75} = 130$$

(b) $\beta = 300$

$$A_{\mathrm{OL}} = \frac{\beta R_{\mathrm{O}}}{R_{\mathrm{O}} + r_{\mathrm{I}}} = \frac{300(5)}{5 + 1.5} = 231$$

For this particular circuit, β_{F} is:

$$\beta_{\mathrm{F}} = \frac{R_{\mathrm{L}}}{R_{\mathrm{B}}} = \frac{5 \text{ k}\Omega}{50 \text{ k}\Omega} = 0.1$$

Then the closed loop gains are:

(a) $\beta = 150$

$$A_{\mathrm{CL}} = \frac{A_{\mathrm{OL}}}{1 + A_{\mathrm{OL}}\beta_{\mathrm{F}}} = \frac{130}{1 + 130(0.1)} = 9.28$$

(b) $\beta = 300$

$$A_{\mathrm{CL}} = \frac{A_{\mathrm{OL}}}{1 + A_{\mathrm{OL}}\beta_{\mathrm{F}}} = \frac{231}{1 + 231(0.1)} = 9.58$$

Thus, the circuit of Fig. 7.7(a) is another example of a transistor amplifier which uses negative feedback to stabilize the gain. Notice again that, although the gain is made more stable, it is reduced considerably; a reduction in the gain is the price which must be paid for removing the capacitor, but we achieve greater stability.

7-6   OVER-ALL FEEDBACK

The feedback amplifiers studied in the previous sections (using emitter bias and collector-to-base bias) incorporated what is known as "local" feedback; i.e., the feedback mechanism was confined to a *single* stage. We saw that, because feedback was used, the gains of these amplifier stages were markedly reduced. We could, of course, make up the loss in gain by cascading the individual stages. However, in this section we see that, rather than using local feedback and then cascading the stages, it is far more efficient to use a method known as "over-all" feedback.

Figure 7.8 shows a two-stage amplifier which incorporates over-all feedback.

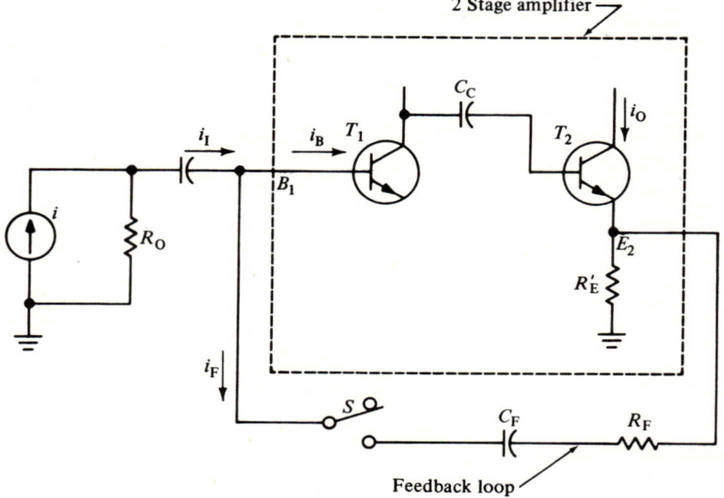

Figure 7.8

The portion of the circuit within the box is an ordinary two-stage R-C coupled transistor amplifier. Note that some of the bias resistors and load resistors have been omitted from the circuit for simplicity. This amplifier is fed from a source consisting of the current generator, i, and an internal resistance, R_O. Up to this point, there is nothing unusual about the circuit. The *new addition* to the circuit is a wire which is connected from the *emitter* terminal (E_2) of the *second* stage, to the series connection of a resistor, R_F, a capacitor, C_F, and a switch, S, and back to the *base* terminal (B_1) of the *first* stage. This portion of the circuit is called the "feedback loop," as indicated in Fig. 7.8. The current which flows through the branch of the circuit consisting of the series connection of R_F, C_F, and S is called the feedback current, i_F.

Now let us first examine the circuit conditions when the switch is *open*. If the switch is open, the current i_F is zero. Since the feedback current is *zero, no* current flows through R_F and C_F, and thus these two circuit elements have *no* effect on the behavior of the circuit. The current, i_I, flows directly into the base of the transistor in the first stage, T_1 (neglecting the effect of the bias resistors of the first stage, which are not shown), because i_F is zero. *Therefore, we can say that $i_I = i_B$ when the switch is open.* The circuit thus behaves as an ordinary two-stage amplifier containing the transistors, T_1 and T_2. The over-all gain of the two-stage amplifier, i_O/i, is equal to the product of the gains of the individual stages:

$$A_{Io} = A_{I_1} \times A_{I_2} = \underbrace{\left[\frac{\beta R}{R + r_I}\right]}_{\text{STAGE 1}} \times \underbrace{\left[\frac{\beta R}{R + r_I + \beta R_E{}'}\right]}_{\text{STAGE 2}}$$

Note that there is some local feedback in stage 2 due to the presence of $R_E{}'$, but there is none in the first stage. *The over-all gain is also called the open loop gain of the amplifier, because this is the gain when the feedback loop is open;* the reader can now appreciate the use of the term "open loop," when applied to a feedback amplifier.

$$A_{Io} = A_{OL}$$

Summarizing, we can say that when the switch in Fig. 7.8 is *open*, the over-all feedback mechanism is *not* in operation, the open loop gain (A_{OL}) is equal to the over-all gain of the two-stage amplifier (A_{Io}), and there is some local feedback present in the second stage due to $R_E{}'$.

Now we can study the situation when the switch is *closed*. A feedback current, i_F, will now flow through the R_F-C_F series combination. If the directions of the currents are as indicated in Fig. 7.8, we can say the i_I is equal to the sum of i_B and i_F:

$$i_I = i_B + i_F$$

$$i_B = i_I - i_F$$

In other words, the current flowing through the base of the first transistor *is less than i_I by the amount flowing through R_F and C_F.* Since i_B is less when the switch is closed, the output current, i_O, will also be less, with the result that the gain of the amplifier will be reduced. The feedback capacitor, C_F, is included to prevent any *dc* current from flowing through the feedback branch. Sometimes, however, C_F, C_C (between the two stages), and the bias resistors are omitted from the circuit, and the feedback branch (consisting now of only R_F) is used to *stabilize the dc, as well as the ac conditions* of the amplifier. We do *not* use the feedback loop to stabilize the dc conditions because of the complexity of the mechanism involved; we use the standard method of *dc* stabilization (emitter bias or collector-to-base bias), and employ the feedback loop to *stabilize the ac operation only.*

Let us now see how the feedback mechanism operates to stabilize the gain of the amplifier when the loop is closed; refer to Fig. 7.9(a). The circuit is drawn with the polarities shown for the instant at which the base-to-emitter voltage of the first stage (v_{BE_1}) and the base current of the first stage (i_{B_1}) are *positive-going;*

we shall be concerned here only with the *ac portions* of the currents and voltages. If v_{BE_1} and i_{B_1} are positive-going, v_{CE_1} is *negative-going* because of the 180 degree phase reversal in the amplifier (see chapter 4); note that the polarity shown for v_{CE_1} in Fig. 7.9(a) indicates that the potential at the collector terminal is *decreasing*

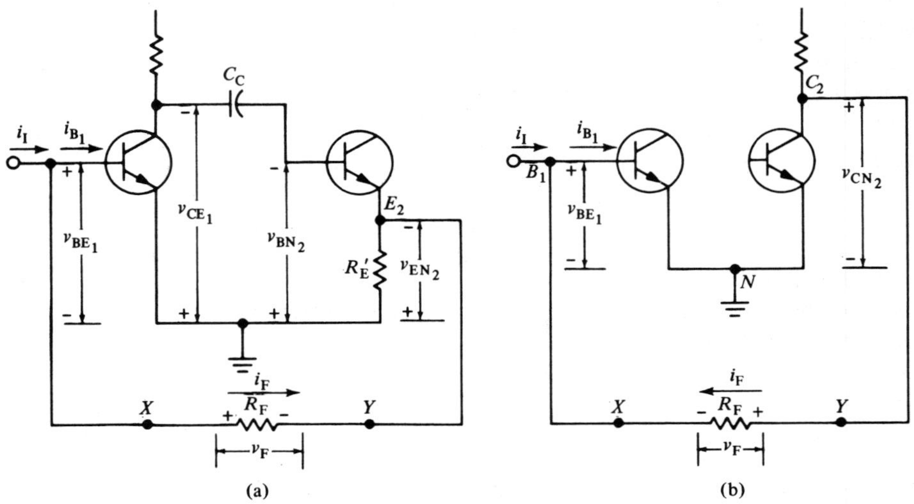

Figure 7.9

with respect to that at the emitter terminal (ground). We know that v_{BN_2} is approximately equal to v_{CE_1} if we neglect the ac voltage drop across the coupling capacitor; therefore, v_{BN_2} is *also* negative-going. Since v_{BN_2} is negative-going, i_{B_2}, i_{E_2} (i_B, i_C, and i_E are all in phase in a transistor), and v_{EN_2} are all negative-going. Now if v_{EN_2} is negative-going, the potential at the *emitter* terminal of the second stage (E_2) is *decreasing* with respect to ground; this is the same terminal (terminal Y) as the *right*-hand terminal of the feedback resistor, R_F. We know that the potential at the *base* terminal of the first stage (B_1) is *increasing* with respect to ground; this is the same terminal (terminal X) as the *left*-hand terminal of R_F. Therefore, if the potential at X is *increasing* with respect to ground, and that at Y is *decreasing*, the potential at X is *increasing* with respect to that at Y, the polarity of v_F is as shown in Fig. 7.9(a), and the direction of the feedback current, i_F, is also as shown.

Suppose that, due to a temperature increase or a transistor replacement, the β of one or both of the transistors increases. The peak-to-peak values of the waveforms of i_{E_2} and v_{EN_2} will *increase*. If the ac portion of v_{EN_2} increases, the potential at terminal E_2 (terminal Y) will *decrease further* with respect to ground, and consequently will become *more negative* with respect to the potential at terminal B_1 (terminal X). If the potential at Y becomes more negative than that at X, the voltage across the feedback resistor ($v_F = v_{XY}$) will *increase* and the feedback current (i_F) will also *increase*. But we know that i_{B_1} is equal to the difference between i_I and i_F.

$$i_{B_1} = i_I - i_F$$

Thus if i_F *tends to increase* due to an increase in β, i_{B_1} will *decrease* to offset the increase. This is another example of negative feedback; a change in the output feeds back information to the input to compensate for the change. The feedback mechanism operates through the loop consisting of R_F and C_F.

The *phase* in which the signals (currents and voltages) are fed back to the input is extremely important. To illustrate this point, let us connect terminal Y of the feedback resistor to the *collector* (C_2) of the second stage, instead of the emitter, as shown in Fig. 7.9(b). The circuit is again drawn for the instant at which v_{BE_1} is *positive-going*. We saw above (and in chapter 4) that there is a *180 degree phase difference* between the input and output voltages in a one-stage transistor amplifier. Therefore, if the voltage undergoes a 180 degree phase shift between the base and collector terminals of stage 1, there will be *another* 180 degree phase shift between the base and collector terminals of stage 2; the voltage at the collector of the second stage (v_{CN_2}) will have *the same phase* as the voltage at the base of the first stage (v_{BE_1}), as shown in Fig. 7.9(b). The potentials both at the base of stage 1 (B_1) and the collector of stage 2 (C_2) are *increasing* with respect to ground. Since the amplifier provides voltage gain, the ac portion of v_{CN_2} is *much larger* than that of v_{BE_1}. Therefore, the potential at $C_2(Y)$ *is increasing with respect to ground at a faster rate* than the potential at B_1 (X). Thus, the potential at Y is *increasing* with respect to that at X, and the polarity of v_F and the direction of i_F are as indicated in Fig. 7.9(b). Note that the direction of i_F in Fig. 7.9(b) is *opposite* to that in Fig. 7.9(a).

Suppose again that the β of the transistors increases. The peak-to-peak value of v_{CN_2} will *increase*, and the potential at terminal C_2 *will increase further* above ground; if the potential at terminal Y becomes *more positive* with respect to that at X, v_F *increases* and i_F *increases*. Notice that, because of the direction of i_F indicated in Fig. 7.9(b), i_{B_1} is now equal to the *sum* of i_I and i_F; i.e.:

$$i_{B_1} = i_I + i_F$$

An increase in i_F due to an increase in β will cause a corresponding *increase* in i_{B_1} [instead of a decrease as in the circuit of Fig. 7.9(a)]. This increase in i_{B_1} will cause a *further increase* in v_{CN_2}, which will, in turn, cause a further increase in i_F and i_{B_1}. The currents and voltages will continue to increase in this cycle; this is an *unstable* situation. If conditions are right (the discussion of these conditions is beyond the scope of this text), the circuit of Fig. 7.9(b) will "break into oscillations." This means that the *circuit itself* will generate an ac waveform (not necessarily sinusoidal) which will oscillate *even if the original ac input source is removed from the circuit*. The frequency of this "self-oscillation" depends upon the transistor parameters and the values of the resistors and capacitors in the circuit. If the circuit generates these oscillations, it is called an "oscillator," not an amplifier. Suppose that the input signal (i_I) is applied to the circuit of Fig. 7.9(b) *while it is oscillating*. Since the oscillator generates a waveform which may *not* be sinusoidal, and the frequency of these oscillations may bear *no* relation to the frequency of i_I, the output signal at the collector of the second stage *will not have the same waveshape as i_I*. From chapter 4 you may recall that one of the requirements of an amplifier is that *the output retain the same waveshape as the input*. This is clearly *not* the case when the circuit of Fig. 7.9(b) is oscillating, and thus this circuit is no longer an amplifier.

Even if the circuit of Fig. 7.9(b) does not oscillate, it still does not meet our requirements because it does not provide negative feedback. Because the feedback was taken from the second collector, an increase in v_{CN_2} (due to an increase in β) caused an increase in i_F which, in turn, caused an *increase* in i_{B_1}, instead of a *decrease* as in Fig. 7.9(a). Thus, instead of stabilizing the gain by *reducing* the value of i_{B_1} [as in the circuit of Fig. 7.9(a)], the feedback used in the circuit of Fig. 7.9(b) aggravates the situation by *increasing the value of* i_{B_1}, thus contributing further to the instability.

The type of feedback used in the circuit of Fig. 7.9(b) is called "positive" feedback. In this case, signals are fed back in such a phase as to *increase* any change in the gain. Positive feedback generally promotes an *unstable* situation. The most important use of positive feedback is in the design of oscillators; we do not discuss oscillators in this text. The type of feedback used in the circuit of Fig. 7.9(a) is called negative feedback. In this case, signals are fed back in such a phase as to *decrease* any change in the gain. Negative feedback generally promotes a *stable* situation. This technique is used to stabilize both the gain and the Q point (see chapter 5) of amplifiers. Summarizing, it is extremely important that the feedback signals be in the proper phase.

The next step is to analyze the amplifier of Fig. 7.9(a); i.e., we should like to develop a method for determining its gain. The circuit is re-drawn in Fig. 7.10(a).

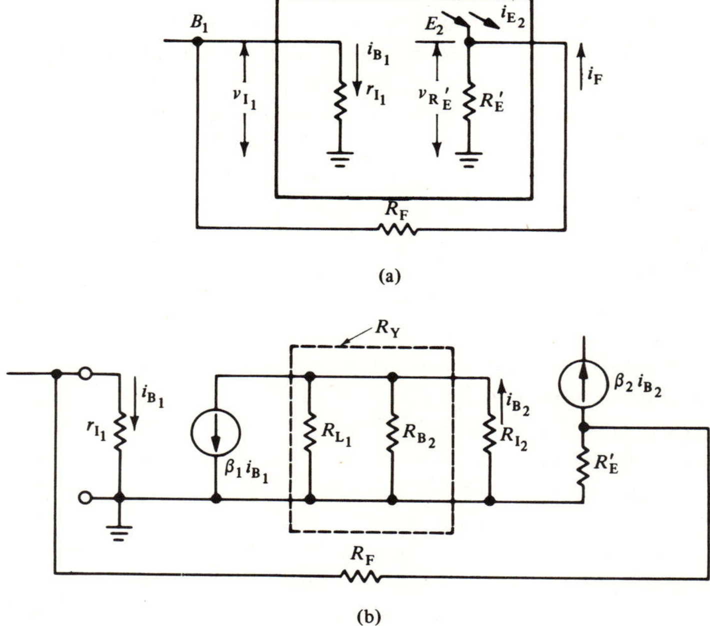

(a)

(b)

Figure 7.10

Note that the feedback resistor, R_F, is connected between E_2 and B_1 (we are neglecting the reactance of the feedback capacitor, C_F). We saw in chapter 6 that if a resistance (or impedance) is connected between two terminals, one of which is the input terminal of the amplifier (B_1), *we can use the Miller effect method to replace this resistor with a new resistance which is in shunt with the input* (base and ground) *terminals.* We used this method to advantage in analyzing the collector-to-base bias circuit for ac operation (see chapter 6). We can thus replace R_F in Fig. 7.10(a) by a new resistance, R_F', whose value is:

$$R_F' = \frac{R_F}{1 + A_V}$$

In this case, A_V must be defined as the ratio of v_{R_E}' to v_{I_1}, since R_F is connected between E_2 and B_1, as shown in Fig. 7.10(a). A_V can be written as follows:

$$A_V = \frac{v_{R_E}'}{v_{I_1}} = \frac{R_E'(i_{E_2} + i_F)}{r_{I_1}i_{B_1}}$$

It can be shown that i_{E_2} is *much larger* than i_F; therefore, A_V can be approximated as follows:

$$A_V \simeq \frac{R_E' i_{E_2}}{r_{I_1}i_{B_1}} \qquad \text{For } i_{E_2} \gg i_F$$

The ac emitter current, i_{E_2}, is approximately equal to the ac collector current, $\beta_2 i_{B_2}$, in the second stage:

$$i_{E_2} \simeq \beta_2 i_{B_2}$$

Using the current divider in Fig. 7.10(b), we can express i_{B_2} as follows:

$$i_{B_2} = \frac{(\beta_1 i_{B_1})R_Y}{R_Y + R_{I_2}} = \frac{(\beta_1 i_{B_1})R_Y}{R_Y + r_{I_2} + \beta_2 R_E'}$$

In this expression, R_Y is equal to the parallel equivalent of R_{L_1} (load resistance of the first stage) and R_{B_2} (parallel equivalent of R_1 and R_2 of the second stage). Thus, we can now write i_{E_2} as follows:

$$i_{E_2} = \beta_2 i_{B_2} = \frac{\beta_2(\beta_1 i_{B_1})R_Y}{R_Y + r_{I_2} + \beta_2 R_E'}$$

$$i_{E_2} = \frac{\beta_1 \beta_2 R_Y i_{B_1}}{R_Y + r_{I_2} + \beta_2 R_E'}$$

A_V is expressed as:

$$A_V = R_E' \frac{\left(\dfrac{\beta_1 \beta_2 R_Y i_{B_1}}{R_Y + r_{I_2} + \beta_2 R_E'}\right)}{r_{I_1}i_{B_1}} = \frac{R_E' \beta_1 \beta_2 R_Y}{r_{I_1}(R_Y + r_{I_2} + \beta_2 R_E')}$$

Since the voltage gain is usually much greater than 1.0, we express the new resistance, R_F', as:

$$R_F' = \frac{R_F}{1 + A_V} \simeq \frac{R_F}{A_V} = \frac{R_F}{\dfrac{R_E'\beta_1\beta_2 R_Y}{r_{I_1}(R_Y + r_{I_2} + \beta_2 R_E')}} = \frac{R_F r_{I_1}(R_Y + r_{I_2} + \beta_2 R_E')}{R_E'\beta_1\beta_2 R_Y}$$

This resistance is now placed in shunt with the input terminals of the amplifier as shown in Fig. 7.11.

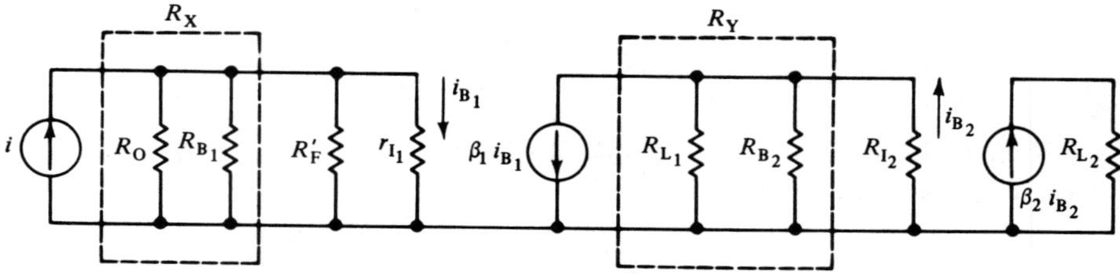

Figure 7.11

Now we are interested in finding the over-all closed loop gain of the amplifier, A_{CL}; this can be expressed as follows:

$$A_{CL} = \frac{\beta_2 i_{B_2}}{i} = \frac{\beta_2 i_{B_2}}{i_{B_1}} \times \frac{i_{B_1}}{i}$$

The ratio, $\beta_2 i_{B_2}/i_{B_1}$, has already been established as:

$$\frac{\beta_2 i_{B_2}}{i_{B_1}} = \frac{\beta_1 \beta_2 R_Y}{R_Y + r_{I_2} + \beta_2 R_E'}$$

The other ratio, i_{B_1}/i, can be written as:

$$\frac{i_{B_1}}{i} = \frac{R'}{R' + r_{I_1}}$$

In the above expression, R' represents the parallel equivalent of R_O (the internal resistance of the source), R_{B_1} (the parallel equivalent of R_1 and R_2 of the first stage), and R_F'. R' can also be expressed as the parallel equivalent of R_X (parallel equivalent of R_O and R_{B_1}) and R_F':

$$R' = \frac{R_X R_F'}{R_X + R_F'} \qquad \text{where } R_X = \frac{R_O R_{B_1}}{R_O + R_{B_1}}$$

If we substitute all of these relationships into the expression for the closed loop gain, we get:

$$A_{CL} = \frac{\beta_2 i_{B_2}}{i_{B_1}} \times \frac{i_{B_1}}{i} = \frac{\beta_1 \beta_2 R_Y}{R_Y + r_{I_2} + \beta_2 R_E'} \times \frac{R'}{R' + r_{I_1}}$$

If the above equation is manipulated algebraically, it can be expressed finally as:

$$A_{\text{CL}} = \frac{\left[\dfrac{\beta_1 R_\text{X}}{R_\text{X} + r_{\text{I}_1}} \times \dfrac{\beta_2 R_\text{Y}}{R_\text{Y} + r_{\text{I}_2} + \beta_2 R_\text{E}'} \right]}{1 + \left[\dfrac{\beta_1 R_\text{X}}{R_\text{X} + r_{\text{I}_1}} \times \dfrac{\beta_2 R_\text{Y}}{R_\text{Y} + r_{\text{I}_2} + \beta_2 R_\text{E}'} \right]\left[\dfrac{R_\text{E}'}{R_\text{F}} \right]} \qquad (7\text{-}4)$$

$$A_{\text{CL}} = \frac{A_{\text{OL}}}{1 + A_{\text{OL}}\beta_\text{F}}$$

In this equation, the open loop gain is equal to the product of the gains of the individual stages, if no feedback is introduced (loop is open):

$$A_{\text{OL}} = A_{\text{I}_1} \times A_{\text{I}_2} = \underbrace{\left(\frac{\beta_1 R_\text{X}}{R_\text{X} + r_{\text{I}_1}} \right)}_{A_{\text{I}_1}} \times \underbrace{\left(\frac{\beta_2 R_\text{Y}}{R_\text{Y} + r_{\text{I}_2} + \beta_2 R_\text{E}'} \right)}_{A_{\text{I}_2}}$$

The feedback factor is:

$$\beta_\text{F} = \frac{R_\text{E}'}{R_\text{F}}$$

Let us now apply this information to the solution of a practical problem. A two-stage amplifier utilizing the feedback scheme that we have been discussing is shown in Fig. 7.12. Notice that both stages are identical, except for the unby-passed

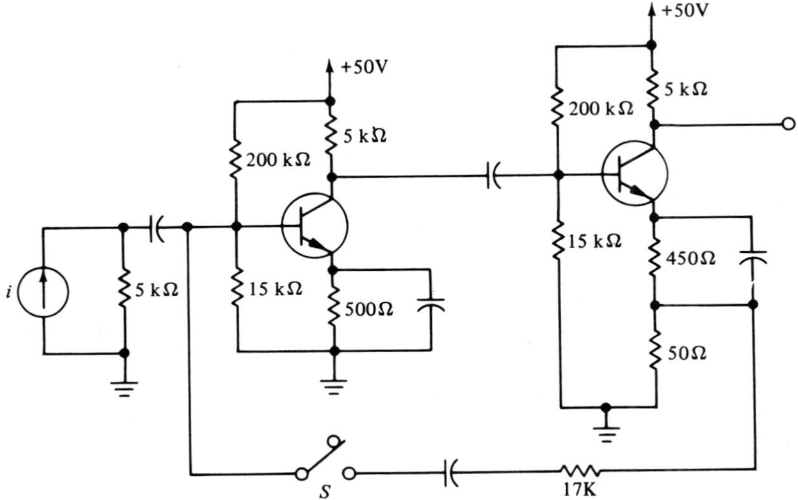

Figure 7.12

resistance ($R_\text{E}' = 50\ \Omega$) in the second stage. Assume that the minimum anticipated β of the transistors in the amplifier of Fig. 7.12 is 100; assume also that the transistors are biased in the center of the load line ($I_{\text{CQ}} = 5$ mA). Then we can write:

$$r_{\text{I}_1} = r_{\text{I}_2} = \frac{\beta}{40 I_{\text{CQ}}} = \frac{100}{40(5)} = 0.5\ \text{k}\Omega$$

The next step is to find the open loop gain of the amplifier. This is done as follows:

$$A_{I_1} = \frac{\beta_1 R_X}{R_X + r_{I_1}} = \frac{100(3.7)}{3.7 + 0.5} = 88$$

Note that these amplifier stages use the same biasing arrangement which was used earlier in the chapter; thus $R_X = R_Y = 3.7$ kΩ (the parallel equivalent of 5 kΩ, 15 kΩ, and 200 kΩ).

$$A_{I_2} = \frac{\beta_2 R_Y}{R_Y + r_{I_2} + \beta_2 R_E'} = \frac{100(3.7)}{3.7 + 0.5 + 100(0.05)} = 40$$

$$A_{OL} = A_{I_1} \times A_{I_2} = (88) \times (40) = 3520$$

Assume that for the desired stability, we should like $A_{OL}\beta_F = 10$. If R_E' is fixed at 50 Ω, we can find the necessary value of R_F as follows:

$$A_{OL}\beta_F = 10; \quad \beta_F = \frac{10}{A_{OL}} = \frac{10}{3520} = \frac{R_E'}{R_F}$$

$$R_F = \frac{3520 R_E'}{10} = 352(0.05) = 17.6 \text{ k}\Omega$$

This means that the closed loop gain should be approximately equal to:

$$A_{CL} \simeq \frac{1}{\beta_F} = \frac{1}{\dfrac{10}{3520}} = 352$$

If we set $R_F = 17$ kΩ, we find the *exact* value of the closed loop gain as follows:

$$A_{CL} = \frac{A_{OL}}{1 + A_{OL}\beta_F} = \frac{3520}{1 + 3520\left(\dfrac{0.05}{17}\right)} = 310$$

From earlier sections in this chapter you recall that when we set $A_{OL}\beta_F = 10$ for a *single-stage* amplifier using *local* feedback ($\beta = 150$), the closed loop gain was approximately equal to 10. If we cascade *two* such identical stages, the gain will be equal to 100. If we use the *same two amplifier stages* (as in Fig. 7.12) in an amplifier incorporating *over-all* feedback with $A_{OL}\beta_F$ again set equal to 10, the closed loop gain is equal to 310. Thus it can be seen that, *for the same number of amplifier stages (two, in this case), and the same desired stability ($A_{OL}\beta_F = 10$, in this case), we obtain a larger closed loop gain (310) using over-all feedback than if we had used local feedback (100).* Since over-all feedback is more efficient, we use it whenever possible; the use of local feedback is limited to single-stage amplifiers, i.e., when the specified gain need not be too large. The use of over-all feedback is generally limited to two stages, because, if more than two stages are included within the feedback loop, there is a danger that the circuit might oscillate; the reason for this phenomenon is beyond the scope of the text.

Suppose that the β's of the transistors in the amplifier of Fig. 7.12 triple, from 100 to 300. Let us compute the new open and closed loop gains. The solution proceeds as follows:

$$r_{\mathrm{I}_1} = r_{\mathrm{I}_2} = \frac{\beta}{40 I_{\mathrm{C_Q}}} = \frac{300}{40 \times 5} = 1.5 \text{ k}\Omega$$

$$A_{\mathrm{I}_1} = \frac{\beta_1 R_{\mathrm{X}}}{R_{\mathrm{X}} + r_{\mathrm{I}_1}} = \frac{300(3.7)}{3.7 + 1.5} = 213$$

$$A_{\mathrm{I}_2} = \frac{\beta_2 R_{\mathrm{Y}}}{R_{\mathrm{Y}} + r_{\mathrm{I}_2} + \beta_2 R_{\mathrm{E}}'} = \frac{300(3.7)}{3.7 + 1.5 + 300(0.05)} = 55$$

$$A_{\mathrm{OL}} = A_{\mathrm{I}_1} \times A_{\mathrm{I}_2} = (213) \times (55) = 11{,}700$$

$$A_{\mathrm{CL}} = \frac{A_{\mathrm{OL}}}{1 + A_{\mathrm{OL}}\beta_{\mathrm{F}}} = \frac{11{,}700}{1 + 11{,}700\,\dfrac{0.05}{17}} = 331$$

Thus, while the open loop gain varies from 3520 to 11,700, the closed loop gain varies from 310 to 331.

The two-stage amplifier employing over-all feedback shown in Fig. 7.12 is not the only type which can be used; there are a number of other circuits. This particular circuit was analyzed to illustrate the major concepts involved in the use of over-all feedback.

7-7 MAXIMUM POWER TRANSFER

Up to this point in the text we have been discussing the various concepts involved in the analysis and design of transistor amplifiers; very little has been said concerning the nature of the original source or final load. You may recall that the primary task of the amplifier is to increase the *amount of ac power* generated in the source, and then deliver this increased power to the load. The energy (or power) in both the original source and the final load is usually in some form *other than electrical*. A device known as a "transducer" converts the energy from this other form into electrical energy; e.g., a microphone is an example of a transducer which converts sound energy into electrical energy at the input to an amplifier. The energy remains in electrical form while in the amplifier. At the output of the amplifier, another transducer re-converts the energy back into its original form; e.g., a speaker is an example of a transducer which converts electrical energy into sound energy.

The nature of the load itself is determined by the type of system in which the amplifier is to be used. Following is a partial list of systems in which an amplifier might be used:

TYPE OF SYSTEM	TYPE OF ENERGY	LOAD
Audio (Radio, Phonograph)	Sound	Speaker
Video (Television, Radar)	Light	Cathode Ray Tube
Control (Inertial Navigation)	Mechanical	Motor
Data Processing (Computer)	Mechanical	Printer

In addition to the types of loads mentioned above, the load may alternately be *another type of electronic circuit.* In many digital systems, the output of the amplifier is fed to the input of a circuit known as a "multivibrator"; in this case, the impedance of the load is merely the input impedance of the multivibrator circuit, and the energy remains in electrical form. The transfer of energy through a system utilizing an amplifier is illustrated in Fig. 7.13; note that the load can take many possible forms, depending upon the type of system.

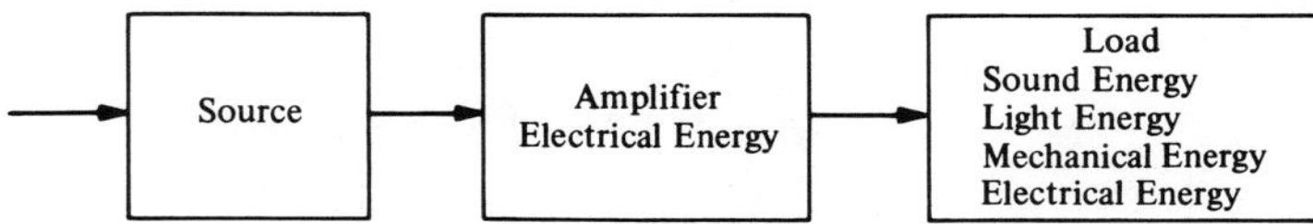

Figure 7.13

Since the task of the amplifier is to deliver a large amount of ac power to the load, we should examine the effect of the load impedance upon the amount of power delivered to it. Figure 7.14(a) shows the final stage of an amplifier ($R_\mathrm{L} =$

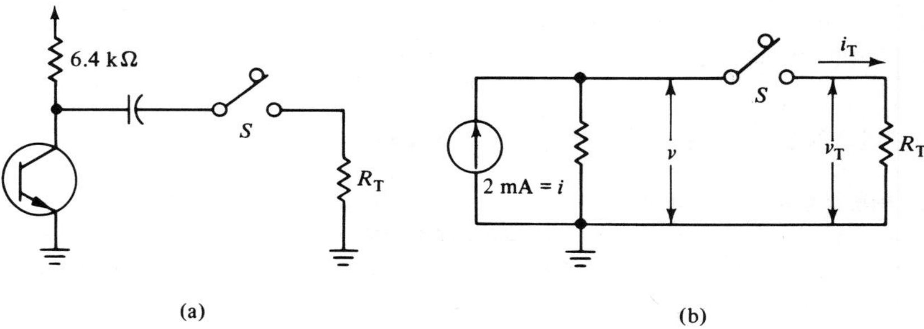

Figure 7.14

6.4 kΩ); the amplifier output is connected through a capacitor and a switch to a resistance, R_T. To avoid confusion, we represent the impedance of the final load by the resistance, R_T; in this case, the "T" stands for "terminal." At this point, we are not concerned with the physical nature of the load; i.e., we don't care whether the load is a speaker, cathode ray tube, motor, etc. R_T represents the value of the electrical resistance (or impedance) of the load. The capacitor is included in the circuit of Fig. 7.14(a) to prevent any dc current from reaching the load, since this is a requirement of many systems.

If the peak-to-peak ac collector current developed in the collector of the final stage is 2 mA, the ac equivalent circuit of Fig. 7.14(a) can be represented as shown in Fig. 7.14(b). With the switch S *open*, no current will flow through R_T. The ac power developed at the output of the amplifier stage *with the load disconnected* can be computed as follows:

$$v = i(6.4 \text{ k}\Omega) = (2 \text{ mA})(6.4 \text{ k}\Omega) = 12.8 \text{ V}$$

$$p = \frac{(v_{\text{P--P}})(i_{\text{P--P}})}{8} = \frac{(12.8)(2)}{8} = 3.2 \text{ mW}$$

In this case, all of the current from the current generator ($i = 2$ mA) flows through the 6.4 kΩ load resistor of the last stage. When the switch is *closed*, a portion of the 2 mA generator current will flow through R_T, and thus the output current, voltage, and power will change.

To determine *the effect of the value of R_T on the amount of power delivered to it*, we must place different values of resistance in the circuit for R_T, and then compute (or measure) the power delivered to the resistance in each case. Refer to the chart shown in Fig. 7.15. In the first column are listed the values of various resistances

R_T (kΩ)	i_T (mA)	v_T (V)	p_T (mW)
0	2	0	0
0.4	1.88	0.75	0.177
1.0	1.73	1.73	0.374
3.2	1.33	4.26	0.710
6.4	1.0	6.40	0.800
10.0	0.78	7.80	0.760
12.8	0.67	8.55	0.710
20.0	0.48	9.70	0.588
∞	0	12.80	0

Figure 7.15

(in kilohms) from zero ohms (short circuit) to infinity (open circuit); these are placed in the circuit individually for R_T. The current which flows through R_T, designated i_T in Fig. 7.14(b), can be computed by the current divider, as follows:

$$i_\text{T} = \frac{i(6.4)}{6.4 + R_\text{T}} = \frac{2(6.4)}{6.4 + R_\text{T}} = \frac{12.8}{6.4 + R_\text{T}}$$

These values are computed and are shown in the second column of the chart. The voltage across R_T, designated v_T in Fig. 7.14(b), is computed by Ohm's law; the values are listed in the third column of the chart:

$$v_\text{T} = (i_\text{T})(R_\text{T})$$

Finally, the power delivered to R_T, which we call "p_T," is computed by the standard method, and the values listed in the last column; we use peak-to-peak values, and thus we must include the "one-eighth" factor described in chapter 4:

$$p_\text{T} = \frac{(v_\text{T})(i_\text{T})}{8}$$

Figure 7.16 shows a graph which is plotted from the data of the chart of Fig.

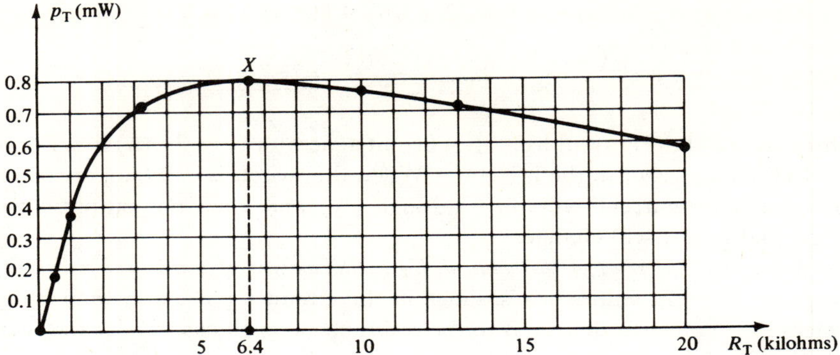

Figure 7.16

7.15; the graph is a plot of the power delivered to the load in milliwatts (p_T) vs. the value of the terminal resistance in kilohms (R_T). Notice that the power *increases* at first as the terminal resistance increases, then reaches a *maximum* at point X, and finally begins to *decrease* with a further increase in the value of R_T. The maximum power which can be delivered to *any* load under these circumstances is 0.8 mW, considerably less than the 3.2 mW available from the amplifier output when the load was disconnected. This 0.8 mW is delivered to the load (R_T) when $R_T =$ 6.4 kΩ as can be seen from the graph of Fig. 7.16; *this is the same value as the load resistance of the final stage* (6.4 kΩ). In other words, if we consider the 2 mA current generator and the 6.4 kΩ load resistor as comprising a source, as shown in Fig. 7.17,

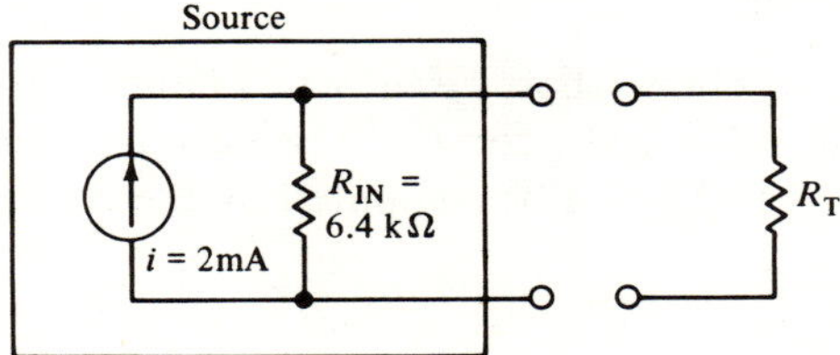

Figure 7.17

then *once the source is fixed, maximum power will be delivered to R_T when R_T is equal to the internal resistance of the source;* in Fig. 7.17, the internal resistance of the source, R_{IN}, is equal to 6.4 kΩ. This is known as the condition of *maximum power transfer.* Therefore, *it is true in general that when the source is fixed* (both i and R_{IN} are fixed), *maximum power will be transferred to R_T when $R_T = R_{IN}$.*

For the condition of maximum power transfer, we say that *the load is matched to the source;* i.e., $R_T = R_{IN}$. Observation of the graph of Fig. 7.16 reveals that when R_T is *not* equal to R_{IN}, p_T is *less* than the maximum (less than 0.8 mW); this condition is known as a *mis-match.* Note that the value of p_T increases rapidly for values of R_T *below* 6.4 kΩ, but then decreases slowly for values *above* 6.4 kΩ. The value of p_T is zero for both $R_T =$ zero and $R_T =$ infinity. This is obvious because in order for power to be delivered to R_T, there must be both voltage across, and current through the resistance; when R_T = zero (short circuit), v_T = zero and when R_T = infinity (open circuit), i_T = zero.

The situation pictured in Fig. 7.17 should not be confused with that of interstage coupling. You may recall from chapter 6 that the object in coupling two stages together is to transfer as much power as possible on to the next stage. In Fig. 7.18,

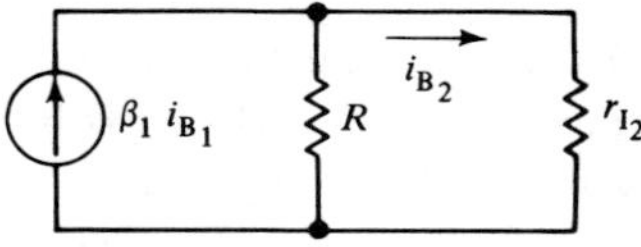

Figure 7.18

r_{I_2} represents the load, R_T, R represents the internal resistance of the source, R_{IN}, and $\beta_1 i_{B_1}$ represents the current generator, i. In this case, r_{I_2} *is fixed once the transistor and the Q point have been chosen.* We *can*, however, vary the value of R (remember that R is the parallel equivalent of R_B and R_O); we can increase the current (i_{B_2}, in Fig. 7.18), and hence the power delivered to r_{I_2} by *increasing* the value of R, as explained in chapter 6. The situation pictured in Fig. 7.17 is *different* in that, in this case, *the values of both i and R_{IN} are fixed;* the only way to achieve maximum power is to match R_T to R_{IN}. The closer we make the value of R_T to R_{IN}, the greater is the power delivered to R_T, and the closer we come to a matched condition; this is illustrated clearly in Fig. 7.16.

7-8 IMPEDANCE MATCHING

We saw in the previous section that the load should be matched to the source in order to achieve maximum power transfer. However, in most instances, the *source* impedance (output impedance of the final amplifier stage) is *fixed* by the requirements of the *amplifier*, and the *load* impedance (R_T) is *fixed* by the requirements of the *system*. Suppose that the output impedance of the final amplifier stage is again equal to 6.4 kΩ, as shown in Fig. 7.19(a), and that R_T = 400 Ω. It is

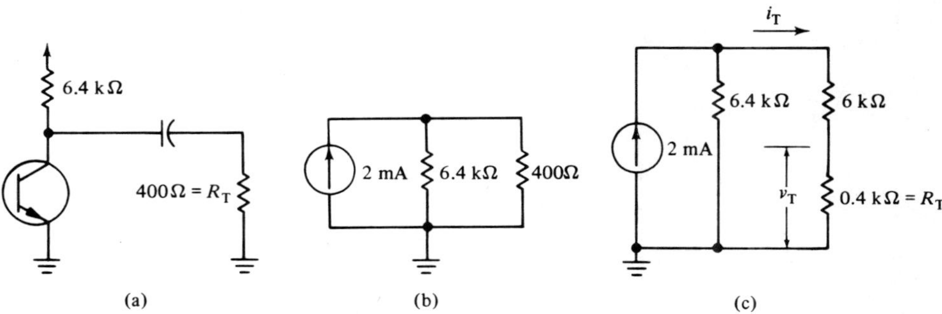

(a)   (b)   (c)

Figure 7.19

desired that we supply the maximum amount of power to R_T. It should be obvious from the ac equivalent circuit of Fig. 7.19(b) that we have a condition of *mis-match;* i.e., because the load is *not* matched to the source, considerably *less* than the maximum amount of power will be delivered to the 400 Ω load. Reference to the chart of Fig. 7.15 (or the graph of Fig. 7.16) reveals that, while the power delivered to the load under matched conditions is 0.8 mW (when $R_T = 6.4$ kΩ), only 0.177 mW is delivered to the 400 Ω load. In other words, the transfer of power from the amplifier to the load in the circuit of Fig. 7.19(a) is highly inefficient.

The next question to ask is: Is there any way of transferring the maximum amount of power to the 400 Ω load? If we could somehow make the load impedance equal to 6.4 kΩ (the proper impedance for matching), we could transfer the maximum amount of power. Suppose a 6 kΩ resistor were placed in series with the 0.4 kΩ R_T, as shown in Fig. 7.19(c). It should be obvious that maximum power is now transferred *to the equivalent series combination* (6.4 kΩ). We are interested, however, in delivering the maximum amount of power *to the 400 Ω load*, not to the series combination of the 6 kΩ and the 400 Ω. Let us analyze the circuit of Fig. 7.19(c) to see how much power is delivered to R_T. Since the impedances of the two branches are equal (both are 6.4 kΩ), i_T is equal to one-half of the generator current, or 1 mA. The voltage across R_T (v_T) and the power delivered (p_T) are computed as follows:

$$v_T = (i_T)(R_T) = (1 \text{ mA})(0.4 \text{ kΩ}) = 0.4 \text{ V}$$

$$p_T = \frac{(v_T)(i_T)}{8} = \frac{(0.4)(1)}{8} = 0.05 \text{ mW}$$

Thus the power delivered to R_T is only 0.05 mW, even less than the 0.177 mW delivered when no 6 kΩ resistor was included in series with R_T. The problem here, of course, is that, since the 6 kΩ is in *series* with the 400 Ω, *most* of the voltage is dropped across the

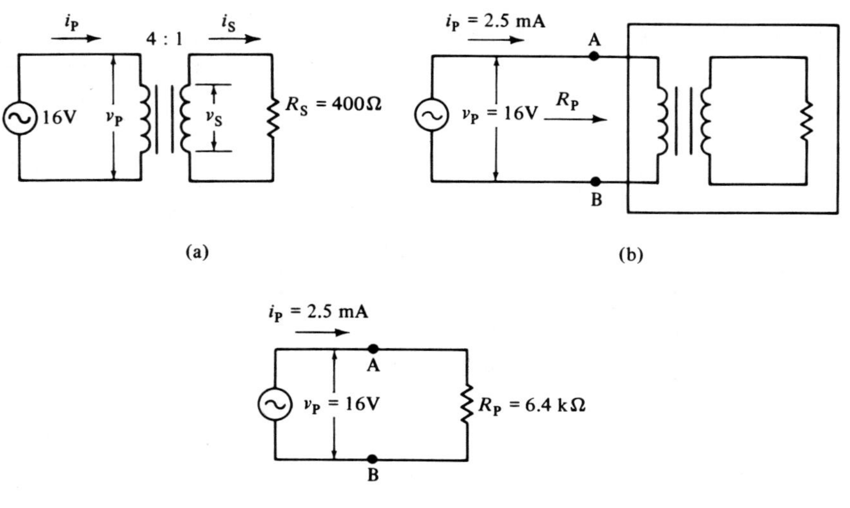

Figure 7.20

6 kΩ resistor, *not* the 400 Ω load. Therefore, the method of including a resistor in series with R_T actually produces a worse situation (less power delivered to R_T) than when no resistor is included; thus this solution will not work.

Figure 7.20(a) shows a 16 V ac source connected to the primary terminals of a 4:1 step-down transformer; the secondary terminals are connected to a 400 Ω resistor, R_S. In this circuit, v_P is the primary voltage (16 V, in this case), v_S is the secondary voltage, i_P is the primary current, and i_S is the secondary current. The ratio of the primary voltage to the secondary voltage is equal to the ratio of the number of turns on the primary to the number of turns on the secondary, i.e.:

$$\frac{v_P}{v_S} = \frac{N_P}{N_S} = \frac{4}{1} = 4$$

In this case, N_P is the number of turns on the primary, and N_S is the number of turns on the secondary. Thus we can solve for the secondary voltage, v_S, since we know that $v_P = 16$ V:

$$v_S = \frac{v_P}{\left(\dfrac{N_P}{N_S}\right)} = \frac{16 \text{ V}}{4} = 4 \text{ V}$$

Since v_S is the voltage across the 400 Ω resistor, we can solve for i_S (i_S is the current through the 400 Ω resistor) as follows:

$$i_S = \frac{v_S}{R_S} = \frac{4 \text{ V}}{400 \text{ }\Omega} = 10 \text{ mA}$$

Assuming no losses, the power delivered to the primary coil is equal to the power delivered to the secondary; thus, if the *voltage* is stepped-*down* 4:1 from primary to secondary, the *current* is stepped-*up* in the same ratio; this is illustrated as follows:

$$p_P = p_S$$
$$v_P i_P = v_S i_S$$
$$\frac{v_P}{v_S} = \frac{i_S}{i_P}$$
$$\frac{i_P}{i_S} = \frac{v_S}{v_P} = \frac{N_S}{N_P} = \frac{1}{\left(\dfrac{N_P}{N_S}\right)} \tag{7-5}$$

We can now compute the value of the primary current, i_P, as follows:

$$i_P = \frac{i_S}{\left(\dfrac{N_P}{N_S}\right)} = \frac{10 \text{ mA}}{4} = 2.5 \text{ mA}$$

Now suppose that we wanted to determine the input resistance to the transformer between the *primary terminals* [terminals A and B in Fig. 7.20(b)]. In other words, we should like to replace both the primary and secondary coils and R_S [portion of circuit within the box in Fig. 7.20(b)] *by a single resistance*. We call this resistance R_P, the resistance "as seen from the primary." You recall from earlier chapters that this was done by dividing the voltage across the two terminals ($v_{AB} = v_P = 16$ V) by the current flowing into the two terminals ($i_P = 2.5$ mA):

$$R_P = \frac{v_P}{i_P} = \frac{16 \text{ V}}{2.5 \text{ mA}} = 6.4 \text{ k}\Omega$$

This means that, as far as the primary source (16 V generator) is concerned, it makes no difference whether we connect the transformer and R_S between terminals A and B as in Fig. 7.20(b), or whether we connect a 6.4 kΩ resistor between the same two terminals as in Fig. 7.20(c); in both cases 2.5 mA is drawn from the 16 V generator.

The 400 Ω impedance has been "transformed" to a 6.4 kΩ impedance; i.e., because of the transformer, the source now "sees" a 6.4 kΩ impedance. Thus we can consider the transformer to be an *impedance* transformer, in addition to a *voltage* and *current* transformer. The impedance has been transformed from 400 Ω to 6.4 kΩ; the ratio of transformation is:

$$\frac{R_P}{R_S} = \frac{6400}{400} = \frac{16}{1}$$

The turns ratio of the step-down transformer is 4:1; this indicates that the impedance has been transformed by the *square of the turns ratio*. This can be written as follows:

$$R_P = \left(\frac{N_P}{N_S}\right)^2 R_S \tag{7-6}$$

In this expression, R_S is the resistance connected across the secondary terminals, N_P/N_S is the transformer turns ratio, and R_P is the resistance as seen from the primary terminals. R_P is also called the "reflected resistance"; the transformer *reflects* the 400 Ω in the secondary circuit into 6400 Ω in the primary circuit.

In general, an impedance, Z_S, is connected across the secondary terminals, and thus Z_P may not be purely resistive. The above expression is expressed in a more general form as follows:

$$Z_P = \frac{v_P}{i_P} = \frac{v_S \left(\dfrac{N_P}{N_S}\right)}{i_S \dfrac{1}{\left(\dfrac{N_P}{N_S}\right)}} \qquad\qquad \frac{v_P}{v_S} = \left(\frac{N_P}{N_S}\right)$$

$$Z_P = \left(\frac{v_S}{i_S}\right) \frac{N_P}{N_S} \times \frac{N_P}{N_S} \qquad\qquad \frac{i_P}{i_S} = \frac{1}{\left(\dfrac{N_P}{N_S}\right)}$$

$$Z_P = Z_S \left(\frac{N_P}{N_S}\right)^2 \tag{7-7} \qquad\qquad v_S = Z_S i_S$$

The use of a transformer for impedance transformation can be applied to the circuit of Fig. 7.19(a). A 4:1 step-down transformer is inserted between the final

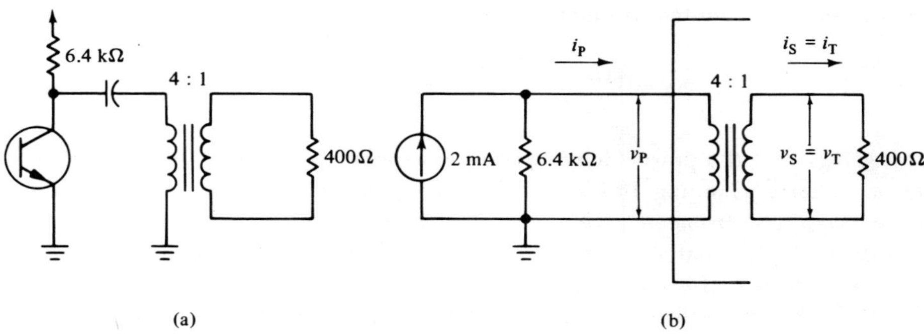

(a)   (b)

Figure 7.21

amplifier stage and the 400 Ω load, as shown in Fig. 7.21(a). We see in chapter 9 that the power can be increased further if the 6.4 kΩ load resistor of the final stage is omitted and the transformer primary coil connected directly in series with the collector of the transistor. The ac equivalent circuit is shown in Fig. 7.21(b). The impedance seen looking into the primary terminals is:

$$Z_P = Z_S \left(\frac{N_P}{N_S}\right)^2 = 0.4(4)^2 = 6.4 \text{ k}\Omega$$

Thus the 4:1 step-down transformer has matched the 400 Ω load to the 6.4 kΩ load resistor of the final stage. The primary current, i_P, is now computed to be:

$$i_P = \frac{(2 \text{ mA})(6.4)}{6.4 + Z_P} = \frac{(2 \text{ mA})(6.4)}{6.4 + 6.4} = 1 \text{ mA}$$

The secondary current, i_S (also equal to i_T), is:

$$i_S = i_P \left(\frac{N_P}{N_S}\right) = 1 \text{ mA}(4) = 4 \text{ mA} = i_T$$

The secondary voltage, v_S (also equal to v_T), is:

$$v_S = i_S Z_S = (4 \text{ mA})(0.4 \text{ k}\Omega) = 1.6 \text{ V} = v_T$$

Finally, the secondary power, p_S (also equal to p_T), is:

$$p_S = \frac{v_S i_S}{8} = \frac{(1.6)(4)}{8} = 0.8 \text{ mW} = p_T$$

Thus the load receives the *full maximum power, 0.8 mW.* The placement of the transformer between the amplifier and the load has matched the load impedance to the amplifier; the use of a transformer for this purpose results in its being called an *impedance matching transformer.* The power delivered to the primary of the transformer is computed as follows:

$$p_P = \frac{v_P i_P}{8} = \frac{(6.4)(1)}{8} = 0.8 \text{ mW} = p_S$$

This shows that the power delivered to the primary is equal to that delivered to the secondary, assuming 100% transformer efficiency (no losses). *Thus the transformer cannot, and does not, provide any power gain* (see chapter 4); it merely reflects an impedance which matches the source, obtains maximum power in the primary, and then transfers this maximum power to the secondary.

A little thought will reveal that transformers can also be used to advantage in coupling amplifier stages; i.e., they can be used in place of capacitor coupling. Figure 7.22(a) shows the ac equivalent circuit of a portion of an amplifier using

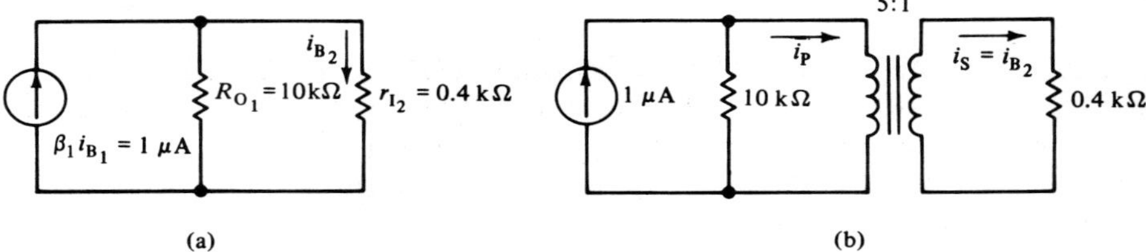

(a)   (b)

Figure 7.22

capacitor coupling. The value of the βi_B current generator of the first stage is 1.0 μA, the output resistance of this stage (R_{O_1}) is 10 kΩ, and the input resistance to the next stage (r_{I_2}) is 0.4 kΩ; we shall neglect the effect of the bias resistors. Using capacitor coupling, we compute the base current flowing in the second stage, i_{B_2}, as:

$$i_{B_2} = \frac{(\beta_1 i_{B_1}) R_{O_1}}{R_{O_1} + r_{I_2}} = \frac{(1 \ \mu\text{A})(10)}{10 + 0.4} = 0.96 \ \mu\text{A}$$

Now suppose we use a matching transformer to couple the stages; i.e., we should like to match the 10 kΩ output impedance of the first stage to the 0.4 kΩ input impedance of the second. We must select a transformer with the proper turns ratio to achieve a match; this is done as follows:

$$Z_P = Z_S \left(\frac{N_P}{N_S}\right)^2$$

$$\frac{N_P}{N_S} = \sqrt{\frac{Z_P}{Z_S}} = \sqrt{\frac{10,000}{400}} = \frac{5}{1}$$

Thus we insert a 5:1 step-down transformer between the two stages, as shown in Fig. 7.22(b). The current, $i_{B_2} = i_S$, can now be computed by the following procedure:

$$i_{\mathrm{P}} = \frac{(\beta_1 i_{\mathrm{B}_1}) R_{\mathrm{O}_1}}{R_{\mathrm{O}_1} + Z_{\mathrm{P}}} = \frac{(1\ \mu\mathrm{A})(10)}{10 + 10} = 0.5\ \mu\mathrm{A}$$

$$i_{\mathrm{S}} = i_{\mathrm{B}_2} = i_{\mathrm{P}} \left(\frac{N_{\mathrm{P}}}{N_{\mathrm{S}}}\right) = (0.5\ \mu\mathrm{A})(5) = 2.5\ \mu\mathrm{A}$$

This indicates that the use of a matching transformer results in a significant improvement over capacitor coupling; $i_{\mathrm{B}_2} = 2.5\ \mu\mathrm{A}$ when the transformer is used, and only $0.96\ \mu\mathrm{A}$ when capacitor coupling is used. Why, then, do we not always use transformer coupling? Why use capacitor coupling at all? The primary reason is that the size, cost, and bulk of the transformer make its use as a coupling device inefficient; it is generally cheaper, in terms of both cost and space, to use an extra amplifier stage to increase the gain than to insert transformers between the stages.

7-9   THE EMITTER-FOLLOWER AMPLIFIER

The emitter-follower amplifier, shown within the dotted box in Fig. 7.23, is a circuit which can be used in place of an impedance matching transformer, as well

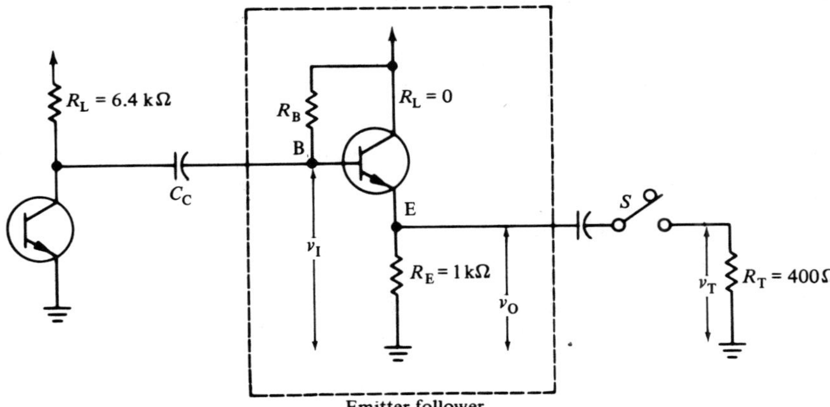

Figure 7.23

as in other applications. Note that the input is fed to the base terminal (just as with any other amplifier), and the biasing arrangement is similar to those discussed in chapter 5. There are *two* major differences, however, between this circuit and the other amplifiers studied in this text: First, there is *no load resistor* in series with the collector terminal of the transistor; i.e., $R_{\mathrm{L}} = 0$ as indicated in the figure. Since the collector is *at ac ground potential* (the collector terminal is tied directly to the positive terminal of V_{CC}), the circuit is sometimes known as the *common* (or grounded) *collector* circuit configuration, to distinguish it from the *common* (or grounded) *emitter* circuit discussed throughout this text. Second, the output is taken from the *emitter* terminal, as shown in Fig. 7.23, *not* the *collector* terminal.

As the input voltage, v_{I}, increases, the base current increases. As the base current increases, the emitter current also increases (recall that all three transistor currents are in phase). But as the emitter current increases, the voltage across the emitter resistor (v_O in Fig. 7.23) will also increase. Therefore, since v_O increases when v_I increases, we can say that *the input and output voltages in an emitter-follower amplifier are in phase* (zero phase shift); contrast this situation with the 180 degree phase shift exhibited by the conventional amplifiers (output taken from the collector) studied earlier in the text. Note that the input voltage (v_I) is developed between the base terminal and ground, and the output voltage (v_O) is developed between the emitter terminal and ground. The difference between these two voltages is the small ac voltage developed between the base and emitter terminals (across the base-emitter diode) of the transistor. Since, as we soon see, the voltage between the base and emitter terminals (v_{BE}) is *much smaller* than either v_I or v_O, we can assume that v_O *is approximately equal to* v_I. Therefore, since the voltage at the emitter (with respect to ground) is approximately equal to that at the base, and since the two voltages (v_I and v_O) are in phase, the potential at the emitter (output voltage) "follows" very closely the potential at the base (input voltage); *because the output voltage follows the input, we call this circuit the emitter-follower*. From this point on, we refer to the circuit simply as the follower.

Now let us study the circuit. There are two different means of viewing the operation of this circuit, and we discuss both. There are *three* quantities which are necessary to describe the performance of the follower circuit: They are the input resistance (impedance), R_I, the voltage gain, A_V, and the output resistance, R_O. In the following analysis, we leave the switch, S, *open*, in Fig. 7.23; thus the voltage, v_T, across the 400 Ω resistor is zero. Dealing first with the input resistance, we know from chapter 6 that the input resistance to a circuit with an emitter resistor can be expressed as follows:

$$R_I = r_I + (\beta + 1)R_E \simeq r_I + \beta R_E$$

Assuming that the β of the follower transistor is 100, and r_I is 1 kΩ, the input resistance is:

$$R_I = r_I + \beta R_E = 1 \text{ kΩ} + 100(1 \text{ kΩ}) = 101 \text{ kΩ} \simeq 100 \text{ kΩ}$$

This indicates that *the input resistance to the follower is quite large*.

The voltage gain of the follower, A_V, is defined as the ratio of the output voltage, v_O, to the input voltage, v_I. The input voltage is developed across the input resistance, R_I, while the output voltage is developed across the emitter resistor, R_E, as shown in Fig. 7.24. Note that the current flowing in the input resistance (which

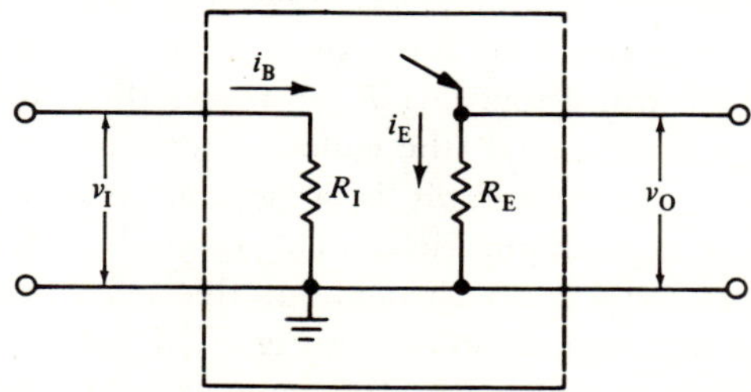

Figure 7.24

is actually a fictitious resistor) is i_B, and the current in the emitter resistor is i_E; thus the voltage gain of the follower can be expressed as follows:

$$A_V = \frac{v_O}{v_I} = \frac{i_E R_E}{i_B R_I} = \frac{(i_B + i_C) R_E}{i_B R_I} = \frac{(i_B + \beta i_B) R_E}{i_B R_I}$$

$$A_V = \frac{i_B(\beta + 1) R_E}{i_B(r_I + \beta R_E)} = \frac{(\beta + 1) R_E}{r_I + (\beta + 1) R_E} \sim \frac{\beta R_E}{r_I + \beta R_E}$$

$$\boxed{A_V = \frac{\beta R_E}{r_I + \beta R_E}} \tag{7-8}$$

The expression indicates that *the voltage gain of the follower is always less than 1.0,* because the numerator of the fraction is always less than the denominator. However, if the factor "βR_E" is much larger than r_I, as is usually the case, the voltage gain is approximately equal to 1.0:

$$A_V = \frac{\beta R_E}{r_I + \beta R_E} \sim \frac{\beta R_E}{\beta R_E} = 1.0 \qquad \text{For } \beta R_E \gg r_I$$

Computing the voltage gain for the follower of Fig. 7.23, we get:

$$A_V = \frac{\beta R_E}{r_I + \beta R_E} = \frac{100(1)}{1 + 100(1)} = 0.99$$

Thus, since $\beta R_E = 100$ kΩ and $r_I = 1$ kΩ, the voltage gain is only *slightly* less than 1.0. The reader may wonder why we use the term emitter follower "amplifier" when the voltage gain is less than 1.0 (the output voltage is actually *less* than the input voltage). We soon see that, although the follower circuit provides no increase in voltage, it *does* provide substantial increases in both the current and the power (both current and power gain are *greater* than 1.0).

The method used for computing the output resistance of the follower involves a considerable amount of algebraic manipulation, as well as the use of some sophisticated circuit theorems. The output resistance of the follower is:

$$R_O = \frac{R_E \left(\dfrac{R + r_I}{\beta} \right)}{R_E + \left(\dfrac{R + r_I}{\beta} \right)} \tag{7-9}$$

This expression represents the parallel equivalent of two resistances: the emitter resistance, R_E, and a resistance equal to $(R + r_I)/\beta$. R is equal to the parallel equivalent of R_B (the bias resistor for the follower circuit) and R_{O_1} (the output resistance of the previous stage). In the majority of cases, R_E is much larger than $(R + r_I)/\beta$, and thus the output resistance of the follower can be approximated as follows:

$$R_O \simeq \frac{R + r_I}{\beta}$$

Let us now compute the output resistance of the follower in the circuit of Fig. 7.23. We assume that the value of the bias resistor, R_B, is large enough to be neglected; thus R can be approximated as:

$$R = \frac{R_B R_{O_1}}{R_B + R_{O_1}} \simeq R_{O_1} = 6.4 \text{ k}\Omega$$

The output resistance of the previous stage (R_{O_1}) is approximately equal to the 6.4 kΩ load resistor of that stage (see Fig. 7.23). Now the output resistance of the follower is computed as:

$$R_O = \frac{R + r_I}{\beta} = \frac{6.4 \text{ k}\Omega + 1 \text{ k}\Omega}{100} = 0.074 \text{ k}\Omega = 74 \text{ }\Omega$$

Note that, since $(R + r_I)/\beta$ (74 Ω, in this case) is much smaller than R_E (1 kΩ, in this case), our approximation is valid; i.e., we can neglect the factor due to the emitter resistor in computing the output resistance. The results of the above computation indicate that *the output resistance of the follower circuit is very small.*

Now that we know how to compute the input resistance, voltage gain, and output resistance, we turn to a method for representing the ac equivalent circuit of the follower; refer to Fig. 7.25. The ac equivalent circuit, consisting of two

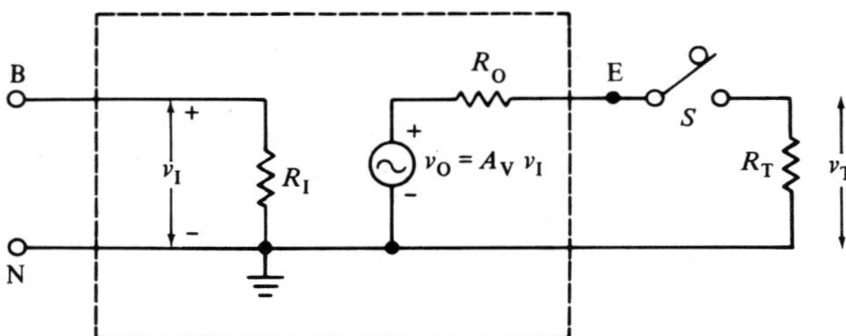

Figure 7.25

resistances (R_I and R_O) and a voltage generator ($v_O = A_V v_I$) is shown within the box in the figure. Between the *input* terminals (terminals B and N), the follower can be represented by the large input resistance, R_I. Between the *output* terminals (terminals E and N), the follower acts like a *voltage* generator equal to $A_V v_I$, in *series* with the small output resistance, R_O; the reader should realize that we could have represented the output circuit alternately by a *current* generator in *parallel* with an internal resistance. Summarizing, we can say that the follower circuit provides a large input resistance, a voltage gain approximately equal to 1.0, and a small output resistance. Since both βR_E and R are generally *much larger* than r_I, we can make some further approximations in the equations for the three quantities, as follows:

$$R_I \simeq \beta R_E$$

$$A_V \simeq 1.0$$

$$R_O \simeq \frac{R}{\beta} \tag{7-10}$$

7-10 THE EMITTER-FOLLOWER AS AN IMPEDANCE MATCHER

Let us now study the use of the follower circuit as an impedance matcher. Before proceeding with a thorough analysis, we eliminate both the coupling capacitor (C_C) and the bias resistor (R_B) in the inter-stage coupling network between the final amplifier stage and the follower in Fig. 7.23. We can accomplish this by *direct coupling* the final amplifier stage to the follower, as shown in Fig. 7.26(a); note that the collector of the final stage is connected *directly* to the base of the follower stage. By direct coupling the two stages, we obtain a number of advantages: First, it is more efficient in terms of both cost and space to eliminate two components from the circuit. Second, R_B shunts the input terminals of the follower under ac conditions; since, as we soon see, one of the major features of the follower is its large input resistance, the shunting effect of R_B will serve only to *reduce* this resistance, and thus adversely affect the performance of the follower. Finally, as we see in chapter 8, the coupling capacitor causes some difficulties at the lower frequencies, and thus we try to exclude it from the circuit wherever possible.

You may recall from chapter 6 that one of the major problems involved in direct coupling two amplifier stages resulted from the fact that the amplifier provided a rather large value of voltage gain; in addition to amplifying *the ac voltage* (which is the purpose of the amplifier), the amplifier *also* amplifies any change in the quiescent collector-to-ground voltage of the previous stage (which is a distinct disadvantage), resulting in the undesirable phenomenon known as drift. However, since the follower provides a voltage gain of *less than 1.0*, the drift problem is negligible when direct coupling to a follower circuit. Thus, direct coupling can usually be used when coupling a follower to a conventional amplifier stage because of the insignificant amount of drift.

Because the *input ac resistance* to the follower is large, we can show that the *input dc resistance* is also large. Therefore, the connection of the base terminal of the follower transistor to the collector terminal of the transistor in the final stage [see Fig. 7.26(a)] produces *negligible loading of the final amplifier stage*. Assume that the quiescent collector-to-ground voltage of the transistor in the final stage of Fig. 7.26(a) is equal to 20 V *before* the follower is connected. If the input resistance to the follower is much larger than the value of the load resistor of the final stage (100 kΩ is much larger than 6.4 kΩ, in this case), V_{CEQ} will drop only a slight amount *after* the follower is connected, say, from 20 V to 19 V (this can be illustrated by means of Thevenin's theorem). Therefore, if $V_{CEQ} = 19$ V for the final amplifier stage, and assuming $V_{BE} = 0.6$ V for both transistors, then V_{ENQ} (the quiescent emitter-to-ground voltage for the follower stage) is equal to V_{CEQ} minus V_{BE}, or 18.4 V. Since this is also the voltage drop across the 1 kΩ emitter resistor in the follower stage, the quiescent emitter (or collector) current of the follower stage can be computed by use of Ohm's law as 18.4 V/1 kΩ, or 18.4 mA. The above method provides a simple means of analyzing the dc behavior of the direct coupled follower circuit.

The next step is an analysis of the circuit of Fig. 7.26(a) to illustrate the use of the follower circuit as an impedance matcher. Note that the final load, R_T, is again equal to 400 ohms. The ac equivalent circuit of Fig. 7.26(a) is shown in Fig. 7.26(b); the follower itself is located within the dotted box. The object here is to compute the ac power delivered to the 400 ohm load. The first step is to compute the input voltage to the follower, v_I. This can be done in one of two ways: Compute the base current by the current divider, and then find the voltage by Ohm's law (v_I equals i_B multiplied by R_I), or find the parallel equivalent of the 6.4 kΩ load resistor of the final stage and R_I, and multiply this parallel equivalent by the value of the current generator (2 mA); we use the latter method. This proceeds as follows:

$$\frac{(6.4 \text{ k}\Omega)(R_I)}{6.4 \text{ k}\Omega + R_I} = \frac{(6.4)(100)}{6.4 + 100} = 6 \text{ k}\Omega$$

$$v_I = (2 \text{ mA})(6 \text{ k}\Omega) = 12 \text{ V}$$

The value of the voltage generator, $A_V v_I$, can now be computed as:

$$A_V v_I = (0.99)(12) = 11.9 \text{ V}$$

The voltage across the 400 Ω load is computed by means of the voltage divider:

$$v_T = \frac{(A_V v_I) R_T}{R_T + R_O} = \frac{(11.9)(400)}{400 + 74} = 10 \text{ V}$$

Finally, we compute both the current through, and the power delivered to the 400 Ω load as follows:

$$i_T = \frac{v_T}{R_T} = \frac{10 \text{ V}}{0.4 \text{ k}\Omega} = 25 \text{ mA}$$

$$p_T = \frac{v_T i_T}{8} = \frac{(10 \text{ V})(25 \text{ mA})}{8} = 31.2 \text{ mW}$$

Note that the power delivered to the load is 31.2 mW; *this is much larger than the power delivered to the same load, 0.8 mW, when transformer matching was used* (see Sec. 7-8). Note also that we have *not* actually achieved an impedance match; the output resistance of the final amplifier stage (6.4 kΩ) is *not* equal to the input resistance of the follower stage (100 kΩ), and the output resistance of the follower stage (74 Ω) is *not* equal to the load resistance (400 Ω). There must be some reason, then, why we deliver *more* power to the load (31.2 mW using the follower, compared with 0.8 mW using the transformer) using the follower circuit when we do not even have an impedance match. *The reason is that, unlike the transformer, the follower circuit provides power gain.*

   This concept can be illustrated more clearly by employing an alternate method of analysis. When the switch, *S*, is closed in Fig. 7.26(a), the 400 Ω load is effectively in parallel with the 1 kΩ emitter resistor as far as ac conditions are concerned (neglecting the reactance of the coupling capacitor). Therefore, the impedance between the emitter terminal and ground is equal to the parallel equivalent of

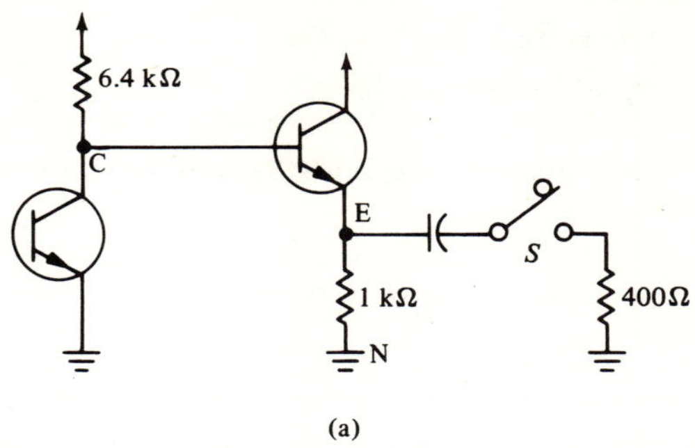

(a)

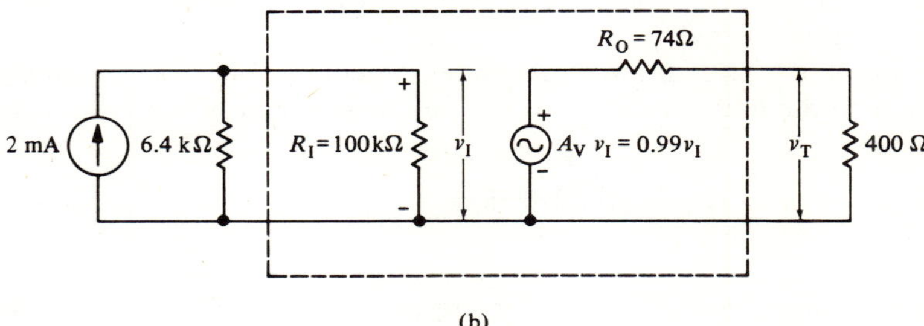

(b)

Figure 7.26

R_E and R_T; we shall call this new resistance R_T'. The value of R_T' for the circuit of Fig. 7.26 is:

$$R_T' = \frac{R_E R_T}{R_E + R_T} = \frac{(1)(0.4)}{1 + 0.4} = 0.286 \text{ k}\Omega$$

Now the input resistance to the follower is expressed as:

$$R_I = r_I + \beta R_T'$$

Its value is:

$$R_I = 1 \text{ k}\Omega + 100(0.286) = 29.6 \text{ k}\Omega$$

Finally, the voltage gain of the follower is:

$$A_V = \frac{\beta R_T'}{r_I + \beta R_T'} = \frac{100(0.286)}{1 + 100(0.286)} = 0.965$$

By considering the impedance between the emitter terminal and ground as being composed of both R_E *and* R_T, we can use the above formulas and thus eliminate the step of finding the output resistance of the follower, R_O.

Let us now apply this approach to the problem of the circuit of Fig. 7.26. We first find i_B and v_I as follows:

$$i_B = \frac{(2\text{ mA})(6.4)}{6.4 + R_I} = \frac{2(6.4)}{6.4 + 29.6} = 0.355\text{ mA}$$

$$v_I = i_B R_I = (0.355\text{ mA})(29.6\text{ k}\Omega) = 10.5\text{ V}$$

Using this method, v_T is the voltage across both R_E and R_T. Therefore, v_T can be found as follows:

$$A_V = \frac{v_T}{v_I}; \qquad v_T = A_V v_I = (0.965)(10.5) = 10\text{ V}$$

Notice that we get the same answer for v_T (10 V) by both methods. This second method is faster because it is not necessary to compute the value of R_O; it is used when we desire a relatively rapid analysis of the circuit. The first method is used to develop an insight into the functioning of the circuit and is discussed further in the following section.

Returning now to a consideration of the power gain of the follower, we compute the input power to the follower circuit as the power delivered to the input resistance, R_I; this is:

$$p_I = \frac{v_I i_B}{8} = \frac{(10.5)(0.355)}{8} = 0.465\text{ mW}$$

Note that this is *less than 0.8 mW* because the output resistance of the final amplifier stage (6.4 kΩ) is *not matched* to the input resistance of the follower (29.6 kΩ). We have already found that the output power ($p_O = p_T$) is equal to 31.8 mW. Thus the power gain is:

$$A_P = \frac{p_O}{p_I} = \frac{31.2\text{ mW}}{0.465\text{ mW}} = 67$$

The amplifier also provides a considerable amount of current gain, since the output current (current in the 400 ohm load) is 25 mA, and the input current (base current of the follower transistor) is 0.355 mA; A_I is:

$$A_I = \frac{25\text{ mA}}{0.355\text{ mA}} = 70$$

Let us now compare the operation of the follower with that of the transformer. We saw in Sec. 7-8 that the transformer reflects an impedance which matches the source (6.4 kΩ, in this case), develops maximum power in the primary (0.8 mW, in this case), and then transfers this maximum power to the secondary; *the transformer provides no power gain itself* ($A_P = 1.0$). The follower reflects an impedance *greater* than the source impedance (29.6 kΩ is greater than 6.4 kΩ), and thus *less* than the maximum power (0.465 mW is less than 0.8 mW) is developed at the input of the follower because of the mis-match. *Unlike the transformer, however, the follower provides power gain* ($A_P = 67$, in this case); the 0.465 mW at the *input* of the follower is multiplied by 67 to yield 31.2 mW at the *output*. In other words, the power gain of the follower more than compensates for the mis-match between the source (final amplifier stage) and the follower input.

The follower is preferred to the transformer as an impedance matcher for a number of reasons:

1. Despite the mis-match, the follower delivers more power to the load than the transformer (31.8 mW compared with 0.8 mW) because it provides power gain.

2. The follower is composed of only one transistor and one resistor, and is therefore lighter, smaller, and cheaper than a transformer.

The disadvantage of the follower lies in the fact that it requires *dc power* to function since it is, basically, an amplifier. Thus the battery or power supply must supply *more* dc current (or power) to the electronic system where followers are used; a transformer does *not* require dc power in order to operate (see chapter 4).

On balance, the follower is far superior to the transformer as an impedance matcher. It is gradually replacing transformers in systems where it is required that a relatively *large* source impedance be matched to a *small* load impedance.

Like the transformer, the follower can be used as an inter-stage coupling device, since it can match the high output impedance of one stage to the relatively low input impedance of another. Figure 7.27(a) shows the output of a common-emitter stage ($R_L = 6.4$ kΩ) *capacitor-coupled* to the input of a *second* common-emitter stage (assume that this second stage uses fixed bias, for simplicity); the ac equivalent circuit is shown in Fig. 7.27(b). Neglecting the effect of the bias resistor, and

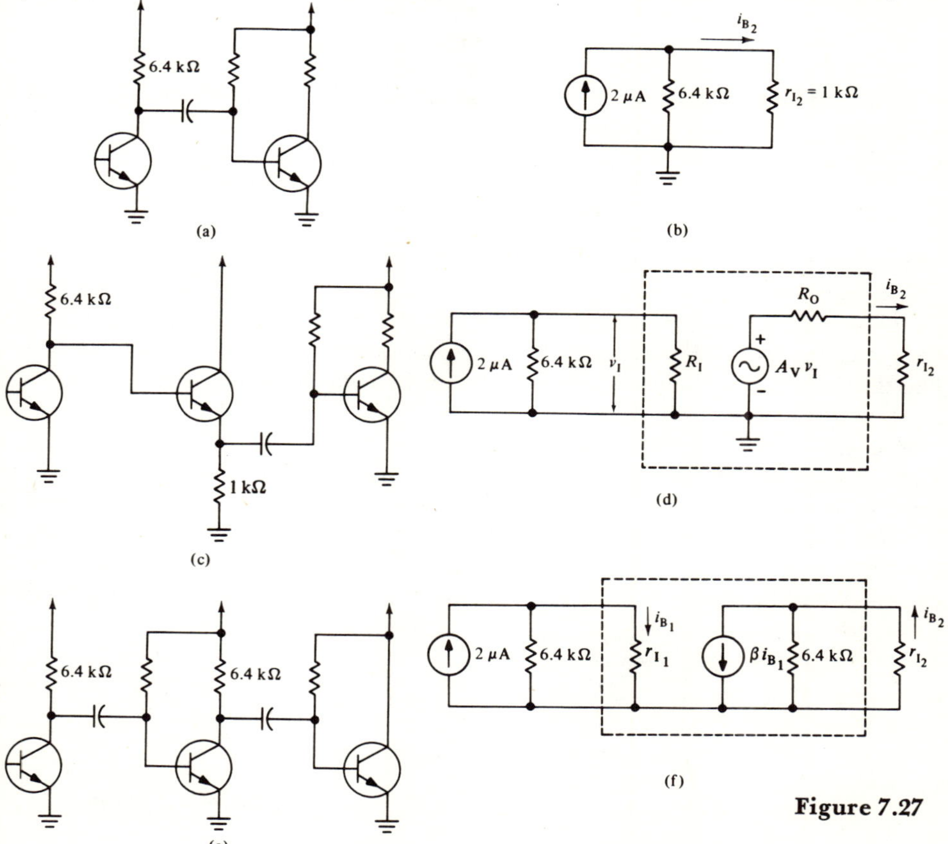

Figure 7.27

assuming that the ac collector current in the first stage is 2 μA peak-to-peak and $r_{I_2} = 1$ kΩ, we find the base current of the second stage to be:

$$i_{B_2} = \frac{(2\ \mu A)(6.4\ k\Omega)}{6.4\ k\Omega + 1\ k\Omega} = 1.73\ \mu A$$

In Fig. 7.27(c), a *follower* ($R_E = 1$ kΩ) has been inserted between the two stages. The ac equivalent circuit is shown in Fig. 7.27(d); the follower is located within the dotted box. Assuming that the β of the transistor in the follower stage is equal to 100, we find the approximate values of R_I, A_V, and R_O for the follower to be:

$$R_I \simeq \beta R_E = 100\ (1\ k\Omega) = 100\ k\Omega$$

$$A_V \simeq 1.0$$

$$R_O \simeq \frac{R}{\beta} = \frac{6.4\ k\Omega}{100} = 64\ \Omega$$

Now proceeding with the analysis of the circuit of Fig. 7.27(d), we should like to find the value of i_{B_2}. We must first find the values of v_I and $A_V v_I$, as follows:

$$6.4\ k\Omega\ ||\ 100\ k\Omega = 6\ k\Omega$$

$$v_I = (2\ \mu A)(6\ k\Omega) = 0.012\ V$$

$$A_V v_I = (1)(0.012\ V) = 0.012\ V$$

To find the value of i_{B_2}, we merely apply Ohm's law to the right-hand portion of the circuit of Fig. 7.27(d), as follows:

$$i_{B_2} = \frac{A_V v_I}{R_O + r_{I_2}} = \frac{0.012\ V}{64\ \Omega + 1000\ \Omega} = 11.2\ \mu A$$

Notice that when a follower is used as an inter-stage coupling device, as in Fig. 7.27(c), *more current* (and hence, more power) is delivered to the input circuit of the following stage; $i_{B_2} = 11.2\ \mu$A when the follower is used, and $i_{B_2} = 1.73\ \mu$A when it is not used. Therefore, as the above computations illustrate, the follower (like the transformer) helps to achieve a better impedance match between amplifier stages.

As long as we provide for the additional space and cost of the follower transistor, however, *it makes more sense to use this transistor in a conventional common-emitter stage,* since the common-emitter configuration provides both current *and* voltage gain. The follower stage is converted into a common emitter stage ($R_L = 6.4$ kΩ) in Fig. 7.27(e); the ac equivalent circuit is shown in Fig. 7.27(f) (the converted stage is located within the dotted box). Assuming the β of the transistor is 100 and $r_{I_1} = 1$ kΩ, we compute the *new* value of i_{B_2} as follows:

$$i_{B_1} = \frac{(2\ \mu A)(6.4\ k\Omega)}{6.4\ k\Omega + 1\ k\Omega} = 1.73\ \mu A$$

$$\beta i_{B_1} = 100\ (1.73\ \mu A) = 173\ \mu A$$

$$i_{B_2} = \frac{(173\ \mu A)(6.4\ k\Omega)}{6.4\ k\Omega + 1\ k\Omega} = 150\ \mu A$$

Notice that more current (and power) is delivered to r_{I_2} when a common emitter stage ($i_{B_2} = 150 \, \mu A$) is inserted in place of the follower stage ($i_{B_2} = 11.2 \, \mu A$). Thus, once the transistor is included in the circuit, *we may as well use it in the common-emitter configuration because of this configuration's relatively large current and voltage gain* (power gain); the follower stage provides only moderate power gain because of its small voltage gain ($A_V = 1$). In other words, it is more efficient, in terms of both cost and space, to tolerate a mis-match by using direct or capacitor coupling between common emitter amplifier stages, than to include impedance matching devices such as transformers and followers. The transformer or follower is generally used at either the input to, or the output of an amplifier.

7-11   THE EMITTER-FOLLOWER AS A BUFFER

Suppose that the final stage of an amplifier is to supply ac power to *several* loads, as illustrated in Fig. 7.28(a). The collector of the transistor in the final stage

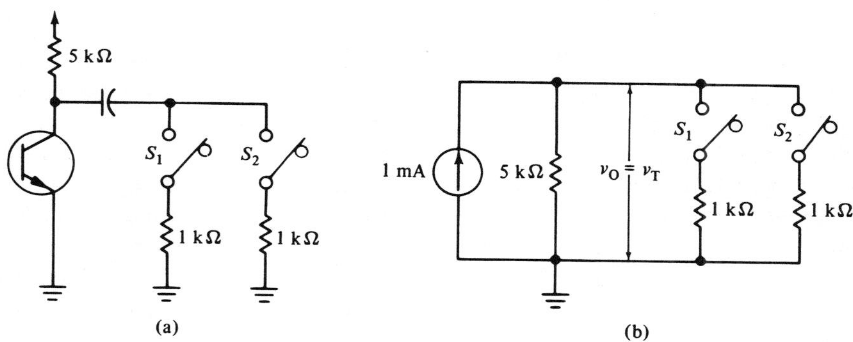

Figure 7.28

($R_L = 5 \, k\Omega$) is capacitor-coupled to *two* 1 kΩ loads through switches S_1 and S_2. Assume that the peak-to-peak current developed in the collector of the final stage is 1.0 mA; the ac equivalent circuit is shown in Fig. 7.28(b). With both switches *open*, the voltage at the output of the amplifier, v_O, is:

$$v_O = (1 \text{ mA})(5 \text{ k}\Omega) = 5 \text{ V}$$

In other words, *with both loads disconnected*, the amplifier output voltage is 5 V.

Now let us close *one* switch, S_1; one of the 1 kΩ resistances is now effectively in parallel with the 5 kΩ load resistor of the final stage. We compute the output voltage, v_O (also equal to the voltage across the load), as follows:

$$\frac{(5 \text{ k}\Omega)(1 \text{ k}\Omega)}{5 \text{ k}\Omega + 1 \text{ k}\Omega} = 0.833 \text{ k}\Omega$$

$$v_O = v_T = (1 \text{ mA})(0.833 \text{ k}\Omega) = 0.833 \text{ V}$$

Next we close *both* switches (S_1 *and* S_2); now both 1 kΩ resistances (parallel equivalent is equal to 500 Ω) are in parallel with the 5 kΩ load resistor of the final stage. The *new* output voltage is:

$$\frac{(5 \text{ k}\Omega)(0.5 \text{ k}\Omega)}{5 \text{ k}\Omega + 0.5 \text{ k}\Omega} = 0.45 \text{ k}\Omega$$

$$v_\text{O} = v_\text{T} = (1 \text{ mA})(0.45 \text{ k}\Omega) = 0.45 \text{ V}$$

Notice first that *when either one or both loads are connected there is a large drop in the output voltage;* the output voltage dropped from 5 V (when the switches were open) to either 0.833 V, or 0.45 V (depending upon which switches were closed). The loads "load down" the amplifier output; i.e., the loads "react back" on the amplifier by reducing the output voltage. This is the same situation encountered in chapter 6 when we cascaded amplifier stages; the output voltage of a stage was loaded down considerably when it was connected to the input of the following stage. The second point to note is that *the output voltage varies considerably* (from 0.45 V to 0.833 V to 5.0 V) *when the various loads are either connected or removed.* For example, when both switches are closed, the output voltage (voltage across both loads) is 0.45 V. Now if S_2 is opened (load in series with S_2 removed), the output voltage (which is still the voltage across the load in series with S_1) *increases* to 0.833 V. In other words, the connection and removal of the load in series with S_2 has a large effect on the voltage across the load in series with S_1.

Both of these features (the large drop in the output voltage, and the variation in this voltage with changes in the load) are *undesirable* in an electronic system. First, it seems wasteful, after having obtained 5 V at the output of the final amplifier stage, to "lose" this voltage by connecting the amplifier to the load. Second, the variation in the voltage across the load in series with S_1 (from 0.45 V to 0.833 V) caused by the connection and removal of the load in series with S_2 can be troublesome if the system requires that the voltage across any particular load remain relatively constant.

The output voltage undergoes a large decrease when the load is connected because the value of the load impedance (500 Ω) is *much smaller* than that of the load resistor of the final amplifier stage (5 kΩ). In Sec. 7-7, we constructed a chart

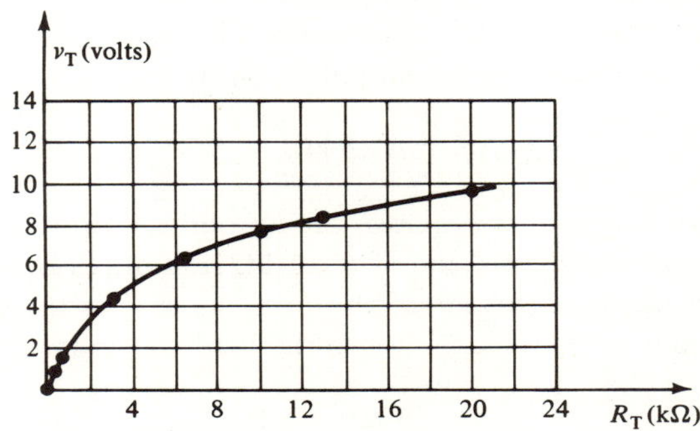

Figure 7.29

(Fig. 7.15) to study the effect of the value of the load impedance upon the amount of power delivered to that load; a graph of p_T (power delivered to the load) vs. R_T (load impedance) was plotted in Fig. 7.16. Figure 7.29 shows a plot of the *voltage across the load, v_T, vs. R_T*; note that the voltage across the load *increases* as R_T *increases*. It can be seen from the graph that, if R_T is very small, v_T will also be very small. This is the same situation encountered in Fig. 7.28 when the two 1 kΩ loads are connected to the amplifier output.

If we could somehow make the value of the load impedance very large, the output voltage would not drop as much when the load was connected. However, the value of the load impedance is usually fixed by the requirements of the system, and is thus beyond the control of the designer. The follower has a very large input impedance and a voltage gain of 1, and is thus an ideal circuit to use in this situation. Let us insert a follower stage between the final amplifier stage and the load, as shown in Fig. 7.30(a). To illustrate the effect of the follower on this circuit,

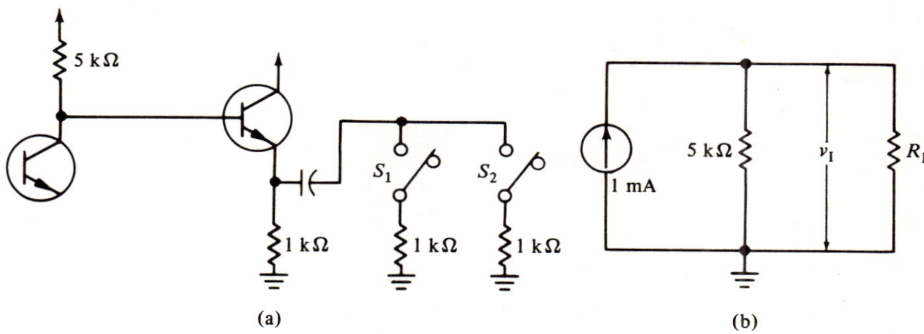

Figure 7.30

we carry through the analysis for *two* conditions: *one* switch closed, and *both* switches closed. Assuming that the β of the follower transistor is 100 and $r_I = 1$ kΩ, we can find the value of R_I as follows:

S_1 CLOSED

$$R_T' = R_T \,||\, R_E = 1 \text{ k}\Omega \,||\, 1 \text{ k}\Omega = 0.5 \text{ k}\Omega$$
$$R_I = r_I + \beta R_T' = 1 + 100(0.5) = 51 \text{ k}\Omega$$

S_1 and S_2 CLOSED

$$R_T' = R_T \,||\, R_E = 0.5 \text{ k}\Omega \,||\, 1 \text{ k}\Omega = 0.33 \text{ k}\Omega$$
$$R_I = r_I + \beta R_T' = 1 + 100(0.33) = 34.3 \text{ k}\Omega$$

Notice that the input resistance is large in both cases. The next step is to find the input voltage to the follower, v_I [see Fig. 7.30(b)].

S_1 CLOSED

$$5 \text{ k}\Omega \parallel 51 \text{ k}\Omega = 4.55 \text{ k}\Omega$$

$$v_I = (1 \text{ mA})(4.55 \text{ k}\Omega) = 4.55 \text{ V}$$

S_1 and S_2 CLOSED

$$5 \text{ k}\Omega \parallel 34.3 \text{ k}\Omega = 4.36 \text{ k}\Omega$$

$$v_I = (1 \text{ mA})(4.36 \text{ k}\Omega) = 4.36 \text{ V}$$

Note that the input voltage to the follower (which is also the output voltage of the final amplifier stage) drops only from 5 V to 4.55 V or 4.36 V. This occurs because the input resistance of the follower is very large, and thus has less effect on the output voltage of the final amplifier stage (see Fig. 7.29). The quantity, v_I, is the *input* voltage to the follower; we are interested in the value of the *output* voltage of the follower, v_T (also equal to the voltage across the load), and thus we must next find the value of the gain:

S_1 CLOSED

$$A_V = \frac{\beta R_T'}{r_I + \beta R_T'} = \frac{100(0.5)}{1 + 100(0.5)} = 0.98$$

S_1 and S_2 CLOSED

$$A_V = \frac{\beta R_T'}{r_I + \beta R_T'} = \frac{100(0.33)}{1 + 100(0.33)} = 0.972$$

The value of gain computed for both conditions is approximately the same because the factor, $\beta R_T'$, is much larger than r_I in both cases. Finally, we compute the load voltage as follows:

S_1 CLOSED

$$v_T = A_V v_I = (0.98)(4.55 \text{ V}) = 4.46 \text{ V}$$

S_1 and S_2 CLOSED

$$v_T = A_V v_I = (0.972)(4.36 \text{ V}) = 4.24 \text{ V}$$

Notice first that the use of the follower between the final amplifier stage and the load helps to maintain a relatively large load voltage; $v_T = 4.46$ V and 4.24 V when the follower was used, and $v_T = 0.833$ V and 0.45 V when *no* follower was used. This occurs because the follower has a large input resistance (51 kΩ and 34.3 kΩ); the output circuit of the final amplifier stage "sees" a *large resistance*, and thus the output voltage of this stage (input voltage of the follower stage) remains relatively large (4.55 V and 4.36 V). Since the voltage gain of the follower is approximately equal to 1 (0.98 and 0.972), the relatively large voltage at the *input* terminals of the follower is *transferred* to the *output* (load) terminals with only a slight loss (4.46 V and 4.24 V).

The second important point concerns the fact that *the load voltage varies very little* (from 4.46 V to 4.24 V) when S_1 (or S_2) is opened and closed. This happens because, although the load impedance varies (from 1 kΩ to 500 Ω), the relatively large input resistance of the follower maintains a large, roughly constant, voltage across the follower input terminals.

The above discussion illustrates the fact that the use of the follower between the final amplifier stage and the load solves both the problem of the large drop in load voltage, and that of the variation of the load voltage with changes in the load. The follower is used here as a "buffer." *A buffer is a device which separates the effect of two things upon each other.* In the circuit of Fig. 7.30(a), the follower is used to separate the load from the amplifier. Recall from one of the earlier paragraphs of this section that the problem with the circuit of Fig. 7.28(a) was that the load "reacted back" on the output of the amplifier. In the circuit of Fig. 7.30(a), the follower effectively "screens" the amplifier output from the load by presenting a large input resistance to the amplifier; the amplifier output voltage thus remains relatively large and stable (4.55 V and 4.36 V) despite the opening and closing of the load switches.

The buffer is usually used at the system "interfaces." *The interfaces are those portions of the system where two or more different types of circuits or equipment interconnect.* Such an interface exists at the connection of the amplifier output to the load, as shown in Fig. 7.31; an interface may also exist between two different types of electronic circuits, such as an amplifier and a multivibrator.

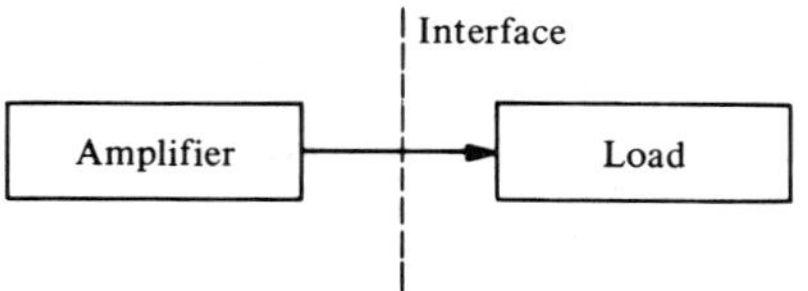

Figure 7.31

For efficient buffer action, we should like the input resistance of the follower to be as large as possible, and the voltage gain as close as possible to unity (1.0); the equations for R_I and A_V are repeated below:

$$R_I = r_i + \beta R_T'$$

$$A_V = \frac{\beta R_T'}{r_i + \beta R_T'} = \frac{1}{\dfrac{r_i}{\beta R_T'} + 1}$$

It can be seen that in order for both R_I and A_V to be large, we should like the factor, $\beta R_T'$, to be as large as possible. This means that we prefer to use a transistor with a large value of β; we should also like the value of R_T' to be as large as possible. We know that R_T' is equal to the parallel combination of R_T and R_E; R_T is the load impedance, and is thus beyond the control of the designer. If we make R_E, the follower emitter resistor, very large, we make the quiescent collector current,

I_{C_Q}, of the follower stage very small. You may recall from chapter 6 that if I_{C_Q} is made very small, we run the risk of reducing the β of the transistor as well as limiting the maximum undistorted output of the follower, i_{C_M}. This latter point is particularly important in this case *since the follower is usually connected at the output of the final amplifier stage*, where the signal swing is quite large; in this case it is usually necessary that i_{C_M} be as large as possible so that the follower can accommodate the large output swing. Therefore, since there are limitations on the size of both R_T' and R_E, we return to the requirement that it is necessary to have a transistor with a large β.

Sometimes, if the load impedance is very small, the buffer may not operate properly even *with* a transistor having a large β. In this case, the designer may use a special cascaded follower arrangement known as the *Darlington* circuit. In other more extreme situations, a technique known as *bootstrapping* is used; the discussion of bootstrapping techniques is beyond the scope of this text.

PROBLEMS

1.   A transistor whose β is 50 is biased at a dc collector current of 5 mA. The value of r_B is 50 Ω. Compute the value of r_I.

2.   Repeat problem 1 if the collector current is 10 mA.
     What effect does the collector current have on the value of r_I?

3.   The circuit of Fig. 7.2 has the following values:

$$V_{CC} = 30 \text{ V} \qquad R_L = 5 \text{ k}\Omega$$
$$R_1 = 50 \text{ k}\Omega \qquad R_E = 1 \text{ k}\Omega$$
$$R_2 = 5 \text{ k}\Omega \qquad \text{Neglect } V_{BE}$$
$$R_O = 5 \text{ k}\Omega$$

(a)   Compute the value of the gain when $\beta = 100$.
(b)   Repeat for $\beta = 200$. Assume that I_{C_Q} remains constant.
What effect does the change in β have upon the gain?

4.   Repeat problem 3 if β varies from 50 to 150.

5.   The circuit of Fig. 7.4 has the following values:

$$V_{CC} = 30 \text{ V} \qquad R_L = 5 \text{ k}\Omega$$
$$R_1 = 50 \text{ k}\Omega \qquad R_E = 1 \text{ k}\Omega$$
$$R_2 = 5 \text{ k}\Omega \qquad R_E' = 250 \ \Omega$$
$$R_O = 5 \text{ k}\Omega$$

(a)   Compute the value of the gain when $\beta = 100$.
(b)   Repeat for $\beta = 200$.
What effect does the change in β have upon the gain?

6.   Repeat problem 5 if $R_E' = 100 \ \Omega$.

7.   Repeat problem 5 by computing A_{OL}, β_F, and A_{CL}.

8.   Repeat problem 6 by computing A_{OL}, β_F, and A_{CL}.

9.   The circuit of Fig. 7.12 uses two identical amplifier stages having the following values:

$$V_{CC} = 30 \text{ V} \qquad R_L = 5 \text{ k}\Omega$$
$$R_1 = 50 \text{ k}\Omega \qquad R_E = 1 \text{ k}\Omega$$
$$R_2 = 5 \text{ k}\Omega \qquad \text{Neglect } V_{BE}$$

R_O of the source is 5 kΩ. $R_E' = 25$ Ω and $R_F = 7.5$ kΩ. β varies from 100 to 300.
  (a)   Compute A_{OL} for both values of β.
  (b)   Compute β_F.
  (c)   Compute A_{CL} for both values of β.
Compare the percentage change in the value of A_{OL} with that of A_{CL}.

10.   Repeat problem 9 if $R_F = 4$ kΩ.
Compare the results with those of problem 9.

11.   Compute the transformer turns ratio necessary to match a 10 kΩ source to a 100 Ω load. What is the impedance seen from the primary, R_P?

12.   Compute the transformer turns ratio necessary to match a 6 kΩ source to a 16 Ω load.

13.   An emitter-follower has the following values:

$$r_I = 1 \text{ k}\Omega$$
$$\beta = 100$$
$$R_E = 2 \text{ k}\Omega$$

  (a)   Compute the input resistance.
  (b)   Compute the voltage gain.
  (c)   Compute the output voltage if the input voltage is 1 mV.

14.   Assume that the values of the two resistors in Fig. 7.28 are each equal to 2 kΩ.
  (a)   Compute v_T with both switches open.
  (b)   Compute v_T with one switch closed.
  (c)   Compute v_T with both switches closed.

15.   Repeat problem 14 if the follower of problem 13 is inserted between the amplifier and the load as in Fig. 7.30.

8

FREQUENCY RESPONSE

It has been shown in the previous chapter that the design and analysis of a transistor amplifier involves a careful consideration of a number of factors. Stabilization of the Q point (chapter 5), designing for cascade connection (chapter 6), and stabilization of the gain (chapter 7), are all major techniques which must be mastered by the technician. In all of this discussion it has been assumed that the ac input to the amplifier was a sinusoidal voltage or current; nothing has been said about the frequency of this signal. We see in this chapter that the gain, in addition to depending on the Q point, method of cascade connection, and transistor, also is a function of the frequency of the input signal. We have intentionally included two capacitors in the amplifier (C_C and C_E) to improve its performance. These capacitors, however, while improving performance in some respects, hinder it in others. In addition, there are capacitances associated with the transistor itself which have not as yet been discussed in this text; these capacitances also have a considerable effect on the performance of the amplifier. The object of this chapter, then, is the study of the effect of all capacitances on the gain of the amplifier at various frequencies, and the methods by which the technician can take these factors into account when working with the amplifier.

8-1 THE CONCEPT OF FREQUENCY RESPONSE

Most electronic circuits, including amplifiers, contain frequency-dependent elements such as capacitances and inductances. The reader will recall from ac circuits theory that the reactance, X, of a capacitor or inductor depends upon the frequency of the applied sinusoidal signal. Since this is true, all of the currents and voltages developed in an amplifier depend upon frequency; thus, in general, the *gain of an amplifier depends upon frequency*. In order to show the relationship between gain and frequency we plot a graph of *gain vs. frequency* for the particular electronic amplifier, as shown in Fig. 8.1(a). Such a graph is called a *frequency response curve*, and the relationship between the gain of an amplifier and the frequency of the input signal is called the frequency response.

Figure 8.1(a) shows the frequency response of an R-C coupled amplifier similar to one we have discussed throughout most of this text. Notice that the gain is relatively constant in the center portion of the graph, but decreases at both the low and high frequencies. We soon see that the gain is actually a complex quantity, so we must examine both the *magnitude* of the complex number ($|A|$) and its *phase*. The graphs of Fig. 8.1 are plots of the magnitude of the gain vs. frequency. We call the gain in the center portion of the graph where it is roughly constant, the *mid-band gain;* in this case the mid-band gain, A_M, is 50 (we drop the subscript "I" and assume we mean current gain). The range of frequencies from zero (dc) to infinite frequency is called the *frequency spectrum;* a range of frequencies within this spectrum is called a frequency *band*. The *mid-frequency band is the band of frequencies at which the gain is greater than or equal to 70.7% of the mid-band gain.* For the graph of Fig. 8.1(a), the mid-band gain is 50; 70.7% of 50 is 35.35. Perpendicular lines

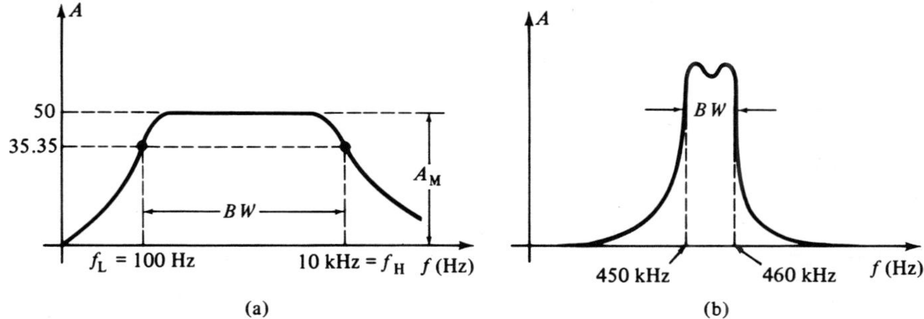

Figure 8.1

are dropped from the curve at the points where the gain is 35.35; the frequencies at which these gains are obtained are labeled f_L and f_H. The frequency f_L is called the *lower cut-off frequency*, and f_H is the *upper* (higher) *cut-off frequency*. For the curve of Fig. 8.1(a), f_L is 100 Hz and f_H is 10 kHz. Instead of the term cut-off, we sometimes use *corner*, or *3 dB*, to describe these frequencies; we explain the use of these terms later on in the chapter.

We define another term, *bandwidth, BW, as the difference between f_H and f_L.*

$$BW = f_H - f_L \tag{8-1}$$

$$BW = f_H - f_L = 10,000 \text{ Hz} - 100 \text{ Hz} = 9900 \text{ Hz}$$

$$BW = 9.9 \text{ kHz}$$

Since in this case f_H is much larger than f_L, we can say that the bandwidth is approximately equal to f_H.

$$BW \simeq f_H \quad \text{for} \quad f_H \gg f_L$$

$$BW \simeq 10 \text{ kHz}$$

Notice that at frequencies below f_L the gain drops off rapidly as the frequency decreases; at frequencies *below* f_L, the gain is less than $(0.707A_M)$, in this case, 35.35. For the particular amplifier whose response curve is shown in the graph, the gain at zero frequency is zero; i.e., the amplifier does not provide gain at dc. We discuss the reasons for this and the implications involved later on in the chapter. The same situation occurs at frequencies *above* f_H; the gain at infinite frequency is also zero. We say that the response is *flat* between f_L and f_H because the curve is flat; this means that the gain is roughly constant in this region.

The response curve pictured in Fig. 8.1(a) is the general curve of an R-C coupled amplifier. There are other types of amplifiers which have response curves of different shape. Figure 8.1(b) is the curve obtained for a radio frequency amplifier using a tuned circuit. We do not discuss tuned circuits in this text; the curve was shown only to point out that there are many different types of response curves. In Fig. 8.1(b), BW is

$$BW = f_H - f_L = 460 \text{ kHz} - 450 \text{ kHz} = 10 \text{ kHz}$$

This shows that the approximation made concerning BW and f_H is not true in general; it holds only when f_H is much greater than f_L. In this case f_H is 460 kHz and f_L is 450 kHz.

The bandwidth desired in an amplifier depends upon the application for which the amplifier is intended. Audio amplifiers are used to amplify signals which are generated by sound waves. Microphones and cartridges are devices which are used to convert sound energy to electrical energy. Usually the electrical signal available from the output of the microphone or cartridge is quite small, a few millivolts. This small signal is applied to the input of an amplifier which then increases the size of the signal. The output of the amplifier is then fed to a speaker which re-converts the electrical energy to sound energy. Thus, the small sound from the cartridge or microphone is increased to a larger level so that it can be heard at greater distances.

The human voice is composed of sound waves whose frequencies fall in the frequency band between 20 Hz and 20 kHz. The waves generated by most musical instruments also fall within this band. The human ear responds to frequencies within the same range. Some animals, such as dogs, can hear frequencies above 20 kHz; this can be proved by blowing a dog whistle which produces a note of frequency greater than 20 kHz. Humans cannot hear this whistle, but dogs can; this can be verified by observing the dog's reactions when the whistle is blown.

Since the range of audio frequencies lies between 20 Hz and 20 kHz, it is desirable that the bandwidth of the amplifier be at least 20 kHz. The amplifier whose response curve is drawn in Fig. 8.1(a) has a bandwidth of approximately 10 kHz. The frequencies above 10 kHz, and below 100 Hz will not be amplified as much as those in the mid-frequency band. This amplifier, then, will not perform adequately as an audio amplifier. If one is listening to a classical music concert, instruments which generate notes below 100 Hz, such as the bass cello, will not be heard as loudly as the violin whose notes produce frequencies which lie in the mid-frequency band. A similar situation will occur with those instruments whose frequencies lie above 10 kHz, such as the piccolo. This loss of gain at both the high and low ends of the frequency spectrum is called *frequency distortion;* the gain is not constant with frequency. The degree to which the output is a faithful representation of the input is called the *fidelity* of the amplifier. If the input remains constant with frequency while the gain decreases, the output will also decrease with frequency. A *high fidelity* amplifier (hi-fi) is one whose gain remains constant with frequency. The greater the bandwidth of an amplifier, the better is the fidelity.

In order for the amplifier whose response curve is shown in Fig. 8.1(a) to be useful as an audio amplifier, the bandwidth must be extended. The value of f_H must be increased from 10 kHz to 20 kHz, and f_L must be decreased from 100 Hz to 20 Hz. In other words, for high fidelity, we should like f_H to be as high, and f_L to be as low as possible. Those amplifiers where it is desirable to have a large bandwidth are called *wideband amplifiers*. Ideally we should like f_H to be infinite and f_L to be zero; this is impossible. It is the job of the designer to come as close as possible to the ideal frequency response curve; see Fig. 8.2(a). This ideal curve

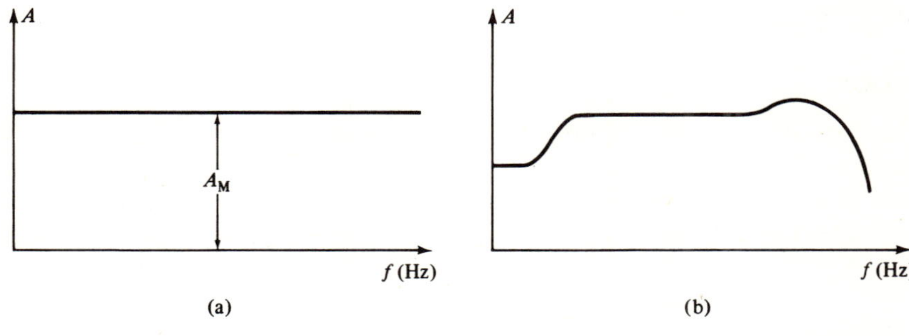

Figure 8.2

shows that the gain remains constant at every frequency in the entire spectrum.

Audio amplifiers must perform over only a limited portion at the lower end of the frequency spectrum; this is because sound waves lie in this frequency band. Video amplifiers are used in systems, such as television and radar, where the energy lies in the frequency band associated with visible light. In both television and radar the output quantity is some type of picture, or graphical representation. The frequency band required for these systems is considerably larger than that required for audio systems; the video band runs into the megahertz range. Thus video amplifiers must be designed for considerably greater bandwidths than audio amplifiers.

Figure 8.2(b) shows a response curve for some amplifier. Notice first that, while the gain decreases at low frequencies, it does *not* drop to zero at dc; this is the response of a dc (direct-coupled) amplifier. The second point to note is that the gain actually increases somewhat at the upper end of the band, and then decreases. This phenomenon is achieved by special design methods and is usually used in video amplifiers. The following sections deal in more detail with the analysis and design of amplifiers, with frequency response as the major point of interest.

8-2   THE DECIBEL

Before proceeding with the analysis of the amplifier for frequency response, it is important that we first define a few new terms. Up to this point in the text we have been describing the gain of an amplifier in numbers, e.g., A_I equals 10. It is useful now to describe the gain in a different way. The gain in decibels is equal to 20 multiplied by the logarithm to the base 10 of the gain.

$$A_{dB} = 20(\log_{10} A) \tag{8-2}$$

The symbol, A_{dB}, will be used to represent the gain of the amplifier expressed in decibels. The technique for computing the gain in decibels is illustrated by the following example:

EXAMPLE 1

The gain of a certain amplifier is 10. Find the gain expressed in decibels, A_{dB}.

SOLUTION

You may recall that the logarithm taken to the base 10 has two components: the characteristic and the mantissa. The characteristic is always one less than the number of digits to the left of the decimal point; the mantissa can be found either from a slide rule or from a table of logarithms. The example proceeds as follows:

$$A_{dB} = 20(\log_{10} A) = 20(1.0) = 20 \text{ dB}$$
$$A_{dB} = 20 \text{ dB}$$

In other words, if the *numerical* gain of an amplifier is 10, its *decibel* gain is 20; or

$$A = 10 \text{ (Numerical)}$$
$$A_{dB} = 20 \text{ (Decibel)}$$

It is important to point out that both numbers (numerical and decibel gain) refer to the *same* amplifier; they are just different ways of expressing the gain.

EXAMPLE 2

Compute A_{dB} of an amplifier whose A is 100.

SOLUTION

Using the same method of example 1,

$$A_{dB} = 20(\log_{10} A) = 20(2) = 40 \text{ dB}$$
$$A_{dB} = 40 \text{ dB}$$

Observation of the results of examples 1 and 2 yield some important results. When the numerical gain was multiplied by 10 (from 10 to 100), the decibel gain increased by 20 decibels (from 20 to 40). This is true in general for the gain of an amplifier. If the gain of the amplifier is again multiplied by 10 (to 1000), the decibel gain will increase by another 20 decibels (to 60); let us see if this is true.

EXAMPLE 3

Compute A_{dB} of an amplifier whose A is 1000.

SOLUTION

$$A_{dB} = 20(\log_{10} A) = 20(3) = 60 \text{ dB}$$
$$A_{dB} = 60 \text{ dB}$$

Example 3 shows again that when the numerical gain of an amplifier is multiplied by 10, its decibel gain increases by 20 decibels. From this point on, we no longer use the term numerical gain; we use just gain. When we speak of the gain expressed in decibels, we say decibel gain, or A_{dB}.

The reader will recall from chapter 6 that the over-all gain of two amplifiers connected in cascade is equal to the product of the individual gains. So if the gains of two amplifiers connected in cascade are 10 and 20, the over-all gain is 200.

EXAMPLE 4

For the two amplifiers mentioned above, find the decibel gains of the individual amplifiers, and also the decibel gain of the over-all cascade amplifier.

SOLUTION

$$A = 10 \qquad A_{dB} = 20 \log_{10} A = 20(1) \quad = 20 \text{ dB}$$
$$A = 20 \qquad A_{dB} = 20 \log_{10} A = 20(1.3) = 26 \text{ dB}$$
$$A = 200 \qquad A_{dB} = 20 \log_{10} A = 20(2.3) = 46 \text{ dB}$$
$$20 \text{ dB} + 26 \text{ dB} = 46 \text{ dB}$$

Notice that if the *decibel gains* of the individual amplifiers are *added*, the sum is the decibel gain of the over-all cascade amplifier. Thus, 20 dB plus 26 dB equals 46 dB. We can say that the *decibel gain of a cascade amplifier is equal to the sum of the decibel gains of the individual amplifier stages*. This comes about because the logarithm of the product of two numbers is equal to the sum of the logarithms of the numbers. The reader should refer to a basic algebra text for a more elaborate explanation. By representing the gains in decibel form, we have made it easier to combine them because all that is necessary is an addition process; admittedly, addition is simpler than multiplication. It is difficult for the reader to appreciate the above statement at this point; the reader will soon see the tremendous benefits to be accrued from representing the gain in decibel form.

EXAMPLE 5

If the gain of an amplifier is 1.0, find A_{dB}.

SOLUTION

The amplifier whose gain is 1.0 is one whose output is equal to its input. This, of course, makes the amplifier useless (except in the case of the emitter-follower) because there is no amplification. It is important, however, that we know its A_{dB}, as we shall see.

$$A_{dB} = 20 \log_{10} A = 20(0) = 0 \text{ dB}$$
$$A_{dB} = 0 \text{ dB}$$

Thus, an amplifier whose gain is 1.0 has a decibel gain of *zero* dB. We say that an amplifier whose decibel gain is zero dB has neither gained any decibels, nor lost any; it has remained the same (output equals input). On the other hand, the amplifier whose gain is 10, has *gained* 20 dB.

EXAMPLE 6

Find A_{dB} for an amplifier whose gain is 0.8.

SOLUTION

This amplifier is one whose output is actually *less* than its input. Taking the logarithm of a number less than 1.0 is a bit complicated; it must first be broken into two factors as follows:

$$A_{dB} = 20 \log_{10} A = 20 \log_{10} (0.8) = 20 \log_{10} [(0.1)(8)]$$

Since it is true that: $\log (ab) = \log (a) + \log (b)$

We can say

$$A_{dB} = 20 \log_{10} [(0.1)(8)] = 20 [\log (0.1) + \log 8]$$

The logarithm of 0.1 is -1; this is because 10^{-1} equals 0.1. Similarly, the logarithm of 0.01 is -2, because 10^{-2} equals 0.01. Thus,

$$A_{dB} = 20[-1 + 0.9] = 20(-0.1)$$
$$A_{dB} = -2 \text{ dB}$$

The logarithm of a number less than 1.0 is always negative, thus the decibel gain of an amplifier whose gain is less than 1.0 is *negative*. We say that the decibel gain of this amplifier is -2 dB; the output is *less* than the input so it has *lost* 2 decibels. There is a simpler way to solve this example. If the gain is less than 1.0, find the A_{dB} of the *reciprocal* of the gain; then put a minus sign in front of the answer.

EXAMPLE 7

Repeat example 6 using the technique just described.

SOLUTION

The first step is to take the reciprocal of the gain, 0.8.

$$\frac{1}{0.8} = 1.25$$

Next, find A_{dB} by the usual method, but put a *minus* sign in front of the answer.

$$A_{dB} = 20 \log_{10} A = -20 \log_{10} (1.25) = -20(0.1)$$
$$A_{dB} = -2 \text{ dB}$$

This is the same answer found in example 6.

We now solve some examples where the decibel gain is given and it is required to solve for the gain. In this case, all that is necessary is to work backwards; this involves finding the anti-logarithm (given the logarithm, find the number).

EXAMPLE 8

Find the gain of an amplifier whose decibel gain is 38 dB.

SOLUTION

$$A_{dB} = 20 \log_{10} A$$
$$38 = 20 \log_{10} A$$
$$\log_{10} A = \frac{38}{20} = 1.9$$

In finding the anti-logarithm of 1.9, we must consider 1.0 as the characteristic and 0.9 as the mantissa. Thus

$$A = 80$$

There are a few ways of checking the results of example 8. One is to work the problem the other way; find the A_{dB}. The second is a rough check. We know that the decibel gain of 10 is 20 dB and that of 100 is 40 dB; thus the decibel gain of 80 should be someplace in between 20 dB and 40 dB, which it is (38 dB).

EXAMPLE 9

Find the gain of an amplifier whose decibel gain is -5 dB.

SOLUTION

In this case proceed as above, but take the reciprocal of the anti-logarithm; this will give the actual gain.

$$A_{\mathrm{dB}} = 20 \log_{10} A$$

$$5 = 20 \log_{10} A$$

$$\log_{10} A = \frac{5}{20} = 0.25; \qquad A = 1.78$$

$$A \text{ (Actual)} = \frac{1}{1.78} = 0.56$$

Notice that the gain turns out to be less than 1.0; this checks with the fact that the decibel gain is negative.

8-3   THE FREQUENCY PLOT

We turn now to a description of a method for plotting graphs in logarithmic form. This method will make it more convenient for the designer to analyze the frequency response of an amplifier. We see in the following sections that the gain of an amplifier may sometimes be represented by the following equation:

$$A = \frac{A_{\mathrm{M}}}{1 + j\dfrac{f}{f_{\mathrm{H}}}}$$

In this equation, A is the gain at any frequency, A_{M} is the mid-band gain, f is the frequency, and f_{H} is the upper cut-off frequency. In other words, this is an algebraic expression for gain as a function of frequency. Since the gain is a complex quantity, it has both magnitude and phase. The magnitude of the gain is given by the following equation:

$$|A| = \frac{A_{\mathrm{M}}}{\sqrt{1 + \left(\dfrac{f}{f_{\mathrm{H}}}\right)^2}}$$

If, as we say, f_{H} is the upper cut-off frequency, the gain should drop to $0.707 A_{\mathrm{M}}$ when f equals f_{H}. Substituting f_{H} for f in the equation,

$$|A| = \frac{A_{\mathrm{M}}}{\sqrt{1 + \left(\dfrac{f_{\mathrm{H}}}{f_{\mathrm{H}}}\right)^2}} = \frac{A_{\mathrm{M}}}{\sqrt{2}} = 0.707 A_{\mathrm{M}}$$

This checks with what has been said in the first section of this chapter; the cut-off frequency is that frequency at which the gain drops to 70.7% of its mid-band value.

It will be found more convenient to express the previous equation in decibel form. Unless otherwise specified, we are always referring to the *magnitude* of the gain. It is known that the logarithm of a fraction is equal to the logarithm of the numerator minus the logarithm of the denominator. The decibel gain can thus be expressed as follows:

$$A_{dB} = 20 \log_{10}\left[\frac{A_M}{\sqrt{1 + \left(\frac{f}{f_H}\right)^2}}\right]$$

$$A_{dB} = 20 \log_{10} A_M - 20 \log_{10}\left(\sqrt{1 + \left(\frac{f}{f_H}\right)^2}\right)$$

$$= A_{MdB} - 20 \log_{10}\left(\sqrt{1 + \left(\frac{f}{f_H}\right)^2}\right)$$

The term A_{MdB} is the decibel value of the mid-band gain. It is instructive to use numbers to demonstrate the following theory, so let us assume that the mid-band gain, A_M, is 10 and the upper cut-off frequency, f_H, is 10 kHz. The equation now reads

$$A_{dB} = 20 - 20 \log_{10}\left(\sqrt{1 + \left(\frac{f}{f_H}\right)^2}\right)$$
$$\overset{}{\underset{}{\longleftarrow 10}}$$

We leave all the frequencies in terms of kHz.

It is desirable that the equation above be plotted on a logarithmic graph. Figure 8.3 shows the graph plotted on a logarithmic scale. The horizontal axis represents

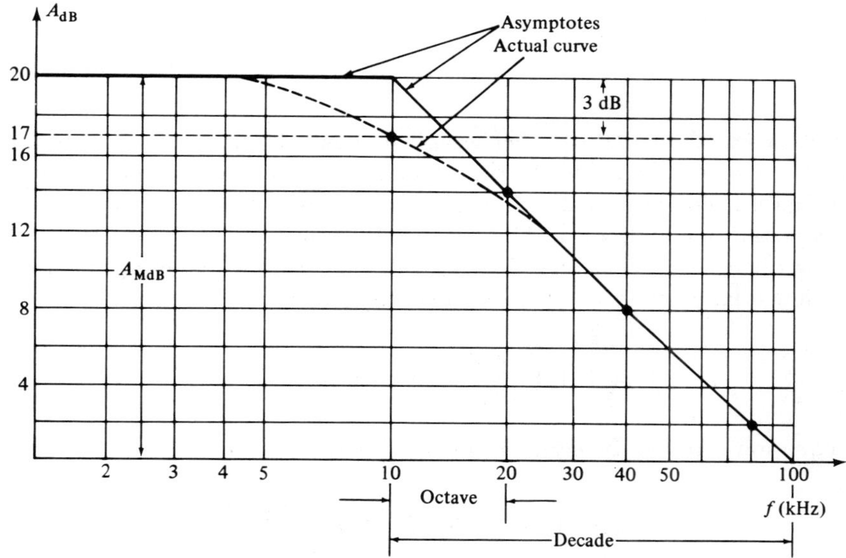

Figure 8.3

frequency (in kHz) and the distances between equal increments of frequency are *not* the same; the distance between 10 and 20 is greater than that between 20 and 30, and so on. The vertical axis represents the gain in decibels. This representation is sometimes called a *decibel frequency plot,* or dB plot. The distance (on the frequency axis) between 1.0 and 10 kHz is the same as that between 10 and 100 kHz. The frequency band over which there has been a ten-fold multiplication of frequency is called a *decade;* that over which the frequency has doubled is an *octave.*

We now turn our attention to the method of plotting the graph shown in Fig. 8.3. Re-stating the decibel equation:

$$A_{\mathrm{dB}} = A_{\mathrm{MdB}} - 20\,\log_{10}\left(\sqrt{1 + \left(\frac{f}{f_{\mathrm{H}}}\right)^2}\right)$$

If the frequency of the input signal to the amplifier is much smaller than 10 kHz (f_{H}), the second term under the square root sign is much smaller than 1.0. For example, suppose f is 1 kHz. Then

$$\sqrt{1 + \left(\frac{1}{10}\right)^2} = \sqrt{1 + 0.01} \simeq 1$$

Thus, for frequencies much less than f_{H} the decibel equation can be *approximated* as

$$A_{\mathrm{dB}} \simeq A_{\mathrm{MdB}} - 20\,\log_{10}(1) \qquad f \ll f_{\mathrm{H}}$$
$$= 20 - 20\,\log_{10}(1)$$

But the logarithm of 1.0 is zero, thus

$$A_{\mathrm{dB}} \simeq A_{\mathrm{MdB}} \qquad f \ll f_{\mathrm{H}}$$
$$A_{\mathrm{dB}} = 20\ \mathrm{dB}$$

In other words the equation when graphed shows the decibel gain constant at A_{MdB} at frequencies below f_{H}. In the particular example we are considering, the decibel gain is constant at 20 dB at frequencies *below* 10 kHz; this is shown by the horizontal line at 20 dB in Fig. 8.3.

For frequencies much larger than f_{H}, the second term under the square root sign is much larger than 1.0. For example, suppose f is 100 kHz. Then

$$\sqrt{1 + \left(\frac{100}{10}\right)^2} = \sqrt{1 + 100} \simeq 10$$

The equation can then be *approximated* as follows:

$$A_{\mathrm{dB}} \simeq A_{\mathrm{MdB}} - 20\,\log_{10}\left(\sqrt{\left(\frac{f}{f_{\mathrm{H}}}\right)^2}\right)$$
$$A_{\mathrm{dB}} = A_{\mathrm{MdB}} - 20\,\log_{10}\left(\frac{f}{f_{\mathrm{H}}}\right) \qquad f \gg f_{\mathrm{H}}$$
$$A_{\mathrm{dB}} = 20 - 20\,\log_{10}\left(\frac{f}{10}\right)$$

The above steps involve some manipulations of logarithms which will not be explained here; the reader is again referred to an algebra text for a review of the basic principles of logarithms. The last equation shows that the decibel gain decreases at higher frequencies. Let us compute the decibel gain at $f = 100$ kHz.

$$\underline{f = 100 \text{ kHz}}$$

$$A_{dB} = 20 - 20 \log_{10}\left(\frac{100}{10}\right) = 20 - 20 \log_{10}(10)$$

$$A_{dB} = 20 - 20 = 0 \text{ dB}$$

Thus, the decibel gain drops to zero dB at 100 kHz; see Fig. 8.3. At frequencies above f_H the decibel gain drops approximately linearly from A_{MdB} toward zero. This is shown by the straight sloping line in Fig. 8.3.

The graph composed of straight line segments is only an approximation to the actual situation; the actual curve is shown dotted in Fig. 8.3. In order to plot the graph of the decibel gain accurately, one would have to substitute into the original equation to obtain a point-by-point plot. The straight line segments are the *asymptotes;* that is, the actual graph approaches these asymptotes at both very low and very high frequencies. Of particular interest is the decibel gain *at $f = f_H$.* In order to find this value we must use the exact form of the decibel equation. This is because neither of the approximations previously made are true; f is neither much larger, nor much smaller than f_H; *it is equal to f_H.* Substituting into the exact equation, we get

$$A_{dB} = A_{MdB} - 20 \log_{10}\left(\sqrt{1 + \left(\frac{f}{f_H}\right)^2}\right)$$

at $f = f_H = 10$ kHz:

$$A_{dB} = 20 - 20 \log_{10}\left(\sqrt{1 + \left(\frac{10}{10}\right)^2}\right)$$

$$= 20 - 20 \log_{10}(\sqrt{2})$$

$$= 20 - 20 \log_{10}(1.414)$$

$$= 20 - 3 = 17 \text{ dB}$$

$$A_{dB} = 17 \text{ dB} \qquad \text{at} \qquad f = f_H$$

This shows that at the upper cut-off frequency, f_H, the decibel gain drops from 20 dB to 17 dB, *for a drop of 3 dB;* this is shown in Fig. 8.3. This is why we sometimes call f_H the *3 dB frequency,* or f_{3dB}. Instead of saying that the gain drops to 70.7% of its mid-band value at the cut-off frequency, we say the *decibel gain drops 3 decibels below the mid-band decibel gain;* in this case it drops from 20 dB to 17 dB.

Reference to Fig. 8.3 shows that the *high-frequency asymptote* (sloping straight line) gives a decibel gain of 8 at f equals 40 kHz, and 2 at f equals 80 kHz. We say that the dB *gain drops 6 dB* (from 8 to 2) *per octave* (frequency doubles from 40 to 80 kHz). When the frequency increases over one decade (from 10 kHz to 100 kHz), the decibel gain drops from 20 dB to zero dB, for a drop of 20 dB; we say the *decibel gain drops at a rate of 20 dB per decade.*

We use the asymptotic plot to represent the high-frequency behavior of an amplifier (straight line graph), bearing in mind that the actual plot appears as the dotted curved line shown in Fig. 8.3. This dotted curve approaches the low frequency asymptote (horizontal line) at frequencies below f_H, and approaches the high frequency asymptote (sloping line) at frequencies above f_H. The terms *corner*, or *break-point* frequency are sometimes applied to the quantity f_H, in addition to cut-off or 3 dB frequency. The terms corner and break-point are used to indicate the fact that the high frequency asymptote *breaks away* from the low frequency asymptote at f_H; the change in the curve leaves a sharp *corner*. The reason for the term 3 dB was already explained; the term cut-off is used because the amplifier is not considered to have much use at frequencies above f_H (the amplifier is cut-off).

EXAMPLE 10

Draw the asymptotic plot for an amplifier whose gain is 43, and whose f_H is 30 kHz.

SOLUTION

We must first find the mid-band decibel gain.

$$A_{MdB} = 20 \log_{10} A = 20 \log_{10} (43) = 20(1.63) = 32.6 \text{ dB}$$
$$A_{MdB} = 32.6 \text{ dB}$$

The asymptotes are then drawn as shown in Fig. 8.4.

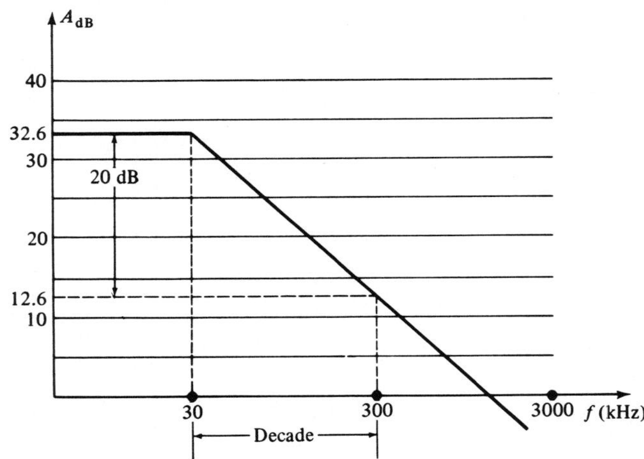

Figure 8.4

EXAMPLE 11

Find the decibel gain of the amplifier of example 10 at $f = 300$ kHz.

SOLUTION

Since the decibel gain falls at the rate of 20 dB per decade, the decibel gain at 300 kHz is 20 dB less than the mid-band value, or 12.6 dB.

$$A_{\mathrm{dB}}\ (f = 300\ \mathrm{kHz}) = 32.6 - 20 = 12.6\ \mathrm{dB}$$

$$A_{\mathrm{dB}} = 12.6\ \mathrm{dB}$$

Figure 8.4 shows that the decibel gain will drop to zero dB somewhere between 300 and 3000 kHz; at frequencies above this value, the dB gain is negative, meaning that the gain is less than 1.0. We could find the value at which the asymptote crosses the frequency axis if the graph were plotted to scale; alternatively, we could find this value analytically by use of equations. In the design of amplifiers, however, we are not concerned with these particular values; we are interested only in f_{H}.

The gain of amplifiers is sometimes given by the equation

$$A = \frac{A_{\mathrm{M}}}{1 + j\dfrac{f_{\mathrm{L}}}{f}}$$

The magnitude of the gain is

$$|A| = \frac{A_{\mathrm{M}}}{\sqrt{1 + \left(\dfrac{f_{\mathrm{L}}}{f}\right)^2}}$$

Using the techniques previously developed, the decibel gain is

$$A_{\mathrm{dB}} = 20\ \log_{10}\left(\frac{A_{\mathrm{M}}}{\sqrt{1 + \left(\dfrac{f_{\mathrm{L}}}{f}\right)^2}}\right)$$

$$A_{\mathrm{dB}} = A_{\mathrm{M\,dB}} - 20\ \log_{10}\left(\sqrt{1 + \left(\dfrac{f_{\mathrm{L}}}{f}\right)^2}\right)$$

In this case f_{L} is the lower cut-off frequency. If f is much *larger* than f_{L}, the decibel gain is roughly constant at $A_{\mathrm{M\,dB}}$.

$$A_{\mathrm{dB}} = A_{\mathrm{M\,dB}} \qquad \text{for} \qquad f \gg f_{\mathrm{L}}$$

If f is much smaller than f_{L}, the decibel gain is given by

$$A_{\mathrm{dB}} = A_{\mathrm{M\,dB}} - 20\ \log_{10}\left(\frac{f_{\mathrm{L}}}{f}\right) \qquad \text{for} \qquad f \ll f_{\mathrm{L}}$$

Thus, the same type of equations apply at the low end of the frequency spectrum, and the decibel plot appears as shown in Fig. 8.5. All that was said concerning asymptotes, decades, octaves, corner frequencies, etc., holds true for the low-frequency decibel plot.

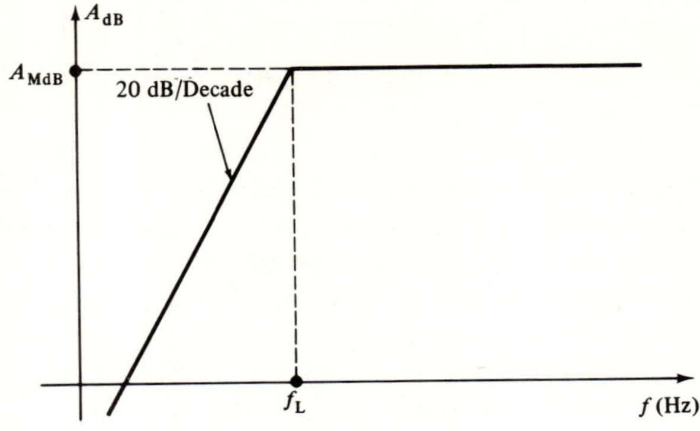

Figure 8.5

Still a final form of the gain equation is given as follows:

$$|A| = \frac{A_M}{k} \frac{\sqrt{1 + \left(\frac{f}{f_1}\right)^2}}{\sqrt{1 + \left(\frac{f}{kf_1}\right)^2}}$$

The decibel gain is given as

$$A_{dB} = 20 \log_{10} \frac{A_M}{k} + 20 \log_{10} \sqrt{1 + \left(\frac{f}{f_1}\right)^2} - 20 \log_{10} \sqrt{1 + \left(\frac{f}{kf_1}\right)^2}$$

In this equation k is a number which depends upon the amplifier circuit (this is discussed later in the chapter). The above equation can be considered as *three separate decibel plots*, which, when added together represent the total graph. From the previous discussion, it can be seen that the second (middle) term *increases* with frequency and has a corner frequency at f_1, whereas the third (last) term *decreases* with frequency and has a corner frequency at kf_1. These two plots are shown separately as dotted lines in Fig. 8.6(a); the sum of these two terms is shown as the

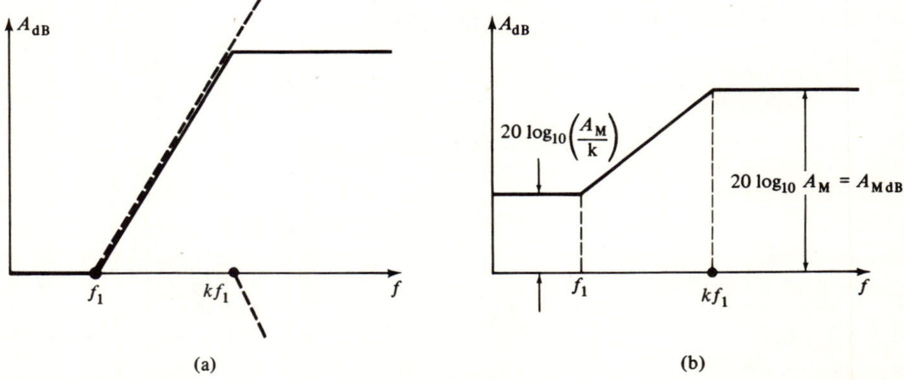

(a)   (b)

Figure 8.6

solid curve. The first term in the decibel equation above is a constant; i.e., it is independent of frequency. If it is added to the sum of the two curves shown in Fig. 8.6(a), the result will be the total graph shown in Fig. 8.6(b). Notice that when this final constant portion of the graph is added to the sum of the other two portions, the effect is that of *shifting the graph of Fig. 8.6(a) up by a level* equal to $20 \log_{10}(A_M/k)$. Remember that this is just an asymptotic plot, an approximation.

The graph of Fig. 8.6(b) checks with the equation since the second and third terms of the equation become zero at frequencies below f_1, leaving just the first, constant term remaining. At frequencies above kf_1 the decibel equation can be approximated as follows:

$$A_{dB} \simeq 20 \log_{10}\left(\frac{A_M}{k}\right) + 20 \log_{10}\left(\frac{f}{f_1}\right) - 20 \log_{10}\left(\frac{f}{kf_1}\right)$$

$$A_{dB} \simeq 20 \log_{10}\left(\frac{A_M}{k}\right) + 20 \log_{10}\left(\frac{\dfrac{f}{f_1}}{\dfrac{f}{kf_1}}\right)$$

$$A_{dB} \simeq 20 \log_{10}\left(\frac{A_M}{k}\right) + 20 \log_{10}(k)$$

$$A_{dB} \simeq 20 \log_{10}\left(\frac{A_M}{k}\, k\right) = 20 \log_{10} A_M = A_{MdB} \qquad f \gg kf_1$$

This shows that the decibel gain at high frequencies (above kf_1) is the mid-band decibel gain, A_{MdB}, as shown in the graph. The reader should now be able to appreciate better the strength of the decibel method of expressing gain. By using this terminology, we can combine many different types of frequency response curves by merely adding them as we have done above. If we tried to follow the same procedure using the numerical gains, we would have to multiply the values of the graphs, point-by-point; this could become quite cumbersome. The decibel plot was introduced to facilitate handling of the various frequency response curves, and also because it is used frequently throughout the engineering field.

8-4 LOW-FREQUENCY RESPONSE: COUPLING CAPACITOR

We now study in some detail the actual causes for the loss of gain at the low and high ends of the frequency spectrum. To do this it is most convenient to analyze the amplifier circuit for one effect at a time. For example, we soon see that both the coupling capacitor, C_C, and the by-pass capacitor, C_E, are responsible for the loss of gain at the low end of the spectrum. To try to study both of these effects at the same time would prove difficult because of the complexity of the equations involved. Thus, this section deals only with the effect of the coupling capacitor on the frequency response.

Figure 8.7(a) shows a one-stage amplifier using emitter bias, with both coupling

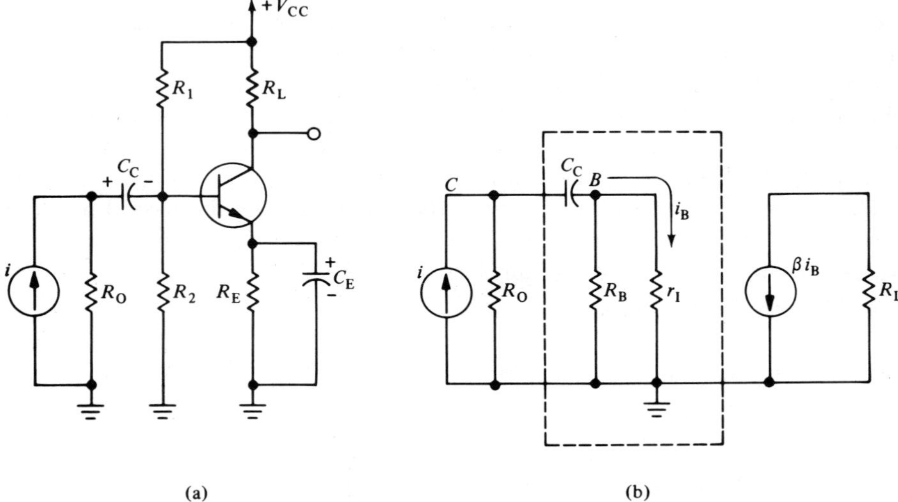

(a)   (b)

Figure 8.7

capacitor and by-pass capacitor. The current source labeled i represents the current developed in the collector of the previous stage, while R_O is the output resistance of that stage (R_O is the parallel combination of R_L and r_O). We assume for the purpose of this section that the emitter resistor is effectively by-passed by C_E at all frequencies, so that the emitter is at ac ground potential. Figure 8.7(b) shows the ac equivalent circuit of Fig. 8.7(a). Notice that the R_E-C_E combination has been omitted in the circuit. You may recall from chapter 6 that the purpose of the coupling capacitor was to separate the dc levels of two stages in cascade so as to eliminate drift problems. It was also important, however, that the reactance of the capacitor be small so that it presented negligible reactance to ac current.

In chapter 6 we found the reactance of a 10 μF capacitor at 1 kHz to be 15.9 Ω. Referring again to Fig. 8.7(b), we can see that we should like most of the current, i, to flow to the base of the transistor (i_B). Thus, we should like both R_O and R_B to be large compared with r_I so that most of the current will go to the base. The current, i, can follow one of two paths: it can flow through R_O, or it can flow through the impedance represented by the reactance of C_C in series with the parallel combination of R_B and r_I. The impedance of this second path is shown within the dotted box in Fig. 8.7(b). If the reactance of C_C is indeed small as it is at 1 kHz (15.9 Ω), C_C has little effect on the operation of the amplifier because 15.9 Ω is quite small in comparison to the values of the resistors (1 kΩ, etc.).

Suppose the frequency of the input signal is decreased to 100 Hz; the reactance of the capacitor is now *10 times as large*, or 159 Ω. This slightly increases the impedance of the dotted box of Fig. 8.7(b). Suppose the frequency is decreased again to 10 Hz; the reactance of the capacitor becomes equal to 1590 Ω, or 1.59 kΩ. This reactance is no longer negligible in comparison to the values of the resistors, and thus must be taken into account when computing values of gain; the equations for gain derived in the previous chapters must now be changed.

Since X_C increases at 10 Hz to such a large value, this X_C must now be added (vectorially) to the parallel combination of R_B and r_I. This will *increase* the impedance of the branch in the dotted box, thus causing *less* current to flow through this branch, and *more* to flow through R_O. This means that now a smaller percentage of i gets to the base of the transistor, with a consequent reduction of gain. If the frequency is further reduced to 1 Hz, X_C becomes equal to 15.9 kΩ; this further increases the impedance of the branch, reduces i_B, and also the gain. If the frequency is reduced to zero Hz (dc), the reactance of the capacitor is infinite (open circuit) and no current gets to the base; the gain is zero. Thus R-C coupled circuits provide no gain (zero output) at dc. The reader can now see that the *coupling capacitor is responsible for the loss of gain at low frequencies.*

We now derive an equation for the gain at low frequencies, and relate this equation to the decibel plots discussed previously. We derive an approximate form of the equation and then give the more exact form, since the derivation of the exact form is quite cumbersome algebraically. We assume that the current flowing through R_B in Fig. 8.7(b) is small in comparison to that flowing through the base, i_B. In other words, assume that R_B is much larger than r_I. We proceed as follows: Using the current-divider formula we obtain

$$\beta i_\mathrm{B} = \frac{\beta i R_\mathrm{O}}{R_\mathrm{O} + r_\mathrm{I} - jX_C} = \frac{\dfrac{\beta R_\mathrm{O}}{R_\mathrm{O} + r_\mathrm{I}}\, i}{1 - j\dfrac{X_C}{R_\mathrm{O} + r_\mathrm{I}}}$$

Using the definition of gain

$$A = \frac{\beta i_\mathrm{B}}{i} = \frac{\dfrac{\beta R_\mathrm{O}}{R_\mathrm{O} + r_\mathrm{I}}}{1 - j\dfrac{X_C}{R_\mathrm{O} + r_\mathrm{I}}} = \frac{\dfrac{\beta R_\mathrm{O}}{R_\mathrm{O} + r_\mathrm{I}}}{1 - j\dfrac{1}{2\pi f C_C(R_\mathrm{O} + r_\mathrm{I})}} \qquad \text{where} \qquad X_C = \frac{1}{2\pi f C_C}$$

Let us make the following definition:

$$f_{\mathrm{L}_1} = \frac{1}{2\pi C_C(R_\mathrm{O} + r_\mathrm{I})}$$

The gain equation can now be written as

$$A = \frac{\dfrac{\beta R_\mathrm{O}}{R_\mathrm{O} + r_\mathrm{I}}}{1 - j\dfrac{f_{\mathrm{L}_1}}{f}}$$

The magnitude of the gain is:

$$|A| = \frac{\dfrac{\beta R_\mathrm{O}}{R_\mathrm{O} + r_\mathrm{I}}}{\sqrt{1 + \left(\dfrac{f_{\mathrm{L}_1}}{f}\right)^2}} = \frac{A_\mathrm{M}}{\sqrt{1 + \left(\dfrac{f_{\mathrm{L}_1}}{f}\right)^2}}$$

It can be seen now that this equation has the same form as the low frequency equation discussed in one of the previous sections. The numerator of the equation is the mid-band gain, A_M; this should be recognized by the reader as the expression for the gain derived in previous chapters *without* considering the effect of the coupling capacitor. The frequency, f_{L_1}, is the corner frequency due to the coupling capacitor.

It was mentioned that this derivation is just an approximation. If the effect of R_B is considered, the actual equations appear as follows:

$$|A| = \frac{A_M}{\sqrt{1 + \left(\dfrac{f_{L_1}}{f}\right)^2}}$$

$$A_M = \frac{\beta R}{R + r_1} \tag{8-3}$$

where

$$R = R_O \parallel R_B$$

$$f_{L_1} = \frac{1}{2\pi C_C (R_O + R_{IE})} \tag{8-4}$$

where

$$R_{IE} = r_1 \parallel R_B$$

and

$$R_O = R_L \parallel r_O$$

These equations can be used to compute the mid-band gain and corner frequency of any R-C coupled amplifier.

EXAMPLE 12

Compute A_M, A_{MdB}, and f_{L_1} for the amplifier with the following values:

$$\begin{aligned}
\beta &= 50 & r_I &= 1 \text{ k}\Omega \\
R_1 &= 90 \text{ k}\Omega & C_C &= 10 \text{ } \mu\text{F} \\
R_2 &= 10 \text{ k}\Omega & &\text{Neglect } r_O \\
R_L &= 1 \text{ k}\Omega & &
\end{aligned}$$

SOLUTION

All that is needed is to substitute into the equations to solve the problem.

$$R_B = \frac{R_1 R_2}{R_1 + R_2} = \frac{90(10)}{90 + 10} = 9 \text{ k}\Omega$$

$$R = \frac{R_O R_B}{R_O + R_B} = \frac{(1)(9)}{1 + 9} = 0.9 \text{ k}\Omega \qquad R_O \simeq R_L = 1 \text{ k}\Omega$$

$$A_M = \frac{\beta R}{R + r_1} = \frac{50(0.9)}{0.9 + 1} = \frac{45}{1.9} = 23.7$$

$$A_{MdB} = 20 \log_{10} A_M = 20 \log_{10} (23.7) = 27.5 \text{ dB}$$

$$R_{IE} = \frac{r_I R_B}{r_I + R_B} = \frac{(1)(9)}{1 + 9} = 0.9 \text{ k}\Omega$$

$$f_{L_1} = \frac{1}{2\pi C_C (R_O + R_{IE})} = \frac{0.159}{(10 \times 10^{-6})(1.9)10^3}$$

$$f_{L_1} = 8.36 \text{ Hz}$$

Thus, this amplifier has a mid-band decibel gain of 27.5 dB, and a corner frequency of 8.36 Hz.

EXAMPLE 13

Repeat example 12, except make C_C equal to 1.0 μF.

SOLUTION

Since neither the resistors nor the transistor have been changed, the mid-band gain remains the same as that found in example 13. The only quantity which will change will be the corner frequency; this is found as follows:

$$f_{L_1} = \frac{1}{2\pi C_C (R_O + R_{IE})} = \frac{0.159}{(1 \times 10^{-6})(1.9 \times 10^3)} = 83.6 \text{ Hz}$$

Notice that the corner frequency has increased 10 times, because C_C has been decreased to one-tenth of its former value. Thus it can be seen that C_C should be made *large* in order to keep the lower corner frequency, f_{L_1}, as *low* as possible. The corner frequency depends upon other values besides C_C, such as R_L, R_B, etc. However, these values are usually either dependent upon the transistor, or are selected to give a particular desired value of gain. Thus the designer usually varies C_C to obtain the particular value of f_{L_1} desired.

In many cases a very low value of f_{L_1} is desired, and thus C_C must be quite large. A large value of C_C is usually physically large. This may limit the corner frequency because the capacitor might be so large that it may be larger than the rest of the amplifier. The reader should bear in mind that one of the main advantages of using transistors instead of vacuum tubes is the enormous savings to be realized in space and weight. Large values of capacitance are available in small physical sizes; these are polarized electrolytic or tantalytic capacitors. Care must be used when placing these capacitors in circuits to use the proper polarity. If the transistors are *npn* as shown in Fig. 8.7(a), the polarity must be as indicated; the reverse is true if the transistors are *pnp*. The polarity is as indicated in Fig. 8.7(a) because the collector of the previous stage is usually at a higher dc potential than the base of the stage shown; in this case higher potential means more positive.

In summary we can say that the coupling capacitor has an important effect on the lower corner frequency; if it is desired that the lower corner frequency be made smaller, the value of the coupling capacitor must be increased.

8-5 LOW-FREQUENCY RESPONSE: BY-PASS CAPACITOR

Figure 8.8 shows the same circuit of Fig. 8.7; Fig. 8.8(a) shows the actual circuit and Fig. 8.8(b) shows the ac equivalent circuit. The coupling capacitor is shown

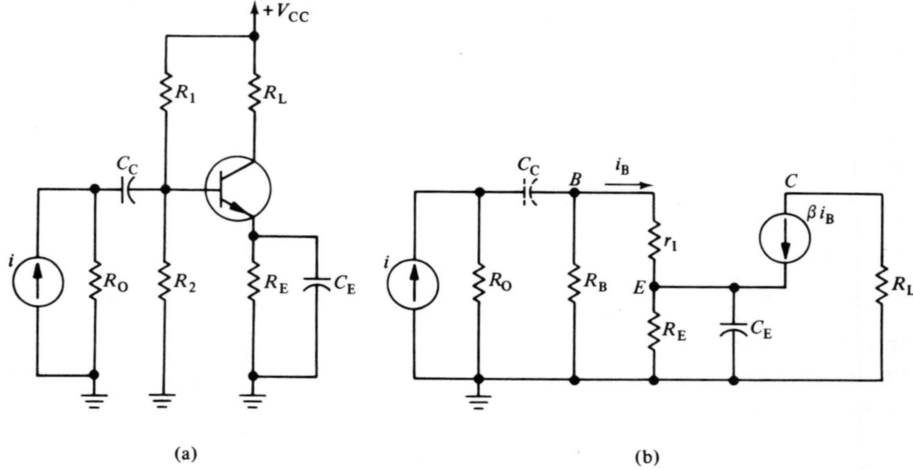

(a)   (b)

Figure 8.8

dotted in Fig. 8.8(b) to indicate that it is actually in the circuit, but we assume for the duration of this section that it is not present. We now consider the effect of the by-pass capacitor on the low-frequency response.

The reader will recall from chapter 7 that the by-pass capacitor was included in the circuit for a very important reason. We saw that without the capacitor the input impedance to an amplifier stage with an emitter resistor is approximately equal to βR_E. This large impedance decreases the amount of current fed to the stage, thus decreasing the gain. The by-pass capacitor is designed to have a small reactance so that the parallel combination of R_E and X_{CE} is very small; C_E *by-passes* the ac portion of the current. If C_E is large enough, X_{CE} will be small enough so that R_E has little effect on the gain. To show the effect of C_E on the gain we consider the following examples:

EXAMPLE 14

Compute the current gain of the amplifier of Fig. 8.8 at $f = 1$ kHz if the values are as follows:

$$\beta = 50 \qquad r_I = 1 \text{ k}\Omega$$
$$R_1 = 90 \text{ k}\Omega \qquad R_E = 250 \ \Omega$$
$$R_2 = 10 \text{ k}\Omega \qquad C_E = 100 \ \mu\text{F}$$
$$R_L = 1 \text{ k}\Omega$$

SOLUTION

We must first find the reactance of C_E at 1 kHz; this is:

$$X_{C_E} = \frac{1}{2\pi f C_E} = \frac{0.159}{10^3 \times 100 \times 10^{-6}} = 1.59 \ \Omega$$

Thus, the parallel combination of R_E and X_{C_E} is about 1.5 Ω. The input resistance to the transistor is:

$$R_I = r_I + \beta Z_E = 1 \text{ k}\Omega + 50(1.5) = 1 \text{ k}\Omega + 0.075 \text{ k}\Omega \simeq 1 \text{ k}\Omega$$

Thus, the 100 μF capacitor effectively by-passes the emitter resistor. The gain is:

$$A_I = \frac{\beta R}{R + R_I} = \frac{50(0.9)}{0.9 + 1} = 23.7$$

EXAMPLE 15

Repeat example 14 at $f = 1$ Hz.

SOLUTION

First compute X_{C_E}.

$$X_{C_E} = \frac{1}{2\pi f C_E} = \frac{0.159}{1 \times 100 \times 10^{-6}} = 1590 \ \Omega$$

It can be seen that at this frequency the 100 μF capacitor no longer bypasses the emitter resistor because the reactance has become too large. Thus the input resistance now is

$$R_I \simeq r_I + \beta Z_E = 1 \text{ k}\Omega + 50(250 \ \Omega) = 13.5 \text{ k}\Omega$$

The gain is

$$A_I = \frac{\beta R}{R + R_I} = \frac{50(0.9)}{0.9 + 13.5} = \frac{45}{14.4} = 3.13$$

The results of examples 14 and 15 illustrate a number of important concepts. If the frequency is large enough (1 kHz), the capacitive reactance is small enough to by-pass the emitter resistor, and thus the gain is unaffected by the presence of R_E. On the other hand at low enough frequencies (1.0 Hz), the capacitive reactance becomes so large that it no longer by-passes R_E. In this case the input resistance increases considerably (from 1 kΩ to 13.5 kΩ) with a consequent reduction of gain (from 23.7 to 3.13). Someplace between 1 Hz and 1 kHz the gain decreases from its relatively large mid-band value to a much smaller value. Thus the by-pass capacitor causes the frequency response to follow a curve similar to the one shown in Fig. 8.6 and repeated in Fig. 8.9 for convenience. There are *two* corner fre-

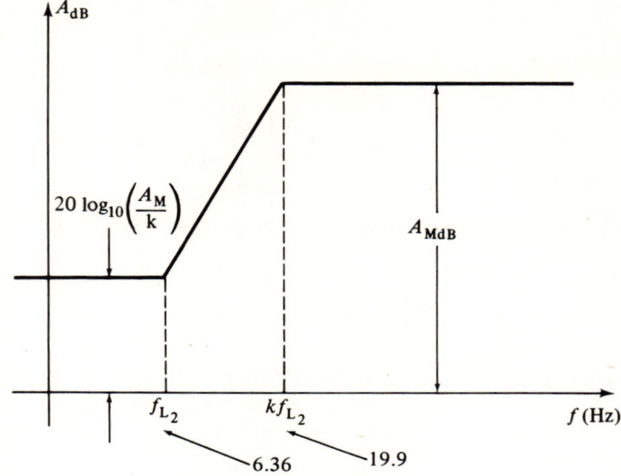

Figure 8.9

quencies associated with the frequency response curve due to the by-pass capacitor: f_{L_2} and kf_{L_2}.

There is an algebraic expression for the gain in terms of frequency similar to the one derived in the previous section for the coupling capacitor. The derivation of this expression is rather involved algebraically and will not be shown here. The results are given as follows:

$$A = \frac{A_M}{k} \frac{\sqrt{1 + \left(\dfrac{f}{f_{L_2}}\right)^2}}{\sqrt{1 + \left(\dfrac{f}{kf_{L_2}}\right)^2}}$$

where

$$A_M = \frac{\beta R}{R + r_I} \tag{8-5}$$

$$k = 1 + \frac{\beta R_E}{R + r_I} \tag{8-6}$$

$$f_{L_2} = \frac{1}{2\pi R_E C_E} \tag{8-7}$$

EXAMPLE 16

Using the values given in example 14, compute A_M, k, A_M/k, f_{L_2}, and kf_{L_2}.

SOLUTION

To solve this problem, merely substitute into the previously derived equations.

$$A_M = \frac{\beta R}{R + r_I} = \frac{50(0.9)}{0.9 + 1} = 23.7$$

$$k = 1 + \frac{\beta R_{\mathrm{E}}}{R + r_{\mathrm{I}}} = 1 + \frac{50(.25)}{0.9 + 1} = 1 + \frac{12.5}{1.9} = 7.57$$

$$\frac{A_{\mathrm{M}}}{k} = \frac{23.7}{7.57} = 3.13$$

$$f_{\mathrm{L_2}} = \frac{1}{2\pi R_{\mathrm{E}} C_{\mathrm{E}}} = \frac{0.159}{(.25 \times 10^3)(100)(10^{-6})} = 6.36 \text{ Hz}$$

$$kf_{\mathrm{L_2}} = (3.13)(6.36) = 19.9 \text{ Hz}$$

The results of example 16 are shown on the graph of Fig. 8.9. This shows that the amplifier is useful at frequencies above $kf_{\mathrm{L_2}}$, in this case, 19.9 Hz. The graph also bears out what we said at the beginning of this section: namely that the corner frequencies associated with the by-pass capacitor lie somewhere between 1 Hz and 1 kHz. The mid-band gain is 23.7 and the gain at the low end of the spectrum is 3.13.

In the previous section it was shown that the lower cut-off frequency due to the coupling capacitor ($f_{\mathrm{L_1}}$) was 8.36 Hz when C_{C} was 10 μF. With the same amplifier the cut-off frequencies due to the by-pass capacitor are in the same range (6.36 Hz and 19.9 Hz), *but the value of C_{E} is 100 μF*. In other words, the design conditions for the by-pass capacitor are much more stringent than those for the coupling capacitor; C_{E} must be made much larger than C_{C} in order that the corner frequencies be in the same range (same order of magnitude).

EXAMPLE 17

Repeat example 16, except make C_{E} equal to 10 μF.

SOLUTION

The values of A_{M} and k are independent of C_{E}; thus the gains remain the same. The only values that will change are $f_{\mathrm{L_2}}$ and $kf_{\mathrm{L_2}}$.

$$f_{\mathrm{L_2}} = \frac{1}{2\pi R_{\mathrm{E}} C_{\mathrm{E}}} = \frac{0.159}{(0.25 \times 10^3)(10 \times 10^{-6})} = 63.6 \text{ Hz}$$

$$kf_{\mathrm{L_2}} = (k)(f_{\mathrm{L_2}}) = (3.13)(63.6) = 199 \text{ Hz}$$

Decreasing C_{E} to one-tenth its previous size results in a ten-fold increase in both corner frequencies.

We now examine the effects of both capacitors simultaneously.

EXAMPLE 18

Using the same amplifier discussed throughout the last few sections, with C_{E} equal to 100 μF and C_{C} equal to 10 μF, draw the decibel plot of the amplifier at low frequencies.

SOLUTION

We have already found the corner frequencies:

$$f_{L_2} = 6.36 \text{ Hz} \qquad f_{L_1} = 8.36 \text{ Hz} \qquad kf_{L_2} = 19.9 \text{ Hz}$$

The mid-band decibel gain has already been found as:

$$A_{\text{MdB}} = 20 \log_{10} (23.7) = 27.5 \text{ dB}$$

Now all that is needed is to draw the plot. The graph will be the sum of the effects of the two individual curves as shown in Fig. 8.10(a). The gain begins to fall at

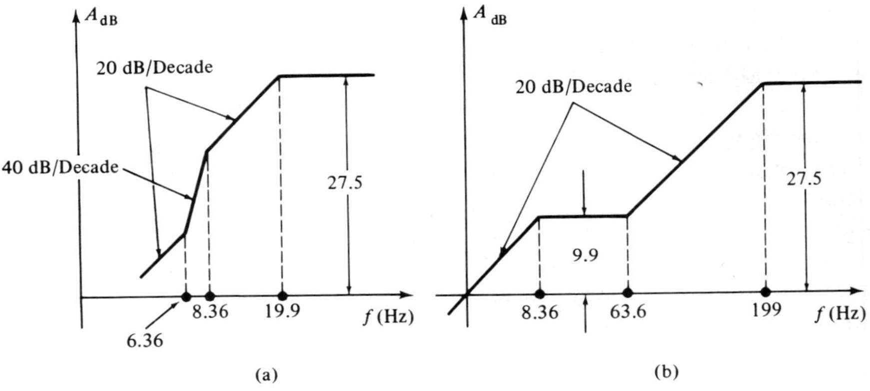

Figure 8.10

the largest of the corner frequencies, in this case, kf_{L_2} (19.9 Hz), at the rate of 20 dB per decade. The corner frequency due to C_C, f_{L_1} (8.36 Hz), occurs between f_{L_2} and kf_{L_2}. At this point the gain drops at the rate of *40 dB per decade* (12 dB per octave); this is because the gain is falling as a result of two separate effects (the effects caused by C_E *and* C_C). When f_{L_2} is reached, the gain falls again at 20 dB per octave. The straight line segments of Fig. 8.10(a) are just asymptotes; the actual curve is more rounded.

EXAMPLE 19

Repeat example 18, except make C_E equal to 10 μF.

SOLUTION

Since C_C remains the same at 10 μF, f_{L_1} also remains the same at 8.36 Hz. The change in C_E causes a change in f_{L_2} and kf_{L_2}.

$$f_{L_1} = 8.36 \text{ Hz} \qquad f_{L_2} = 63.6 \text{ Hz} \qquad kf_{L_2} = 199 \text{ Hz}$$

$$20 \log_{10} \left(\frac{A_M}{k}\right) = 20 \log_{10} (3.13) = 9.9 \text{ dB}$$

The frequency response plot appears in Fig. 8.10(b). Notice that the shape of the curve is different from that in Fig. 8.10(a) because the corner frequency due to C_C occurs below both of the corner frequencies due to C_E.

It can be seen from the results of this section that it is important that both C_C *and* C_E be made as large as possible for good low-frequency response; C_E, like C_C, is usually polarized. The loss of gain due to C_C can be eliminated altogether by direct coupling amplifier stages. You may recall from chapter 6 that this involves the problem of drift. The loss of gain due to C_E can be eliminated by not using emitter bias. This can be done only when the temperature range over which the amplifier must perform is limited, or when the β spread of production units is small. If collector-to-base feedback is used, the same type of problems are encountered.

8-6 HIGH-FREQUENCY RESPONSE

Having handled the low-frequency problems, we now turn our attention to the study of the high-frequency behavior of the transistor amplifier. You will recall from chapter 3 that the emitter-base junction is forward-biased and the collector-base junction, reverse-biased. Figure 8.11 shows an appropriately biased *npn* tran-

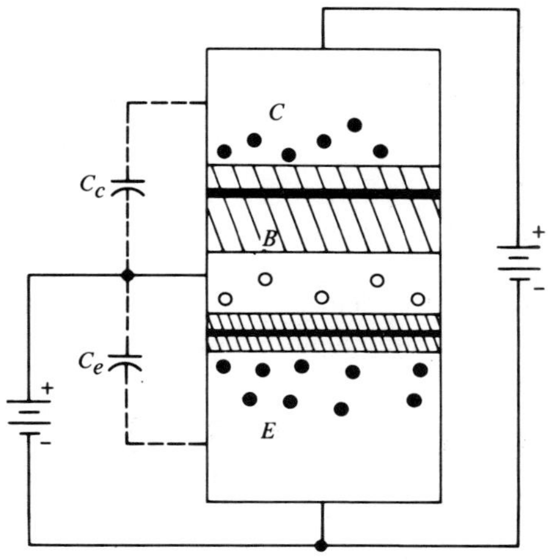

Figure 8.11

sistor. Notice that because of the polarity of the biasing batteries, the depletion region at the emitter-base junction is much narrower than that at the collector-base junction. There are carriers on either side of both junctions, but none (or very few) within the depletion regions. Since there are no carriers within these depletion regions, the regions can be considered to be poor conductors, or good insulators.

The reader will recall from basic dc circuits theory that a capacitor was formed by separating two conductors by an insulator. This is similar to the situation pictured in Fig. 8.11. The carriers on either side of the junction represent the conductors, and the depletion region between the carriers is the insulator. In other words, we can say that *there are capacitances associated with the junctions of a transistor.* These capacitances are indicated by the dotted lines in Fig. 8.11 to show that they are not placed into the circuit intentionally as is the case with the coupling and by-pass capacitors; they exist as part of the transistor. Just as with the other transistor parameters, like r_I, these junction capacitances cannot be seen or touched; as far as the circuit designer is concerned the effect is as if an *actual capacitor were connected across the junctions.*

Figure 8.11 shows that there are two capacitances, one for each junction. The capacitance across the emitter-base junction is labeled C_e, and that across the collector-base junction, C_c. We use lower case subscripts to avoid confusion between these capacitances and the coupling and by-pass capacitors, C_C and C_E. You will again recall from basic dc circuits theory that the size of the capacitor depends upon the dimensions of the plates, as well as thickness and material of the dielectric, or insulator. In general, a thinner dielectric will yield a higher value of capacitance. Note in Fig. 8.11 that the depletion region (dielectric) is much thinner at the emitter-base junction; thus the emitter junction capacitance is generally considerably larger than the collector junction capacitance. Stated alternately, we say that C_e *is much larger than* C_c.

It should be mentioned at this point that the previous discussion was considerably oversimplified. There are actually two components to each of the junction capacitances: a diffusion capacitance and a transition capacitance. The discussion of the differences between these various quantities involves a deeper understanding of the physical behavior of the transistor, along with a more sophisticated mathematical treatment. It is not necessary that the reader be familiar with the intricacies of semiconductor physics in order to comprehend the major concepts of transistor high-frequency behavior. It is necessary for the purposes of this text only that the reader know that there are capacitances associated with the transistor junctions, and that these capacitances depend upon the transistor itself.

Figure 8.12(a) shows the ac equivalent circuit of a one-stage transistor amplifier

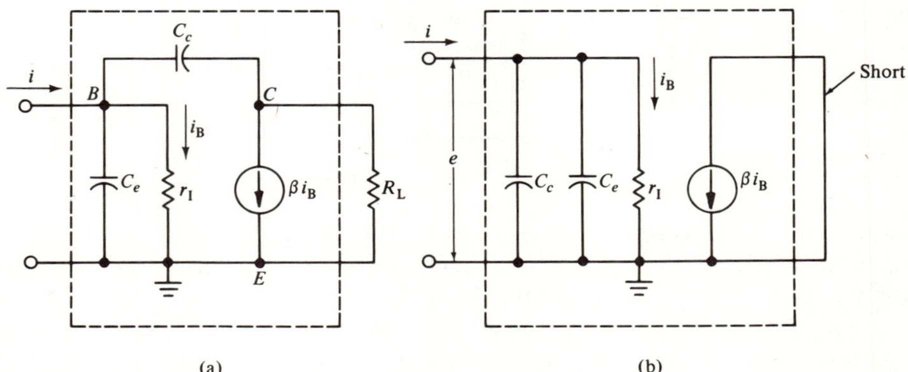

Figure 8.12

with the junction capacitances included. C_e is connected across the emitter-base terminals and C_c is across the collector-base terminals. The transistor itself is inside the dotted box, and the bias resistors are omitted for simplicity. The circuit pictured in Fig. 8.12(a) will be used throughout the remainder of this chapter to illustrate the high-frequency behavior of the transistor amplifier. It should be mentioned that this gives a good approximation to the actual results; however, there is a more exact equivalent circuit, known as the hybrid-pi, which can be used to study the high-frequency response. The hybrid-pi circuit is slightly more complex than the one shown in Fig. 8.12(a), and gives more accurate information. However, for the purposes of this text, the results obtained from a study of the circuit of Fig. 8.12(a) give a more than adequate indication of high-frequency performance; the additional accuracy obtained from a study of the hybrid-pi circuit does not warrant a detailed study of this circuit at this point.

We begin the study of high-frequency response by shorting the output terminals (collector-to-emitter) of the amplifier; i.e., we set the load resistor, R_L, equal to zero. This means that no ac voltage is developed across the collector and emitter terminals; alternatively, we say that the collector and emitter terminals are at the *same ac potential*. If the collector is at the same potential as the emitter, the right hand terminal of C_c in Fig. 8.12(a) can be considered to be connected to the emitter (or ground) as shown in Fig. 8.12(b). Thus, for the condition that the output is shorted, both junction capacitances appear in parallel shunting the input (base-emitter) terminals. The transistor is again inside the dotted box indicating that the capacitances are part of the transistor. It was mentioned previously that C_e is much larger than C_c; thus the equivalent parallel combination (the sum of the values of C_e and C_c) is approximately equal to C_e.

EXAMPLE 20

C_e of a particular transistor is 100 pF and C_c is 5 pF. Compute the total equivalent capacitance shunting the input terminals for the conditions of Fig. 8.12(b).

SOLUTION

Since the equivalent capacitance of two capacitors connected in parallel is equal to the sum of the values of the two capacitors, all that is required is to add the values of C_e and C_c.

$$C_T = C_e + C_c = 100 \text{ pF} + 5 \text{ pF} = 105 \text{ pF} \simeq 100 \text{ pF}$$

Thus, for the conditions of Fig. 8.12(b), we can neglect the effect of C_c.

The input current, i, in Fig. 8.12(b) flows into the transistor. Part of this current goes to the input resistor, r_I, and part of it goes to the junction capacitors. We know that the current developed in the collector is equal to the product of β and i_B. In other words, *only that portion of the input current which enters r_I (i_B) is multiplied by β*. A current divider is formed at the input terminals consisting of r_I and the junction capacitances. For maximum gain we should like most of the input current to enter r_I so that it will then be multiplied by β. That means that it is preferable that r_I be much smaller than the reactance of the junction capacitances. In order that the reactance of the junction capacitances be large, we should like the capacitances themselves to be as small as possible, and the frequency also to be small. As mentioned earlier, the frequency of the input signal is usually specified as part of the design, so this is beyond the designer's control. A few examples will illustrate the problems involved.

EXAMPLE 21

For the transistor described in example 20, C_T (total equivalent capacitance) is approximately equal to 100 pF, and r_I is 1 kΩ. Compute the reactance of the capacitor at a frequency of 10 kHz.

SOLUTION

Using the expression for computing the capacitive reactance:

$$X_{C_T} = \frac{1}{2\pi f C_T} = \frac{0.159}{(10^4)(100 \times 10^{-12})} = 159 \text{ k}\Omega$$

Since X_C is 159 kΩ and r_I is 1 kΩ, it is clear that at 10 kHz most of the input current flows through the 1 kΩ input resistance, and a negligible portion flows through the junction capacitances.

EXAMPLE 22

Repeat example 21 if the frequency is increased to 1 MHz.

SOLUTION

$$X_{C_T} = \frac{1}{2\pi f C_T} = \frac{0.159}{10^6(100 \times 10^{-12})} = 1.59 \text{ k}\Omega$$

Notice that at the higher frequency, the reactance of the junction capacitances has dropped considerably to a value of 1.59 kΩ. This means that the input current will now split up. Since less current flows through r_I at 1 MHz, less current flows through the collector; thus the gain of the amplifier is reduced considerably.

Herein lies the significance of the junction capacitances. At frequencies in the mid-band range, and below, the frequency is low enough so that the reactance of these capacitances is much larger than r_I; thus, most of the input current goes to the resistance, r_I. As the frequency of the input signal is increased to larger values, the reactance decreases and more current is drawn by the capacitances; thus, less current is available for r_I, and the gain decreases. Just as C_C and C_E cause problems at the lower frequencies and can be neglected at mid-band and higher frequencies, the junction capacitances affect the high-frequency behavior but can be neglected at mid-band and lower frequencies.

We now proceed to derive a few expressions which are important in the consideration of high-frequency response. The current flowing through the collector of the circuit of Fig. 8.12(b) is given by the following equation:

$$ i_C = \beta i_B = \beta \frac{e}{r_I} = \frac{\beta}{r_I} iZ $$

In this case e is the voltage across the base-emitter (input) terminals, i is the input current to the amplifier, and Z is the impedance of the parallel combination of r_I and X_C. Substituting the appropriate terms into the equation, and performing some algebraic manipulations,

$$ A = \frac{i_C}{i} = \frac{\beta}{r_I}(Z) = \frac{\beta r_I(-jX_C)}{r_I(r_I - jX_C)} = \frac{\beta(-jX_C)}{r_I - jX_C} $$

$$ A = \frac{\beta}{1 + j\dfrac{r_I}{X_C}} = \frac{\beta}{1 + j2\pi f r_I C_e} $$

Utilizing the following definition

$$ f_\beta = \frac{1}{2\pi r_I C_e} $$

The equation can now be stated as

$$ A = \frac{\beta}{1 + j\dfrac{f}{f_\beta}}; \qquad |A| = \frac{\beta}{\sqrt{1 + \left(\dfrac{f}{f_\beta}\right)^2}} $$

The above expression shows that the mid-band current gain is equal to β, and that there is a corner frequency at f_β. *The frequency f_β is defined as the β cut-off frequency.*
Figure 8.13 shows a decibel plot for the above equation; the sloping asymptote

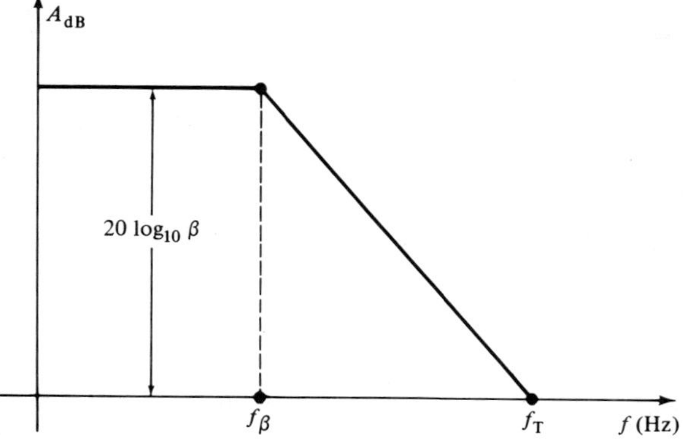

Figure 8.13

drops, as usual, at the rate of 20 dB per decade. We should like to know the frequency at which the short circuit current gain (remember that the output is shorted) drops to zero dB, or 1.0. This frequency is designated f_T as shown in the figure. To find f_T, we set the equation equal to 1.0 and solve, as follows:

$$A = 1 = \frac{\beta}{\sqrt{1 + \left(\frac{f_T}{f_\beta}\right)^2}}; \qquad \beta = \sqrt{1 + \left(\frac{f_T}{f_\beta}\right)^2}$$

$$\beta^2 - 1 = \left(\frac{f_T}{f_\beta}\right)^2$$

$$f_T = f_\beta \sqrt{\beta^2 - 1}$$

The term β^2 is usually much larger than 1.0, so the following approximation can be made.

$$f_T \simeq \beta f_\beta \tag{8-8}$$

Many manufacturers list f_T on the specification sheet for the particular transistor; this is the frequency at which the short circuit current gain drops to 1.0. We soon see that this is a very important factor when we consider the useful high-frequency range of an amplifier.

The above discussion was necessary in order to illustrate the effect of the junction capacitances on the high-frequency response. We turn now to a consideration of a practical amplifier; i.e., one in which the output is *not* shorted. The ac equivalent circuit is shown in Fig. 8.14(a); since the output is not shorted, C_c is connected

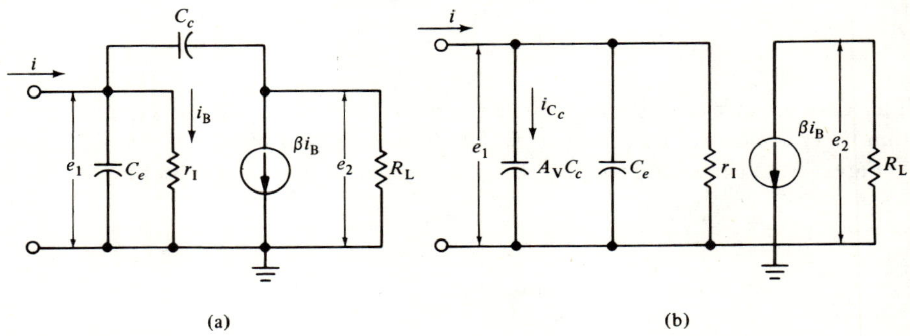

(a)   (b)

Figure 8.14

as shown. We proceed now with an analysis of this circuit. The current through C_c can be expressed as the voltage across C_c divided by the capacitive reactance of C_c; thus:

$$i_{C_c} = \frac{e_{C_c}}{X_{C_c}} = 2\pi f C_c e_{C_c}$$

The voltage across C_c can be expressed as the difference between the input voltage, e_1, and the output voltage, e_2.

$$e_{C_c} = e_1 - e_2 \qquad i_{C_c} = 2\pi f C_c (e_1 - e_2)$$

But the ratio of e_2 to e_1 is the voltage gain of the amplifier, A_V.

$$A_V = \frac{e_2}{e_1}; \qquad e_2 = A_V e_1$$

$$i_{C_c} = 2\pi f C_c (e_1 - A_V e_1) = 2\pi f C_c e_1 (1 - A_V)$$

$$|i_{C_c}| \simeq 2\pi f C_c e_1 A_V \qquad \text{since } A_V \gg 1$$

Thus the ratio of e_1 (the voltage across the input terminals of the amplifier) to the current through C_c is given by

$$\frac{e_1}{i_{C_c}} = \frac{1}{2\pi f C_c A_V} = \frac{1}{2\pi f (A_V C_c)}$$

Since e_1 is the voltage across the input terminals, the equation shows that a capacitance with a value equal to $A_V C_c$ is connected across the input terminals. In other words, instead of considering C_c connected between base and collector as shown in Fig. 8.14(a), we can assume that a capacitance equal to $A_V C_c$ is connected across the input terminals as shown in Fig. 8.14(b). The circuits of Figs. 8.14(a) and (b) are equivalent; we use the circuit of Fig. 8.14(b) because it is easier to work with algebraically. The phenomenon illustrated by the previous equations and Figs. 8.14(a) and (b) is known as the *Miller effect*. Whenever some impedance is connected between the output and the input terminals, the circuit can be changed into a more convenient form. We say that C_c is reflected into the input circuit and multiplied by A_V; this is the same principle which applies in the case of the emitter resistor's being reflected into the base circuit.

Using the results derived above, we analyze a simple one-stage transistor amplifier for high-frequency behavior. Figure 8.15(a) is a schematic diagram of the

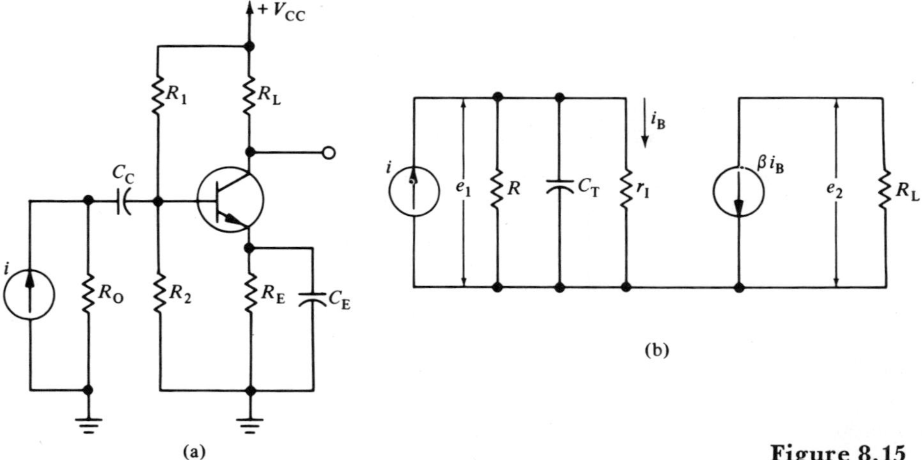

(a)

(b)

Figure 8.15

amplifier and Fig. 8.15(b) shows the ac equivalent circuit at high frequencies. Notice that both the coupling and by-pass capacitors are omitted since they act like short circuits at these frequencies. Note also that C_e is now in parallel with $A_V C_c$; the total equivalent capacitance is now equal to the sum of these two values:

$$C_T = C_e + A_V C_c$$

Thus C_c no longer contributes a negligible effect to the total capacitance as it did in the case where the output was shorted; it is now multiplied by A_V. The resistor R is as usual the parallel combination of the output resistance of the source, R_O, and the bias resistors, R_B. Let us now derive an expression for the current gain from the equivalent circuit of Fig. 8.15(b).

$$i_C = \beta i_B = \frac{\beta e_1}{r_I} = \frac{\beta}{r_I} iZ$$

$$A_I = \frac{\beta i_B}{i} = \frac{\beta}{r_I} \frac{R_T(-jX_{C_T})}{R_T - jX_{C_T}} = \frac{\beta}{r_I} \frac{R_T}{1 + j\dfrac{R_T}{X_{C_T}}}$$

where

$$R_T = \frac{Rr_I}{R + r_I}$$

The resistance, R_T, is the parallel combination of R and r_I.

$$A_I = \frac{\beta}{r_I} \frac{\dfrac{Rr_I}{R + r_I}}{1 + j\dfrac{R_T}{X_{C_T}}} = \frac{\dfrac{\beta R}{R + r_I}}{1 + j\dfrac{R_T}{X_{C_T}}}$$

$$A_I = \frac{\dfrac{\beta R}{R + r_I}}{1 + j2\pi f C_T R_T} = \frac{\dfrac{\beta R}{R + r_I}}{1 + j\dfrac{f}{f_H}} \qquad f_H = \frac{1}{2\pi C_T R_T}$$

$$|A_I| = \frac{\dfrac{\beta R}{R + r_I}}{\sqrt{1 + \left(\dfrac{f}{f_H}\right)^2}}$$

$$\boxed{A_M = \frac{\beta R}{R + r_I}} \tag{8-9}$$

$$\boxed{f_H = \frac{1}{2\pi R_T C_T}} \tag{8-10}$$

This shows that the mid-band gain is the same as that derived in previous sections; it also shows that the upper corner frequency is a function of R_T and C_T. As explained earlier in the chapter, we should like the upper corner frequency, f_H, to be as large as possible. R_T depends upon the resistors in the circuit; these are usually specified on the basis of gain and stability requirements. Thus, the designer must try to minimize C_T in order to obtain a wide bandwidth. Since C_T depends primarily upon the transistor itself, the designer must select a transistor with small junction capacitances in order to achieve a wide bandwidth.

There is another factor upon which the high-frequency performance depends which has not yet been discussed; this is the *stray capacitance*. Figure 8.16 shows a portion of a transistor amplifier circuit. The lead from the junction of the two resistors, R_1 and R_2, to the base of the transistor is a metal wire encased in insulation, as is the ground wire. Both of these leads are therefore conductors separated by insulation; the insulation in this case is composed of the actual wire covering (if there is any) and the air space between the wires. When two conductors are separated by an insulator, the effect is as if there were a capacitor present. Thus, there are capacitances present between the leads as shown by the dotted lines of Fig. 8.16. There is a certain amount of capacitance per foot of length; this depends

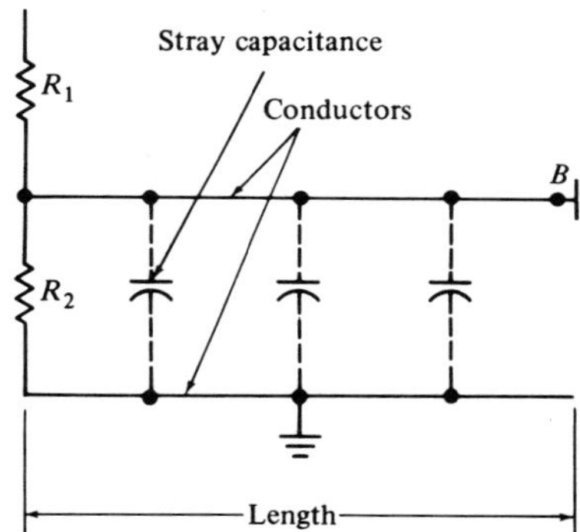

Figure 8.16

upon the dimensions and material of the wire, the type of insulation, and the distance between the two conductors.

The capacitance present in the circuit due to the wires is called *stray capacitance*. Because these capacitances exist between base (or collector) and ground, they are effectively *in parallel* with the junction capacitances. When capacitances are in parallel they *add;* thus the effect of the stray capacitance is to increase the total capacitance in shunt with the input and output terminals of the amplifier. An increase in C_T causes a decrease in the upper corner frequency, f_H, with a consequent reduction in bandwidth. Thus it is desirable to minimize stray capacitance. This is generally done by using very short leads in wiring the actual amplifier circuit; careful wiring layout is of prime importance in the design of transistor amplifiers for wide bandwidth. Thus, it can be seen that while the low-frequency response is a function of C_C and C_E, high-frequency response depends upon the transistor itself and the wiring layout.

The expression for computing the high-frequency corner was given in terms of R_T and C_T; C_T is in terms of C_e and C_c. Most manufacturers do not give this information directly in the specification sheets, but give it instead in another form. It is, therefore, necessary that we manipulate the expression algebraically to achieve a more useful equation. This proceeds as follows:

$$f_{\mathrm{H}} = \frac{1}{2\pi R_{\mathrm{T}} C_{\mathrm{T}}} = \frac{1}{2\pi \left(\dfrac{R r_{\mathrm{I}}}{R + r_{\mathrm{I}}}\right)(C_e + A_{\mathrm{V}} C_c)}$$

$$= \frac{R + r_{\mathrm{I}}}{2\pi R r_{\mathrm{I}}(C_e + A_{\mathrm{V}} C_c)} = \frac{1}{r_{\mathrm{I}}(2\pi)}\left(\frac{1 + \dfrac{r_{\mathrm{I}}}{R}}{(C_e + A_{\mathrm{V}} C_c)}\right)$$

$$f_{\mathrm{H}} = \frac{1}{2\pi r_{\mathrm{I}} C_e}\left(\frac{1 + \dfrac{r_{\mathrm{I}}}{R}}{1 + A_{\mathrm{V}}\dfrac{C_c}{C_e}}\right); \qquad f_{\beta} = \frac{1}{2\pi r_{\mathrm{I}} C_e}$$

$$f_{\mathrm{H}} = f_{\beta}\left(\frac{1 + \dfrac{r_{\mathrm{I}}}{R}}{1 + A_{\mathrm{V}}\dfrac{C_c}{C_e}}\right) \tag{8-11}$$

We have yet to show how we calculate the voltage gain, A_{V}. We know that the voltage gain is equal to the ratio of the output voltage to the input voltage. Thus the following manipulations lead to an expression for the voltage gain:

$$e_2 = i_{\mathrm{C}} R_{\mathrm{L}} = \beta i_{\mathrm{B}} R_{\mathrm{L}}; \qquad e_1 = i_{\mathrm{B}} r_{\mathrm{I}}$$

$$A_{\mathrm{V}} = \frac{e_2}{e_1} = \frac{\beta i_{\mathrm{B}} R_{\mathrm{L}}}{i_{\mathrm{B}} r_{\mathrm{I}}} = \frac{\beta R_{\mathrm{L}}}{r_{\mathrm{I}}}$$

$$A_{\mathrm{V}} = \frac{\beta R_{\mathrm{L}}}{r_{\mathrm{I}}} \tag{8-12}$$

Some manufacturers specify f_{T} and C_{OB}. C_{OB} is the capacitance between collector and base with the emitter open; this is usually specified at some value of v_{CB} near the suggested Q point. The value of C_{OB} will decrease as v_{CB} is increased since an increase in the reverse bias will increase the depletion region, thus decreasing the capacitance. This value of C_{OB} specified by the manufacturer is the *same* as C_c. We illustrate the use of the above formulas by means of a few examples.

EXAMPLE 23

A transistor amplifier has the following values associated with it:

$R_1 = 90 \ \text{k}\Omega$		$\beta = 50$
$R_2 = 10 \ \text{k}\Omega$		$r_{\mathrm{I}} = 1 \ \text{k}\Omega$
$R_0 = 1 \ \text{k}\Omega$ (previous stage)		$f_{\mathrm{T}} = 20 \ \text{MHz}$
$R_{\mathrm{L}} = 1 \ \text{k}\Omega$		$C_{\mathrm{OB}} = 7 \ \text{pF}$

Compute the mid-band gain, A_{M}, and the upper corner frequency, f_{H}.

SOLUTION

The expression for A_M is as follows:

$$A_M = \frac{\beta R}{R + r_I}$$

We must first compute R, then A_M

$$R_B = \frac{R_1 R_2}{R_1 + R_2} = \frac{(90)(10)}{90 + 10} = 9 \text{ k}\Omega$$

$$R = \frac{R_O R_B}{R_O + R_B} = \frac{1(9)}{1 + 9} = 0.9 \text{ k}\Omega$$

$$A_{IM} = \frac{\beta R}{R + r_I} = \frac{50(0.9)}{0.9 + 1} = 23.7$$

To find f_H we must first find f_β.

$$f_\beta = \frac{f_T}{\beta} = \frac{20 \text{ MHz}}{50} = 400 \text{ kHz} = 0.4 \text{ MHz}$$

We must then find A_V.

$$A_V = \frac{\beta R_L}{r_I} = \frac{50(1)}{1} = 50$$

Finally, we find C_e.

$$f_\beta = \frac{1}{2\pi r_I C_e}; \qquad C_e = \frac{1}{2\pi f_\beta r_I}$$

$$C_e = \frac{0.159}{0.4 \times 10^6 \times 10^3} = 0.397 \times 10^{-9} \simeq 400 \text{ pF}$$

Thus f_H is found as follows:

$$f_H = f_\beta \left(\frac{1 + \dfrac{r_I}{R}}{1 + A_V \dfrac{C_c}{C_e}} \right) = 0.4 \, \frac{1 + \dfrac{1.0}{0.9}}{1 + 50 \dfrac{(7)}{400}} = 0.4 \, \frac{2.11}{1.875}$$

$$f_H = 450 \text{ kHz}$$

This example illustrates that f_H for a common-emitter circuit is usually nowhere near the value of f_T. In this case it is roughly equal to f_β. In general, it can be either larger or smaller than f_β. As a very rough approximation we can consider the f_H of a single stage amplifier to be of the order of f_β. The value of f_H computed in this example is probably higher than would be measured in practice. This is because we have not taken into account the problem of stray capacitance. It is extremely difficult to predict the values of stray capacitance; the only way to ascertain the actual bandwidth is to build the amplifier and make measurements. If the value of f_H computed in this problem is not large enough for a particular application, the designer would have to go to a different transistor; he would choose one with a larger f_T and a smaller C_{OB}. Since f_T depends upon f_β which in turn depends upon C_e, we see again that the junction capacitances must be small for good high frequency response.

Other manufacturers give β (h_{fe}) measured at some specific high frequency. Since β is the short circuit current gain, we can use one of the previously derived equations to compute the value of f_β.

EXAMPLE 24

Repeat example 23 using a different transistor whose specifications are given as follows:

$$\beta = 50 \qquad\qquad \beta = 4.0 \ @ \ f = 100 \text{ MHz}$$
$$r_I = 1 \text{ k}\Omega \qquad C_{OB} = 5 \text{ pF}$$

SOLUTION

The mid-band gain is independent of the high-frequency behavior of the transistor; thus A_M is still equal to 23.7 because β of the new transistor is also 50. We must find f_β from the given information. We can use the following expression:

$$A = \frac{\beta}{\sqrt{1 + \left(\dfrac{f}{f_\beta}\right)^2}}$$

In this expression A represents the short circuit current gain at some frequency, in this case, 100 MHz; β represents the short circuit current gain at mid-frequency, 50. Substitution into this equation will yield f_β.

$$\sqrt{1 + \left(\frac{f}{f_\beta}\right)^2} = \frac{\beta}{A}$$

$$\frac{f}{f_\beta} = \sqrt{\left(\frac{\beta}{A}\right)^2 - 1} \simeq \frac{\beta}{A} \qquad \text{since } \beta \gg A$$

$$\boxed{f_\beta = f\,\frac{A}{\beta}} = 100\,\frac{(4.0)}{50} = 8 \text{ MHz} \qquad\qquad (8\text{-}13)$$

$$C_e = \frac{1}{2\pi f_\beta r_I} = \frac{0.159}{8 \times 10^6 \times 10^3} = 0.0199 \times 10^{-9}$$

$$C_e \simeq 20 \text{ pF}$$

$$f_H = f_\beta \left(\frac{1 + \dfrac{r_I}{R}}{1 + A_V\dfrac{C_c}{C_e}}\right) = 8\,\frac{1 + 1.11}{1 + 50\,\dfrac{5}{20}} = 8\,\frac{2.11}{13.5}$$

$$f_H = 1.25 \text{ MHz}$$

The bandwidth (upper corner frequency) is larger for the amplifier of example 24 (1.25 MHz) than for the one in example 23 (0.45 MHz). The only change which was made was replacement of the transistor; this accounted for the change in bandwidth. The transistor used in example 24 had much smaller junction capacitances (20 and 5 pF) than the one used in example 23 (400 and 7 pF). Again, the stray capacitances were not taken into account, so the value of 1.25 MHz is overoptimistic. Examples 23 and 24 were solved to bring out two major ideas. The first is that the transistor itself is the major cause of poor high-frequency response. The second is that manufacturers state the specifications in different ways; the reader should be able to interpret the information given on the data sheet.

8-7 FREQUENCY RESPONSE OF MULTI-STAGE AMPLIFIERS

Up to this point in the chapter, we have been considering only single-stage amplifiers. We know, of course, that amplifiers are usually cascaded to achieve greater gain. In order to extend the theory of this chapter to multi-stage amplifiers, we would have to become involved with considerably more complex mathematics. For example, it was mentioned in the previous section that f_H depends upon the voltage gain of the stage, A_V. But we know from earlier chapters that the voltage gain depends strongly upon the input impedance of the following stage; i.e., we must consider the loading effect of the next stage. However, the input impedance of this stage is also a function of frequency since there is a junction capacitance in shunt with the input. The calculations and algebra become involved and thus will not be considered in this text. We develop a number of approximate relationships which should suffice for a practical design situation.

Figure 8.17 shows a decibel plot of the frequency response of a few amplifiers.

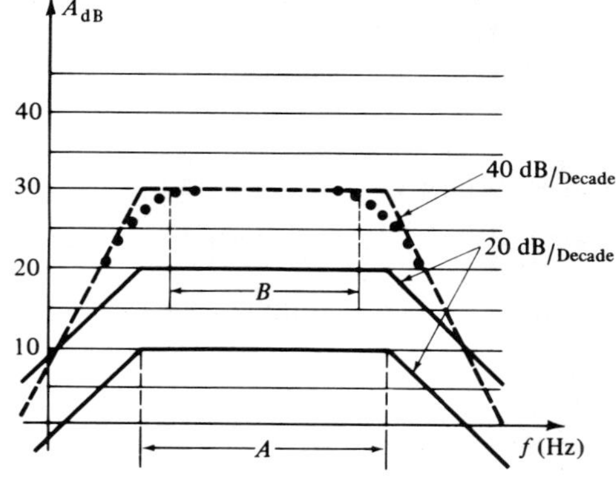

Figure 8.17

The solid lines are the plots of two single-stage amplifiers, one having a decibe gain (mid-band) of 10 and the other, 20. Both of these amplifiers have the sam

f_L and f_H, and thus have the same bandwidth, labeled A in the figure. Notice that the asymptotes of both show the gain decreasing at 20 dB per decade. We know from earlier chapters that the over-all gain is equal to the product of the gains of the individual stages. The advantage of working with decibel plots is that we can *add* the plots of the individual amplifier stages to find the decibel plot of the over-all amplifier (in this case, a two-stage amplifier). If the two solid curves are added, the result is the *dashed* curve shown; this curve shows the asymptotes of the curve of the over-all amplifier. Notice that the mid-band decibel gain is 30 dB (10 plus 20); however, if the addition is carried out very carefully, it can be seen that the asymptotes of the over-all response curve fall at the rate of *40 dB per decade*. This is because over any decade, the gain of each individual amplifier falls 20 dB, so the over-all gain falls 40 dB.

Because of the rapidly falling asymptotes, the shape of the *actual* curve (not the asymptotes) will be different from curves where the asymptotes fall at the rate of 20 dB per decade. The *dotted* curve of Fig. 8.17 shows the actual response curve of the over-all gain. If this is plotted carefully it can be seen that f_L and f_H for the over-all amplifier are not the same as those values for the individual amplifiers. The f_H will *decrease* and f_L will *increase* as shown in the figure; the actual over-all bandwidth decreases, and is labeled as B in the figure. This is true in general: the *over-all bandwidth of cascaded stages is usually less than the bandwidth of the individual stages.*

This concept can be illustrated in a different way. Figure 8.18 shows two cas-

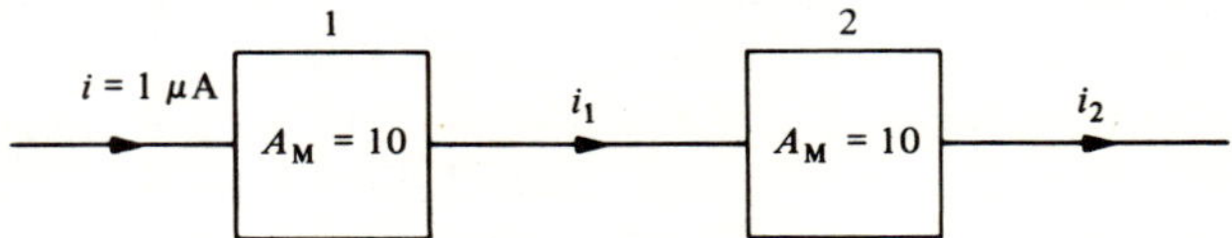

Figure 8.18

caded amplifiers each with a mid-band gain of 10.0. Let us assume that f_H for each individual amplifier is 100 kHz. If the input current, i, is 1.0 μA, the output of stage 1 (input to stage 2) is 10 μA (i_1) and the output of stage 2 is 100 μA (i_2) at mid-band.

$$i_1 = A_M i = 10(1\ \mu\text{A}) = 10\ \mu\text{A}$$

$$i_2 = A_M i_1 = 10(10\ \mu\text{A}) = 100\ \mu\text{A} \qquad \textit{At mid-band}$$

Thus the over-all gain of the cascaded amplifier at mid-band is 100.

$$\frac{i_2}{i} = \frac{100\ \mu\text{A}}{1\ \mu\text{A}} = 100$$

We know that the gain drops to 70.7% of its mid-band value at f_H. Thus at 100 kHz, i_1 is 7.07 μA (gain of stage 1 is 7.07) and i_2 is 50 μA (gain of stage 2 is also 7.07).

$$i_1 = Ai = [0.707(10)](1\ \mu A) = 7.07\ \mu A$$

$$i_2 = Ai_1 = (7.07)(7.07\ \mu A) = 50\ \mu A \qquad \textit{At 100 kHz}$$

Thus the over-all gain at 100 kHz is 50.

$$\frac{i_2}{i} = \frac{50\ \mu A}{1\ \mu A} = 50$$

This means that the over-all gain dropped to *50%* of its mid-band value at f_H; the f_H of the cascaded amplifier must be *smaller* than 100 kHz. In other words, we have again shown that f_H of the over-all amplifier is less than f_H of the individual stages.

It can be shown by algebraic manipulation that the following equations are true:

$$f_H(n) = f_H\sqrt{2^{1/n} - 1}$$

$$f_L(n) = \frac{f_L}{\sqrt{2^{1/n} - 1}}$$

In these equations n stands for the number of stages connected in cascade, $f(n)$ is the corner frequency of the cascaded amplifier, and f is the corner frequency of a single stage. These equations show that f_H is decreased, and f_L is increased for a cascaded amplifier.

EXAMPLE 25

Apply the above equations to the problem just discussed. Find the f_H of the cascaded amplifier if f_H of each individual stage is 100 kHz.

SOLUTION

Applying the above equation for f_H, we get

$$f_H(2) = f_H\sqrt{2^{1/2} - 1} = 100\sqrt{1.414 - 1} = 100\sqrt{0.414}$$

$$f_H(2) = 64\ \text{kHz}$$

This proves that the bandwidth deteriorates when amplifier stages are cascaded. It should be mentioned that the above equations hold only if the amplifier stages are identical; if they are not, the equations become quite complex. However, these equations give a good approximation even if the stages are not identical. If the designer is interested in knowing the exact bandwidth of a cascaded amplifier with dissimilar stages, it is frequently easier to build the circuit and measure the bandwidth directly; we discuss this further in chapter 10.

It should be mentioned in passing that the incorporation of negative ac feedback results in an *improvement* in the bandwidth of an amplifier. Since the equations become quite involved, we shall not undertake an extensive analysis at this point. It can be shown, however, that the bandwidth of the feedback amplifier is *increased* by approximately the same factor that the gain is *decreased*.

PROBLEMS

1.   An amplifier has $f_L = 30$ Hz and $f_H = 50$ kHz.
Compute the bandwidth.

2.   Repeat problem 1 if $f_L = 200$ kHz and $f_H = 500$ kHz.

3.   Compute the values of decibel gain for the amplifiers whose gains are:
    (a)   80
    (b)   120
    (c)   1
    (d)   0.7

4.   Compute the values of gain for the amplifiers whose decibel gains are:
    (a)   31 dB
    (b)   70 dB
    (c)   0 dB
    (d) -8 dB

5.   Sketch the asymptotic plot for an amplifier whose mid-band gain is 37, whose $f_L = 30$ Hz, and whose $f_H = 80$ kHz.

6.   Repeat problem 5 if the gain is 130, $f_L = 70$ Hz, and $f_H = 200$ kHz.

7.   The circuit of Fig. 8.7 has the following values:

$$\beta = 100 \qquad r_I = 1 \text{ k}\Omega$$
$$R_1 = 80 \text{ k}\Omega \qquad C_C = 5 \ \mu\text{F}$$
$$R_2 = 20 \text{ k}\Omega \qquad R_E = 500 \ \Omega$$
$$R_O = 5 \text{ k}\Omega \qquad C_E = 25 \ \mu\text{F}$$

Compute the values of:
    (a) A_M
    (b) f_{L_1}
    (c) f_{L_2}
    (d) kf_{L_2}
    (e)   Sketch the asymptotic plot.
    (f)   What is the lowest frequency at which the amplifier is useful?

8.   Repeat problem 7, except make $C_C = 0.5 \ \mu\text{F}$.
What effect does this have on the value of f_{L_1}?

9.   Repeat problem 7, except make $C_E = 10 \ \mu\text{F}$.
What effect does this have on the values of f_{L_2} and kf_{L_2}?

10.   The circuit of Fig. 8.7 has the following values:

$$\beta = 70 \qquad r_I = 1 \text{ k}\Omega$$
$$R_1 = 100 \text{ k}\Omega \qquad R_E = 1 \text{ k}\Omega$$
$$R_2 = 15 \text{ k}\Omega$$
$$R_O = 3 \text{ k}\Omega$$

It is desired that the amplifier be useful to frequencies as low as 20 Hz. Select the values of C_C and C_E.

11.   Repeat problem 10 for 50 Hz.

12.   A transistor amplifier has the following values associated with it:

$$R_1 = 100 \text{ k}\Omega \qquad \beta = 80$$
$$R_2 = 20 \text{ k}\Omega \qquad r_I = 1 \text{ k}\Omega$$
$$R_O = 2 \text{ k}\Omega \qquad f_T = 100 \text{ MHz}$$
$$R_L = 5 \text{ k}\Omega \qquad C_{OB} = 10 \text{ pF}$$

Compute the values of:
(a) A_M
(b) f_H
(c)   Is the value of f_H computed in part b the same value which would be measured in the laboratory? Explain.

13.   Repeat problem 12, except make $f_T = 50$ MHz.

14.   Two identical stages, each having $f_L = 50$ Hz and $f_H = 50$ kHz, are connected in cascade.
Compute the values of f_L and f_H for the cascaded amplifier.

15.   Repeat problem 14 if three stages are used.

9

POWER AMPLIFIERS

It has already been shown that there is a progressive increase in the ac signal (voltage and current) as we move from stage-to-stage in the cascaded amplifier chain. The signal has its maximum value in the *final* amplifier stage. This final stage, which is known as the *power stage*, delivers a relatively large amount of ac power to the load. The problems which are peculiar to this particular amplifier stage form the basis of study for this chapter.

There are a number of problems concerned with obtaining large amounts of ac power from the power stages. In particular, we shall see that there is a definite limit to the amount of ac power which can be obtained from a given stage. In addition, it is frequently necessary to incorporate a transformer to match the transistor to the load.

Since the signals (voltages and currents) are quite large in the power stage, the major problem is one of amplitude distortion; i.e., there is a tendency for the output waveforms to become *nonsinusoidal* due to the geometry of the input and output characteristics of the transistor. Therefore, a considerable portion of this chapter is devoted to the study of methods used to minimize amplitude distortion, just as the major problem was dc stability in chapter 5, cascading in chapter 6, gain stability in chapter 7, and frequency response in chapter 8.

Finally, we shall see that it is necessary to connect the circuit in a different fashion in order to minimize amplitude distortion. This new configuration, known as the *push–pull amplifier*, contributes to a definite reduction in the amount of amplitude distortion.

9-1 MAXIMUM OUTPUT POWER

Since it is generally desired that we supply a large amount of ac power to the load, it is logical at this point to determine just how much ac power *is available* from a particular transistor amplifier stage. We know that the ac output power is proportional to the product of the ac output voltage (v_{CE}) and the ac output current (i_C); in particular, this output power is given by the following expression:

$$P_O = \frac{v_{CEP\text{-}P}\, i_{CP\text{-}P}}{8}$$

It can be seen from this equation that, in order to obtain a large amount of power, we must provide both a large voltage *and* a large current. Figure 9.1(a) shows a

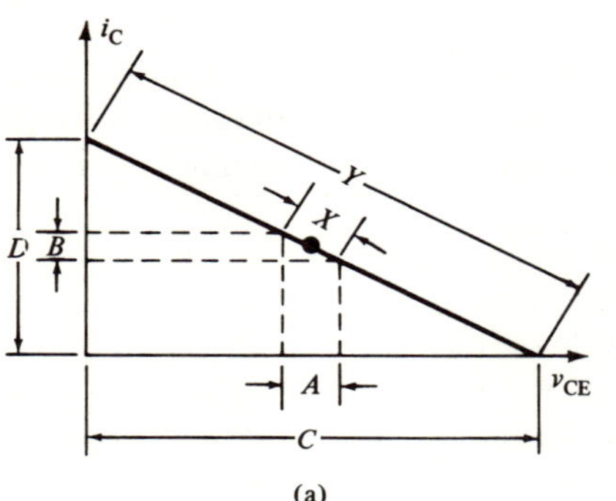

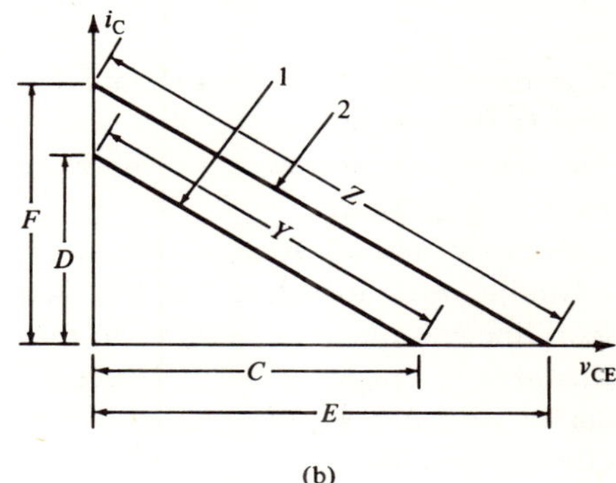

(a)   (b)

Figure 9.1

load line drawn on a set of characteristic axes. If the input signal swing (peak-to-peak ac base current) is relatively *small* (indicated by X in Fig. 9.1(a)), the peak-to-peak output voltage (A) and the peak-to-peak output current (B) will also be relatively small. Consequently, the output power will be small; in this particular case it can be expressed as follows:

$$P_O = \frac{(A)(B)}{8}$$

Suppose that the input signal swing is increased to a *much larger* value (indicated by Y in Fig. 9.1(a)). In this case, the output voltage will increase (from A to C) and the output current will increase (from B to D); the output power will also increase to

$$P_O = \frac{(C)(D)}{8}$$

Obviously, we obtain more output power with a *large* input signal swing. In the situation pictured in Fig. 9.1(a), the maximum output power will be obtained when the input signal swing is that indicated by Y; with the load line shown in Fig. 9.1(a), the maximum undistorted peak-to-peak collector-to-emitter voltage is equal to C, and the maximum undistorted peak-to-peak collector current is equal to D. Of course, in order to obtain this maximum output power, we must provide an input signal swing equal to Y.

The next question to ask is: Is it possible to obtain a value of output power *greater* than $CD/8$? It was mentioned above that the maximum output power was equal to $CD/8$ *using the load line shown in Fig. 9.1(a)*. Suppose we move the load line some distance to the right, as shown in Fig. 9.1(b). With the *original* load line (load line 1), the maximum output power was equal to $CD/8$, when the input signal swing was equal to Y. With the *new* load line (load line 2), it can be seen that the maximum output power is

$$P_{O_M} = \frac{(E)(F)}{8}$$

The maximum output power available $(EF/8)$ using load line 2 is *greater* than that $(CD/8)$ which can be obtained using load line 1; of course, in order to obtain this larger output power, we must provide a larger input signal swing (indicated by Z in Fig. 9.1(b)).

Suppose that we wish to obtain a value of output power *even greater* than that available using load line 2 in Fig. 9.1(b). What is to prevent us from moving the load line *even farther to the right* (to the right of load line 2 in Fig. 9.1(b)) to obtain this greater output power? Recall from chapter 3 that each transistor has associated with it a set of maximum ratings; in particular, these are the maximum allowable collector current (I_{C_M}), the maximum allowable collector-to-emitter voltage (V_{CE_M}), and the maximum allowable collector power dissipation (P_{C_M}). If one or more of these ratings are exceeded, there is a possibility that the transistor will fail. Therefore, the *allowable region of operation* on the collector characteristics is indicated by the shaded area in Fig. 9.2. This area is bound at the top by I_{C_M}, at

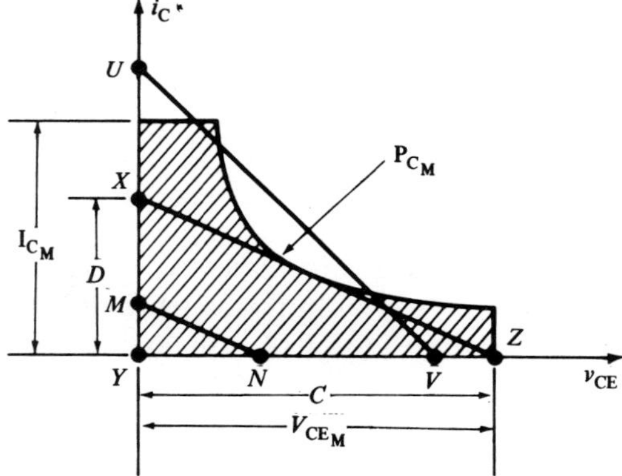

Figure 9.2

the right by V_{CE_M}, and is also bound by the maximum power dissipation hyperbola. Any load line which is used with the transistor must fit *entirely within* the shaded region; this is true of load line X–Z in Fig. 9.2. In other words, a load line used with the transistor must contain *no points which fall outside the shaded region*. For example, load line U–V *cannot* be used with the transistor whose allowable region of operation is shown in Fig. 9.2 because *it contains points which are not within the shaded region.* Therefore, there is a *limit* to how far we can move the load line to the right because its position is limited by the maximum ratings of the transistor. Consequently, there is a definite limit to the maximum output power which can be obtained from an amplifier using a particular transistor.

It has already been mentioned that the output power depends upon both the output voltage *and* the output current. Let us consider load line X–Z in Fig. 9.2. Using this load line, the maximum output power is equal to $CD/8$; i.e., the maximum power is *proportional to the product of C and D*. Notice in Fig. 9.2 that load line X–Z, together with portions of the vertical and horizontal axes, forms a *right triangle*, triangle X–Y–Z. Recall from plane geometry that the area of a triangle is proportional to the product of the base and the altitude. Thus the area of triangle X–Y–Z is equal to $CD/2$; i.e., the area of the right triangle is *proportional to the product of the lengths of the two legs, C and D*. Since the maximum output power ($CD/8$) and the area of the triangle ($CD/2$) are *both* proportional to the product of C and D, we can say that *the maximum output power is directly proportional to the area of the right triangle formed by the intersection of the load line with the vertical and horizontal axes*. Thus, there is a definite procedure to be followed when attempting to determine the maximum output power available from an amplifier using a particular transistor: We draw a series of load lines, ensuring that they all fall entirely within the shaded region shown in Fig. 9.2; the load line which results in the maximum area of the right triangle is the one which will yield the maximum output power. Obviously load line X–Z will yield a larger maximum output power than load line M–N, since triangle X–Y–Z has a larger area than triangle M–Y–N (see Fig. 9.2).

Figure 9.3 shows the characteristic curves of a particular power transistor. It can

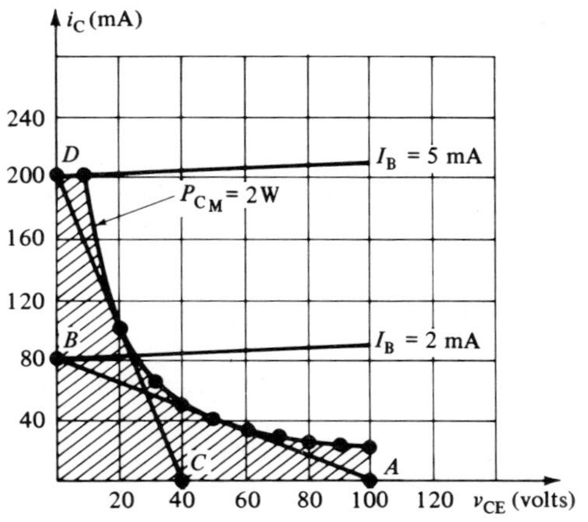

Figure 9.3

be seen from the boundaries of the shaded region that $I_{C_M} = 200$ mA, $V_{CE_M} = 100$ V, and $P_{C_M} = 2$ W. We shall now draw some load lines and attempt to determine which of these lines will result in the maximum output power for this particular transistor. Load line A–B is drawn by fixing one end of the line at the lower right-hand corner of the shaded region (point A), and then drawing the line so that it is *tangent* to the maximum power dissipation hyperbola; if the line is drawn in this manner, it will intersect the vertical axis at point B ($i_C = 80$ mA). You will recall from chapter 1 that the slope of the load line is equal to the reciprocal of the load resistance; thus, load line A–B corresponds to a load resistance of

$$R_{\mathrm{L}} = \frac{\Delta v_{\mathrm{CE}}}{\Delta i_{\mathrm{C}}} = \frac{100\ \mathrm{V}}{80\ \mathrm{mA}} = 1.25\ \mathrm{k\Omega}$$

The maximum output power available from an amplifier using this transistor with a load resistance of 1.25 kΩ is computed as follows:

$$P_{\mathrm{O_M}} = \frac{(100\ \mathrm{V})(80\ \mathrm{mA})}{8} = 1000\ \mathrm{mW} = 1\ \mathrm{W}$$

Load line $C\!-\!D$ is drawn by fixing one end at the upper left-hand corner of the shaded region (point D), and then drawing the line tangent to the hyperbola; the load line intersects the horizontal axis at point C ($v_{\mathrm{CE}} = 40$ V). Load line $C\!-\!D$ corresponds to a load resistance of

$$R_{\mathrm{L}} = \frac{\Delta v_{\mathrm{CE}}}{\Delta i_{\mathrm{C}}} = \frac{40\ \mathrm{V}}{200\ \mathrm{mA}} = 200\ \Omega$$

The maximum output power available from an amplifier using this transistor with a load resistance of 200 Ω is

$$P_{\mathrm{O_M}} = \frac{(40\ \mathrm{V})(200\ \mathrm{mA})}{8} = 1.0\ \mathrm{W}$$

Notice that load lines $A\!-\!B$ and $C\!-\!D$ *both* provide a maximum output power of 1.0 W. It can be shown that there are many load lines which can provide the same maximum output power of 1.0 W.

The next question to ask is: since both load lines $A\!-\!B$ and $C\!-\!D$ can provide the *same maximum output power of 1.0 W*, which load line is preferred? The answer is that we prefer to use load line $A\!-\!B$ for *two* important reasons: First, it is important to bear in mind that we have been discussing methods for obtaining the *maximum* output power from a particular power amplifier stage; this maximum output power will be obtained *only if the input signal swing* (ac base current) *is sufficiently large.* If the input signal swing is relatively *small* (see Fig. 9.1(a)), the output power will be *much less than the maximum.* With this in mind, refer again to Fig. 9.3. In order to obtain maximum output power (1.0 W) with load line $A\!-\!B$, we must provide an input signal swing of 2 mA peak-to-peak; to obtain the *same output power* with load line $C\!-\!D$, it is necessary to supply a 5 mA peak-to-peak signal to the base of the transistor in the power stage. In other words, we require a *larger* ac input (or "drive") from the *previous* amplifier stage (the stage which feeds the power stage, sometimes referred to as the "driver") to obtain maximum output power with load line $C\!-\!D$ (5 mA peak-to-peak) than we do to obtain the *same output power* using load line $A\!-\!B$ (2 mA peak-to-peak); this means that the over-all current gain of the pre-amplifier (all of the amplifier stages *minus* the power stage) *need not be as large* when load line $A\!-\!B$ is used as when load line $C\!-\!D$ is used. The above explanation illustrates one reason why we prefer to use load line $A\!-\!B$ (the one with the largest resistance, or smallest slope). The second reason is concerned with the problem of amplitude distortion, and is explained shortly.

Summarizing, we can say that the maximum output power available from a particular amplifier stage depends upon both the maximum ratings of the transistor, and the placement of the load line. For each transistor, there is maximum output power available, and a corresponding optimum value of load resistance. We turn next to the study of a practical power amplifier.

9-2   THE SINGLE-ENDED CLASS A POWER AMPLIFIER

Let us now consider a practical problem concerning a power amplifier stage. Suppose that we desire to supply 1.0 W to a 12.5 Ω load. We know that the particular transistor studied in the previous section is capable of providing a maximum output power of 1.0 W, so we shall utilize this transistor in our circuit. The first problem which presents itself concerns the nature of the load. Suppose that this 12.5 Ω load is connected directly *in series with the collector* of the transistor; i.e., the load resistance (R_L) for this amplifier stage is equal to 12.5 Ω. Since the slope of the load line is *inversely* proportional to the value of the load resistance, the load line which corresponds to this relatively *small* value of load resistance will be quite *steep*, as shown in Fig. 9.4(a). If the load line is steep, it is possible to obtain

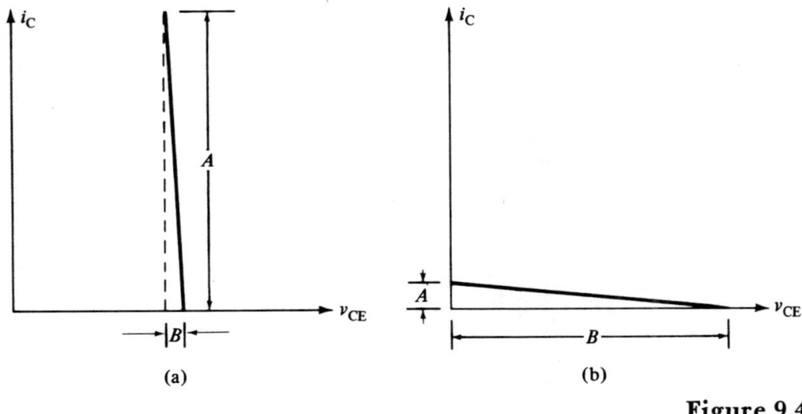

Figure 9.4

a *large* peak-to-peak output current (indicated by A in the figure), but only a *small* peak-to-peak output voltage (indicated by B). Recall that in order to obtain a large output power, we must have both a large current *and* a large voltage. Similarly, if the load resistance were very large, the slope of the load line would be very small, as shown in Fig. 9.4(b). In this case, we could achieve a large value of output voltage (B), but only a small value of output current (A).

We know from the results of the discussion in the previous section that the optimum value of load resistance for this particular transistor is 1.25 kΩ; i.e., we saw that, with this load resistance, we can obtain the maximum output power (1.0 W) using the minimum amount of input drive (2 mA peak-to-peak). Thus we are faced with the following problem: We know that the transistor would like to "see" a 1.25 kΩ load resistance, but the value of the actual load resistance is only 12.5 Ω. Here is an ideal application for the matching transformer. If a transformer is inserted in the amplifier circuit between the transistor and the load, we can select the value of turns ratio which will make the 12.5 Ω load *appear as a 1.25 kΩ resistance to the transistor*. Recall from chapter 7 that we can express the reflected resistance, R_P (the resistance "as seen from the primary"), as follows:

$$R_P = \left(\frac{N_P}{N_S}\right)^2 R_S$$

In this equation, N_P/N_S is the value of the turns ratio of the transformer, and R_S is the value of the actual load resistance (connected across the secondary terminals). Solving the equation for the turns ratio, we get

$$\frac{N_P}{N_S} = \sqrt{\frac{R_P}{R_S}}$$

If we substitute into the equation, setting $R_P = 1.25$ kΩ and $R_S = 12.5$ Ω, we find the required value of turns ratio to be

$$\frac{N_P}{N_S} = \sqrt{\frac{R_P}{R_S}} = \sqrt{\frac{1250}{12.5}} = 10{:}1$$

This means that if we connect a 10:1 step-down transformer between the transistor and the load (primary connected to the transistor and secondary to the load), the 12.5 Ω load will appear to the transistor as a 1.25 kΩ resistance.

The actual amplifier circuit is shown in Fig. 9.5; the reason for the name

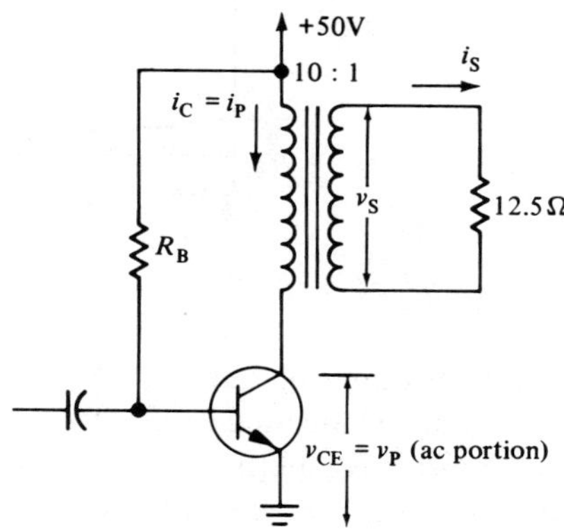

Figure 9.5

"single-ended class A" power amplifier is explained later in the chapter. Note that the primary of the transformer is connected directly between the collector of the transistor and the positive terminal of the 50 V dc supply (we shall soon see the reason for the value of 50 V). The transformer secondary is connected to the 12.5 Ω load. We have used fixed bias for simplicity; however, the reader should bear in mind that, because of the *large signals* present in the power amplifier stage, it is *particularly important* that this stage have a very stable Q point (a low value of stability factor).

Let us now analyze the operation of this circuit. We consider *first the dc behavior* of the amplifier. The matching transformer reflects a resistance equal to 1.25 kΩ into the collector circuit of the transistor *only under ac conditions,* since the transformer is an ac device. *As far as dc conditions* are concerned (assuming for the moment that no ac is applied to the base of the transistor through the coupling capacitor shown in Fig. 9.5), the collector of the transistor "sees" *only the dc resistance of the primary coil of the transformer;* i.e., *the dc load resistance* is equal to the dc resistance of the coil. Since the dc resistance of a coil is relatively small, the dc load line will be quite steep. If we assume that the dc resistance of the coil is approximately equal to zero ohms, the slope of the dc load line will be infinite; i.e., the dc load line will appear vertical, as shown in Fig. 9.6(a). Since we have assumed that the dc resistance of the primary coil is zero ohms, there will be a *zero volt dc drop across the primary coil.* Therefore, if we designate the dc voltage drop across the primary coil as V_P, we can compute the value of the quiescent collector-to-emitter voltage, V_{CEQ}, as follows:

$$V_{CEQ} = V_{CC} - V_P = 50 - 0 = 50 \text{ V}$$

In other words, if the dc coil resistance is approximately equal to zero ohms, V_{CEQ} *is approximately equal to* V_{CC} (50 V, in this case), as shown in Fig. 9.6(a).

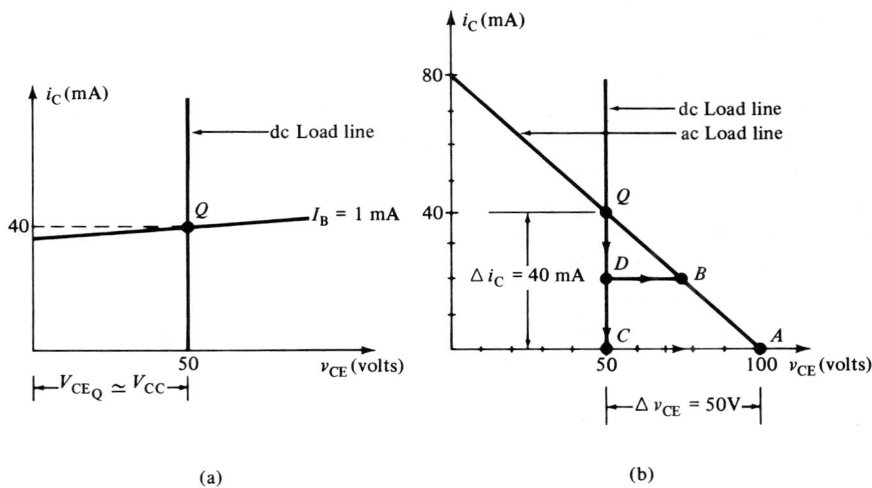

Figure 9.6

The next problem is the determination of the Q point. We know that the maximum allowable power dissipation for this transistor, P_{CM}, is 2 W; we also know that $V_{CEQ} = 50$ V. Therefore, I_{CQ} cannot exceed

$$I_{CQ} \leq \frac{P_{CM}}{V_{CEQ}} = \frac{2 \text{ W}}{50 \text{ V}} = 40 \text{ mA}$$

If we make $I_{C_Q} = 40$ mA, we shall need to supply a quiescent base current, I_{B_Q}, of 1 mA, as shown in Fig. 9.6(a). This means that the value required for the bias resistor, R_B, is (neglecting the base-emitter voltage drop)

$$R_B = \frac{V_{CC}}{I_{B_Q}} = \frac{50 \text{ V}}{1 \text{ mA}} = 50 \text{ k}\Omega$$

Therefore, if we select $R_B = 50$ kΩ, the Q point will be located as shown in Fig. 9.6(a). Of course, this means that the transistor is dissipating 2 W at the Q point (50 V multiplied by 40 mA equals 2 W). In practice, the Q point is selected to be somewhat *lower* than that shown in Fig. 9.6(a), in order that there be some safety factor in the design. We can now appreciate another reason why it is important that this power stage have a stable Q point. It requires only a small change in temperature to shift the Q point into the region of excessive power dissipation (greater than 2 W), resulting in a possible transistor failure.

We turn next to the *study of the ac operation* of this amplifier. We know that under ac conditions (ac current entering the stage via the coupling capacitor shown in Fig. 9.5), the transistor sees a resistance equal to the reflected resistance, R_P; we have already selected the transformer turns ratio such that *this ac load resistance* is equal to 1.25 kΩ. This means that the slope of the *ac load line* will be equal to the reciprocal of *the ac load resistance*, just as the slope of *the dc load line* was equal to the reciprocal of *the dc load resistance*. In other words, for dc currents, the transistor sees a resistance equal to the dc resistance of the primary coil (approximately zero ohms), while for ac currents it sees a resistance equal to R_P (1.25 kΩ, in this case). We have already plotted the dc load line in Fig. 9.6(a). Let us now draw the ac load line.

In order to draw any straight line, we need *two* points, or *one* point plus the *slope* of the line. We know from earlier chapters that all load lines must pass through the Q point; thus, we already have one point. We also know that the slope of this ac load line is equal to the reciprocal of the ac load resistance; thus, we have enough data to plot the ac load line. We begin at the Q point, as shown in Fig. 9.6(b). Since the slope of the ac load line, m, is equal to the reciprocal of R_P, we can write the following expression:

$$m = \frac{1}{R_P} = \frac{\Delta i_C}{\Delta v_{CE}}$$

Solving for Δv_{CE}, we get

$$\Delta v_{CE} = \Delta i_C (R_P)$$

Since we know the position of one point through which the ac load line passes (the Q point), and we also know the slope of this line ($1/R_P$), our task now becomes one of drawing a line through the Q point with a slope equal to $1/R_P$; this line is the ac load line. We begin by selecting *any desired change* in i_C, Δi_C; this choice is entirely up to the one who is analyzing the problem. Let us select a change in i_C of 40 mA. Since the Q point is located at $I_C = 40$ mA, a *decrease* (change) of 40 mA will bring us down to *zero* mA; i.e., a change of 40 mA will move us from point Q ($I_C = 40$ mA) to point C ($I_C = 0$ mA), as shown by the direction of the arrow in Fig. 9.6(a). Using the equation for the slope, let us compute the value of Δv_{CE} which corresponds to a Δi_C of 40 mA:

$$\Delta v_{CE} = \Delta i_C(R_P) = (40 \text{ mA})(1.25 \text{ k}\Omega) = 50 \text{ V}$$

This says that, with an ac load resistance (R_P) of 1.25 kΩ, a 50 V change in v_{CE} corresponds to a 40 mA change in i_C. The 40 mA change in i_C moved us from point Q to point C (from 40 mA to 0 mA); the 50 V change in v_{CE} moves us from point C to point A (from 50 V to 100 V), as indicated by the direction of the arrow in Fig. 9.6(b). Point A is the *second* point through which the ac load line passes; if point A is connected to point Q and extended (it intersects the vertical axis at 80 mA), the resulting line is the ac load line, as shown in Fig. 9.6(b).

What we have actually done in Fig. 9.6(b) is to construct a line (the ac load line) through the Q point with a slope equal to $1/R_P$. Refer to triangle Q–C–A in Fig. 9.6(b). The ratio of the *vertical* side ($QC = 40$ mA) to the *horizontal* side ($CA = 50$ V) is equal to the slope of the ac load line; but this ratio is equal to the reciprocal of R_P, as follows:

$$\frac{QC}{CA} = \frac{40 \text{ mA}}{50 \text{ V}} = \frac{1}{1.25 \text{ k}\Omega} = \frac{1}{R_P}$$

In constructing this ac load line, it was not necessary to select a Δi_C of 40 mA. Suppose we had selected $\Delta i_C = 20$ mA. A decrease of 20 mA brings us from 40 mA (point Q) to 20 mA (point D). The corresponding value of Δv_{CE} is

$$\Delta v_{CE} = (\Delta i_C)R_P = 20 \text{ mA}(1.25 \text{ k}\Omega) = 25 \text{ V}$$

Thus we move *horizontally* from point D an amount equal to 25 V, which brings us to point B at $V_{CE} = 75$ V (follow the arrows in Fig. 9.6(b)). Note that point B ($V_{CE} = 75$ V, $I_C = 20$ mA) *also* falls on the ac load line; this indicates that we can select any value we wish for Δi_C. We could even have selected an *increase* in the value of i_C; in this case we would have to move to the *left* after we found the corresponding value of Δv_{CE}.

It can be seen from Fig. 9.6(b) that the Q point is exactly in the center of the ac load line. This explains the reason for selecting $V_{CC} = 50$ V. With the Q point as indicated in Fig. 9.6(b), the collector-to-emitter voltage can swing between zero and 100 V; you will recall that the maximum voltage for this transistor, V_{CEM}, is equal to 100 V. In this way, we can use the transistor efficiently by providing for a maximum undistorted peak-to-peak collector-to-emitter voltage equal to 100 V. Since we can obtain a maximum peak-to-peak collector-to-emitter voltage of 100 V, and a collector current of 80 mA (see Fig 9.6(b)), the maximum output power from the amplifier of Fig. 9.5 is

$$P_{OM} = \frac{(100 \text{ V})(80 \text{ mA})}{8} = 1 \text{ W}$$

This maximum ac power is delivered to the transformer primary. Since the efficiencies of most transformers are in the vicinity of 98%, we can say that approximately *98% of the ac power delivered to the primary is transferred to the secondary;* since the 12.5 Ω load is connected across the secondary terminals, it receives 98% of the 1.0 W, or 0.98 W.

Figure 9.7 shows the waveforms of the various voltages and currents when a 2 mA

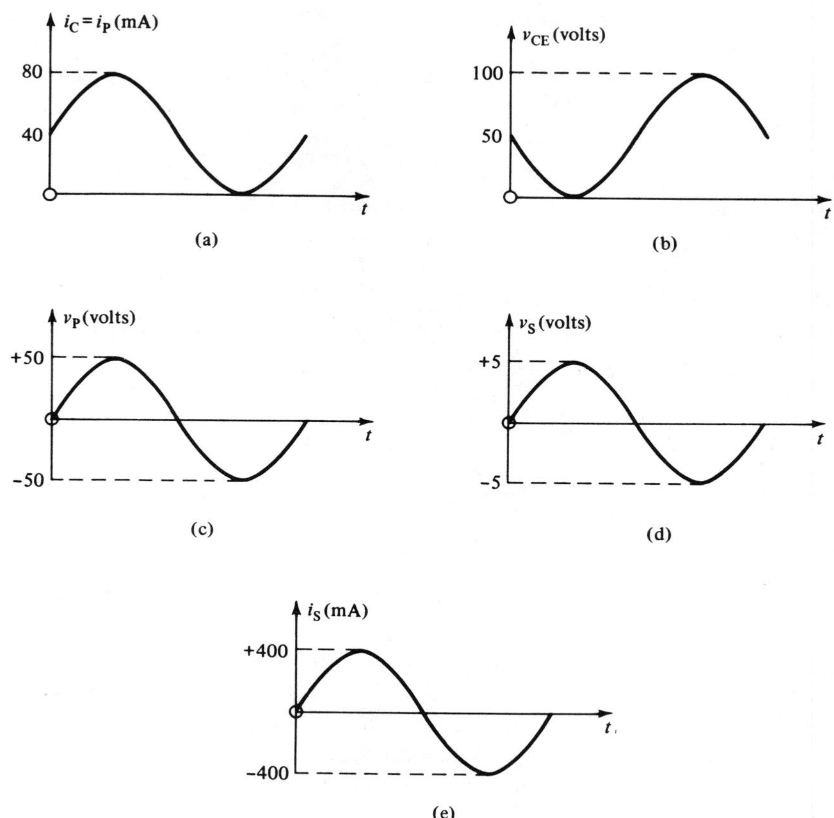

Figure 9.7

peak-to-peak base current is applied to the input of the power stage. In drawing these waveforms, we have assumed that all of the transistor characteristics (both input and output) are perfectly *linear* and *uniformly spaced;* we shall see in the following section that this is *not* generally true, and that because of the nonlinearity and nonuniform spacing, the phenomenon known as amplitude distortion results. Figures 9.7(a) and (b) show the waveforms of i_C and v_{CE}, respectively; the collector current (i_C) is the *same* as the transformer primary current (i_P), as can be seen from an examination of Fig. 9.5. Figure 9.7(c) shows the waveform of v_P, the voltage across the primary coil of the transformer. Note that v_P has the *same ac portion* as v_{CE} (100 V peak-to-peak), but has *a dc level of zero volts.* There is *no dc voltage drop* across the primary coil because we assumed a value of zero ohms for the dc resistance of this coil in our analysis. Figures 9.7(d) and (e) show the waveforms of the secondary voltage (v_S) and the secondary current (i_S), respectively. Note that both v_S and i_S have dc levels of zero volts; i.e., they are *pure sine waves*, not complex. The reason for this behavior, of course, is that *the dc voltages and currents* in the primary cannot be transferred into the secondary. This situation is entirely satisfactory as we usually desire that there be no dc voltages or currents in the load. Note also that, since we are using a 10:1 step-down transformer, the 100 V peak-to-peak primary voltage is stepped-*down* to 10 V peak-to-peak in the secondary (Fig. 9.7(d)), and the 80 mA peak-to-peak primary current is stepped-*up* to 800 mA peak-to-peak in the secondary (Fig. 9.7(e)).

The next step is the study of the power distribution in the collector circuit of the power stage. There are actually *three* distinct quantities which concern us at this point: The first is the power supplied to the collector circuit by the battery, P_{CC}. P_{CC} is equal to the product of the battery voltage, V_{CC}, and the quiescent collector current, I_{C_Q},

$$P_{CC} = V_{CC}I_{C_Q}$$

This is "dc" power. The second quantity is the actual output power, P_O, supplied by the collector circuit to the transformer primary (and then transferred to the secondary). This quantity is expressed as follows:

$$P_O = \frac{v_{CE_{P-P}}i_{C_{P-P}}}{8}$$

This is ac power. The third and final quantity is the power dissipated in the collector of the transistor, P_C.

The distribution of power in the collector circuit can be viewed as follows: The battery supplies power (P_{CC}) to the collector circuit; this power is then divided between output power supplied to the transformer (P_O), and power dissipated in the collector of the transistor (P_C). Thus we can write the following relationship:

$$P_{CC} = P_O + P_C$$

Solving the equation for P_C, we get

$$P_C = P_{CC} - P_O$$

This states that the transistor dissipates an amount of power equal to the difference between the power supplied by the battery to the circuit (P_{CC}) and the power supplied by the circuit to the load (P_O).

Let us compute the values of these quantities for the circuit of Fig. 9.5; it will be necessary for the reader to refer to Figs. 9.5, 9.6, and 9.7. The value of P_{CC} is

$$P_{CC} = V_{CC}I_{C_Q} = (50 \text{ V})(40 \text{ mA}) = 2 \text{ W}$$

We have already computed the value of the *maximum* output power (P_{OM}) to be 1 W. Therefore, the value of the collector dissipation is

$$P_C = P_{CC} - P_O = 2 \text{ W} - 1 \text{ W} = 1 \text{ W}$$

Since the collector of the transistor is dissipating only 1 W, and the maximum allowable collector dissipation for this particular transistor (P_{CM}) is 2 W (see Fig. 9.3), it is evident that we are operating this transistor well within its limits; i.e., we are dissipating only 50% (1 W) of the maximum allowable power (2 W). This would seem to indicate that we need not use a transistor whose power rating is as large as 2 W. In other words, why can we not replace this transistor with one whose P_{CM} is 1 W, and thus take advantage of the resulting savings in both cost and space?

The situation which we have just examined applies to the condition of maximum output power. The battery supplies 2 W to the collector circuit; 1 W is delivered to the load and 1 W is dissipated in the transistor. Let us suppose that the input drive to the base of the power transistor is reduced momentarily to zero; i.e., there is no ac input but the battery is still connected to the circuit. If the input drive is zero (no ac base current), *the ac output power will also be zero;* i.e., $P_O = 0$ W. Under these conditions, the value of the collector dissipation can be computed as follows:

$$P_C = P_{CC} - P_O = 2 \text{ W} - 0 = 2 \text{ W}$$

This indicates that, if the input drive is *reduced* from 2 mA peak-to-peak to zero, the ac output power will *decrease* from 1 W to zero, and the collector dissipation will *increase* from 1 W to 2 W. Thus, if we had used a 1 W transistor ($P_{CM} = 1$ W), *the transistor would have failed* when the input drive was reduced to zero, since the collector dissipation increased from 1 W to 2 W. In other words, we can say that a *decrease* in the base drive causes a corresponding *decrease* in the ac output power, and an *increase* in the collector dissipation.

It can be seen from the preceding discussion that the collector dissipation attains its *maximum* value when the input base drive is *zero*. Under these conditions, $P_C = P_{CC}$; i.e., the collector dissipation is equal to the power supplied by the battery (2 W, in this case). We know that the maximum output power available from the amplifier of Fig. 9.5 is 1 W. Therefore, since the transistor must be able to dissipate at least 2 W under the "worst case conditions" (the worst case occurring when the input drive is zero), we must select a transistor whose maximum dissipation rating is *at least twice* the value of the maximum output power available from the amplifier stage. In other words, if the maximum output power computed for a single-ended class A amplifier stage is 1 W, we must select a transistor whose P_{CM} is at least 2 W.

This discussion also illustrates that, since the collector dissipation attains its *maximum* value when the input drive is zero, it attains its *minimum* value when the input drive is a maximum (2 mA peak-to-peak, in this case); i.e., when the output power is at a *maximum* ($P_O = 1$ W), the collector dissipation is at a *minimum* ($P_C = 1$ W). In other words, the metal surface of the transistor is actually *cooler* (less power dissipation) when the amplifier is delivering its maximum output than when no output is being delivered.

It is the function of the power stage to *convert the dc power* supplied by the battery

(P_{CC}) *to ac power* supplied to the load (P_O). The ratio of *the ac output power* to *the dc battery power* (multiplied by 100) is defined as the *collector circuit efficiency;* this is designated by the Greek letter eta (η):

$$\eta = \frac{P_O}{P_{CC}} \times 100 \ (\%)$$

Computing the value of collector circuit efficiency for the condition of maximum output power, we get

$$\eta = \frac{P_O}{P_{CC}} \times 100 = \frac{1 \ \text{W}}{2 \ \text{W}} \times 100 = 50\%$$

If the output power is *less* than the maximum value of 1 W, P_{CC} will still be equal to 2 W, and thus the efficiency will be *less* than 50%. We conclude, then, that *the maximum value of the collector circuit efficiency* for a single-ended class A power amplifier stage *is 50%.*

This can be illustrated with the aid of Fig. 9.8. If the Q point is selected in the

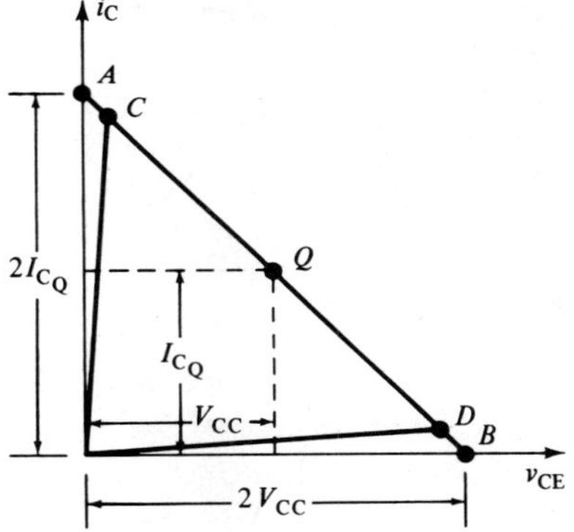

Figure 9.8

center of the ac load line (line A–B) as shown, the maximum undistorted peak-to-peak collector current is equal to $2I_{C_Q}$, and the maximum undistorted peak-to-peak collector-to-emitter voltage is $2V_{CC}$; thus the maximum output power can be computed as

$$P_{O_M} = \frac{(2I_{C_Q})(2V_{CC})}{8} = \frac{I_{C_Q}V_{CC}}{2}$$

The value of P_{CC} is

$$P_{CC} = V_{CC}I_{C_Q}$$

Therefore, the maximum value of the collector circuit efficiency is

$$\eta_M = \frac{P_{O_M}}{P_{CC}} \times 100 = \frac{\dfrac{V_{CC}I_{C_Q}}{2}}{V_{CC}I_{C_Q}} \times 100 = 50\%$$

Strictly speaking, the maximum value of the collector circuit efficiency will be *slightly less* than 50%. It can be seen from Fig. 9.8 that, in computing the maximum value of efficiency, we assumed that the input base drive could move the *entire length* of the load line (from point B to point A). Actually, because of the geometry of the characteristics, we can drive only from point D to point C (see Fig. 9.8), and thus the maximum efficiency will be somewhat less than 50%. In addition, the efficiency of the matching transformer will be less than (although close to) 100%; thus this is another factor tending to reduce the over-all efficiency below 50%.

To summarize some of the concepts developed in this section, let us analyze the situation pictured in Fig. 9.9. Using the same transistor and load, we shall compute

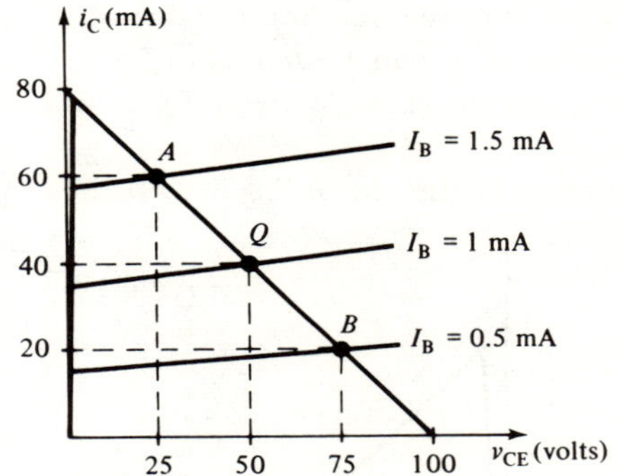

Figure 9.9

the value of the various powers when the input base drive is 1 mA peak-to-peak; i.e., the ac input varies the base current from 1.5 mA (point A) to 0.5 mA (point B). The output power will obviously be less than the maximum because the input base drive is less than the maximum of 2 mA peak-to-peak. The ac output power is

$$P_O = \frac{v_{CEP\text{-}P}\, i_{CP\text{-}P}}{8} = \frac{(75\text{ V} - 25\text{ V})(60\text{ mA}\text{--}20\text{ mA})}{8} = 250\text{ mW} = 0.25\text{ W}$$

The value of P_{CC} *is still equal to 2 W* because neither the battery nor the Q point has been changed. The collector dissipation is

$$P_C = P_{CC} - P_O = 2 - 0.25 = 1.75\text{ W}$$

The transistor is operating within its limits because the maximum dissipation rating is 2 W. The collector circuit efficiency is

$$\eta = \frac{P_O}{P_{CC}} = \frac{0.25}{2} = 0.125\ (12.5\%)$$

Note that the efficiency is less than the theoretical maximum of 50% because the output power is less than the maximum (0.25 W is less than 1 W).

9-3 AMPLITUDE DISTORTION

Up to this point in the text we have assumed that, if the waveforms of i_C and v_{CE} are not limited by either cut-off or saturation, then the ac portions of these waveforms are pure sinusoids. In other words, the only type of distortion discussed thus far has been the flattening of one or both ends of the waveform due to cut-off and/or saturation. This assumption was extended to the previous section, as evidenced by the "clean" sinusoidal waveforms sketched in Fig. 9.7. In order to obtain these sinusoidal waveforms, it is necessary to make *two* assumptions: These are that the input characteristic of the transistor is perfectly linear, and that the collector (output) characteristics are linear and uniformly spaced. If these two conditions are met, a sinusoidal input will result in a sinusoidal output. However, we know from chapters 3 and 4 that the input characteristic is *non*linear and the collector characteristics are *not* uniformly spaced. These two factors are responsible for the undesirable phenomenon known as amplitude distortion. We shall examine the effects of each of these two factors *separately*, taking first the problem resulting from the nonlinear input characteristic.

Figure 9.10 shows the input characteristic of some transistor. Note that the char-

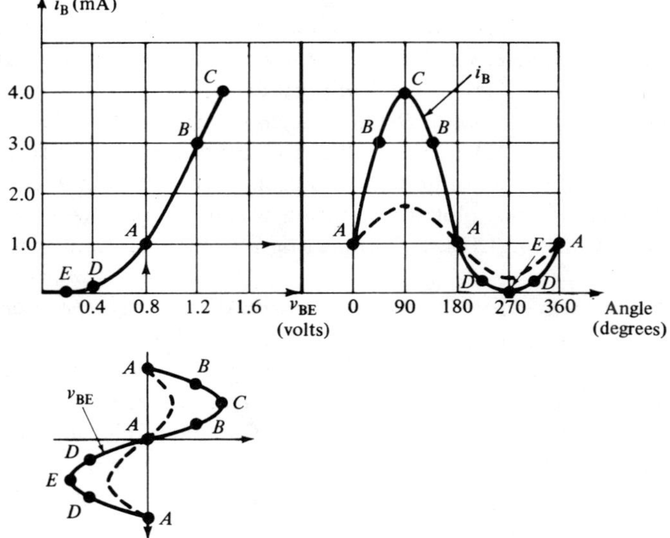

Figure 9.10

acteristic is relatively straight (linear) at values of base current *above* 1 mA (between points A and C), but has a considerable amount of curvature at values of base current *below* 1 mA (between points A and E). Let us assume that the transistor is biased at a quiescent base current of 1 mA (point A in Fig. 9.10), and that an ac input voltage (v_I) of 1.2 V peak-to-peak is applied to the input (base) circuit of the amplifier; for simplicity, we have assumed that the internal resistance of the source which generates this 1.2 V sine wave is zero. If Fig. 9.10 is rotated 90 degrees counter-clockwise, it can be seen that the waveform of v_{BE} (the 1.2 V peak-to-peak *ac* plus a 0.8 V *dc*) is plotted to scale on a set of axes; v_{BE} is the *solid* (as distinguished from the dashed) curve which extends from a maximum value of 1.4 V to a minimum of 0.2 V.

There is a useful graphical technique which we can apply to this problem to determine the waveform of the base current, i_B. Notice that there is a heavy vertical line separating the left-hand portion of Fig. 9.10 from the right-hand portion. The left-hand portion contains a set of perpendicular axes (i_B and v_{BE}) on which is plotted the input characteristic. In the right-hand portion of the figure, the vertical axis again represents the base current (i_B), but the horizontal axis now represents "angle" in degrees. To determine the waveshape of i_B, we proceed as follows: We first obtain the value of v_{BE} when the input sine wave is at its "zero" value; this occurs at angles of 0, 180, and 360 degrees. At all of these angles, v_{BE} is equal to its dc value, 0.8 V. We locate $V_{BE} = 0.8$ V on the v_{BE}=axis (left-hand portion of Fig. 9.10), travel *upward vertically* (follow the arrow) until we reach the input characteristic at point A ($I_B = 1$ mA), then travel *horizontally to the right* (follow the arrow) into the right-hand portion of Fig. 9.10, and plot *three* points (A) in the right-hand portion of Fig. 9.10 at angles of 0, 180, and 360 degrees. Let us repeat this procedure for some other angle, say 45 degrees. Returning to the left-hand portion of Fig. 9.10, we can see that v_{BE} has a value of approximately 1.2 V (point B) at both 45 *and* 135 degrees (the sine wave is symmetrical about the 90 degree axis). We again move vertically up from $V_{BE} = 1.2$ V until we reach the input characteristic at point B ($I_B = 3$ mA); traveling horizontally to the right, we plot *two* points (B) in the right-hand portion of Fig. 9.10 at angles of 45 and 135 degrees. If we continue with this procedure, selecting angle increments of 45 or 30 degrees, we can plot the resulting waveform of i_B (solid curve in the right-hand portion of Fig. 9.10), by connecting all of the plotted points (A, B, C, D, and E).

Notice that this waveform is *severely distorted*. While the *upper* portion (from $I_B = 1$ mA to $I_B = 4$ mA) appears to be the top half of a sine wave, the *lower* portion (from $I_B = 0$ mA to $I_B = 1$ mA) is severely flattened. Since the upper and lower portions do not "match up," the base current waveform is *nonsinusoidal*. The reason for this phenomenon is the *curvature* of the input characteristic which is present at low values of base current. At large values of base current (greater than 1 mA), the slope of the input characteristic is steep; this means that the input resistance of the transistor is small, resulting in relatively large values of base current. At smaller values of base current (less than 1 mA), the slope of the characteristic is smaller; in this region, the input resistance of the transistor is larger, resulting in smaller values of base current. It is this variation in the input resistance of the transistor which is responsible for the unequal upper and lower portions of the base current waveform.

The phenomenon illustrated in Fig. 9.10 is referred to as *amplitude distortion;* i.e., the amplitude (or "height") of the waveform is *distorted*, or "twisted" from its proper shape. You will recall from chapter 4 that the function of an amplifier is not only to increase (or "stretch") the amplitude of the waveform, but also *to retain the same waveshape.* Obviously, the amplifier is not performing properly in the situation pictured in Fig. 9.10, because the base current waveform is distorted. A distorted base current waveform, such as that pictured in Fig. 9.10, will result in both a distorted collector *current* waveform and collector-to-emitter *voltage* waveform. Since the upper portion of the base current waveform is *much larger* than the lower portion (see Fig. 9.10), the input drive will vary the operating point from point A to point C in Fig. 9.11 when the base current is *positive*-going, but only from point A to point E when the base current is *negative*-going; the resulting waveforms of i_C and v_{CE} will also be distorted, as shown in Fig. 9.11.

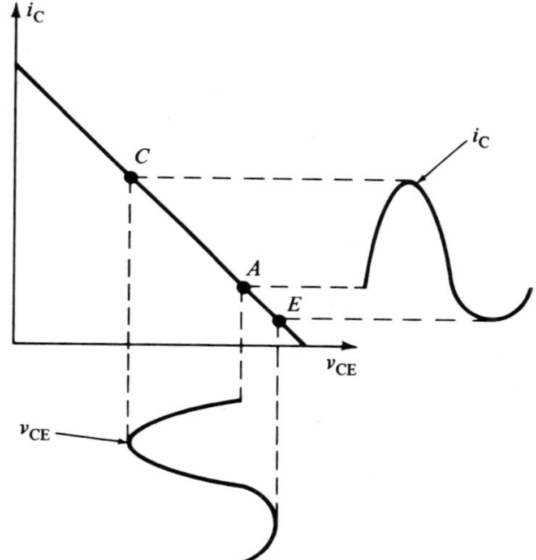

Figure 9.11

This distortion of the amplitude of the waveform is undesirable in all types of electronic systems. In an audio system, the nonlinearity of the input characteristic will result in a "garbled" sound. In a video system, such as a television set, the result will be a distorted picture on the screen; e.g., a person's head may appear much larger than the rest of his body if the distortion is very severe.

In plotting the waveforms of Fig. 9.10, we illustrated an extreme case; i.e., we assumed that the internal resistance of the ac input source was zero. It can be shown that the base current distortion will be reduced considerably if the input to the stage is fed from a high impedance source. In any case, the distortion of the base current waveform caused by the nonlinear input characteristic will still prove to be a problem in any practical electronic system. Of course, we could reduce the amount of distortion by reducing the value of the input drive. The *dashed* input waveform of Fig. 9.10 has an ac value of only 0.4 V peak-to-peak (extends from a maximum value of 1.0 V to a minimum value of 0.6 V). If the previously described graphical procedure is followed, the resulting base current waveform (dashed) appears as shown in Fig. 9.10. Note that there is *much less distortion* in the dashed waveform than in the solid one. The reason for this behavior is that, with the smaller input drive (0.4 V peak-to-peak), we have kept away from the "high curvature" region of the input characteristic. The price which must be paid, however, for this lower distortion is a reduction in the amount of ac output power, since we have reduced the amount of input drive. It appears, at this point, as though we are faced with a choice between large output power, along with a correspondingly large amount of distortion, and smaller output power with a smaller amount of distortion.

Let us turn now to the problem caused by the nonuniform spacing of the collector characteristics. Since we are studying the effects of the nonlinear input characteristic and the nonuniformly spaced collector characteristics *separately*, we shall assume now that the input characteristic is *perfectly linear;* i.e., we shall assume that the ac portion of the base current waveform is a pure sinusoid. The collector characteristics of the transistor discussed in the previous sections are shown in Fig. 9.12. Notice that the curves are *nonuniformly spaced;* there is a

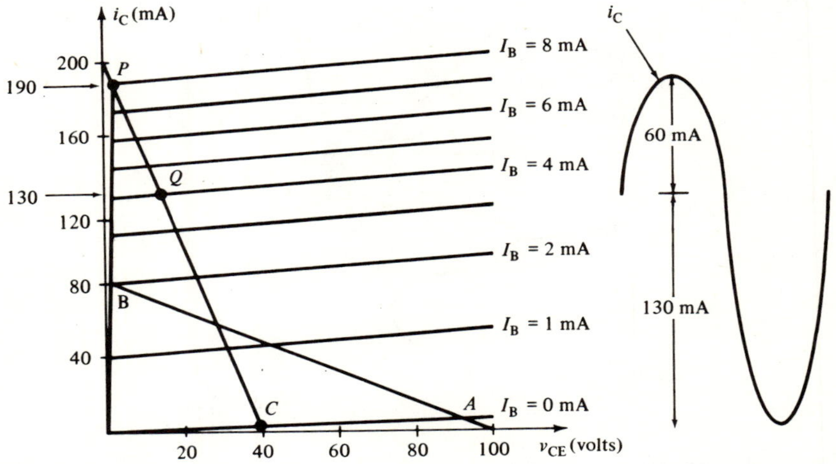

Figure 9.12

smaller spacing (using a base current increment of 1 mA between curves) at the *larger* values of collector current. This indicates that the β of the transistor is largest at values of collector current *below* 100 mA, where the spacing is largest. Suppose for a moment that we use load line C–D, and it is required that we deliver the maximum output power to the load. This would require a peak-to-peak base current drive of 8 mA; we must drive from $I_B = 0$ mA at one end of the load line (point C), to $I_B = 8$ mA at the other end (point D). Assuming that the ac portion of the base current waveform is a pure sinusoid, as previously mentioned, the dc value of the base current must be equal to 4 mA, and thus the Q point will appear as shown in Fig. 9.12. If the 8 mA peak-to-peak signal is now driven into the base of the transistor, the resulting collector current waveform will appear *severely distorted* as shown in Fig. 9.12. The reason for the occurrence of this distortion is explained as follows: When the base current swings from 4 mA to 8 mA, the collector current swings from 130 mA to 190 mA, *for a total swing of 60 mA;*

however when the base current swings from 4 mA to 0 mA, the collector current swings from 130 mA to 0 mA, *for a total swing of 130 mA.* In other words, the nonuniform spacing of the curves results in the upper and lower portions of the collector current waveform being unequal; thus this is a second cause of amplitude distortion.

The reader can now appreciate a *second* reason why we prefer to use a *flatter* ac load line, such as load line $A–B$, rather than load line $C–D$. You will recall from Sec. 9-1 that one reason why load line $A–B$ was preferred to load line $C–D$ was that a smaller drive (2 mA peak-to-peak, in this case) was required to yield maximum output power; with load line $C–D$ we require a drive of 8 mA peak-to-peak. Not only can we achieve maximum output power with a smaller drive using load line $A–B$, but, since the curves are more uniformly spaced in this region, there will be much less distortion. In other words, the use of the flatter load line keeps us away from the region of the closely spaced curves.

Actually, both input (due to the nonlinear input characteristic) and output (due to the nonuniform curve spacing) distortion prove to be troublesome in power amplifier stages. In some instances, the two types of distortion have a cancelling effect upon each other, resulting in a relatively undistorted output waveform. In general, however, both types of distortion cause serious problems in amplifier design.

As mentioned before, the major cause of amplitude distortion is the geometry of the transistor characteristics. Amplitude distortion is particularly troublesome in the power stage because it is in this stage that *the ac voltage and current swings attain their maximum values.* As these quantities swing over a larger and larger region of the characteristics, the amount of distortion increases due to the nonlinearities and the nonuniform spacing of the curves. We can reduce the amount of distortion by reducing the drive; however, this entails a corresponding reduction in the amount of ac output power. So it seems again that we are faced with the following choice: We can obtain a large amount of ac output power, but we must be prepared to accept a large amount of distortion. Or we could obtain a more distortion-free waveform by reducing the drive, but at the expense of ac output power.

9-4   CLASS A PUSH–PULL AMPLIFIER

It has been shown that a definite limit is imposed on the amount of power output that can be derived from a single-ended amplifier stage; i.e., a stage using a single transistor. To obtain a greater power output with acceptable distortion, use is made of the so-called Class A push–pull amplifier shown in Fig. 9.13. Class A operation is characterized by having a transistor biased so that it conducts at all times.

Referring to Fig. 9.13, notice that two transformers are used in this particular circuit; namely, an input transformer and an output transformer. As a result of having the center tapped input transformer, ac signals of equal amplitude, but opposite polarity, are fed to the bases of T_1 and T_2 through the coupling capacitors

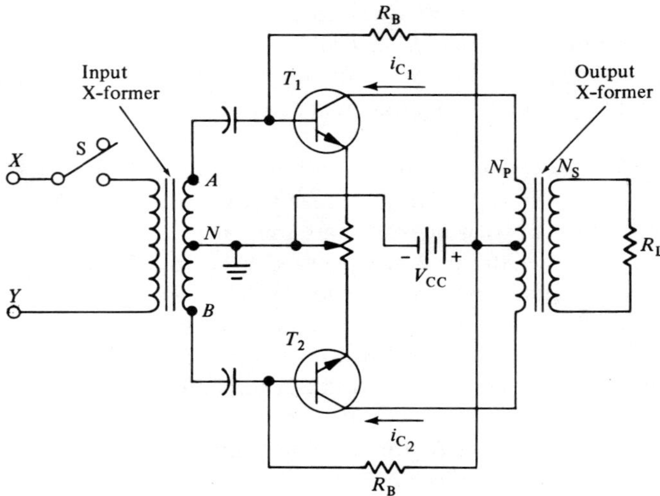

Figure 9.13

which act to keep the dc bias currents from passing through the low-resistance winding of the transformer to ground. When the input signal drives point A positive, the forward bias on T_1 is increased causing increases in base and collector currents. Simultaneously, point B swings negative, driving T_1 toward cutoff so that its base and collector currents decrease. The increase in currents in one transistor accompanied by simultaneous decreases in currents in the other transistor accounts for the name push–pull amplifier.

The arrows of Fig. 9.13 indicate the instantaneous directions of both dc and ac components of collector current. Notice that both of the ac components of collector current, i_{C_1} and i_{C_2}, pass through the primary of the output transformer in opposite directions. The total load current in the secondary is, therefore, proportional to the instantaneous difference in the collector currents of T_1 and T_2.

$$i_{P_{total}} = i_{C_1} - i_{C_2} \tag{9-1}$$

When the input signal changes polarity on the next alternation, T_1 is driven toward cutoff and T_2 toward saturation. Thus, we have alternate conduction, divided between the two transistors.

Referring back to Fig. 9.13, notice that each amplifier stage in this particular example uses fixed bias, supplied through the resistors marked R_B. The collector of each transistor is connected to one end of the primary of the center-tapped output transformer, and the center tap is connected to $+V_{CC}$. The emitters of T_1 and T_2 are connected to opposite ends of a potentiometer, the slider of which is connected to ground. The values of the bias resistors are selected to allow a specific amount of dc base (and consequently, collector) current to flow through the transistors. It is necessary for the proper operation of this circuit that *the two power amplifier stages be identical*. This requires that the two transistors, T_1 and T_2, have *perfectly matched characteristic curves*. While present industrial techniques permit the manufacture of matched pairs, the percentage difference between the corresponding parameters of these matched units is rarely less than 10%. It is for this reason that the potentiometer has been included in the emitter circuits of the two amplifiers. This potentiometer is adjusted until *equal dc currents* flow in the two transistors. When this is done, the amplifier is said to be *balanced*. In other words, the potentiometer is used to compensate for any imbalance due to the inherently different characteristics of the two transistors.

Now, from earlier studies, recall that a dc current tends to saturate the core of a transformer and reduce the primary inductance at signal frequencies. In the push–pull arrangement, the dc components of collector current are 180° out of phase with one another and the dc magnetization effect one might expect is canceled. This assumes, of course, a balanced condition, as noted previously.

The cancellation effect noted above also extends to ripple components resulting from insufficient filtering of the power supply.

Because each stage of the Class A push–pull circuit is conducting at all times, the power dissipation is high, and efficiency low. These are the major disadvantages of Class A operation. For the moment, however, let us turn our attention to an analysis of the harmonics produced by the push–pull configuration.

Because the transfer characteristic of a transistor is nonlinear, harmonic components appear in the output. The relationship between the output current and the input-signal voltage may be expressed by the power series

$$i_C = a_0 + a_1 e_S + a_2 e_S^2 + a_3 e_S^3 + a_4 e_S^4 + \cdots \tag{9-2}$$

This equation may be written, of course, with whatever number of terms is needed to fit a specific curve. The terms a_0, a_1, a_2, etc., are constants that relate voltage to current and have different values for each type of transistor. In the analysis of a push–pull circuit, a balanced condition is assumed so that the power series with its coefficients apply to each transistor. The input signals, which are 180 degrees out of phase with one another, may be expressed as

$$e_{S_1} = E \sin \omega t \tag{9-3}$$

and

$$e_{S_2} = - E \sin \omega t \tag{9-4}$$

If these expressions are substituted in the power series, the two collector currents become

$$i_{C_1} = A_0 + A_1 \sin \omega t + A_2 \sin 2\omega t + A_3 \sin 3\omega t + \cdots \tag{9-5}$$

and

$$i_{C_2} = A_0 - A_1 \sin \omega t + A_2 \sin 2\omega t - A_3 \sin 3\omega t + \cdots \tag{9-6}$$

Subtracting these two equations gives the total primary current as

$$i_{P_{total}} = i_{C_1} - i_{C_2} \tag{9-7}$$

$$= 2A_1 \sin \omega t + 2A_3 \sin 3\omega t + \cdots$$

The dc term, A_0, and the even harmonics, A_2, A_4, etc., are, therefore, eliminated in push–pull operation. Accordingly, there is no dc flux in the primary of the output transformer and saturation cannot occur.

In the equation for $i_{P_{total}}$, the term $2A_1$ is the amplitude of the fundamental signal frequency, and is twice the value for a single-ended stage. Thus, the push–pull stage develops about twice the power of a single-ended stage. Moreover, the ratio of the third harmonic to the fundamental frequency is the same as for a single-ended amplifier. Thus, push–pull operation does not affect the percentage of odd-harmonic distortion.

10

AMPLIFIER DESIGN

This chapter is devoted to the design of a complete transistor amplifier. Because of the limitations imposed by a textbook, we can attempt only a "paper" design; i.e., we can proceed no further than the specification of the circuit configuration and the values of the components on a sheet of paper. The actual construction and trouble-shooting of a prototype of the circuit (known as "breadboarding"), which is the most critical part of any design process, cannot be achieved in a textbook.

There are two major objectives to be achieved in this chapter: First, we shall illustrate the steps involved in the design of a practical, functioning amplifier circuit. This begins with the specifications of the input and output quantities, frequency response, temperature range, and so on, and proceeds through the selection of the circuit configuration, type of transistors, and values of the various resistors and capacitors, to the final construction and testing of the actual prototype circuit. The object here is for the reader to obtain some familiarity with the design process. The second objective is a review of some of the major concepts discussed in chapters 1 through 9. We shall attempt to integrate these concepts by applying them to the design of an amplifier.

In addition, we shall also discuss some related topics, such as the difference between a circuit and a system, and the difference between analysis and synthesis, or design. It is important to emphasize that the paper design is a "ball-park" design; i.e., it just gets us into the ball-park, or the *vicinity* of the final objective since many of the equations used in the design are only approximations. The design is usually finalized after the breadboard has been built; we have more to say about this point later on in the chapter. It is extremely important to emphasize that it is *not* the purpose of this chapter to mold the reader into a circuit designer. Expertise in circuit design comes only after years of experience with different types of circuits under practical laboratory conditions. The object here is to give the reader some idea of the problems involved in electronic circuit design.

10-1   SYSTEMS

An electronic *system* is generally composed of *many circuits* and transducers; i.e., a circuit is one small part of a larger system. Some systems are composed of many different *types* of circuits, such as amplifiers, multivibrators, sweep circuits, logic circuits, differentiators, etc. These various circuits are interconnected in some manner to comprise a system, and the system performs some useful task. For example, an audio system is composed of a multi-stage amplifier and two transducers, a cartridge and a speaker; these components are interconnected to amplify the low sound level coming from the cartridge to provide a much higher level at the speaker. Video systems (television and radar) are composed of amplifiers and different types of pulse circuits; amplifiers and motors are some of the components to be found in control systems. As can be seen from the above discussion, amplifiers are to be found in most types of systems.

A system is generally represented by a *block diagram;* Fig. 10.1 shows the block

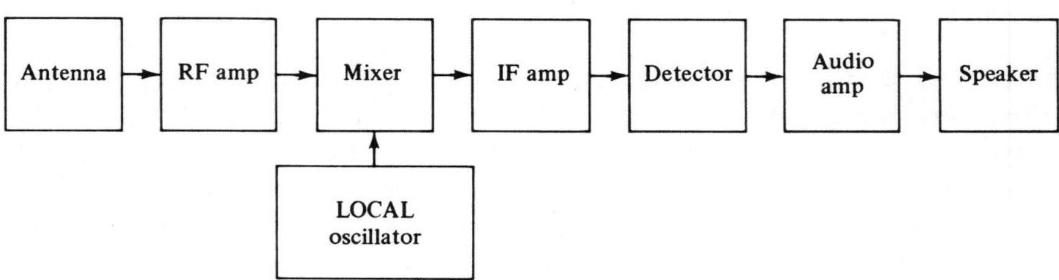

Figure 10.1

diagram of a superheterodyne AM radio receiver. Each block in the system represents a different circuit or transducer; these blocks are interconnected as shown to form the entire system. Notice that this particular system requires three different amplifiers: an rf (radio frequency) amplifier, an if (intermediate frequency) amplifier, and an audio amplifier. In addition, other types of circuits (mixer, local oscillator, and detector) also form a part of the system.

In the early stages of system design, the block diagram is usually constructed by an experienced engineer, one who has had approximately ten years of design experience. He has designed many of the individual circuit blocks earlier in his career, and, thus, he knows the ability of each block to perform its task as a part of the entire system. When a tentative block diagram has been established, the designs of the various circuit blocks are assigned to younger, less experienced engineers or technicians. These engineers are given the input and output requirements of their particular circuit block, along with any other pertinent information. They then set about designing the circuit, bearing in mind that, when finished, the circuit must fit into its proper place in the system.

Up to this point in the text, we have been *analyzing* transistor amplifier circuits. In an analysis, the given information includes the circuit configuration, the values of the different components (resistors, capacitors, and so on) and the transistor parameters (β, r_I, etc.), and the value of any voltage or current applied to the circuit, such as V_{CC}, and v_I. The object then is to compute the values of the various voltages, currents, voltage and current gains, or frequency response. Therefore, *in an analysis, the circuit is already in existence;* the configuration and component values have already been specified. The task involves obtaining information (such as current gain) concerning an existing circuit.

Synthesis, or *design,* is the reverse process. In this case, the given information includes the desired value of gain, frequency response, or other values, of some particular circuit. The task here is to construct the circuit so that it will provide the desired values of gain and frequency response. This involves selection by the designer of the circuit configuration, the number of stages necessary, the type of transistor to be used, and the values of the various resistors and capacitors. In other words, *the designer must create a circuit which did not exist previously;* i.e., he must select the transistors, resistors, and capacitors, and interconnect them such that the resulting circuit meets the given specifications.

In an analysis, the solution to the problem is *unique.* Once the circuit is in existence, there can be *one and only one value of current* flowing through a particular resistor. In design, the solution is *not* unique. It is entirely possible that *more than one* amplifier circuit can meet the given specifications. The task of the designer is to select that circuit which is most economical in terms of cost, space, and reliability. It can now be seen why the circuit designer relies very heavily upon past experience.

10-2   THE STATEMENT OF THE DESIGN PROBLEM

Let us now consider a practical design problem. Suppose that the output circuit of the previous block in the block diagram, which we call the source, can be illustrated as shown in Fig. 10.2(a). Between terminals A and B, the source can

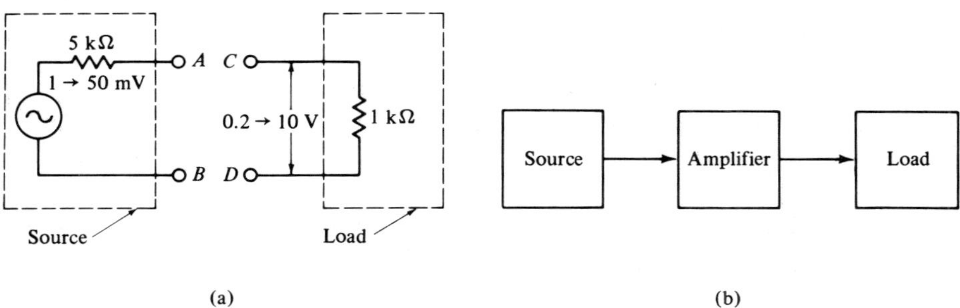

(a)　　　　　　　　　　　　　　　　　(b)

Figure 10.2

be represented by an ac sinusoidal voltage generator whose peak-to-peak value can be anywhere between 1 mV and 50 mV, in series with a 5 kΩ internal resistance. It is desired that a peak-to-peak sinusoidal voltage whose value varies between 0.2 V and 10 V be supplied to a 1 kΩ resistive load (terminals C and D), as shown in the figure; it is further specified that *no dc power* be developed in the load. In other words, when the source generator voltage is 1 mV, we should like the load voltage to be 0.2 V (200 mV), and when the source generator voltage is 50 mV, we should like the load voltage to be 10 V. In order to increase the value of the ac voltage (from 1 mV to 200 mV, and from 50 mV to 10 V), it will be necessary to insert an amplifier between the source and the load, as shown in Fig. 10.2(b). The voltage gain required for this amplifier is

$$A_\text{V} = \frac{\text{output voltage}}{\text{input voltage}} = \frac{v_\text{T}}{v_\text{I}} = \frac{200 \text{ mV}}{1 \text{ mV}} = \frac{10 \text{ V}}{50 \text{ mV}} = 200$$

The above information is necessary to establish the value of the amplifier voltage gain. We now proceed with some further specifications. The frequency of the incoming signal is expected to vary between 100 Hz and 50 kHz. Since the system (of which this amplifier is to be a part) is to be placed inside an orbiting satellite, it is specified that the amplifier be able to operate over an ambient temperature range (temperature of the surrounding environment) of -55 to $+95$ degrees C. Because of limitations of available space inside the satellite, we should like the physical size of the finished amplifier circuit to be as small as possible. The system can tolerate no more than a plus or minus 10% variation in the gain due to temperature change or replacement of transistors. In other words, if the source generator voltage is 50 mV, the load voltage can vary between 10% *above* (11 V) and 10% *below* (9 V) the "nominal" value of 10 V; the nominal value is the average, or center value for the design. Finally, the system is to be powered by a 50 V supply whose negative terminal is to be connected to the system ground.

The above information is given to the circuit designer by the systems engineer. Before proceeding with the actual design, we shall summarize the specifications. An amplifier is to be designed which is to operate according to the following requirements:

(1)   Voltage gain: 200, + or − 10%.
(2)   Frequency response: greater than 100 Hz to 50 kHz.
(3)   Temperature range: −55 to +95 degrees C.
(4)   Physical size: as small as possible.
(5)   Power supply: 50 V dc, negative ground.

10-3 DETERMINATION OF THE CIRCUIT CONFIGURATION

Now that the specifications for the amplifier have been set down, the next step is to make a decision as to the type of circuit configuration to be used. In doing this, we generally work *backward*, from the desired output to the given input quantities. We shall design the amplifier using the *maximum* input (50 mV) and output (10 V) quantities, since this is the most critical condition; i.e., we must make sure that the final amplifier stage can accommodate the maximum peak-to-peak output signal (no distortion due to saturation or cut-off). If the amplifier operates satisfactorily under the condition of maximum input signal, it will perform equally well for smaller inputs (from 1 mV to 50 mV) and outputs (from 200 mV to 10 V). The phenomenon known as "noise" may cause trouble at the lower level inputs, but the discussion of this topic is beyond the scope of the text.

We know that the desired output voltage is 10 V. Therefore, it might be a good idea to use an emitter follower as the final stage (load connected to the follower output terminals) to avoid loading of the amplifier by the 1 kΩ load impedance. Since the voltage gain of the follower is slightly *less* than 1.0, we require that the *input* voltage to the follower be slightly *greater* than 10 V. Let us include a *safety factor* by requiring that the input voltage to the follower be 15 V; this is the same voltage at the collector of the previous amplifier stage. Let us assume for a moment that the load resistor of this previous amplifier stage is 5 kΩ, as shown in Fig. 10.3(a). Notice that the *output* voltage of this stage (15 V_{P-P}) is the *input* voltage to the follower stage. We shall assume that the ac collector current of the stage (i_{C_2}, in Fig. 10.3(a)) is approximately equal to the current in the 5 kΩ load resistor;

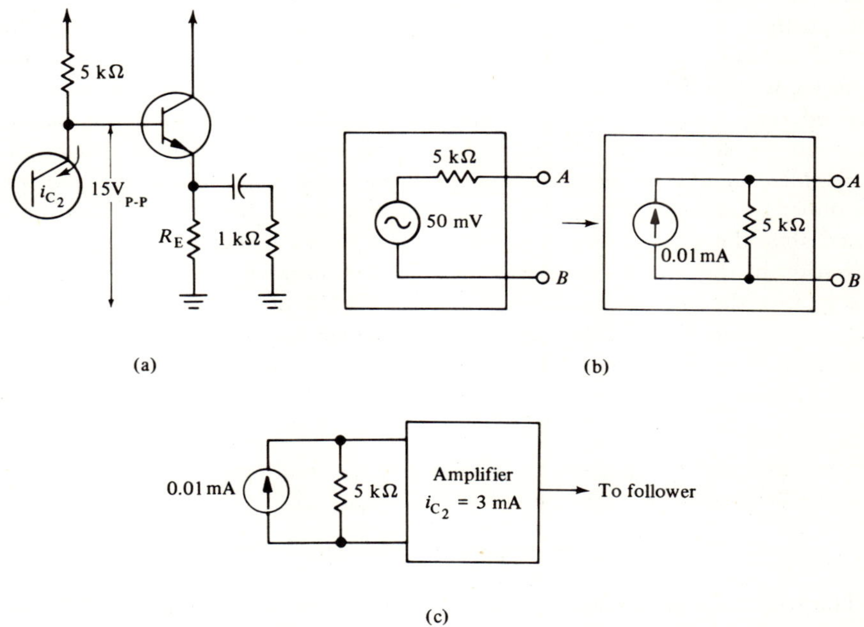

Figure 10.3

we can make this assumption because very little current flows into the base of the follower transistor due to the high input impedance of the follower stage. Therefore, since the ac collector-to-ground voltage (15 V_{P-P}) is equal to the ac voltage across the 5 kΩ load resistor (see chapter 4), we can solve for the desired value of i_{C_2} as follows:

$$i_{C_2} \simeq i_{5\ k\Omega} = \frac{v_{5\ k\Omega}}{5\ k\Omega} = \frac{15\ V}{5\ k\Omega} = 3\ mA$$

For convenience, we shall convert the input voltage source (50 mV in series with 5 kΩ) to a current source (a current generator in parallel with 5 kΩ), as shown in Fig. 10.3(b). The value of the current generator is

$$i = \frac{50 \text{ mV}}{5 \text{ k}\Omega} = \frac{50 \times 10^{-3}}{5 \times 10^{3}} = 10 \text{ } \mu\text{A} = 0.01 \text{ mA}$$

Thus, that portion of the amplifier *exclusive of the follower stage* must have an output current (i_{C_2}) of approximately 3 mA and an input current of 0.01 mA. The current gain of this portion of the amplifier is

$$A_{\mathrm{I}} = \frac{i_{C_2}}{i} = \frac{3 \text{ mA}}{0.01 \text{ mA}} = 300$$

We shall now concentrate primarily on the design of the portion of the amplifier exclusive of the follower, and return to the design of the follower stage later on in the chapter. For convenience, we shall refer to this portion of the *entire* amplifier simply as "the amplifier," even though the entire amplifier is composed of both this portion *and* the follower stage. Thus the circuit can be represented as shown in Fig. 10.3(c); the input source, consisting of a 0.01 mA current generator in parallel with a 5 kΩ resistance, is applied to the input of an amplifier whose desired output current is 3 mA, and the output of this amplifier is connected through the follower to the load. Let us attempt the design of this amplifier.

Recall that one of the specifications limited the variation in the gain of the amplifier to plus or minus 10% of the nominal value. Another specification indicated that the temperature range was -55 to $+95$ degrees C. Because of the variation in the β of the transistors due to temperature change, the variation in the gain of the amplifier will easily exceed 10% if no negative feedback is used. Therefore, our first decision must be to incorporate negative feedback in our amplifier. The closed loop gain of this feedback amplifier is 300:

$$A_{\mathrm{CL}} = 300$$

Since reasonably good stability is required, we shall specify that the open loop gain be *10 times larger* than the closed loop gain, or

$$A_{\mathrm{OL}} = (10)A_{\mathrm{CL}} = (10)(300) = 3000$$

The next question to ask is: can we obtain an open loop gain of 3000 with *one* amplifier stage? This would require *a transistor whose β is greater than 3000;* at the time of this writing, transistors with β's of this size have not been manufactured. Can we obtain an open loop gain of 3000 with *two* stages? Assuming for a moment that the stages have *equal* gains, we can write

$$A_{\mathrm{I}_1} \times A_{\mathrm{I}_2} = A_{\mathrm{OL}}$$

If $A_{\mathrm{I}_1} = A_{\mathrm{I}_2}$:
$$A_{\mathrm{I}}^2 = A_{\mathrm{OL}}$$

The gain of *each* stage would then have to be

$$A_{\mathrm{I}} = \sqrt{A_{\mathrm{OL}}} = \sqrt{3000} = 55$$

Transistors with β's in excess of 55 are readily available from manufacturers these days. Therefore, the amplifier shown in Fig. 10.3(c) is composed of *two* stages; counting the follower stage, this makes a total of *three* stages for the entire amplifier.

We employ *over-all* (as opposed to local) feedback since, as we saw in chapter 7, this is the most efficient scheme. We use as our circuit configuration the two-stage feedback amplifier, with feedback taken from the emitter of the second stage to the base of the first stage, shown in Fig. 7.12. It should be strongly emphasized that it is entirely possible for *some other type* of circuit configuration to fulfill the specifications of the design, perhaps even more efficiently than the one we are using. The ability of the designer to select the most economical amplifier for a particular set of specifications is developed only after years of practical experience.

Recall from chapter 7 that, because of the existence of the feedback loop, it was necessary to leave unbypassed a portion of the emitter resistance of the second stage; the unby-passed portion is designated R_E'. Since this resistance is unby-passed, some local feedback is introduced into the second stage, and, since *no* such local feedback is present in the first stage, *the gain of the first stage will always be greater than that of the second* (assuming the same type transistor is used in both stages); refer to Sec. 7-6 for a review of these concepts. Let us assume, as a rough approximation, that the gain of stage one is *twice* the gain of stage two (under open loop conditions). Then we can write the following:

$$A_{I_1} = 2A_{I_2}$$

We know that the open loop gain is equal to the product of the gains of the individual stages, since the stages are connected in cascade. Thus we can solve for the gains of the individual stages as follows:

$$A_{I_1} \times A_{I_2} = 3000 = A_{OL}$$

$$2A_{I_2} \times A_{I_2} = 2(A_{I_2}{}^2) = A_{OL}$$

$$A_{I_2} = \sqrt{\frac{A_{OL}}{2}} = \sqrt{\frac{3000}{2}} = 38.8 \simeq 40$$

$$A_{I_1} = 2A_{I_2} = 2(40) = 80$$

Now that we have an approximate indication as to the values of the current gains of the individual stages (under open loop conditions), we can attempt to specify the type of transistor required for our design. Since the desired value of gain for the first stage is 80, and since no local feedback is present in this stage, we must use a transistor whose β is somewhat *greater* than 80, say 100. Because the β of a transistor generally has its lowest value at the lower temperatures, we must use a transistor *whose β is at least 100* at -55 degrees C. The next step, then, is to search through the various transistor manuals and specification sheets to find a transistor type which meets the above requirements.

In this section, we have translated the specifications listed in Sec. 10-2 into an actual circuit. We have specified that three stages are required in the design, that the first two stages comprise a feedback amplifier, and that the third stage is an emitter follower. We turn next to the problem of selecting an appropriate transistor to use in the amplifier.

10-4 SELECTION OF THE TRANSISTOR TYPE: THE SPECIFICATION SHEET

Now that we have decided on a circuit configuration, the next step in the design process is the selection of the type of transistor to be used in the amplifier circuit. Generally speaking, there are *five* factors concerning the transistor which must be known by the designer before he can make a decision as to its use; these are

(1) the transistor β
(2) the upper cut-off frequency
(3) the maximum allowable collector-to-emitter voltage
(4) the maximum allowable power dissipation
(5) the storage temperature range.

Let us say that we have searched the various transistor manuals and decided that the Texas Instruments type 2N1574 shows promise in meeting the requirements of the design. The specification sheet is usually in the form of a folder, as shown.* Let us now study the specification sheet in some detail to determine, if, indeed, we can use this transistor in our design.

On page 1, in the upper right-hand corner, the transistor type is listed (TYPE 2N1574). Just below this, we see that the transistor is an NPN DOUBLE-DIF-FUSED SILICON MESA TRANSISTOR; the words "double-diffused" and "mesa" are descriptive of the process whereby this particular transistor type is manufactured. Next, we see written the words, *GENERAL-PURPOSE TRAN-SISTOR*, and beneath this appear four items: The first, *A–C Beta from 80 to 200*, gives us an idea of the β of the transistor; this is the guaranteed β spread of the transistor *at room temperature*. The second item specifies that the transistor has a *High Breakdown Voltage;* we have more to say about this later. Third, *600 MW Free Air Dissipation*, tells us the maximum power dissipation allowable with *no heat sink* (free air). Finally, the fourth item mentions that there is a *Guaranteed Low-Temperature Beta* for this transistor type. These four items contain general information which aids the designer as he skims through the various specification sheets.

Continuing on page 1, there is some information concerning the *environmental tests* to which all of the 2N1574 units are subjected before they pass inspection; since the transistors are often placed inside aircraft and missiles, they must be able to withstand a considerable amount of shock and vibration. In addition, there is some *mechanical data* listed, such as that the transistor is enclosed in a TO-5 package; this gives us information as to its physical size, as indicated by the drawing.

At the bottom of page 1, note the listing entitled, *absolute maximum ratings at 25 degrees C ambient (unless otherwise noted)*. This tells the designer the values of the maximum voltage, current, power, temperature, etc.; thus he knows that he must keep his design within these ratings in order that the transistors enjoy a long life. The first rating states that the maximum *Collector-Emitter Voltage* is 80 V; we shall

---

* The following 3 pages of specification sheet 2N1574 are supplied through the courtesy of Texas Instruments, Inc., Dallas, Texas.

TYPE 2N1574

NPN DOUBLE-DIFFUSED SILICON MESA TRANSISTOR

TYPE 2N1574
BULLETIN NO. DL-S 60404, AUGUST 1960

GENERAL-PURPOSE TRANSISTOR

- A-C Beta From 80 to 200
- High Breakdown Voltage
- 600 MW Free Air Dissipation
- Guaranteed Low-Temperature Beta

environmental tests

Each unit is heat cycled from $-65°C$ to $+175°C$ for ten cycles. A rigorous tumbling test subjects each unit to 12 mechanical shocks of up to 500 G's to ensure mechanical reliability. Each unit is thoroughly tested to determine the electrical characteristics. Production samples are life tested at regularly scheduled periods to ensure maximum reliability under extreme operating conditions.

mechanical data

The transistor is in a JEDEC TO-5 hermetically sealed, welded package with glass-to-metal hermetic seal between case and leads. Approximate weight is 1.0 gram.

THE COLLECTOR IS IN ELECTRICAL CONTACT WITH THE CASE

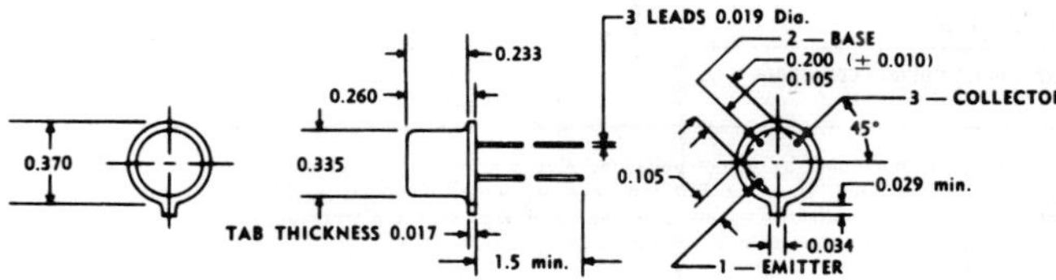

DIMENSIONS ARE MAXIMUM IN INCHES UNLESS OTHERWISE SPECIFIED

absolute maximum ratings at 25°C ambient (unless otherwise noted)

Collector-Emitter Voltage (see note 1)	80 v
Emitter-Base Voltage	5 v
Total Device Dissipation at 25°C Case Temperature (see note 2)	1.2 w
Total Device Dissipation at 25°C Ambient Temperature (see note 3)	0.6 w
Collector Junction Temperature	175°C
Storage Temperature Range	$-65°C$ to $+200°C$

Note 1: The voltage at which h_{FE} approaches one when the emitter-base diode is open circuited. This value can be exceeded in applications where the dc circuit resistance (R_{BE}) between base and emitter is a finite value.

Note 2: Derate linearly to 175°C case temperature at the rate of 8.0 mw/°C.

Note 3: Derate linearly to 175°C ambient temperature at the rate of 4.0 mw/°C.

TYPE 2N1574

NPN DOUBLE-DIFFUSED SILICON MESA TRANSISTOR

electrical characteristics at 25°C ambient temperature (unless otherwise noted)

SYMBOL	PARAMETER	TEST CONDITIONS	MIN	TYP	MAX	UNIT
I_{CBO}	Collector Reverse Current	$V_{CB} = 40$ v $I_E = 0$			1	μa
I_{CBO}	Collector Reverse Current	$V_{CB} = 40$ v $I_E = 0$ $T_A = 150°C$			100	μa
I_{EBO}	Emitter Reverse Current	$V_{EB} = 5$ v $I_C = 0$			10	μa
BV_{CBO}	Collector-Base Breakdown Voltage	$I_C = 10 \mu a$ $I_E = 0$	125			v
$BV_{CEO}{}^*$	Collector-Emitter Breakdown Voltage	$I_C = 10$ ma $I_E = 0$	80			v
$V_{CE(sat)}{}^*$	Collector-Emitter Saturation Voltage	$I_B = 2$ ma $I_C = 10$ ma			1	v
h_{fe}	A-C Common-Emitter Forward-Current Transfer Ratio	$V_{CE} = 5$ v $I_E = -1$ ma $f = 1$ kc	60			
h_{fe}	A-C Common-Emitter Forward Current Transfer Ratio	$V_{CE} = 5$ v $I_E = -5$ ma $f = 1$ kc	80		200	
h_{fe}	A-C Common-Emitter Forward Current Transfer Ratio	$V_{CE} = 5$ v $I_E = -5$ ma $T_A = -55°C$ $f = 1$ kc	40			
h_{ie}	A-C Common-Emitter Input Impedance	$V_{CE} = 5$ v $I_E = -5$ ma $f = 1$ kc		1000	1800	ohm
h_{oe}	A-C Common-Emitter Output Admittance	$V_{CE} = 5$ v $I_E = -5$ ma $f = 1$ kc		95		μmho
h_{re}	A-C Common-Emitter Reverse Voltage Transfer Ratio	$V_{CE} = 5$ v $I_E = -5$ ma $f = 1$ kc		1.3×10^{-4}		
$\|h_{fe}\|$	A-C Common-Emitter Forward Current Transfer Ratio	$V_{CE} = 5$ v $I_E = -5$ ma $f = 30$ mc	2	5		
C_{ob}	Common-Base Output Capacitance	$V_{CB} = 5$ v $I_E = 0$ $f = 1$ mc		5	10	pf

*Semiautomatic testing is facilitated by using pulse techniques to measure these parameters. A 300-microsecond pulse (approximately 2% duty cycle) is utilized. Thus, the unit can be tested under maximum current conditions without a significant increase in junction temperature. The parameter values obtained in this manner are particularly pertinent for switching-circuit design and, in general, indicate the true capabilities of the device.

TYPICAL CHARACTERISTICS

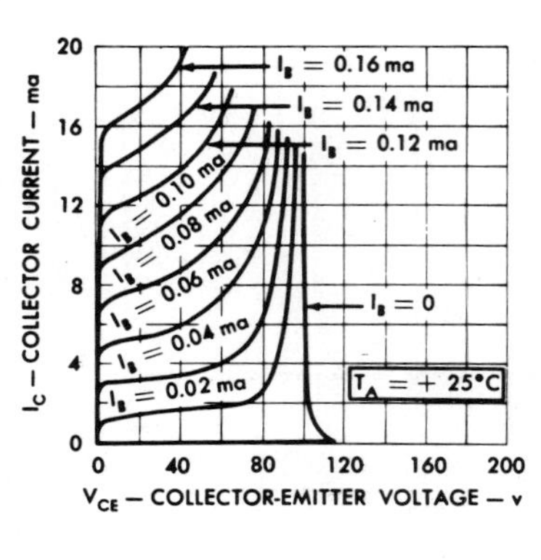

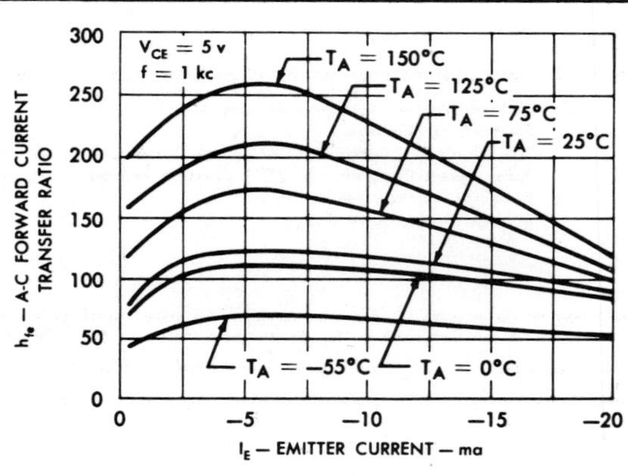

TYPE 2N1574
NPN DOUBLE-DIFFUSED SILICON MESA TRANSISTOR

TYPICAL CHARACTERISTICS

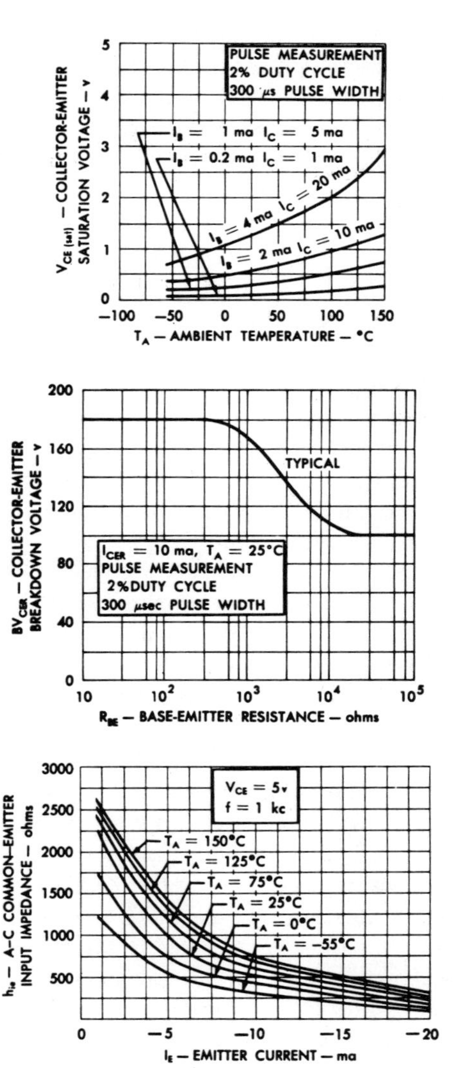

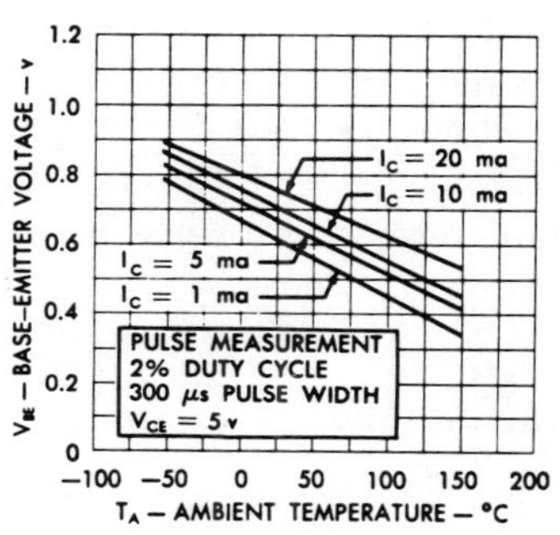

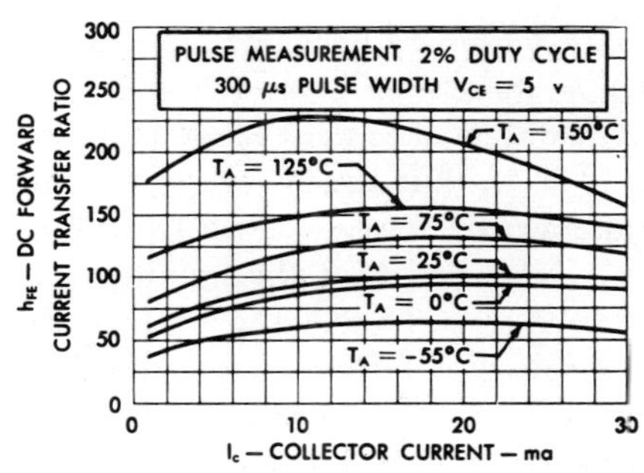

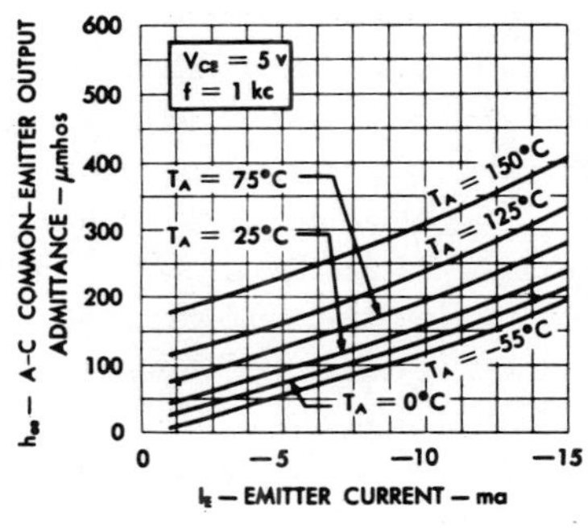

have more to say about this later. The second rating says that the maximum
Emitter-Base Voltage is 5 V. Since this is the maximum rating for a *reverse* biased
emitter-base junction, we are not concerned with this particular rating in our
design; we are designing a class A amplifier (emitter-base junction always forward-
biased). Skipping now to the fourth item, we note that the *Total Device Dissipation
at 25 degrees C Ambient Temperature* is 0.6 W; this is the maximum allowable dissipation at *room* temperature. Note 3, at the very bottom of page 1, states that the
maximum allowable power dissipation must be derated linearly to 175 degrees C
ambient temperature at the rate of 4.0 mW/degree C. In our particular design,
the maximum temperature anticipated is $+95°C$; therefore, the ambient temperature will change from 25°C to 95°C, for a total *change* of 70°C. Since we must
derate at the rate of 4.0 mW/°C for the 2N1574 transistor, the *change* in the
maximum allowable dissipation is

$$\Delta P_{\text{CM}} = (4.0 \text{ mW/°C}) \times (70°\text{C}) = 280 \text{ mW}$$

Thus the maximum allowable power dissipation at $+95°C$ is

$$P_{\text{CM}}(95°\text{C}) = 600 \text{ mW} - 280 \text{ mW} = 320 \text{ mW}$$

This means that the 2N1574 transistor cannot dissipate more than 320 mW at
$+95°C$ without a heat sink. The sixth item in the list states that the *Storage Temperature Range* is $-65°C$ to $+200°C$.

Turning now to page 2, we note that there is a plot of h_{fe} (β) vs. I_E (or I_C) for a
typical 2N1574, in the lower right-hand corner; these curves are plotted for ambient
temperatures between $-55°C$ and $+150°C$. (T_A specifies "ambient" temperature.) Note first that the maximum β for the 2N1574 is obtained at a collector
current between 5 mA and 7.5 mA; this indicates that we should select the Q point
for our design somewhere within this range. Second, the typical β at room temperature ($T_A = 25°C$) and $I_C = 5$ mA is approximately equal to 125. In the lower
left-hand corner is shown a typical set of collector characteristics for the 2N1574 at
room temperature.

The upper portion of page 2 is occupied by a chart entitled, *electrical characteristics at 25°C ambient temperature* (*unless otherwise noted*). This chart contains information concerning leakage currents, breakdown voltages, transistor parameters, and
high frequency performance. In the eighth row of the chart, it is mentioned that
the β (h_{fe} = the ac common-emitter forward current transfer ratio) of the 2N1574
varies between 80 and 200 at $I_C = 5$ mA and room temperature. In the ninth row, it
is specified that the *minimum β* at a temperature of $-55°C$ is 40; this is what is
meant by a "guaranteed low-temperature beta" (see page 1 of the specification
sheet). Recall from Sec. 10-3 that our design requires a transistor whose minimum β at $-55°C$ is 100; therefore, *the 2N1574 does not meet our specifications since its
minimum β is only 40.*

At this point we can do one of *three* things: First, we can continue to search the
specification sheets for a transistor type which has a minimum β of 100 at $-55°C$.
Let us assume, however, that production techniques have not yet progressed to

the point where the manufacturer can guarantee a low-temperature β of 100 for *each one* of a large number of mass-produced units. The second alternative involves *increasing* our original estimate of the number of stages required in the feedback amplifier from two to three. In this case, the gain required for each individual stage within the amplifier would be correspondingly *smaller*, with the result that the amplifier would probably operate satisfactorily if the β dropped to 40 at $-55°C$. It was mentioned, however, in Sec. 10-2 that our amplifier is to be placed inside an orbiting satellite, where space is at a premium; therefore, the designer should think twice before allocating space for an additional amplifier stage. Finally, we could specify that the manufacturer deliver *only those 2N1574 units whose β is greater than 100 at $-55°C$*. Of course, this would increase the price of each individual unit, because the manufacturer must take the time and effort to select the particular units required. Let us assume that this last alternative proves to be the most economical in terms of both cost and space. Therefore, when ordering the transistors from the manufacturer, we shall specify that the minimum β at $-55°C$ be equal to 100.

Now that we have selected a transistor type on the basis of its β (the first of the five factors listed at the beginning of this section), we must check to see if this particular type meets the other requirements, namely, those of frequency response, maximum voltage, maximum power, and temperature. You will recall from chapter 8 that the upper cut-off frequency, f_H, is partially dependent upon the transistor itself; it also depends somewhat on the values of the various circuit components and the distributed wiring capacitances. On row 13, page 2 of the specification sheet, it is mentioned that the *magnitude* of h_{fe} ($|h_{fe}|$) is equal to 5 at a frequency of 30 MHz. Recall also from chapter 8 that the *magnitude of the short circuit current gain*, $|A|$, can be expressed as follows:

$$|A| = \frac{\beta}{\sqrt{1 + \left(\dfrac{f}{f_\beta}\right)^2}}$$

In this expression, $|A|$ represents the magnitude of the short circuit current gain at some frequency, f; $|A|$ corresponds directly to $|h_{fe}|$. The equation states that $|A|$ is equal to β (h_{fe}) at the mid-band frequencies, where f is *much smaller* than f_β; however, $|A|$ decreases as f increases, dropping to a value of 5 at $f = 30$ MHz. Solving the above equation for f_β, we get

$$f_\beta = f_{h_{fe}} = \frac{f}{\sqrt{\left(\dfrac{\beta}{A}\right)^2 - 1}} \simeq \frac{f(|A|)}{\beta}$$

In this equation, f is the frequency (30 MHz) at which the magnitude of the short circuit current gain ($|A|$) drops to 5; we use the typical β of the transistor (125). Substituting into the equation, we find the value of f_β to be

$$f_\beta = \frac{f(|A|)}{\beta} = \frac{30 \text{ MHz}(5)}{125} = 1.2 \text{ MHz}$$

The value of f_β depends *only* upon the transistor. You will recall from chapter 8 that f_β can be expressed as follows:

$$f_\beta = \frac{1}{2\pi r_{\mathrm{I}} C_e}$$

Assuming that the r_{I} of the transistor is approximately equal to 1 kΩ, we can solve for the value of the emitter junction capacitance, C_e:

$$C_e = \frac{1}{2\pi r_{\mathrm{I}} f_\beta} = \frac{0.159}{10^3 \times 1.2 \times 10^6} = 132.5 \text{ pF}$$

On row 14, page 2 of the specification sheet, it is stated that C_{OB} has a typical value of 5 pF; recall from chapter 8 that this is an alternate method of expressing the collector junction capacitance, C_c. It was shown in chapter 8 that the upper cut-off frequency, f_{H}, for a single-stage amplifier can be expressed as follows:

$$f_{\mathrm{H}} = f_\beta \left[\frac{1 + \dfrac{r_{\mathrm{I}}}{R}}{1 + A_{\mathrm{V}} \dfrac{C_c}{C_e}} \right]$$

We should like to estimate the value of the upper cut-off frequency obtainable in an amplifier using a 2N1574 transistor; in this way we can determine if this particular transistor type meets the frequency response specification set forth in Sec. 10-2, namely, that f_{H} be *greater* than 50 kHz. In the above expression, we already know the values of f_β, C_c, and C_e; however, we do *not* know the values of r_{I}, R, and A_{V} because the amplifier has not yet been designed. Therefore, we shall have to make some rough assumptions; let us assume that r_{I} is approximately equal to 1 kΩ, $R = 5$ kΩ, and $A_{\mathrm{V}} = 300$. Substituting into the equation for f_{H}, we get

$$f_{\mathrm{H}} = f_\beta \left[\frac{1 + \dfrac{r_{\mathrm{I}}}{R}}{1 + A_{\mathrm{V}} \dfrac{C_c}{C_e}} \right] = 1.2 \left[\frac{1 + \dfrac{1}{5}}{1 + 300 \left(\dfrac{5}{132.5} \right)} \right] = 117 \text{ kHz}$$

This equation predicts that the upper cut-off frequency for a *single*-stage amplifier using a 2N1574 transistor is somewhere in the neighborhood of 117 kHz. The upper cut-off frequency of a multi-stage amplifier will be *smaller* than that of a single-stage amplifier; this seems to indicate that the f_{H} for our *three*-stage amplifier will be somewhat *less* than 117 kHz. However, because we are incorporating negative feedback into our amplifier to stabilize the gain, we shall also achieve, as a by-product, a considerable *improvement* (increase) *in the bandwidth of the amplifier.* Since the equations used for the computation of the upper cut-off frequency of a single-stage amplifier yield only very rough results (we assumed values for r_{I}, R, and A_{V}), and since it is very difficult to estimate the value of the total stray wiring capacitance, we shall proceed no further with the analysis of the high-frequency performance at this point. We have to depend upon the designer's experience; let us assume that the designer has found in the past that the incorporation of negative feedback (with $A_0\beta_\mathrm{F} = 10$) in an amplifier has contributed to a *ten-fold* increase in the upper cut-off frequency. We can be reasonably confident, then, that the upper cut-off frequency of our three-stage amplifier will be considerably greater than 50 kHz; i.e., we conclude that the 2N1574 transistor can be used in our design. We see later in this chapter that the upper cut-off frequency of the entire three-stage amplifier can be checked after the amplifier design has been completed. We have thus disposed of the second of the five factors (listed at the beginning of this section) concerning the suitability of the transistor for use in this particular design.

We now check to see if the breakdown voltage rating of the 2N1574 is large enough to permit its use in the design. On page 3 of the specification sheet, in the first column, second row, is a graph of BV_CER-*COLLECTOR-EMITTER BREAK-DOWN VOLTAGE* vs. R_BE-*BASE-EMITTER RESISTANCE*. Since we are using a 50 V power supply in our design, the voltage across the collector and emitter terminals will never exceed 50 V (we are using a resistive load), and thus the 2N1574 is suitable for our design as far as breakdown voltage is concerned.

We established that the maximum allowable collector dissipation for this transistor is 320 mW without a heat sink; this occurs at the maximum temperature of $+95$ degrees C. This means that the transistor cannot dissipate more than 320 mW at $+95$ degrees C. Although the 2N1574 can dissipate 600 mW at room temperature, we must design for the *worst possible case* (this is known as "worst case design"); the worst case occurs at the high temperature where the transistor can dissipate only 320 mW. It should be mentioned that the specification given concerning power dissipation applies to *total device dissipation* (see page 1 of the specification sheet); however, since the base circuit of the amplifier dissipates much *less* power than the collector circuit, we can assume that the collector dissipation is approximately equal to the total device dissipation (see chapter 3). If we select our quiescent collector current at 5 mA (region of maximum β), and if we establish the Q point in the center of the load line ($V_\mathrm{CEQ} = 25$ V), the collector dissipation can be computed as follows:

$$P_\mathrm{C} = I_\mathrm{CQ} \times V_\mathrm{CEQ} = (5 \text{ mA})(25 \text{ V}) = 125 \text{ mW}$$

Since the computed value of 125 mW is much less than the maximum of 320 mW, we conclude that the 2N1574 can be used in the design.

Finally, we come to the fifth of the five factors listed at the beginning of this section, namely, the storage temperature range. The 2N1574 is a silicon transistor, and as such, it is ideally suited to high-temperature operation. Since the maximum storage temperature is $+200°\mathrm{C}$ (see page 1 of the specification sheet), and the maximum temperature specified for the design is $+95°\mathrm{C}$, the 2N1574 should operate satisfactorily in the design.

We conclude, then, that the 2N1574 transistor shows reasonable promise of satisfactory performance in our design. This conclusion is based on a study of the β, f_β, BV_CER, P_CM, and storage temperature range of the transistor. If, in the previous discussions, we found that the 2N1574 did *not* meet one or more of the requirements (e.g., if the breakdown voltage were 20 V), we would have had to search for a different transistor type. Now that we have decided to use the 2N1574, we turn next to the actual design of the first amplifier stage.

10-5   THE DESIGN OF STAGE 1

The first step in the design of an amplifier stage is the selection of the Q point. The graph in the lower right-hand corner of page 2 of the specification sheet indicates that the maximum β is achieved at values of collector current between 5 mA and 7.5 mA. Let us select a quiescent collector current of 5 mA. If it is desired that the Q point be exactly in the center of the load line, the collector saturation current must be twice as large as the quiescent value, or 10 mA. The desired value of load resistance is now found as follows:

$$I_{Cs} = \frac{V_{CC}}{R_L}; \; R_L = \frac{V_{CC}}{I_{Cs}} = \frac{50 \text{ V}}{10 \text{ mA}} = 5 \text{ k}\Omega$$

Since the signal swing is relatively small in the first stage, it is not necessary that we select the Q point exactly in the center of the load line; i.e., with a very small signal swing, the possibility of distortion due to saturation or cut-off is minimal. Therefore, we can select a slightly *larger* value for R_L; this will result in the Q point's being closer to the saturation value (to the *left* of the center of the load line). You will recall from chapter 6 that a large value of R_L will contribute to an increase in the gain by allowing more current to flow into the base of the transistor in the second amplifier stage. Since there is only one power supply available in the system in which our amplifier is to function, we use emitter bias, rather than collector-to-base bias (which may require *two* power supplies), in our design. We select $R_L = 6.2$ kΩ and $R_E = 470$ Ω; these odd values of resistance are chosen because 5% carbon resistors are manufactured for only certain specific values (multiples of 1.0, 1.2, 1.5, 1.8, 2.2, 2.7, 3.3, 3.9, 4.7, 5.1, 5.6, 6.2, 6.8, 7.5, 8.2, 9.1).

The next step is the design of the biasing network; i.e., the selection of the values for the bias resistors, R_1 and R_2. Since the stability factor is dependent upon the values of these resistors (as well as upon the values of R_E and β), we must first obtain an estimate of the desired stability factor for the amplifier stage. This means that we must determine the range of variation of those quantities which are temperature sensitive, namely, I_{CO}, β, and V_{BE}. In the second row of page 2 of the specification sheet, it is stated that the maximum value of the collector reverse current, I_{CBO} (the reverse current from *collector* to *base* with the emitter *open*, also referred to as I_{CO}), is 100 μA at 150°C. Taking one-half of each successive value for each 10°C decrease, we find that the maximum anticipated value of I_{CO} at +95°C is approximately 3 μA. Since we assume that I_{CO} is approximately equal to zero at room temperature, we can say that the expected *change* in I_{CO} is

$$\Delta I_{CO} = 3 \ \mu\text{A} - 0 \ \mu\text{A} = 3 \ \mu\text{A}$$

Turning next to the variation in β, recall that we specified that the minimum β of the transistors used in the design must be 100. Reference to the graph in the lower right-hand corner on page 2 of the specification sheet reveals that the β of a *typical* 2N1574 (room temperature β equal to 125) is approximately equal to 185 at $+95°C$; this is determined by interpolating between the curves for $T_A = +75°C$ and $T_A = +125°C$. Thus we can say that the ratio of the β at $+95°C$ to that at room temperature is

$$\frac{\beta(+95°C)}{\beta(+25°C)} = \frac{185}{125} = 1.48$$

On row 8, page 2 of the specification sheet, it is mentioned that the maximum value of β for *any* 2N1574 unit at *room* temperature is 200. Since the ratio of the β at $+95°C$ to that at room temperature is 1.48, we can calculate that *the maximum value of β for any 2N1574 unit at $+95°C$ is 1.48* multiplied by 200, or 296; let us round this off to 300. This means that 300 is the largest value anticipated for β under *any* circumstances. If the minimum and maximum values of β are 100 and 300 respectively, the anticipated *change* in β is

$$\Delta\beta = 300 - 100 = 200$$

Finally, we come to the variation in the value of V_{BE}. In the upper right-hand corner of page 3 of the specification sheet, there is a graph of V_{BE}-BASE-EMITTER VOLTAGE vs. T_A-AMBIENT TEMPERATURE, for different values of collector current. Let us study the curve for $I_C = 5$ mA, from $-55°C$ to $+95°C$. It can be ascertained from the graph that V_{BE} varies from 0.5 V (at $+95°C$) to approximately 0.85 V (at $-55°C$); this is a *change* of 0.35 V over the temperature range. If the graph were not given, we could have found the change in V_{BE} by making use of the relationship which says that V_{BE} decreases 2.5 mV for each degree centigrade increase in temperature. Let us assume that the total change in V_{BE} is 0.4 V. We can summarize the various changes in the temperature-sensitive quantities, as follows:

$$
\left.
\begin{array}{ll}
(1) & \Delta I_{CO} = 3 \ \mu A \\
(2) & \Delta\beta = 200 \\
(3) & \Delta V_{BE} = -0.4 \text{ V}
\end{array}
\right\} -55°C \text{ to } +95°C
$$

Let us begin the design of the bias network by assuming that the stability factor, S, is 50. We can now find the change in collector current due to changes in leakage current, β, and V_{BE}; these changes can then be summed to find the total change in collector current. These operations proceed as follows:

$$\Delta I_{C_L} = S(\Delta I_{CO}) = 50(3 \ \mu A) = 150 \ \mu A = 0.15 \text{ mA}$$

$$\Delta I_{C_\beta} = \frac{I_{C_1}(\Delta\beta)S}{\beta_1\beta_2} = \frac{5(200)(50)}{(100)(300)} = 1.67 \text{ mA}$$

$$\Delta I_{C_V} = \frac{-(\Delta V_{BE})S}{R_B + R_E} = \frac{-(-0.4)50}{10 + 0.47} = 1.91 \text{ mA}$$

$$\Delta I_{C_T} = \Delta I_{C_L} + \Delta I_{C_\beta} + \Delta I_{C_V}$$
$$= 0.15 + 1.67 + 1.91 = 3.73 \text{ mA}$$

Note that we assumed a value of 10 kΩ for R_B in the above computations. Strictly speaking, *not all* of this change in I_C (3.73 mA) will be *above* the room temperature value of 5 mA; because the change is computed for the entire temperature range, a portion of the 3.73 mA change will tend to *increase* I_C (above 5 mA), and some portion will tend to *decrease* it (below 5 mA). As a safety factor, let us assume that the *entire 3.73 mA change tends to increase* I_C; i.e., we are assuming that the situation is worse than it actually is. Therefore, if $I_{CQ} = 5$ mA at room temperature, and a stability factor of 50 predicts a 3.73 mA increase in I_C, the value of I_{CQ} at $+95°C$ is 5 mA $+$ 3.73 mA, or 8.73 mA. However, the collector saturation current for this amplifier stage is

$$I_{Cs} = \frac{V_{CC}}{R_L + R_E} = \frac{50 \text{ V}}{6.2 \text{ k}\Omega + 0.47 \text{ k}\Omega} = 7.5 \text{ mA}$$

Therefore, the collector current will never reach a value of 8.73 mA because *the transistor saturates at $I_C = 7.5$ mA;* i.e., if $I_{CQ} = 5$ mA at room temperature, we can tolerate a change in collector current of *only* 2.5 mA (7.5 mA $-$ 5 mA), as shown in Fig. 10.4. Thus the transistor will saturate at $+95°C$, resulting in distortion of the ac collector current waveform (see chapter 4).

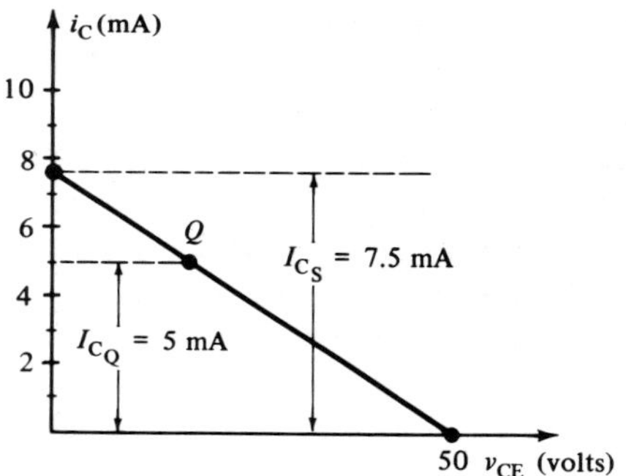

Figure 10.4

The above discussion indicates that a stability factor of 50 results in an excessive Q point shift. Let us try a stability factor of 25. Since the various changes in collector current (due to changes in I_{CO}, β, and so on) are directly proportional to S, a reduction in the value of S from 50 to 25 should result in a *halving* of the value of ΔI_{CT}; thus the total change in collector current, with $S = 25$, is one-half of 3.73 mA, or 1.86 mA. Since a change of 1.86 mA is *less* than the allowable change of 2.5 mA for this amplifier stage, we conclude that a stability factor of 25 is small enough to prevent an excessive Q point shift.

Before proceeding with the actual design of the bias network, we should make sure that the ac collector current swing of this stage is small enough, so that even with a 1.86 mA increase in the value of I_{CQ}, the saturation point is not reached. We know that the current gain for the *second* stage can be written as follows:

$$A_{I_2} = \frac{\beta_2 R}{R + r_{I_2} + \beta_2 R_E'} \simeq \frac{i_{C_2}}{i_{C_1}}$$

This current gain is approximately equal to the ratio of i_{C_2} (ac collector current in the second stage) to i_{C_1} (ac collector current in the first stage). Solving for i_{C_1}, we get

$$i_{C_1} = \frac{i_{C_2}(R + r_{I_2} + \beta_2 R_E')}{\beta_2 R}$$

It is our intention to estimate the ac collector current in stage 1 (i_{C_1}) in order to determine if this swing is small enough to prevent distortion when the Q point shifts up by 1.86 mA. Since neither stage has yet been designed, we do not know the values of R, r_{I_2}, and R_E'. We know that R is equal to the parallel equivalent of r_O (stage 1), R_L (stage 1), and R_B (stage 2). Neglecting r_O, we know that $R_L = 6.2$ kΩ, and we assume that $R_B = 10$ kΩ; thus R is approximately equal to

$$R \simeq R_L \parallel R_B = 6.2 \text{ k}\Omega \parallel 10 \text{ k}\Omega \simeq 4 \text{ k}\Omega$$

We assume further that $r_{I_2} = 1$ kΩ, $R_E' = 50$ Ω, and $\beta_2 = 125$ (typical value). We also know that the anticipated value of i_{C_2} is 3 mA (peak-to-peak), since this is one of the specifications upon which we based our design. Substituting into the equation, we find an approximate value for i_{C_1} as follows:

$$i_{C_1} = \frac{i_{C_2}(R + r_{I_2} + \beta_2 R_E')}{\beta_2 R} = \frac{3(4 + 1 + 125(0.05))}{125 \times 4} \simeq 0.07 \text{ mA}$$

It can be seen from the illustration in Fig. 10.5 that, when the Q point shifts from

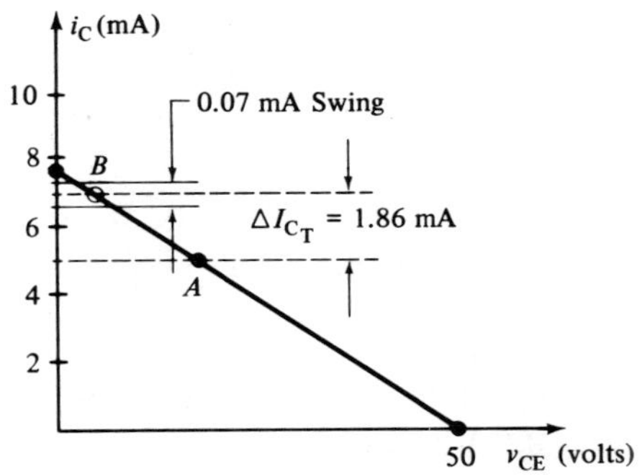

Figure 10.5

point A ($I_{C_Q} = 5$ mA) to point B ($I_{C_Q} = 6.86$ mA), the 0.07 mA peak-to-peak collector current swing is small enough to prevent distortion due to saturation. In other words, the 0.07 mA collector current swing will oscillate about a quiescent collector current of 6.86 mA at $+95°$C, instead of about 5 mA at room temperature. We conclude, finally, that a stability factor of 25 is appropriate for the dc design of the first stage.

Our task now is to select the proper values for R_1 and R_2. R_B (the parallel equivalent of the two resistors) is fixed by the desired value of S. Recall from chapter 5 that S can be written as follows:

$$S = \frac{1 + \dfrac{R_E}{R_B}}{\dfrac{1}{\beta} + \dfrac{R_E}{R_B}}$$

Solving the equation for R_B, we get

$$R_B = \frac{\beta R_E(S - 1)}{\beta - S}$$

We know that $\beta = 300$ (the maximum value), $R_E = 0.47$ kΩ (470 Ω), and the desired value of S is 25. Substituting into the equation, we find the required value of R_B to be

$$R_B = \frac{\beta R_E(S - 1)}{\beta - S} = \frac{300(0.47)(24)}{300 - 25} = 12.3 \text{ k}\Omega$$

Notice that in design, as opposed to analysis, the procedure is *reversed;* here we determine the desired value of S, and then work backward to find the required value of R_B. Although we now know the value of R_B required in the design, we do not know the individual values of R_1 and R_2. These values are selected to maintain the desired quiescent collector current (5 mA).

Recall from chapter 5 that we can express the quiescent collector current for the emitter bias circuit as follows:

$$I_{C_Q} = \frac{\beta(V_B - V_{BE})}{R_B + \beta R_E}$$

Solving for V_B, we get

$$V_B = \frac{I_{C_Q}(R_B + \beta R_E)}{\beta} + V_{BE}$$

In this equation, we desire that $I_{C_Q} = 5$ mA and $R_B = 12.3$ kΩ; using $\beta = 125$, $R_E = 0.47$ kΩ, and $V_{BE} = 0.6$ V, we find the required value of V_B to be

$$V_B = \frac{I_{C_Q}(R_B + \beta R_E)}{\beta} + V_{BE} = \frac{5[12.3 + 125(0.47)]}{125} + 0.6 = 3.44 \text{ V}$$

Essentially, we have *two* unknown quantities, R_1 and R_2. We now also have *two* equations relating these unknowns, as follows:

$$R_B = \frac{R_1 R_2}{R_1 + R_2} = 12.3 \text{ k}\Omega$$

$$V_B = \frac{R_2 V_{CC}}{R_1 + R_2} = \frac{R_2 50}{R_1 + R_2} = 3.44 \text{ V}$$

We now proceed to solve these equations simultaneously for R_1 and R_2:

$$R_1 + R_2 = \frac{R_1 R_2}{R_B}; \qquad R_1 + R_2 = \frac{R_2 V_{CC}}{V_B}$$

$$\frac{R_1 R_2}{R_B} = \frac{R_2 V_{CC}}{V_B}; \qquad R_1 = \frac{R_B V_{CC}}{V_B} = \frac{(12.3)(50)}{3.44} = 179 \text{ k}\Omega$$

$$R_2 = \frac{R_1}{\dfrac{V_{CC}}{V_B} - 1} = \frac{179}{\dfrac{50}{3.44} - 1} = 13.3 \text{ k}\Omega$$

In order to specify commercially available values, we make $R_1 = 180$ kΩ and $R_2 = 12$ kΩ. Let us now check the values of both the quiescent collector current and the stability factor:

$$R_B = \frac{R_1 R_2}{R_1 + R_2} = \frac{(180)(12)}{180 + 12} = 11.25 \text{ k}\Omega$$

$$V_B = \frac{R_2 V_{CC}}{R_1 + R_2} = \frac{12(50)}{180 + 12} = 3.12 \text{ V}$$

$$I_{CQ} = \frac{\beta(V_B - V_{BE})}{R_B + \beta R_E} = \frac{125(3.12 - 0.6)}{11.25 + 125(0.47)} = 4.5 \text{ mA}$$

$$S = \frac{1 + \dfrac{R_E}{R_B}}{\dfrac{1}{\beta} + \dfrac{R_E}{R_B}} = \frac{1 + \dfrac{0.47}{11.25}}{\dfrac{1}{300} + \dfrac{0.47}{11.25}} = 23.1$$

Note first that $I_{CQ} = 4.5$ mA, and not 5 mA. The reason for this result is the selection of a 12 kΩ resistor for R_2, instead of a 13.3 kΩ resistor. If the situation develops where it is necessary to have a 13.3 kΩ resistor for R_2, it is a simple matter to order one from the resistor manufacturer; in this case we might have to pay a slightly higher price. A glance, however, at the graph in the lower right-hand corner of page 2 of the specification sheet reveals that the situation is not quite critical enough to warrant the use of a 13.3 kΩ resistor. Note that the curve for $T_A = +25°$C is relatively *flat* for values of collector current from approximately 3 mA to 10 mA. This means that the transistor β at 4.5 mA is approximately the same as at 5 mA, and thus a value of I_{CQ} of 4.5 mA should in no way impair the performance of the amplifier stage. The second point to note is that the stability factor is actually *less* than 25; i.e., the Q point shift will be *smaller* than 1.86 mA.

We have now completed the design of the first stage of the amplifier (except for the value of the by-pass capacitor, which is designed later); the schematic diagram of the first stage is shown in Fig. 10.6. Before leaving this stage, let us compute the value of the minimum gain (minimum β) to see if it meets our original specification; recall that we specified that the minimum gain of the first stage be equal to 80. The gain (open loop) of the stage is

$$A_{I_1} = \frac{\beta R}{R + r_I}$$

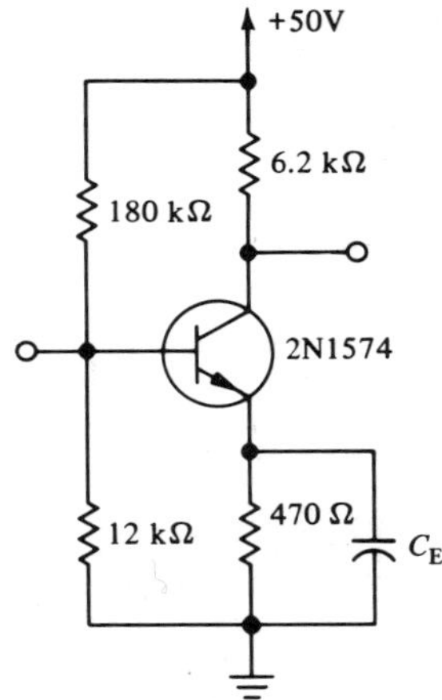

Figure 10.6

R is equal to the parallel equivalent of the source resistance (5 kΩ) and R_B (11.25 kΩ):

$$R = \frac{(5)(11.25)}{5 + 11.25} = 3.46 \text{ k}\Omega$$

The value of r_I is

$$r_I \simeq \frac{\beta}{40 I_C} = \frac{100}{40 \times 4.5} = 0.55 \text{ k}\Omega$$

We add a few hundred ohms to account for the base spreading resistance (see chapter 7), and assume that r_I is approximately equal to 800 Ω. The value of r_I can be found in another manner. In the lower left-hand corner of page 3 of the specification sheet, there is a graph of h_{ie}-AC COMMON-EMITTER INPUT IMPEDANCE vs. I_E-EMITTER CURRENT. It can be seen from the graph that the value of h_{ie} (which is the same as r_I) is approximately equal to 600 Ω at $I_E = 5$ mA and $T_A = -55°C$; this agrees reasonably well with our estimated value of 800 Ω. Note that the value of h_{ie} *decreases* with increasing emitter (collector) current; note also that h_{ie} *increases* with increasing temperature because of the increase in β at the higher temperatures. The minimum gain of the first stage can now be computed as follows:

$$A_{I_1} = \frac{\beta R}{R + r_I} = \frac{100(3.46)}{3.46 + 0.8} = 81.2$$

Notice that this is slightly greater than the minimum requirement. This means that our original estimation of the required minimum β for the transistor was correct; i.e., we estimated that the minimum β of the transistor be equal to 100 in order to obtain a minimum gain of 80.

10-6   THE DESIGN OF STAGE 2

You will recall from Sec. 10-3 that we assumed a value of 5 kΩ for the load resistance of the second stage; thus a 3 mA ac current in the collector of this stage will produce a 15 V signal at the output. Let us then select $R_L = 4.7$ kΩ; we also select $R_E = 1$ kΩ. Before designing the bias network, we must check to see if we can tolerate a stability factor of 25 for this stage. You will recall from the previous section that an S of 25 will result in a total change in I_{CQ} of 1.86 mA over the temperature range. This means that I_{CQ} will change from 5 mA (at point A in Fig. 10.7) to 6.86 mA (at point B). Since we are anticipating a 3 mA peak-to-peak

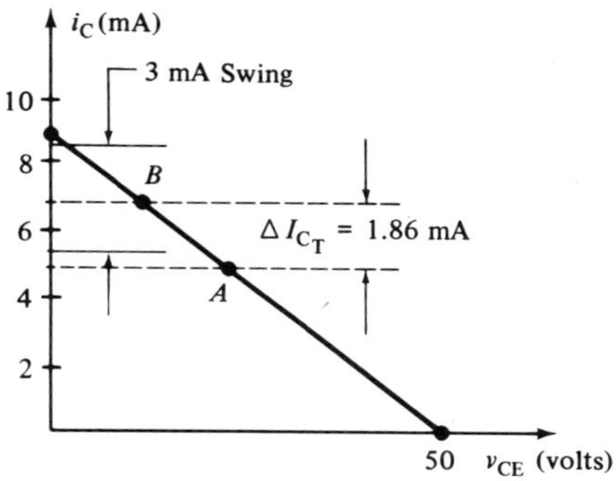

Figure 10.7

current swing in the collector of this stage (much larger than the 0.07 mA swing in the first stage), the *instantaneous collector current at +95 degrees C* will vary from 8.36 mA (1.5 mA *above* 6.86 mA) to 5.36 mA (1.5 mA *below* 6.86 mA). The value of the collector saturation current for this stage is

$$I_{Cs} = \frac{V_{CC}}{R_L + R_E} = \frac{50 \text{ V}}{4.7 \text{ k}\Omega + 1 \text{ k}\Omega} = 8.77 \text{ mA}$$

Since I_{Cs} (8.77 mA) is *greater* than the maximum instantaneous value of collector current (8.36 mA), there will be no problem of distortion due to saturation at the high temperature; this is illustrated in Fig. 10.7. Therefore, we can again use a stability factor of 25. It should be mentioned that we are including a large safety factor in our design by assuming that the total Q point shift occurs *above* the 5 mA room temperature value; as previously noted, a portion of the 1.86 mA change serves to decrease I_{CQ} *below* 5 mA. Since the Q point (5 mA) has been selected slightly closer to saturation (8.77 mA) than to cut-off, we can be reasonably confident that the waveform will not distort due to cut-off. The situation pictured in both Figs. 10.5 and 10.7 is over-pessimistic for another reason: Recall from chapter 6 that, when the output of an amplifier stage is connected to another circuit, the signal swing moves along the ac load line, *not* the dc load line shown in Figs. 10.5 and 10.7. A study of Fig. 6.25 will reveal that the instantaneous collector current can actually increase to values *greater* than I_{Cs} (8.77 mA), depending upon the value of *the ac load resistance, R_L'*. For these reasons, we can be confident that our amplifier will operate well within the specifications.

Now that we have decided upon a safety factor of 25, we can begin the design of the bias network. Following are the computations used to determine the values for R_1 and R_2:

$$R_B = \frac{\beta R_E(S - 1)}{\beta - S} = \frac{300(1)(24)}{300 - 25} = 26.2 \text{ k}\Omega$$

$$V_B = \frac{I_{CQ}(R_B + \beta R_E)}{\beta} + V_{BE} = \frac{5[26.2 + 125(1)]}{125} + 0.6 = 6.65 \text{ V}$$

$$R_1 = \frac{R_B V_{CC}}{V_B} = \frac{26.2(50)}{6.65} = 197 \text{ k}\Omega$$

$$R_2 = \frac{R_1}{\dfrac{V_{CC}}{V_B} - 1} = \frac{197}{\dfrac{50}{6.65} - 1} = 30.3 \text{ k}\Omega$$

We select $R_2 = 27$ kΩ and $R_1 = 200$ kΩ, even though some manufacturers do not produce 5% 200 kΩ resistors in large quantities (see Sec. 10-5); i.e., we may have to pay a slightly higher price for a 200 kΩ resistor. Let us now check the values of both I_{CQ} and S.

$$R_B = \frac{R_1 R_2}{R_1 + R_2} = \frac{200(27)}{200 + 27} = 23.8 \text{ k}\Omega$$

$$V_B = \frac{R_2 V_{CC}}{R_1 + R_2} = \frac{27(50)}{200 + 27} = 5.95 \text{ V}$$

$$I_{CQ} = \frac{\beta(V_B - V_{BE})}{R_B + \beta R_E} = \frac{125(5.95 - 0.6)}{23.8 + 125(1)} = 4.5 \text{ mA}$$

$$S = \frac{1 + \dfrac{R_E}{R_B}}{\dfrac{1}{\beta} + \dfrac{R_E}{R_B}} = \frac{1 + \dfrac{1}{23.8}}{\dfrac{1}{300} + \dfrac{1}{23.8}} = 23.2$$

Note that $I_{CQ} = 4.5$ mA and $S = 23.2$; these values are well within our design specifications.

We select $R_E' = 10$ Ω; this is the unbypassed portion of the emitter resistance of the second stage (the point from which the feedback loop is returned to the first stage). R_E' is generally selected to be a *small* value of resistance in order that the open loop gain of the second stage be relatively *large*. The second amplifier stage is shown in Fig. 10.8. Notice that the *total dc emitter resistance* is actually 1010 Ω (the 1 kΩ in series with 10 Ω); however, we neglect the effect of the 10 Ω resistor *on the dc operation* of the circuit because it is much smaller than the 1 kΩ resistor.

The next step is to check the minimum value of the open loop gain of the second stage. We know that the gain is given by the following equation:

$$A_{I_2} = \frac{\beta R}{R + R_I} = \frac{\beta R}{R + r_I + \beta R_E'}$$

R is equal to the parallel equivalent of R_L (stage 1), R_B (stage 2), and r_O (stage 1).

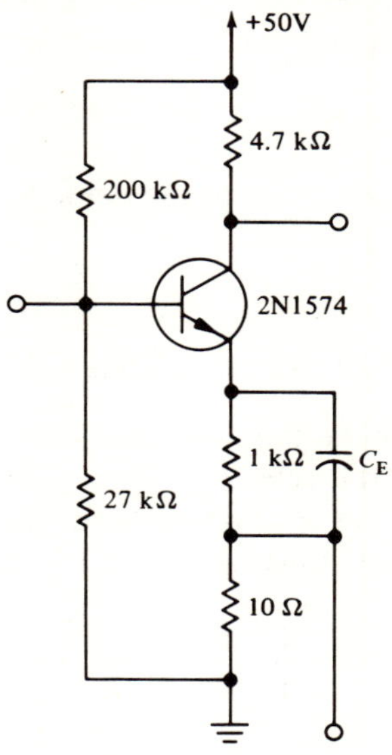

Figure 10.8

We also know that $R_L = 6.2$ kΩ and $R_B = 23.8$ kΩ; the parallel equivalent o
these two resistances is

$$\frac{(6.2)(23.8)}{6.2 + 23.8} = 4.92 \text{ k}\Omega$$

In the lower right-hand corner of page 3 of the specification sheet, there is a grapl
of h_{oe}-AC COMMON-EMITTER OUTPUT ADMITTANCE vs. I_E-EMITTEI
CURRENT. At $T_A = +25°C$ and $I_E = 5$ mA, h_{oe} is approximately equal to 10
μmhos; therefore, $1/h_{oe}$ (which we have been referring to as r_O) is

$$r_O = \frac{1}{h_{oe}} = \frac{1}{100 \ \mu\text{mhos}} = 0.01 \text{ M} = 10 \text{ k}\Omega$$

Thus, R is equal to

$$R = \frac{(4.92)(10)}{4.92 + 10} = 3.29 \text{ k}\Omega$$

Finally, the minimum value of gain (minimum β is equal to 100) is computed :
follows:

$$A_{I_2} = \frac{\beta R}{R + r_{\mathrm{I}} + \beta R_{\mathrm{E}}'} = \frac{100(3.29)}{3.29 + 0.8 + 100(0.01)} = 64.7$$

This is considerably larger than the minimum required value of 40; recall from Sec. 10-3 that we estimated that the minimum required value of A_{I_2} is 40. The *minimum* open loop gain of our two-stage amplifier is then equal to

$$A_{\mathrm{OL}} = A_{I_1} \times A_{I_2} = (81.2)(64.7) = 5260$$

This should be *more than enough* to provide the required value of closed loop gain at the required level of stability.

10-7   THE DESIGN OF THE FOLLOWER STAGE

We turn now to the design of the follower stage. This is a relatively simple task because the design involves the selection of only *one* resistor, R_{E}. Recall from chapter 7 that it is more efficient to *direct-couple* the follower to the collector of the transistor in the previous stage; thus we use this method, as shown in Fig. 10.9.

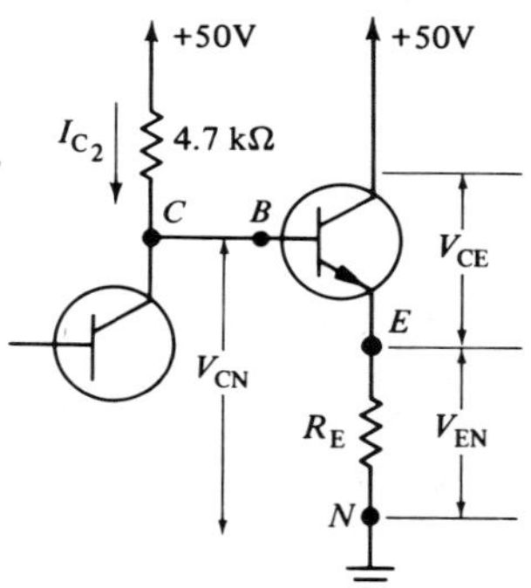

Figure 10.9

The dc base-to-ground voltage of the follower is equal to the collector-to-ground voltage of the previous stage, V_{CN}. Neglecting the *dc* loading effect of the follower on the previous stage, we can express V_{CN} as follows:

$$V_{\mathrm{CN}} = V_{\mathrm{CC}} - I_{C_2}R_{L_2}$$
$$V_{\mathrm{CN}} = 50 - (5)(4.7) = 26.5 \text{ V}$$

Let us assume that the base-to-emitter voltage drop for the follower transistor, V_{BE}, is equal to 0.5 V. Then the dc emitter-to-ground voltage drop for the follower, V_{EN} (the drop across R_{E}), is

$$V_{\mathrm{EN}} = V_{\mathrm{CN}} - V_{\mathrm{BE}} = 26.5 - 0.5 = 26 \text{ V}$$

The collector-to-emitter voltage of the follower transistor is

$$V_{CE} = V_{CC} - V_{EN} = 50 - 26 = 24 \text{ V}$$

It was shown earlier in the chapter that the maximum allowable power dissipation, P_{CM}, for the 2N1574 transistor (without a heat sink) is 320 mW. Therefore, since $V_{CE} = 24$ V and $P_{CM} = 320$ mW, we can compute the value of the *maximum allowable quiescent collector current* of the follower as follows:

$$I_{C_{maxQ}} = \frac{P_{CM}}{V_{CE}} = \frac{320 \text{ mW}}{24 \text{ V}} = 13.3 \text{ mA}$$

Assuming that the collector current is approximately equal to the emitter current, we can say that the value of the emitter resistance is given as follows:

$$R_E = \frac{V_{EN}}{I_{CQ}} = \frac{26 \text{ V}}{13.3 \text{ mA}} = 1.95 \text{ k}\Omega$$

This is the *minimum* value which can be selected for the emitter resistance; the selection of a *smaller* value for R_E will result in the transistor's exceeding the 320 mW maximum allowable power dissipation. Let us select $R_E = 2.2$ kΩ; the follower circuit appears in Fig. 10.10.

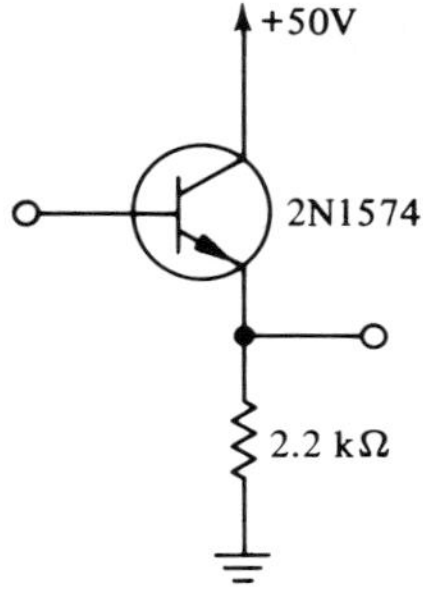

Figure 10.10

Our next task is to determine whether or not the follower can accommodate the anticipated output signal. Figure 10.11(a) shows both the dc and ac load lines drawn on the collector characteristics of the follower transistor. The dc load line is drawn with a load resistance of 2.2 kΩ (R_E). The ac load line is drawn using the ac load resistance of the follower transistor, R_T'; R_T' is equal to the parallel equivalent of the 2.2 kΩ emitter resistor of the follower (R_E) and the 1 kΩ load resistance (R_T):

$$R_T' = \frac{R_E R_T}{R_E + R_T} = \frac{2.2(1)}{2.2 + 1} = 0.687 \text{ k}\Omega$$

The actual value of I_{CQ} for the follower stage is

$$I_{CQ} = \frac{V_{EN}}{R_E} = \frac{26 \text{ V}}{2.2 \text{ k}\Omega} = 11.8 \text{ mA}$$

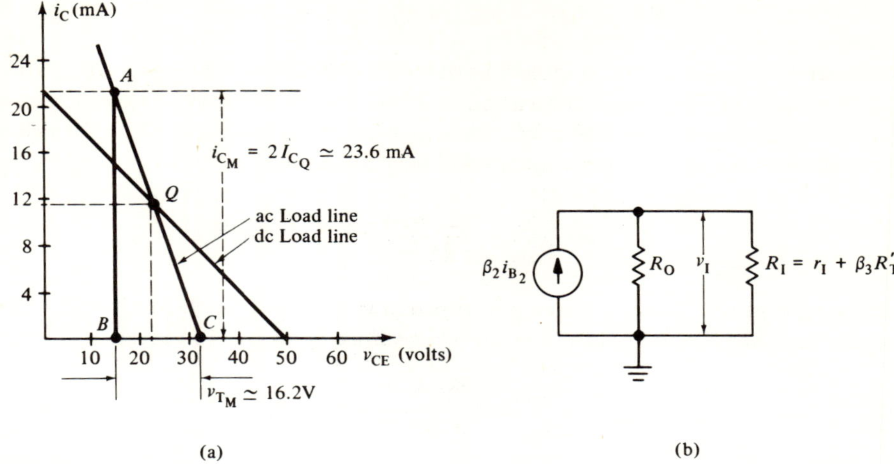

Figure 10.11

A glance at the diagram in Fig. 10.11(a) reveals that the maximum undistorted peak-to-peak collector current available from the follower stage is equal to $2I_{C_Q}$.

$$i_{C_M} = 2I_{C_Q} = 2(11.8 \text{ mA}) = 23.6 \text{ mA}$$

The maximum undistorted peak-to-peak load voltage, v_{T_M}, is equal to the product of i_{C_M} and R_T'; this can be shown as follows: We know that the slope of the ac load line, m, is equal to the reciprocal of the ac load resistance; in this particular case, the ac load resistance is equal to R_T':

$$m = \frac{1}{R_T'}$$

It can be seen from triangle ABC in Fig. 10.11(a) that the slope of the ac load line (line A–C) is equal to the ratio of line A–B (i_{C_M}) to line B–C (v_{T_M}):

$$m = \frac{A-B}{B-C} = \frac{i_{C_M}}{v_{T_M}}$$

Setting the two expressions for the slope equal to each other, we get

$$m = \frac{1}{R_T'} = \frac{i_{C_M}}{v_{T_M}}$$

$$v_{T_M} = (i_{C_M})(R_T')$$

Thus the value of v_{T_M} is found as follows:

$$v_{T_M} = (i_{C_M})(R_T') = (23.6 \text{ mA})(0.687 \text{ k}\Omega) = 16.2 \text{ V}$$

Since v_{T_M} (16.2 V peak-to-peak) is *greater* than our maximum anticipated peak-to-peak sinusoidal output voltage of 11 V (see Sec. 10-2), we can safely say that the follower stage *can* accommodate the maximum output; i.e., the computations predict that there will be no distortion due to saturation or cut-off.

Before proceeding with the design of the feedback loop, we find it advantageous to derive an expression relating the output voltage of the follower (v_T) to the output current of the second amplifier stage. Figure 10.11(b) shows the *output* circuit of the second amplifier stage (consisting of a current generator, $\beta_2 i_{B_2}$, in parallel with the output resistance of the stage, R_O) connected to the *input* circuit of the follower stage (consisting of the input resistance, R_I); the input voltage to the follower stage is represented by v_I. This voltage can be expressed as the product of the current generator ($\beta_2 i_{B_2}$) and the parallel equivalent of R_O and R_I:

$$v_I = \frac{(\beta_2 i_{B_2})(R_O)(r_I + \beta_3 R_T')}{R_O + r_I + \beta_3 R_T'}$$

We know that the voltage gain of the follower is given by the following expression:

$$A_V = \frac{v_T}{v_I} = \frac{\beta_3 R_T'}{r_I + \beta_3 R_T'}$$

Therefore, the output voltage of the follower (the load voltage, v_T) can be expressed as follows:

$$v_T = A_V v_I = \frac{\beta_3 R_T'}{r_I + \beta_3 R_T'} \times \frac{(\beta_2 i_{B_2})(R_O)(r_I + \beta_3 R_T')}{R_O + r_I + \beta_3 R_T'}$$

$$v_T = \frac{(\beta_3 R_T')(\beta_2 i_{B_2} R_O)}{R_O + r_I + \beta_3 R_T'} = \frac{\beta_2 i_{B_2} R_O}{1 + \dfrac{R_O + r_I}{\beta_3 R_T'}}$$

We now have an equation which relates the output load voltage of the follower stage (v_T) to the ac current generator in the second stage ($\beta_2 i_{B_2}$). In this equation, β_2 is the β of the transistor in the second stage, β_3 is the β of the transistor in the third (follower) stage, R_O is the output resistance of the second stage, r_I is the base-to-emitter resistance of the follower transistor, and R_T' is the parallel equivalent of the follower emitter resistor (R_E) and the load resistance (R_T). This equation will prove useful in the following section; it is written as follows:

$$v_T = \frac{(\beta_2 i_{B_2})R_O}{1 + \dfrac{R_O + r_I}{\beta_3 R_T'}} \tag{10-1}$$

10-8 THE DESIGN OF THE FEEDBACK LOOP

We have already shown (Sec. 10-6) that theoretically we have more than enough open loop gain to meet the specifications of the design. Our present task is to select the value of R_F which will result in an output load voltage of 10 V when the original source voltage is 50 mV; i.e., when the original source current is 0.01 mA (see Fig. 10.3(b)). To do this, we work *backward* from the load. In the last section, we derived the following equation:

$$v_T = \frac{\beta_2 i_{B_2} R_O}{1 + \dfrac{R_O + r_I}{\beta_3 R_T{}'}}$$

Solving this equation for the second stage current generator ($\beta_2 i_{B_2}$), we get

$$\beta_2 i_{B_2} = \frac{v_T}{R_O}\left(1 + \frac{R_O + r_I}{\beta_3 R_T{}'}\right)$$

We should like to compute the value of $\beta_2 i_{B_2}$ necessary to maintain an output load voltage (v_T) of 10 V. R_O is *approximately* equal to the parallel equivalent of R_L (4.7 kΩ) and r_O (10 kΩ) of the second stage; strictly speaking, the situation is considerably more complicated due to the presence of the 10 Ω unby-passed emitter resistance ($R_E{}'$) and the feedback loop. However, we can ascertain a reasonably good idea of the value of R_O by taking the parallel equivalent of R_L and r_O; thus R_O is

$$R_O = \frac{R_L r_O}{R_L + r_O} = \frac{(4.7)(10)}{4.7 + 10} = 3.2\ \text{k}\Omega$$

Since 10 V is the *center* value for the design, we use the *typical β* of the transistor, 125. We can compute the value of r_I for the follower as follows:

$$r_I = \frac{\beta}{40 I_{C_Q}} = \frac{125}{40(11.8\ \text{mA})} = 0.265\ \text{k}\Omega$$

Adding a few hundred ohms for the base spreading resistance, we approximate r_I as 500 Ω. Substituting into the equation for $\beta_2 i_{B_2}$, and setting $\beta_3 = 125$, we get

$$\beta_2 i_{B_2} = \frac{v_T}{R_O}\left(1 + \frac{R_O + r_I}{\beta_3 R_T{}'}\right) = \frac{10}{3.2}\left(1 + \frac{3.2 + 0.5}{125(0.687)}\right) = 3.26\ \text{mA}$$

We know that the *closed* loop current gain for the *first two* amplifier stages is

$$A_{CL} = \frac{\beta_2 i_{B_2}}{i} = \frac{3.26\ \text{mA}}{0.01\ \text{mA}} = 326$$

The closed loop gain is also given by the following equation:

$$A_{CL} = \frac{A_{OL}}{1 + A_{OL}\beta_F}$$

Solving this equation for β_F (the feedback factor), we get

$$\beta_F = \frac{\dfrac{A_{OL}}{A_{CL}} - 1}{A_{OL}}$$

Now we know that the *required* value of the closed loop gain is 326; we also know that $R_E' = 10\ \Omega$. Therefore, if we compute the value of the open loop gain for a typical type 2N1574 transistor ($\beta = 125$), we can substitute into the above equation to find the required value of β_F, and from this, the value of R_F.

Let us first find the open loop gain of stage one. The value of r_I (with $\beta = 125$) is

$$r_I = \frac{\beta}{40 I_{C_Q}} = \frac{125}{40(4.5\ \text{mA})} = 0.694\ \text{k}\Omega$$

Rounding off r_I to $900\ \Omega$ (taking into account the base spreading resistance), we find the gain of the first stage to be

$$A_{I_1} = \frac{\beta R}{R + r_I} = \frac{125(3.46)}{3.46 + 0.9} = 99$$

The gain of the second stage is

$$A_{I_2} = \frac{\beta R}{R + r_I + \beta R_E'} = \frac{125(3.29)}{3.29 + 0.9 + 125(0.01)} = 75.5$$

The open loop gain is

$$A_{\text{OL}} = A_{I_1} \times A_{I_2} = 99 \times 75.5 = 7470\ \cdot$$

Substituting into the equation for the feedback factor, we find the required value for β_F to be

$$\beta_F = \frac{\dfrac{A_{\text{OL}}}{A_{\text{CL}}} - 1}{A_{\text{OL}}} = \frac{\dfrac{7470}{326} - 1}{7470} = 0.00293$$

The required value for R_F is

$$\beta_F = \frac{R_E'}{R_F}; \qquad R_F = \frac{R_E'}{\beta_F} = \frac{0.01}{0.00293} = 3.41\ \text{k}\Omega$$

Thus we select a $3.41\ \text{k}\Omega$ resistor for R_F.

The next step is to check to see if the *minimum* specifications are met; i.e., we should like to know if *the output voltage is somewhere between 9 V and 10 V when the β of the transistors is at the minimum value of 100.* You will recall from Sec. 10-6 that the *minimum* value of the open loop gain ($\beta = 100$) was computed to be 5260. Working now from the *input* of our amplifier toward the *output*, we first find the value of the closed loop gain

$$A_{\text{CL}} = \frac{A_{\text{OL}}}{1 + A_{\text{OL}}\beta_F} = \frac{5260}{1 + 5260(0.00293)} = 321$$

The value of the $\beta_2 i_{B_2}$ current generator is

$$\beta_2 i_{B_2} = (A_{\text{CL}})i = 321(0.01\ \text{mA}) = 3.21\ \text{mA}$$

Finally, the value of the output load voltage is

$$v_T = \frac{(\beta_2 i_{B_2})R_O}{1 + \dfrac{R_O + r_I}{\beta R_T'}} = \frac{(3.21)(3.2)}{1 + \dfrac{3.2 + 0.4}{100(0.687)}} = 9.77 \text{ V}$$

Since this value (9.77 V) is *greater* than the minimum specified value of 9.0 V, we have met the minimum specification.

Next we shall check the *maximum* load voltage; i.e., we must check to make sure that *the output load voltage is somewhere between 10 V and 11 V when the β of the transistors is at the maximum value of 300*. The value of r_I ($\beta = 300$) is

$$r_I = \frac{\beta}{40I_{C_Q}} = \frac{300}{40 \times 4.5 \text{ mA}} = 1.67 \text{ k}\Omega$$

Rounding this off to 1.7 kΩ to account for the base spreading resistance, we repeat the above procedure as follows:

$$A_{I_1} = \frac{\beta R}{R + r_I} = \frac{300(3.46)}{3.46 + 1.7} = 201$$

$$A_{I_2} = \frac{\beta R}{R + r_I + \beta R_E'} = \frac{300(3.29)}{3.29 + 1.7 + 300(0.01)} = 123.5$$

$$A_{OL} = A_{I_1} \times A_{I_2} = (201)(123.5) = 24{,}800$$

$$A_{CL} = \frac{A_{OL}}{1 + A_{OL}\beta_F} = \frac{24{,}800}{1 + (24{,}800)(0.00293)} = 336$$

$$\beta_2 i_{B_2} = (A_{CL})i = 336(0.01 \text{ mA}) = 3.36 \text{ mA}$$

$$v_T = \frac{(\beta_2 i_{B_2})R_O}{1 + \dfrac{R_O + r_I}{\beta R_T'}} = \frac{3.36(3.2)}{1 + \dfrac{3.2 + 0.7}{300(0.687)}} = 10.55 \text{ V}$$

In the above computations, we assumed that r_I of the follower transistor varied from 400 Ω ($\beta = 100$) to 500 Ω ($\beta = 125$) to 700 Ω ($\beta = 300$). Since the output voltage (10.55 V) is *less* than the maximum specified value of 11.0 V, we have met the maximum specification. Notice that, according to our computations (which are somewhat approximate), *we are well within the specified design limits*. Note also that, while the *open* loop gain varies from 5260 ($\beta = 100$) to 24,800 ($\beta = 300$), the *closed* loop gain varies only from 321 to 336; in other words, this small variation in the gain is responsible for maintaining the output load voltage between 9.77 V and 10.55 V.

10-9   THE SELECTION OF THE RESISTOR POWER RATINGS

Before we build and test our amplifier, we must determine the necessary power ratings for the resistors in the circuit. We generally proceed by computing the dc power delivered to the resistor, and then as a safety factor, we specify a resistor whose power rating is *at least twice* as much as the computed value. For example, let us consider the case of the 6.2 kΩ load resistor in the first stage. We know that the quiescent collector current of the transistor in the first stage is 4.5 mA; thus the dc voltage across the 6.2 kΩ resistor is

$$V_{\mathrm{RL}} = (I_{\mathrm{CQ}})R_{\mathrm{L}} = (4.5 \text{ mA})(6.2 \text{ k}\Omega) = 27.9 \text{ V}$$

The dc power delivered to the resistor is

$$P = (V_{\mathrm{RL}})I_{\mathrm{CQ}} = (27.9 \text{ V})(4.5 \text{ mA}) = 125.5 \text{ mW}$$

In this case, we shall specify a *one-half watt* (500 mW) resistor.

This process can be repeated for all of the other resistors in the circuit. The feedback resistor, R_{F}, will dissipate *no dc power* because there is a capacitor, C_{F}, in series with it. Therefore, we must determine the amount *of ac power* delivered to R_{F} before we can specify its power rating. It can be seen from Fig. 10.12 that the

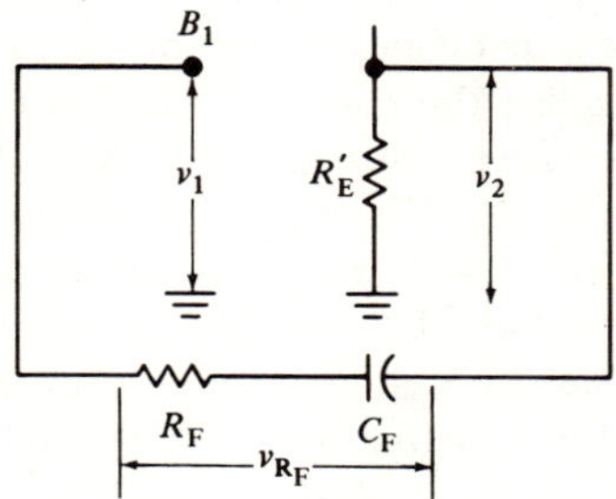

Figure 10.12

ac voltage across R_{F} (neglecting the capacitive reactance of C_{F}) is equal to the difference between the voltage across R_{E}' (v_2) and the voltage across the input terminals of the amplifier (v_1). It can be shown that the voltage v_1 is much smaller than v_2; thus we can say that:

$$v_{\mathrm{R_F}} = v_2 - v_1 \simeq v_2$$

The voltage v_2 is approximately equal to the product of $i_{\mathrm{C_2}}$ (the peak-to-peak collector current in the second stage) and R_{E}' (10 Ω); thus v_2 is:

$$v_2 \simeq (i_{\mathrm{C_2}})R_{\mathrm{E}}' = (3.21 \text{ mA})(0.01 \text{ k}\Omega) = 0.0321 \text{ V}$$

The ac current in R_{F} is:

$$i_{\mathrm{R_F}} = \frac{v_{\mathrm{R_F}}}{R_{\mathrm{F}}} = \frac{0.0321 \text{ V}}{3.41 \text{ k}\Omega} = 0.00942 \text{ mA}$$

The ac power delivered to R_{F} can be computed as follows (remember that $v_{\mathrm{R_F}}$ and $i_{\mathrm{R_F}}$ are peak-to-peak values, and thus we must divide by a factor of 8 when computing the power):

$$P = \frac{(v_{R_F})(i_{R_F})}{8} = \frac{(0.0321 \text{ V})(0.00942 \text{ mA})}{8} = 0.0377 \text{ } \mu\text{W}$$

In this case we can select a one-tenth watt resistor.

You will recall from chapter 7 that, when the factor "$A_{OL}\beta_F$" is much larger than 1.0, we can approximate the closed loop gain as

$$A_{CL} \simeq \frac{1}{\beta_F} = \frac{1}{\dfrac{R_E'}{R_F}} = \frac{R_F}{R_E'}$$

Substituting into the equation, we get

$$A_{CL} \simeq \frac{R_F}{R_E'} = \frac{3.41 \text{ k}\Omega}{0.01 \text{ k}\Omega} = 341$$

This compares favorably with the *actual* values of 321 and 336 computed in Sec. 10-8. The negative feedback incorporated in the amplifier circuit is responsible for this relatively constant gain; the constant gain, in turn, is responsible for the fact that the output voltage varies only from 9.77 V to 10.55 V, even though the transistor β varies from 100 to 300. Since it is important that the closed loop gain remain as constant as possible, and since the value of this closed loop gain is highly dependent upon the values of R_F and R_E', *we shall specify that these two resistors have a tolerance of 1% or better*. The actual determination as to whether or not the resistor tolerances should be better than 1% involves a considerable number of additional computations, which we shall not undertake. Of course, 1% resistors cost more than 5% units, but the computations will reveal that we must use the *closer tolerance* units in order to maintain the desired level of stability (output varies no more than plus or minus 10%) at the output of the amplifier.

Following is a list of all of the resistors used in the amplifier, together with their power ratings and tolerances.

STAGE 1

R_L	6.2 kΩ	0.5 W	5%
R_E	470 Ω	0.1 W	5%
R_1	180 kΩ	0.1 W	5%
R_2	12 kΩ	0.1 W	5%

STAGE 2

R_L	4.7 kΩ	0.5 W	5%
R_E	1 kΩ	0.1 W	5%
R_1	200 kΩ	0.1 W	5%
R_2	27 kΩ	0.1 W	5%

FOLLOWER STAGE

R_E	2.2 kΩ	1 W	5%

FEEDBACK NETWORK

R_E'	10 Ω	0.1 W	1%
R_F	3.41 kΩ	0.1 W	1%

10-10 THE SELECTION OF THE CAPACITORS

The final step in the paper design is the selection of the values for the various capacitors in the amplifier circuit. You will recall from chapter 8 that these capacitors (the coupling and by-pass capacitors) are responsible for the decrease in gain at the low frequencies. Three of the capacitors are located in that portion of the amplifier circuit which is enclosed *within the feedback loop;* these are the coupling capacitor between stages one and two, and the two by-pass capacitors (strictly speaking, C_F is also located within the loop, but this is a separate problem which will be handled later). Because these capacitors are located within the loop, the lower cut-off frequency for the amplifier (f_L) will be *better* (smaller) than if *no* feedback were incorporated (see chapter 8). Since, however, the equations used for the computation of the improvement in f_L due to the incorporation of feedback are extremely complex (and as such they are beyond the scope of this text), we have to rely upon the designer's experience, or refer to a more advanced text. Let us say that there is a *ten-fold* decrease in the value of f_L due to the incorporation of the amount of feedback present in our amplifier. Since the *lowest* frequency anticipated for our particular design is 100 Hz (see Sec. 10-2), *the individual corner frequencies due to these three capacitors can be made approximately ten times as large as 100 Hz;* i.e., we can select these capacitors for corner frequencies of 1000 Hz. The selection of corner frequencies on the order of 1 kHz will result in the use of *smaller* values of capacitance, with a consequent saving in physical space. We now select the values of these three capacitors.

Let us first select the value of the inter-stage coupling capacitor (between stages one and two). We know that the expression for the corner frequency due to this capacitor is given by the following equation:

$$f_{L_1} = \frac{1}{2\pi C_C (R_O + R_{I_E})}$$

Solving for C_C, we get

$$C_C = \frac{1}{2\pi f_{L_1}(R_O + R_{I_E})}$$

R_O is equal to the parallel equivalent of R_L and r_O (stage 1); this is

$$R_O = \frac{R_L r_O}{R_L + r_O} = \frac{(6.2)(10)}{6.2 + 10} = 3.83 \text{ k}\Omega$$

R_{I_E} is equal to the parallel equivalent of R_B and R_I (stage 2); this is

$$R_I = r_I + \beta R_E' = 0.8 + 125(0.01) = 2.05 \text{ k}\Omega$$

$$R_{IE} = \frac{R_B R_I}{R_B + R_I} = \frac{(23.8)(2.05)}{23.8 + 2.05} = 1.89 \text{ k}\Omega$$

Substituting into the equation for C_C, and setting $f_{L_1} = 1$ kHz, we get

$$C_C = \frac{1}{2\pi f_{L_1}(R_O + R_{IE})} = \frac{0.159}{10^3(3.83 + 1.89) \times 10^3} = C_C = 0.0278 \ \mu\text{F}$$

As a safety factor, we select a value of 0.1 μF for this coupling capacitor. In addition to the value, we must also specify the *voltage rating* of the capacitor. The dc voltage across this coupling capacitor, V_{C_C}, is equal to the difference between V_{CN_1} (the dc collector-to-ground voltage of stage 1) and V_{BN_2} (the dc base-to-ground voltage of stage 2); the computations proceed as follows:

$$V_{CN_1} = V_{CC} - I_{CQ}R_L = 50 - (4.5)(6.2) = 22.1 \text{ V}$$

$$V_{BN_2} = V_{BE} + I_{CQ}R_E = 0.6 + (4.5)(1) = 5.1 \text{ V}$$

$$V_{C_C} = V_{CN_1} - V_{BN_2} = 22.1 - 5.1 = 17 \text{ V}$$

Strictly speaking, if one or both of the transistors in the first two stages should fail during operation, the voltage across this coupling capacitor could increase to a value close to 50 V. We assume, however, that, because of our careful design, there is little chance of transistor failure, and thus we specify a voltage rating of 20 V for this capacitor.

The second of the three capacitors within the feedback loop is the by-pass capacitor in stage 1. Recall from chapter 8 that there are *two* corner frequencies associated with the by-pass capacitor, f_{L_2} and kf_{L_2}. Since kf_{L_2} is the *larger* of these two corner frequencies, we must set this quantity equal to 1 kHz; in other words, f_{L_2} (the smaller of the corner frequencies) will have a value *less* than 1 kHz. The value of f_{L_2} is computed as follows:

$$k = 1 + \frac{\beta R_E}{R + r_I} = 1 + \frac{125(0.47)}{3.46 + 0.8} = 14.8$$

$$f_{L_2} = \frac{kf_{L_2}}{k} = \frac{1000 \text{ Hz}}{14.8} = 67.5 \text{ Hz}$$

We can now solve for the required value of C_E:

$$f_{L_2} = \frac{1}{2\pi R_E C_E}$$

$$C_E = \frac{1}{2\pi f_{L_2} R_E} = \frac{0.159}{(67.5)470} = 5 \ \mu\text{F}$$

We specify a value of 10 μF. The dc voltage drop across C_E is equal to the dc drop across the emitter resistor of stage 1; this is:

$$V_{RE} = (I_{CQ})R_E = (4.5 \text{ mA})(0.47 \text{ k}\Omega) = 2.12 \text{ V}$$

We specify a voltage rating of 5 V.

Using the same procedure, it can be shown that the by-pass capacitor in stage 2 (the third of the three capacitors within the feedback loop) should be 10 μF with a 10 V rating.

The fourth capacitor, C_F, while technically within the feedback loop, does not have the same effect upon the low-frequency response as do the coupling and by-pass capacitors. C_F was included within the loop for the purpose of preventing an *interaction* between the dc quantities in stages one and two via the feedback loop. As mentioned in chapter 7, the feedback capacitor could be eliminated, and the feedback resistor (R_F) could be used for dc, as well as ac stabilization. However, we have elected to include the feedback capacitor, and we have used the emitter bias method for dc stabilization of our amplifier stages. We know that C_F is in series with R_F (in the feedback loop). In order that the reactance of C_F have a negligible effect upon the amplifier performance, we should like this reactance to be *very small* in comparison with the value of R_F over the entire frequency range of operation. If we make X_{C_F} very small at the *lowest* frequency (100 Hz), it will be *even smaller at the higher frequencies;* this is another example of "worst case design," since X_{C_F} has its largest value (worst case) at the lowest frequency. Therefore, we make X_{C_F} equal to *one-tenth* of the value of R_F at $f = 100$ Hz; we can then solve for the value of C_F:

$$X_{C_F} = \frac{1}{10} R_F = \frac{1}{10} (3.41 \text{ k}\Omega) = 341 \ \Omega \ @ \ f = 100 \text{ Hz}$$

$$C_F = \frac{1}{2\pi f X_{C_F}} = \frac{0.159}{100 \times 341} = 4.66 \ \mu\text{F}$$

We select a 5 μF capacitor. The dc voltage drop across this capacitor is equal to the difference between V_{BN_1} (the dc base-to-ground voltage of stage 1) and V_{R_E}' (the dc drop across R_E'):

$$V_{BN_1} = V_{BE} + I_{C_Q}R_E = 0.6 + (4.5)(0.47) = 2.72 \text{ V}$$

$$V_{R_E}' = I_{C_Q}R_E' = (4.5)(0.01) = 0.045 \text{ V}$$

$$V_{C_F} = V_{BN_1} - V_{R_E}' = 2.72 - 0.045 = 2.68 \text{ V}$$

We use a 5 V rating for C_F.

The fifth capacitor to be selected is the capacitor which couples the output of the follower to the load. Since this is a coupling capacitor, its value can be found by the usual equation:

$$C_C = \frac{1}{2\pi f_{L_1}(R_O + R_{I_E})}$$

Since this capacitor, however, is *outside* the feedback loop, there will be *no* improvement in the low-frequency response due to the negative feedback, and consequently we must set $f_{L_1} = 100$ Hz. In the above equation, R_O represents the output resistance of the follower and $R_{\mathrm{I_E}}$ is the load resistance. R_O is the parallel equivalent of R_E (of the follower) and:

$$\frac{R + r_\mathrm{I}}{\beta}$$

(See Fig. 10.13(a).)

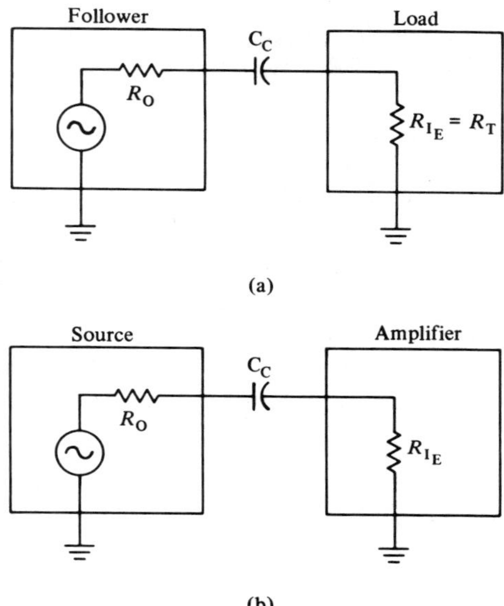

Figure 10.13

In this equation, R is the parallel equivalent of R_L (stage 1), r_O (stage 1), and R_B (stage 2); we have already found that this is equal to 3.29 kΩ. The resistance $r_\mathrm{I} = 0.5$ kΩ (for the follower). We can now compute the output resistance of the follower:

$$R_\mathrm{O} = R_\mathrm{E} \left\|\left(\frac{R + r_\mathrm{I}}{\beta}\right)\right. = 2.2 \left\|\left(\frac{3.29 + 0.5}{125}\right)\right. = 2.2 \text{ k}\Omega \,\|\, 0.03 \text{ k}\Omega$$

$$R_\mathrm{O} \simeq 0.03 \text{ k}\Omega = 30 \ \Omega$$

$R_{\mathrm{I_E}}$ is equal to the load resistance, R_T (1 kΩ). Substituting into the equation for C_C, we get

$$C_\mathrm{C} = \frac{1}{2\pi f_{\mathrm{L_1}}(R_\mathrm{O} + R_{\mathrm{I_E}})} = \frac{0.159}{100(30 + 1000)} = 1.54 \ \mu\mathrm{F}$$

This equation indicates that the connection of a 1.54 μF coupling capacitor between the follower output and the load will result in a corner frequency (due to this capacitor only) of 100 Hz. Recall from chapter 8 that the gain drops to 70.7% of its mid-band value at the corner frequency. It was specified, however, in

Sec. 10-2 that the output load voltage cannot be *less* than 9, nor *more* than 11 V (when the input source voltage is 50 mV) for frequencies between 100 Hz and 50 kHz. In other words, *we cannot tolerate a 29.3% (100% minus 70.7%) drop in the gain.* If a 1.54 μF capacitor were connected into the circuit, the gain at 100 Hz would drop to 70.7% of its mid-band value, and *the output voltage would be 7.07 V instead of 10 V, less than the minimum value of 9 V.* For this reason, we shall select the value of this capacitor (as we have done with most of the others) to be considerably *larger* than the computed value of 1.54 μF; let us use a value of 15 μF. Since we anticipate *no* dc voltage at the load, the dc voltage drop across this capacitor is equal to the dc voltage across the 2.2 kΩ emitter resistor in the follower circuit; this is:

$$V_{C_C} = V_{\mathrm{RE}} = I_{C_Q}R_{\mathrm{E}} = (11.8 \text{ mA})(2.2 \text{ k}\Omega) = 26 \text{ V}$$

We specify a voltage rating of 35 V.

The sixth and final capacitor to be selected is the one which couples the source to the base of the first amplifier stage; the value of this capacitor can be computed using the following equation:

$$C_C = \frac{1}{2\pi f_{\mathrm{L_1}}(R_O + R_{\mathrm{IE}})}$$

In this equation, R_O represents the internal resistance of the source (5 kΩ) and R_{IE} represents the input resistance to the entire amplifier (see Fig. 10.13(b)). Recall from chapter 7 that the effect of the feedback loop on the input circuit of the amplifier can be represented by a small resistance, R_F', in shunt with the input terminals. Using some of the equations developed in chapter 7, it can be shown that the presence of R_F' causes the input resistance to the entire amplifier (R_{IE}) to be *less than 30* Ω. Since R_{IE} (30 ohms) is much smaller than R_O (5 kΩ), we can neglect the effect of R_{IE} on the low-frequency response; substituting into the equation for C_C, we get:

$$C_C \simeq \frac{1}{2\pi f_{\mathrm{L_1}}R_O} = \frac{0.159}{100 \times 5000} = 0.318 \ \mu\text{F}$$

We again make the actual value larger than the computed value; we select $C_C = 2 \ \mu$F. The dc voltage drop across the capacitor depends upon the dc level at the output of the source terminals. Assuming that this dc level is zero volts, the dc voltage drop across C_C is equal to $V_{\mathrm{BN_1}}$, or:

$$V_{C_C} = V_{\mathrm{BN_1}} = V_{\mathrm{BE}} + I_{C_Q}R_{\mathrm{E}} = 0.6 + 4.5(0.47) = 2.72 \text{ V}$$

We select a 5 V rating for this capacitor. We have now finished with the selection of the capacitors. It is important to note that this selection was based upon a number of rough assumptions. In the following section, we discuss some of the methods which can be used to check these assumptions.

10-11   TROUBLESHOOTING THE CIRCUIT

We have now finished the paper design of the circuit; the schematic diagram is shown in Fig. 10.14. Notice that the feedback loop is returned to the base of the

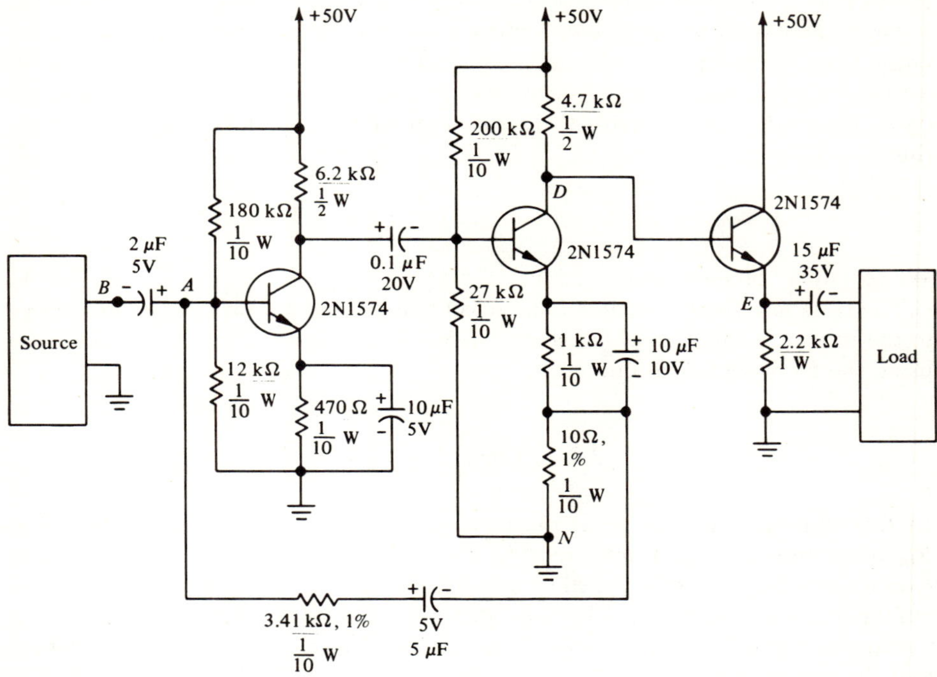

Figure 10.14

transistor in the first stage (the *right*-hand terminal of the 2 µF coupling capacitor); this is point *A* in Fig. 10.14. If the loop were returned instead to point *B* (the *left*-hand terminal of the capacitor), *the 2 µF coupling capacitor would effectively be included within the feedback loop.* According to the reasoning advanced in chapter 8, and again in Sec. 10-10, the inclusion of the capacitor within the feedback loop will contribute to an improvement in the low-frequency response of the amplifier. Stated alternately, if we include the capacitor within the loop (connection of loop return to point *B*), we can use a *smaller* value for this capacitor and obtain the *same* low-frequency response achieved when a 2 µF capacitor is used and the capacitor is *not* included within the loop (connection of loop return to point *A*). The amount of improvement obtained by including the capacitor within the loop depends upon the value of the internal resistance of the source (5 kΩ, in this case); the equations used for the computation of the amount of improvement are quite complex and will not be derived here. It can be shown, however, that the smaller the value of the internal source resistance, the greater will be the improvement in the low-frequency response when the loop return is connected to point *B* instead of point *A*. In practice, the designer could compute the improvement in the low-frequency response, or he could measure it directly in the laboratory.

The next step in the design process is to "breadboard" the circuit; i.e., the amplifier is built on a flat, thin, perforated plastic board commonly known as a breadboard. Small metal terminals are inserted into the holes at various points on the board, and the circuit components (the transistors, resistors, and capacitors shown in Fig. 10.14) are wired between these terminals. Since the high-frequency response of the amplifier is adversely affected by stray wiring capacitances, every effort is made to keep the physical lay-out compact, and the leads kept as short as possible. It is important to emphasize that the breadboard construction and testing is the *intermediate* step in the design process between the paper design and the construction of the final prototype. At this point, the designer has an opportunity to test his circuit to see if it meets the specifications; he measures the dc voltages, gain, stability, distortion, and frequency response of the amplifier. Since many of the decisions made in the paper design are based upon rough assumptions, *it is highly probable that the original circuit* (the circuit in Fig. 10.14) *will not function perfectly when first tested.* In this case the designer will have to make some changes in the values of the components, and possibly, the physical lay-out. In extreme cases, if the performance of the circuit is very poor, it may be necessary to discard the design completely and begin anew; this situation will occur only if the designer makes major miscalculations in the paper design as a result of inexperience or incompetence. The process of building, testing, and revising the breadboarded circuit is known as "troubleshooting." It is in the troubleshooting process that the designer combines his knowledge of both the theory and practical laboratory techniques.

It is senseless to carry out the paper design in more detail than we have done (such as using *all four* h-parameters in the computations), since the values of these parameters vary considerably with Q point, temperature, and production unit. In fact, many designers might feel that our paper design (Secs. 10-2 through 10-10) was too "analytical"; i.e., it may be their opinion that it was not necessary to carry through as many computations as we *did.* These designers would prefer to make rougher estimates of the circuit performance on paper, and spend more time perfecting the design by the use of troubleshooting techniques. This practice can be carried to an extreme, however, by some designers who make practically no use of the circuit theory. They breadboard the circuit almost immediately, including in the breadboarded circuit a number of potentiometers. They then vary these potentiometers, observing the readings of voltmeters and oscilloscopes, until they obtain the performance desired. The trouble with this "hit or miss" technique is that the designer has not provided for changes in temperature or replacement of transistors. It is entirely possible (more likely, probable) that a slight increase in the ambient temperature will result in a malfunction of the circuit. Thus, it can be seen that a reasonable balance between paper design and laboratory troubleshooting leads to the most efficient design. The designer who combines these two qualities in the proper balance is the most skillful worker. Of course, some circuits lend themselves more readily to paper, and others to laboratory designs; the knowledge of which circuits should be designed by which methods is gained through experience.

Let us suppose that we have built the circuit of Fig. 10.14 on a breadboard; we use a laboratory power supply for the 50 V dc. Before applying any ac signal to the amplifier, we must first check the dc levels. We know that V_{CN_1} (the collector-to-ground voltage of stage 1) should be:

$$V_{CN_1} = V_{CC} - I_{C_Q}R_L = 50 - (4.5)(6.2) = 22.1 \text{ V}$$

Let us say that we measure it (using a voltmeter or an oscilloscope) to be 30 V. This means that the quiescent collector current, I_{C_Q}, in the first stage is:

$$I_{C_Q} = \frac{V_{CC} - V_{CN_1}}{R_L} = \frac{50 - 30}{6.2} = 3.23 \text{ mA}$$

Knowing that the transistor has its maximum β at values of collector current between 5 mA and 7.5 mA, we may feel that 3.23 mA is *too small* a value of collector current for our particular design; thus we make a decision to *increase* the value of I_{C_Q} in the first stage. The expression for the value of I_{C_Q} can be written as:

$$I_{C_Q} = \frac{\beta(V_B - V_{BE})}{R_B + \beta R_E} \qquad V_B = \frac{R_2 V_{CC}}{R_1 + R_2}$$

The simplest way to *increase* the value of I_{C_Q} is to *increase* the value of V_B; this can be seen from the equation for I_{C_Q}. To increase the value of V_B, we must change the voltage divider consisting of R_1 and R_2; we can either *increase R_2 or decrease R_1*. If we increase R_2 (to a value larger than 12 kΩ), we effectively increase R_B, and consequently we increase the value of the stability factor; this means that the Q point shift may be *greater* than the value anticipated (1.86 mA) in the previous sections, possibly resulting in distortion. If we decrease R_1 (to a value smaller than 180 kΩ), we effectively decrease R_B, and thus the stability factor is improved (decreased). A decrease in the value of R_B, however, will cause a corresponding *decrease* in the open (and hence the closed) loop gain. Since we have included some amount of safety factor in our design, the situation is not as serious as it appears. Let us say that, as a result of his past experience, the designer concludes that he has provided a greater amount of safety factor in the stability factor; thus he decides to increase the value of R_2 to 15 kΩ. He makes the change, using long-nose pliers and soldering iron, and then re-measures the value of V_{CN_1}. If the dc voltmeter reads 22 V, the designer concludes that the quiescent collector current is approximately 4.5 mA, and so he continues with the troubleshooting process. This process (that of checking the dc voltages) is repeated for the second stage and the follower.

After we are convinced that the dc operation of the circuit is satisfactory, we turn to the testing of the ac performance. We simulate the source voltage using an ordinary audio oscillator (whose frequency can be varied between 100 Hz and 50 kHz) in series with a resistance. The audio oscillator has an internal resistance of its own, and so the value of the external series resistance must be adjusted such that the output resistance of the voltage source consisting of the *oscillator-external resistance combination* is 5 kΩ. For example, if the output resistance of the oscillator (R_O) is 600 Ω, the external series resistance should be equal to 4.4 kΩ (5 kΩ minus 0.6 kΩ). We must also make sure that the open circuit voltage of the oscillator can be varied between 1 mV and 50 mV (peak-to-peak); this may necessitate the use of some type of resistive voltage divider. The load can be simulated by an ordinary 1 kΩ resistor. The circuit is connected as shown in Fig.10.15. The switch, S, is

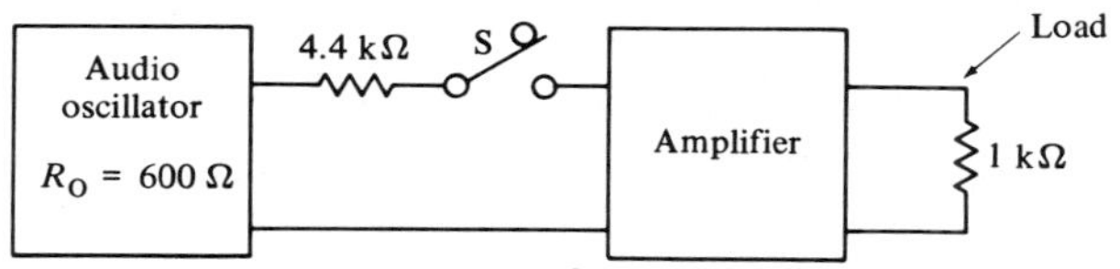

Figure 10.15

opened and the open circuit voltage of the oscillator is adjusted first to 1 mV and then to 50 mV; these two positions are then marked on the voltage control dial of the oscillator. When the switch is closed, the voltage will drop by a very large amount because of the very small input resistance of the amplifier.

To study the ac performance of the amplifier, we adjust the input source voltage to the *maximum* position (open circuit voltage equal to 50 mV), we set the frequency to 1 kHz, and we observe the waveform of the output load voltage, v_T, with an oscilloscope. We first check to see if the waveform is sinusoidal. Suppose the waveform is *flattened* at the positive peak, as shown in Fig. 10.16(a); this amount

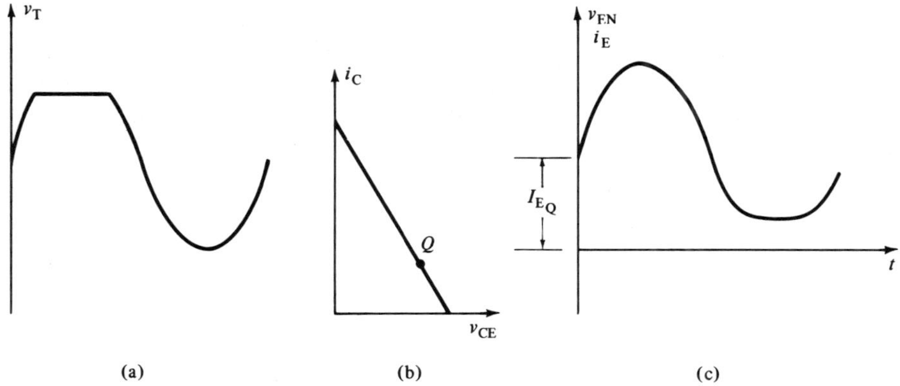

Figure 10.16

of distortion cannot be tolerated in the amplifier. The next step is to trace back through the circuit (using the oscilloscope) to determine the source of this distortion. Suppose we observe the waveform of v_{CN_2} (the collector-to-ground voltage of stage 2) and notice that the waveshape is similar to that in Fig. 10.16(a); this can be accomplished by connecting the oscilloscope leads to terminals D and N in Fig. 10.14. Suppose also that we observe the waveform at the collector of the *first* stage (with respect to ground) and see *no distortion*. We must conclude that the source of the distortion is the *second* stage; in other words, there is some problem in the second stage which is causing distortion in the output waveform. The fact that the waveform flattens at the *positive* peak indicates that the transistor is probably biased too close to *cut-off*, as shown in Fig. 10.16(b) (see chapter 4). This means that we must re-measure the dc voltages and currents in the second stage to check our initial assumption, and then change the values of one or both bias resistors accordingly (increase R_2 or decrease R_1).

Suppose that we connect the scope to measure the waveform of v_{EN} (the voltage across the 2.2 kΩ emitter resistor of the follower), and the resulting waveform appears as shown in Fig. 10.16(c); note that both the current (i_E) and the voltage (v_{EN}) have the *same waveshape* because they are developed at a resistor. The severe distortion at the *negative* peak occurs because the quiescent emitter current, I_{EQ} (or I_{CQ}), of the follower transistor is too small. This current can be expressed as follows:

$$I_{\mathrm{EQ}} = \frac{V_{\mathrm{CN_2}} - V_{\mathrm{BE}}}{R_{\mathrm{E}}}$$

Since the value of $V_{\mathrm{CN_2}}$ is fixed by the Q point requirements of stage 2, the only way to increase I_{EQ} is to decrease the value of R_{E}. If, however, the value of R_{E} is decreased (below 2.2 kΩ), *two* adverse effects may result: First, a decrease in the value of R_{E} will cause a corresponding decrease in R_{T}' (the parallel equivalent of R_{E} and R_{T}) which will, in turn, cause a decrease in both the input resistance and the voltage gain of the follower. If both the input resistance and the voltage gain of the follower are decreased, the load voltage will *also* decrease, perhaps below the 9 V minimum. This can be rectified by increasing the value of R_{F} (we return to this point later). Second, and more serious, a decrease in R_{E} will cause an increase in I_{CQ}; this increase in I_{CQ} may result in *the transistor's exceeding the maximum power dissipation*. For example, suppose that we reduce R_{E} to 1.5 kΩ; then the power dissipation in the transistor can be computed as follows:

$$I_{\mathrm{CQ}} \simeq \frac{V_{\mathrm{CN_2}}}{R_{\mathrm{E}}} = \frac{26 \text{ V}}{1.5 \text{ k}\Omega} = 17.3 \text{ mA}$$

$$V_{\mathrm{CEQ}} \simeq V_{\mathrm{CC}} - V_{\mathrm{CN_2}} = 50 - 26 = 24 \text{ V}$$

$$P_{\mathrm{C}} = (I_{\mathrm{CQ}})V_{\mathrm{CEQ}} = (17.3 \text{ mA})(24 \text{ V}) = 416 \text{ mW}$$

This is *greater* than the maximum rating of 320 mW, which means that *the transistor will fail* at a temperature of $+95$ degrees C. In this case, we can do one of *two* things: We can search the specification sheets for another transistor type, one whose maximum power dissipation is greater than 600 mW (at room temperature); of course, it should also have a minimum β of 100. We use this new transistor type for the follower stage *only*, retaining the two 2N1574 transistors in the first two amplifier stages. The alternative to selecting a different transistor type for the follower stage involves the application of some type of *heat sink* to the 2N1574 follower transistor. Time and cost will determine which of these two methods will be used.

The next step in the troubleshooting procedure is to check the gain. The open circuit input voltage is varied between 1 mV and 50 mV in definite increments, and the output voltage is measured at each step. If a graph is plotted of output voltage vs. open circuit input voltage, a *straight line* should result; the slope of this straight line should be equal to the over-all voltage gain, 200 (see Fig. 10.17).

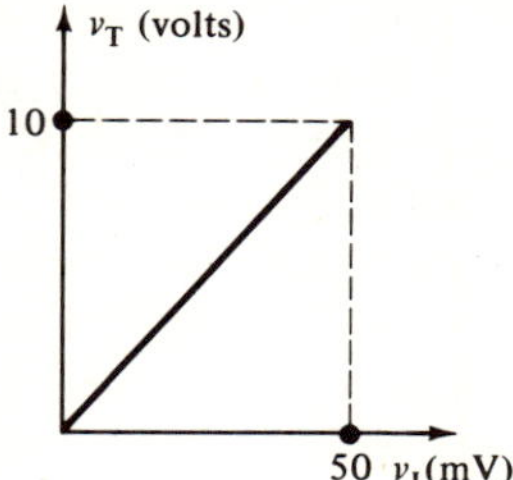

Figure 10.17

The straight line is an indication of the fact that *the output voltage is directly proportional to the input;* i.e., *the gain is constant.* This straight line relationship *does not apply at much larger voltages* because the transistor outputs will be limited due to saturation and cut-off. When the input voltage is 50 mV, the output voltage should be between 9.0 V and 11.0 V. Suppose that, when the input is 50 mV, the output is 13 V; i.e., we have *too much gain* and thus we should like to reduce it. We know that the closed loop current gain of the two-stage feedback amplifier is approximately equal to the ratio of R_F to R_E'. Therefore, to reduce the gain, we should reduce the value of R_F; the reverse situation will require an *increase* in the value of R_F.

To measure the frequency response of the amplifier, we must vary the frequency of the audio oscillator signal over a range extending from 100 Hz to 50 kHz; the gain is measured at appropriate intervals (see chapter 8). When the input is 50 mV, we must check to see that *the output voltage is greater than 9.0, but less than 11.0 V for each and every frequency between 100 Hz and 50 kHz.* Suppose we find that the output voltage is equal to 8.0 V at a frequency of 125 Hz; in other words, we have *not* met the low-frequency specification. Evidently, the value of one or more of the six capacitors (three coupling, two by-pass, and one feedback) is *too small.* Recall from chapter 8 that there are one or two corner frequencies associated with each of these capacitors. We can view the low-frequency response as a *chain,* and each one of the corner frequencies as a *link* in this chain. The *highest* corner frequency is the *weakest* link in the chain, and thus it is our task to determine which of the capacitors is associated with this corner frequency. We can replace *each* of the six capacitors (one at a time) with a capacitor whose value is *five to ten times* the original value; in each case, we note the effect on the output voltage at 125 Hz. The capacitor change which causes the *largest increase* in the output voltage is probably the weakest link. Let us assume that this capacitor turns out to be the 15 μF capacitor which couples the follower to the load. In this case, we reduce the oscillator frequency to 100 Hz (the output will drop below 8.0 V). We then place successively *larger* values of capacitance into the circuit until the output voltage at 100 Hz increases above 9.0 V; the smallest value of capacitance which meets the specification is the correct value for the design.

To measure the high-frequency response, the oscillator frequency is set to 50 kHz, and the output load voltage is measured when the open circuit input voltage is 50 mV; this output voltage should, of course, be between 9.0 and 11.0 V. The rough computations made in Sec. 10-4 indicated that the upper cut-off frequency of our amplifier will probably be *much larger* than 50 kHz; thus we should anticipate no trouble at this end of the frequency spectrum. If the output load voltage *is* found to be less than 9.0 V at $f = 50$ kHz, the effect can be interpreted in either one of *two* ways: The first is that we may have made a major miscalculation in Sec. 10-4 when we attempted to determine the approximate high-frequency response of an amplifier using 2N1574 transistors; in this case we would have to *scrap the design and begin anew* by selecting a different transistor type. This first possibility is highly unlikely, particularly if the designer has had some prior experience with this type of amplifier circuit. The second possibility concerns the wiring lay-out of the breadboarded circuit; recall from chapter 8 that the stray wiring capacitances tend to *decrease* the upper cut-off frequency of the amplifier. Therefore, we can attempt to re-wire the breadboarded circuit, such as to *reduce* the stray wiring capacitances (by using shorter lead lengths), and consequently, *increase* the upper cut-off frequency.

The final step in the troubleshooting process is the examination of the amplifier performance under varying environmental conditions. If the amplifier is to meet military specifications, the environmental tests might include temperature, humidity, shock and vibration, sand and dust, salt spray, and other types. Since transistors are highly sensitive to changes in temperature, it is wise to examine the performance of the *breadboarded* amplifier circuit at temperatures ranging from $-55°$ to $+95°$C; i.e., it is important to check this particular environmental factor (temperature) *before* the final prototype is built. The breadboarded circuit is placed inside a temperature chamber, and leads are brought out to the oscillator, power supply, voltmeter, and oscilloscope. The temperature inside the chamber is varied from $-55°$ to $+95°$C and the amplifier performance is observed and measured over this entire temperature range to see if the specifications are met at all temperatures. Suppose that the output load voltage *falls below 9.0 V at* $-55°C$; this implies one of *two* possibilities: The β of the transistor *is less than 100* (minimum specified value) at $-55°$C, in which case the manufacturer must be contacted. The second possibility is that we have made a miscalculation in the value of the minimum β required for the transistors in our design; in this case we must recheck our computations, and if necessary, specify a *larger* value for the minimum β. Suppose that we observe distortion of the waveform of the output load voltage at $+95°$C. This probably indicates that the Q point has shifted an excessive amount from its position at room temperature. In other words, the stability factor for one or both amplifier stages is *too large*, and thus we must re-design for a smaller value of S.

After the designer is satisfied that the amplifier circuit meets the specifications, he begins to consider the construction of the final prototype. At this point, the amplifier circuit may look somewhat different from the schematic shown in Fig. 10.14 (our original paper design); this is due to the changes which we have made in the course of the troubleshooting process. *The final prototype is one completely con-*

structed unit of the entire system; in many instances we have to manufacture thousands of these systems for distribution throughout the world. Our amplifier is now built and carefully wired on a *circuit card;* an electronic system may be composed of thousands of these circuit cards, each one containing a circuit which is part of the system. If a portion of the system (into which our amplifier is to be placed) is already built, we can insert the amplifier into its proper position and recheck its performance; it is entirely possible that further changes in the design will be necessitated at this point.

It should be emphasized again that the purpose of this chapter is *not* to teach the reader to be a circuit designer. The wide variety of available circuit configurations and the rapid change in the pace of technology both make this task nearly impossible for any textbook. Our purpose here is to indicate to the reader the many problems involved in the design of a circuit to meet a set of specifications, and the various steps by which the designer proceeds from these specifications to the final design.

INDEX